Molecular Interactions

Molecular Interactions

Concepts and Methods

David A. Micha
University of Florida, Gainesville, FL, USA

WILEY

This edition first published 2020

Registered Office
John Wiley & Sons, Inc., 111 River Street, Hoboken, NJ 07030, USA

Editorial Office
111 River Street, Hoboken, NJ 07030, USA

For details of our global editorial offices, customer services, and more information about Wiley products visit us at www.wiley.com.

Wiley also publishes its books in a variety of electronic formats and by print-on-demand. Some content that appears in standard print versions of this book may not be available in other formats.

Library of Congress Cataloging-in-Publication Data applied for
ISBN: 9780470290743

Cover image: © Photobank gallery/Shutterstock
Cover design by Wiley

Set in 10/12pt Warnock by SPi Global, Pondicherry, India

Printed in the United states of America
V10015059_110619

Contents

Preface

The temperatures and densities of the chemical elements present in our natural world lead to the formation of molecules, which interact to create complex many-atom systems. Many materials and the components of living organisms are made up of aggregates of atomic and molecular units. Fluids, molecular and atomic solids, polymers, and proteins are examples of those aggregates. The properties of these objects can be described from first principles of quantum mechanics and statistical mechanics, after their composition in terms of electrons and atomic nuclei have been specified.

Molecular encounters are governed by intermolecular forces, which can be measured and calculated. This book presents concepts and methods needed to obtain energies of a molecular system, as they change with interatomic distances to provide intermolecular forces. Once the interaction forces in a molecular system are known, the equations of motion of classical or quantum mechanics, implemented with physical boundary conditions, can be used to obtain thermodynamic equilibrium and nonequilibrium properties from first principles, and the response of the system to external factors, such as light. Interactions of molecular species can be described by their electromagnetic properties and chemical reactivity. The reverse is also true. Information about molecular interaction energies can be derived from measurements of thermodynamic, kinetic, and electromagnetic properties of matter, and from experiments specifically devised to extract interaction energies, such as crossed molecular beam and photodissociation experiments. This information has been incorporated into the selection of concepts and methods for the book.

The advent of quantum theory and its applications during the first half of the twentieth century has led to many of the quantitative concepts about molecular structure and properties employed today, and how to calculate them, as well as ways to describe their interaction energies. Molecular dynamics and statistical mechanics have also provided thermodynamical and kinetics values of properties to be compared with experimental measurements. Comparisons and tests of theoretical and experimental results require extensive computational work. Computational methods, software and utilities, and the ever-increasing speed and data storage power of computers have been essential for these purposes. Some of these methods are covered in the following chapters.

Intermolecular forces are essential in many applications of molecular and materials properties to technologies contributing to the needs of society. To illustrate the enormous impact of the subject, some of their subjects (and their applications) are storage of hydrogen in solids (fuel cells), storage and transport of ions in solids (batteries), synthesis of thermally stable and conducting surfaces (solar energy devices), delivery of compounds through biological cell membranes (pharmacology), catalysis and photocatalysis in electrochemical cells (sustainable fuel production), atmospheric reactions (environmental sciences), efficient fuel combustion (transportation and energy), and solvation and lubricants (machinery). Furthermore, as the quantitative tools of chemistry and physics in this book have become more useful and common in biology, pharmaceutics, and medicine, its contents should also be of interest in these new areas of applications.

This book is based on the author's lectures given over many years at the University of Florida in Chemical Physics courses taken by graduate and advanced undergraduate students in Chemistry, Physics, Chemical Engineering, and Materials Sciences, with working knowledge of quantum and statistical mechanics as usually covered in Physical Chemistry or Modern Physics courses. The lectures have emphasized concepts and methods used in calculations with realistic models, to be compared with empirical data. They have been expanded to cover recent developments with advanced theoretical methods, as presented by researchers at conferences and particularly at many Sanibel Symposia on the quantum treatments of atoms, molecules, and materials, which this book's author has helped to co-organize.

The text contains hundreds of citations, but these are only a small portion of the tens of thousands of relevant publications on the subject of molecular interactions. To compensate, much can be done to gain additional knowledge, including physical and chemical data, by searching for information in the World Wide Web. The concepts and methods described here are meant to also provide a broad vocabulary as needed to search the Web. To keep the book length within reasonable bounds, its contents have been restricted to cover the energetics of molecular interactions and the potential energy surfaces that are generated as interatomic distances are varied. A short introduction has been given about how these interactions affect the spectroscopy, dynamics, and kinetics of molecular interactions, but these subjects are not covered in detail here and have been left for possible future treatments. References have been limited to fundamental publications, monographs, reviews, and comprehensive reports. Apologies are advanced here to the authors of many relevant publications that could not be mentioned due to length limitations.

Chapters in this book have been organized to cover first the concepts and methods for simpler systems, and next to cover more advanced subjects for complex systems. Each chapter begins with qualitative aspects before these are treated in quantitative fashion. The first four chapters include the well-established concepts and quantitative aspects of long-range (electrostatic, induction, and dispersion) forces and how they extend to intermediate and short ranges, for ground and excited states. They are followed by chapters dealing with recent developments including electronically nonadiabatic interactions, correlated many-electron treatments, generalized density functional theory, decomposition and embedding of molecular fragments for large systems, and very recent developments using artificial intelligence with network training for many-atom systems. The first four chapters and the first sections of the following four chapters can provide an introductory course on molecular interactions, for students with a knowledge of the fundamentals of quantum and statistical mechanics and of molecular structure. A more advanced course can be based on material starting with the fourth chapter and continuing with the following ones.

This author thanks his science teachers at the Physics Institute, Universidad de Cuyo, Bariloche (Argentina), presently Instituto Balseiro, and at the Quantum Chemistry Group, Uppsala University (Sweden), and particularly Dr. Jose Balseiro in Bariloche for his introduction to quantum theory and Dr. Per-Olov Lowdin in Uppsala, for his mentoring in quantum chemistry. Much was learned from colleagues and researchers at the University of Wisconsin (Madison) and at the University of California, San Diego, and at institutions visited during sabbaticals at Gothenburg University in Sweden, Harvard University, Max Planck Institutes for Stroemungsforschung and for Astrophysics in Germany, Imperial College in London, England, Institute of Theoretical Physics at the University of California Santa Barbara, JILA at the University of Colorado in Boulder, the Weizmann Institute in Rehovot, Israel, Florida State University, the Ecole Normal Superieur, Paris, France, and the Institute for Mathematics and its Applications, University of Minnesota, Minneapolis. This author is grateful also to the many graduate and undergraduate students who did research under his direction and to postdoctoral associates and visiting scientists working with him at the University of Florida. They are too numerous to be listed here but their names can be found at the website https://people.clas.ufl.edu/Micha.

This author is greatly appreciative of the financial support by the Foundations that provided essential funds for his research and collaborations: the Swedish International Development Agency, the Alfred P. Sloan Foundation, the National Science Foundation of the USA for many years of Principal Investigator research and also for support of US-Latin American workshops, NASA, the Alexander von Humboldt Foundation Program of U.S. Senior Scientists Awards twice, and the Dreyfus Foundation.

Thanks are also due to reviewers at the early stages when this book was being planned, particularly Bernard Kirtman, George Schatz, and Victor Batista. Several of the author's research colleagues have been kind enough to read and comment on some of the book's chapters, in particular Rod Bartlett, Adrian Roitberg, John Stanton, Akbar Salam, Pilar de Lara-Castells, and Benjamin Levine. All of them had helpful comments at early and later times, but they are not to be blamed for any shortcomings in this book!

This author appreciates the patience and support of the editors at the publisher, John Wiley and Sons, Inc., over the years it took to have this book completed. As the times for its production and publication approach, special thanks go to the project editor, Ms. Aruna Pragasam, and the senior editor Jonathan T. Rose for their guidance. A special thanks is also due to the author's wife, Rebecca A. Micha, for her help keeping the English grammar right in many paragraphs and for her unwavering support.

David A. Micha,
Gainesville, FL, USA

December 2018

1

Fundamental Concepts

CONTENTS

Molecular Interactions: Concepts and Methods, First Edition. David A. Micha.

1.1 Molecular Interactions in Nature

Many materials and the components of living organisms in nature are made up of aggregates of atomic and molecular units. Fluids, molecular and atomic solids, polymers, and proteins are examples. The properties of these objects can be described from first principles of quantum mechanics and statistical mechanics, after their composition in terms of electrons and nuclei have been specified. This requires a theoretical framework to describe the structure of atoms and molecules and the way the atoms and molecules interact. In some cases, the interactions affect but do not change the conformations of molecules, which can then be taken as the basic building units of the objects being studied.

Once the interaction forces in a molecular system are known, the equations of motion of classical or quantum mechanics, implemented with physical boundary conditions, can be used to derive thermodynamic equilibrium and nonequilibrium properties from first principles. The response of the system to external factors, such as light and interacting species, can be described by their electromagnetic and chemical reactivity properties. The reverse is also true. Information about molecular interaction energies can be derived from measurements of thermodynamic, kinetic, and electromagnetic properties of matter, and from experiments specially devised to extract interaction energies, such as crossed molecular beam and photodissociation experiments.

Early treatments that proceed from atomic and molecular structure to the calculation of intermolecular forces and properties of molecular systems have been covered in several books going back over 50 years [1–4] and have been expanded to incorporate results of more extensive calculations [5–13] made possible by continuous improvements in computational power. The present work introduces and updates theoretical concepts and methods needed to model structures and properties and provides links to the more recent computational developments.

A chronological Table 1.1 follows with some important early discoveries on molecular interactions. They start with the acceptance of the existence of molecules and understanding of how they interact, as early as the nineteenth century. The advent of quantum theory and its applications during the first half of the twentieth century has then lead to many of the quantitative concepts about molecular structure and properties and how to calculate them, as well as ways to describe their interaction energies. Statistical mechanics has also provided thermodynamical and kinetics values of properties to be compared with experimental measurements.

Table 1.1 Chronology of some early discoveries on molecular interactions.

Year	Authors	Subject
1857	R. Clausius	Distance dependence of interaction potentials
1868	J.C. Maxwell	Simple molecular transport theory
1872	L. Boltzmann	Transport theory for fluids
1873	J.D. van der Waals	Equation of state for real gases
1905	P. Langevin	Ion–molecule interaction potential
1912	P. Debye	Dielectric properties of fluids
1924	J.E. Lennard-Jones	Analytic interaction potentials
1925	J. Franck and E.U. Condon	Molecular photoexcitations
1927	W. Heitler and F. London	Chemical bonding in H2
1927	M. Born and J.D. Oppenheimer	Quantum theory of molecules
1930	F. London	Quantal calculation of dispersion forces
1931	J.C. Slater and J.G. Kirkwood	Variational calculation of dispersion forces
1932	J.H. Van Vleck	Electric and magnetic susceptibilities of molecules
1933	P.K.L. Drude	Optical properties of fluids
1939	R. Feynman	Hellmann–Feynman or force theorem
1943	B.M. Axilrod and E. Teller	Three-atom interaction potentials
...	...	...

The link between electrons and nuclei and potential energy functions related to molecular structure is provided by quantum chemistry, and thermodynamical properties follow by using statistical mechanics. Steady-state (or transport) properties as well as nonequilibrium properties, including reactivity, can be obtained from molecular dynamics, as shown in the block diagram illustrated in Figure 1.1. Spectroscopic properties follow from the electrodynamics of molecular systems. Transport in gases follow from molecular collision cross sections.

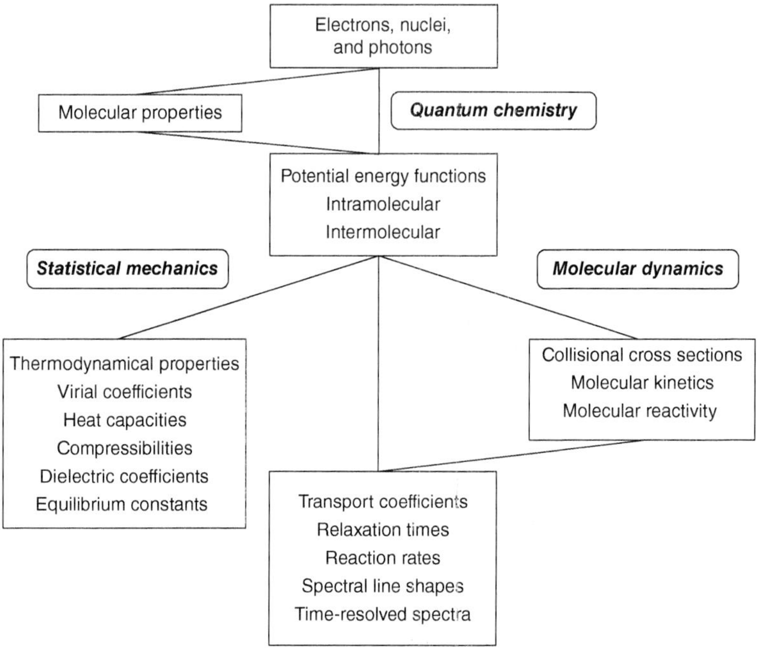

Figure 1.1 Properties derived from interacting electrons, nuclei, and photons.

1.2 Potential Energies for Molecular Interactions

1.2.1 The Concept of a Molecular Potential Energy

The large difference between electron and nuclear masses leads to a qualitative difference between electronic and nuclear motions. Nuclei move slowly compared to electronic motions and this allows the formation of molecules, with positively charged nuclei held together by the negative electron distributions. This leads to stable molecular systems when total energies are not too high compared to electronic binding energies. We consider electrons bound by Coulomb forces inside a finite region of space, and separately deal with (i) bound nuclei, where all particles are restricted by their own Coulomb forces to a finite region of space, and (ii) unbound nuclei with attached electrons, where some of the particles (atoms, molecules, etc.) are involved in a collision event.

For molecules occupying a finite region of space, electrons and nuclei move under their own Coulomb forces. Since the magnitude of electron and nuclear charges are comparable, their Coulomb forces are similar so that $F_e \approx F_n$ and in terms of masses and accelerations, $m_e \Delta v_e / \Delta t \approx m_n \Delta v_n / \Delta t$. In the time interval Δt one therefore finds comparable momentum changes, $m_e \Delta v_e \approx m_n \Delta v_n$. Provided electronic and nuclear momenta are comparable to begin with and a system is observed over short times, one finds that over time $p_e = m_e v_e \approx p_n = m_n v_n$. Hence, since $m_n > 2000\ m_e$, one finds that velocities satisfy

$$v_e = \frac{p_e}{m_e} \ll v_n = \frac{p_n}{m_n}$$

and to a first approximation nuclei can be assumed to be at rest while electrons move around them. Fixing the nuclear positions, the molecular energies become functions of the nuclear coordinates and provide the potential energies for the nuclear motions; hence, they can be referred to as molecular potential energies. This is the Born–Oppenheimer picture of molecular structure [14, 15]. These authors showed, using quantum mechanical perturbation theory, that for bound molecular states, the potential energy correction due to nuclear motion goes as $(m_e/m_n)^{1/2}$, while the correction to molecular wavefunctions goes as $(m_e/m_n)^{1/4}$, and therefore are relatively small and acceptable for most applications.

To be more specific, we can compare electronic and nuclear kinetic energies; for the electronic kinetic energy K_{el} we find, assuming hydrogenic states for an electron moving in a framework of atomic cores with an effective (screened) positive charge, and since then $K_{el} \approx -E_{el}$ for the bound states,

$$K_{el} = \frac{m_e v_e^2}{2} \approx \frac{1}{2}\frac{Z_{scr}^2}{n^2}\frac{e^2}{a_B} = \frac{1}{2}\frac{Z_{scr}^2}{n^2} 27.2\,\text{eV}$$

where e, a_B, $Z_{scr}e$, and n are the electron charge in esu units, the atomic Bohr radius, the screened nuclear charge around which the electron orbits, and the principal quantum number of the orbital, respectively. The energy units conversion is that 1 au(E) = 27.2116 eV, and 1 eV = 1.60218×10^{-19} J = 96.485 kJ mol^{-1} for future reference in this work.

The magnitudes of nuclear kinetic energies depend on whether the nuclei are bound or not. Letting $v_n \leq v_e/10$ be our criterion for separation of electronic and nuclear motions, we find for the nuclear kinetic energy,

$$K_n = \frac{m_n v_n^2}{2} \leq \frac{m_n (v_e/10)^2}{2} = \frac{1}{100}\frac{m_n}{m_e}\frac{1}{2}\frac{e^2}{a_B}\frac{Z_{scr}^2}{n^2}$$

where $m_n \approx 2000 m_n \bar{A}$ in terms of an average atomic mass number $\bar{A}$. This expression shows that as the electronic state is excited, and n increases, the upper energy bound for motion separation decreases, and it becomes more difficult to satisfy the inequality; the separation of electronic and nuclear motions is more doubtful for electronically excited states.

For ground and low excited electronic states, we can separately estimate nuclear kinetic energies for molecular systems with bound and unbound nuclei.

A) Bound nuclei

The nuclei in molecules vibrate and rotate. We can estimate the vibrational and rotational kinetic energies by means of

$$K_{vib} \approx h\nu_v n_v, \quad \nu_v = (k/\mu_n)^{1/2}/2\pi$$

$$K_{rot} \approx \hbar^2 l(l+1)/(2I)$$

in a system where the vibrational frequencies are of magnitude ν_v (given in terms of a vibrational force constant k and a reduced mass μ_n), I is a moment of inertia, and the vibrational and rotational quantum numbers are n_v and l, respectively. These terms would typically add to about 1.0 eV and would be much smaller than the required upper bound to the nuclear kinetic energy. However, the bound can be exceeded if the vibrational–rotational levels are very high, signifying less accuracy in the separation of electronic and nuclear motions.

B) Unbound nuclei
For colliding atoms and molecules, nuclear momenta include the center of mass components acquired in the collision, which can be very large for example in atomic beams. Potential energy functions can then be defined only provided as before, in eV units, $K_n \leq 272\,\mathrm{eV}.\bar{A}.Z_{scr}^2/n^2$. For a hydrogen atom, a potential energy function can safely be defined for kinetic energies below 272 eV, while for deuterium ($A = 2$) the upper limit would be 544 eV. The potentials can also be defined above these values, but one must consider the possibility that electronic and nuclear motions will be coupled.

1.2.2 Theoretical Classification of Interaction Potentials

Interaction potential energies between species A and B are obtained from the Coulomb interaction of all electrons and nuclei making up the whole system, and their quantal expectation value in given electronic states, for fixed nuclear positions. The nuclear position variables are collected in a set $\{Q_j\} = \mathbf{Q}$, where the Q_j are for example atomic position vectors, or bond coordinates (distances, angles), and the distance R between the centers of mass of the two molecules A and B. The electronic variables (positions and spin) are collected in $\mathbf{X}$. The potential energy $V_\Gamma(\mathbf{Q})$ of the system in a quantum state $\Phi_\Gamma(\mathbf{X};\mathbf{Q})$ is obtained from a quantal expectation value by integrating over electronic variables. It can be treated separating nuclear variables as $\mathbf{Q} = (R, \mathbf{Q}')$ and concentrating first on its dependence on R for a given quantum state.

Total pair potential energies are the sum of long-range, short-range, and intermediate-range contributions, $V(R) = V_{SR}(R) + V_{IR}(R) + V_{LR}(R)$. Short-range potentials follow from electronic charge distributions obtained in quantum chemical calculations or by parametrizing measured properties. This can also be done for the intermediate range potential energies. Long-range potential energies follow from properties of the two species.

A theoretical analysis of the electronic states of a molecular system allows us to classify the interactions in accordance with the separation between interacting fragments A and B. We work in a reference frame where the z-axis goes through the centers of mass of A and B, at relative distance R. Indicating with a_A and a_B the approximate radii of the two fragments, the classification can be made with respect to ranges of $R/(a_A + a_B)$ omitting, to simplify matters here, the dependences on internal and orientation coordinates. In each region, one

Table 1.2 Classification of molecular interaction potentials.

Separation	$R/(a_A + a_B)$	Potential type	R-dependence	Treatment
Small	< 1.0	Repulsive	$A(R)\exp(-bR)$	Variational
Intermediate	1.0–10	Valence	$F(R; R_m, \ldots V_m, \ldots)$	Variational–perturbative
Large	10–50	Electrodynamic	$\sum_n C_n R^{-n}$	Perturbative
Very large	>50	Retardation	$\sum_n D_n R^{-n}$	Perturbative

can consider the types of contributions to the potential energy, its dependence with distance R, and the way it can be calculated with theoretical methods. Table 1.2 summarizes this information.

The calculation of molecular interaction potential energies requires a quantum mechanical treatment starting from the structure of the whole collection of electrons and nuclei. The Schroedinger equation of motion of the whole system can be solved to obtain stationary state solutions $\Phi_\Gamma(\boldsymbol{X}; \boldsymbol{Q})$ where the collection of nuclear position variables $\boldsymbol{Q}$ is considered, to begin with, as fixed at chosen values. The index Γ is a collection of quantum numbers that also labels the energy $E_\Gamma(\boldsymbol{Q})$ of the steady state. These steady states can be calculated with a variety of methods, including variational and perturbation treatments of the equations of motion with given boundary conditions [16–19].

1.2.2.1 Small Distances

Short-range interactions are repulsive due to the so-called Pauli repulsion of closed electron shells, resulting from the requirement of single occupancy of electron orbitals imposed by the antisymmetry of electronic wavefunctions. This leads to distortion of overlapping shells that acquire opposing dipoles so that their repulsion goes as $V_{SR}(R) = C_{SR}R^{-3}$ at short distances $R \geq R_{min}$ and contains larger powers of R at even shorter distances where distortion multipoles are formed. This can be represented by an exponential function $A\exp(-R/a)$, more approximately by an inverse power form C_n/R^n, or instead by a combination $(\sum_n A_n/R^n)\exp(-R/a)$ with some coefficients chosen to reproduce intermediate-region energies. However, for two atoms A and B with nuclear charges Z_a and Z_b and at very short distances that exclude electronic charge in between them, the potential energy goes as $Z_aZ_b/(4\pi\varepsilon_0 R)$ as $R \to 0$.

Variational methods can be used to calculate potential energies at small distances with relatively simple trial electronic wavefunctions of the Hartree–Fock type, including Coulomb and exchange electronic interactions. Contributions are the electronic kinetic, Coulomb, and exchange energies, and the potential can be obtained by minimization of an energy functional of the trial wavefunction. Energy functionals of the electronic density can also provide useful information for short distances.

1.2.2.2 Intermediate Distances

At intermediate distances, the potential can be parameterized in terms of the position R_m and energy V_m at the potential minimum, and other parameters referring, e.g. to the zero of the potential, inflection point, etc. Results can be fit to expansions in powers of $R - R_m$, or can be fit by functions suitable for near equilibrium regions, such as the Morse potential [2] within a finite region, connected by splines to short- and long-range energy functions.

Elaborate electronic wavefunctions are required for trial functions in variational methods, including the effects of electron correlation, for example by using superpositions of configurations, or wavefunctions must be obtained from perturbation expansions properly symmetrized to account for overall antisymmetry with respect to electron exchange. This will be covered in the chapter on many-electron treatments.

1.2.2.3 Large Distances

At large distances, the potential results from the interaction of permanent and fluctuating charge distributions, leading to electrostatic, induction, and dispersion energies,

$$V_{LR} = V_{el} + V_{ind} + V_{disp} = \sum_{n \geq 1} C_n R^{-n}$$

They can be derived from perturbation theory using electronic wavefunctions appropriate for the separated species and can be expressed in terms of molecular properties such as charge, electric multipole, and multipolar dynamical polarizabilities.

1.2.2.4 Very Large Distances

It is convenient to describe very long-range interactions in the language of quantum electrodynamics, with interaction energies resulting from the exchange of photons between the two molecular species. When the propagation time of photons exchanged between the fragments, $t_{prop} = R/c$, with c the speed of light, is comparable to or larger than the periods of charge oscillation $\tau = \lambda_{rad}/c = h/\Delta E$, with λ_{rad} the radiation wavelength and h the Planck constant, or equivalently when the distance R is larger than or equal to the wavelength of light emitted and absorbed in the interaction, one must in addition consider retardation effects. We find, using the smallest excitation energies $(\Delta E)_{A,B}$ of a pair of fragments, that for

$$R \gg \frac{ch}{(\Delta E)_{A,B}}$$

we must include retardation effects in molecular potentials. The result from perturbation theory is also an expansion in R^{-1}, with different powers from before. For two neutral molecules, the leading term goes as R^{-7}, instead of R^{-6} in the absence of retardation [2].

At very large distances, the electromagnetic contributions are very small and may become less important than magnetic (relativistic) contributions coming from spin couplings, and these must be accounted for.

Effects resulting from the electron spin coupled to its orbital angular momentum are in principle present at all distances, particularly for open-shell systems. They can be treated by means of a quantal perturbation treatment for the states $\Phi_\Gamma(\boldsymbol{X}; \boldsymbol{Q})$ due to spin-orbit energy couplings, which lead to splittings of interaction potential energies.

Combining short- and long-range functional forms for two neutral and isotropic species gives model potentials, with two common ones being

$$V(R) = V_m\left[\frac{6}{n-6}\left(\frac{R_m}{R}\right)^n - \frac{n}{n-6}\left(\frac{R_m}{R}\right)^6\right]$$

the Lennard–Jones $n-6$ potential with parameters n, and V_m and R_m for the minimum energy and radius, and

$$V(R) = \frac{6}{\alpha-6}\exp\left[\alpha\left(\frac{R}{R_m}-1\right)\right] - \frac{\alpha}{\alpha-6}\left(\frac{R_m}{R}\right)^6$$

the exponential-6 potential, with parameters α, V_m, and R_m. The Lennard–Jones potential function is usually chosen to have $n = 12$, with $V_m = \epsilon$, and is also written using instead the distance $R_0 = R_m/2^{1/6}$, where the potential energy is zero so that $V(R) = 4\epsilon[(R_0/R)^{12} - [(R_0/R)^6]$.

1.3 Quantal Treatment and Examples of Molecular Interactions

Energies can be obtained within a quantal treatment of molecular states [16–19]. We can to begin with, fix nuclei in space at positions $\{\vec{r}_n\}$ or alternatively using variables $\{Q_j\} = \boldsymbol{Q}$, where the Q_j are bond coordinates (distances, angles) and the distance between centers of mass of two molecules, to define a Hamiltonian operator $\hat{H}_{\boldsymbol{Q}}$ containing all energies except the nuclear kinetic energy. It depends on all the electronic variables (positions and spin) collected in $\boldsymbol{X}$. We have for the total energy Hamiltonian operator, using carats to signify operators on functions of coordinates, and adding the nuclear kinetic energy $\hat{K}_n$,

$$\hat{H} = \hat{K}_n + \hat{H}_{\boldsymbol{Q}}$$

$$\hat{H}_{\boldsymbol{Q}} = \hat{K}_{el} + V_{ne} + V_{ee} + V_{nn}$$

and the equation for stationary molecular states,

$$\hat{H}_Q \Phi(X;\mathbf{Q}) = E(\mathbf{Q})\Phi(X;\mathbf{Q})$$

where $\Phi(X;\mathbf{Q}) = \langle X \mid \Phi(\mathbf{Q})\rangle$ is the electronic eigenstate of energy E, a function of all the electronic variables and dependent on the nuclear coordinates, shown as parameters. Here $\hat{H}_Q$ is the Hamiltonian operator for fixed nuclear positions, so that it includes the electronic kinetic energy, nuclear–electronic, electronic–electronic, and nuclear–nuclear Coulomb charge energies but no nuclear kinetic energy. Alternatively, the energy $E(\mathbf{Q})$ of the molecular system can be obtained as the expectation value of $\hat{H}_Q$, using the Dirac bra-ket notation,

$$E(\mathbf{Q}) = \langle \Phi(\mathbf{Q}) \mid \hat{H}_Q \mid \Phi(\mathbf{Q}) \rangle = \int dX\, \Phi(X;\mathbf{Q})^* \hat{H}_Q \Phi(X;\mathbf{Q})$$

where a normalization $\langle \Phi(\mathbf{Q}) \mid \Phi(\mathbf{Q}) \rangle = 1$ has been used. This energy may be decomposed into contributions from the several terms in the Hamiltonian operator and is the so-called Born–Oppenheimer or adiabatic potential energy. Other properties (such as the electric dipole), represented by an operator $\hat{A}_Q$, are also obtained for fixed nuclei as averages,

$$A(\mathbf{Q}) = \langle \Phi(\mathbf{Q}) \mid \hat{A}_Q \mid \Phi(\mathbf{Q}) \rangle.$$

An accurate treatment of the electronic state must describe the electrons as moving in a self-consistent field shaped by the antisymmetry properties of the wavefunction, as given usually by the Hartree–Fock formulation to incorporate electronic exchange energies [16], and must in addition contain corrections for the correlation energy of electronic motions, spin-orbit coupling, and magnetic (relativistic) effects, so that

$$E(\mathbf{Q}) = E_{el}(\mathbf{Q}) + E_{nn}(\mathbf{Q})$$
$$E_{el}(\mathbf{Q}) = E_{HF} + E_{corr} + E_{SO} + E_{magn}$$

where $E_{nn}(\mathbf{Q})$ is the nuclear–nuclear repulsion energy, and all the terms depend on the nuclear coordinates.

The electronic state also provides information on electronic density functions of electronic positions, $\rho(\vec{r})$; these can be decomposed into atomic core ρ_c and valence ρ_v electron terms, with the first leading to nuclear screening and to core–core interactions. Electron exchange leads to a "Pauli pressure," which explains the distortion and repulsion of atomic cores as they overlap.

The energy contributions can be described for the case of H_2^+ in ground and excited electronic states. Here, we have

$$\left(\hat{K}_{el} + V_{ne}\right)\varphi_\mu(\vec{r};R) = \varepsilon_\mu(R)\varphi_\mu(\vec{r};R)$$

$$E_\mu(R) = \varepsilon_\mu(R) + \frac{e^2}{4\pi\varepsilon_0 R}$$

where $\varepsilon_\mu(R)$ and $\varphi_\mu(\vec{r};R)$ are electronic energy and orbital function at internuclear distance R, and we have added the electronic energy of the electron orbital to the nuclear Coulomb repulsion energy to obtain the total adiabatic potential energy $E_\mu(R)$. Figure 1.2a and b, from [20], shows the electronic orbital energy $\varepsilon_\mu(R)$ and the diatomic potential energy $E_\mu(R) = U_\mu(R)$ versus the interatomic distance R in atomic units of energy (Hartree energy), with 1 au(E) = 4.35979×10^{-18} J, and length (Bohr radius), with 1 au(L) = 5.2918×10^{-11}m = 52.918 pm. The electronic binding energies are negative and increasing as the nuclei move apart, with the limits at $R = 0$ and at infinite R given by the electronic energies for He^+ and $H + H^+$, respectively, and are labeled by molecular point group symmetry symbols. After adding the positive repulsive energy of the nuclei in Figure 1.2b, the lowest (or ground) total potential energy $E_g(R)$ shows a minimum that defines the static equilibrium distance R_e and displays a net positive repulsion energy at short distances.

Similarly, for H_2 in ground and excited electronic states, energies depend only on the internuclear distance R. Figure 1.3 shows the total, electronic plus n–n repulsion contributions for the ground state Σ_g^+. Here, the separated atoms (S.A.) limit is H(1s) + H(1s), while the electronic united atom (U.A.) limit is He(1s^2) and has a lower electronic energy. The contributions are again found to add to a potential with a minimum, attractive at long distances and repulsive at short distances. The contributions to the excited Σ_u^+ electronic state add up instead to a purely repulsive potential and here the united atom limit is He(1s2p), with a larger electronic energy [22].

A different situation arises for LiF, where potential curves show avoided crossings between states $1^1\Sigma^+$ and $2^1\Sigma^+$ as seen in Figure 1.4. The electronic ground state and five electronically excited states have been obtained with a many-electron treatment including electronic correlation energies. Electron transfer occurs as the distance R increases, and couples the state $1^1\Sigma^+$ going asymptotically to Li + F with the state $2^1\Sigma^+$ going to $Li^+ + F^-$. A full treatment of the atomic interactions here requires that one also specifies the interstate coupling energy due to atomic displacements, leading to transitions between the two nearly degenerate potential energies.

Potential energies for more than two atoms require special treatments to account for energy functions of many atomic position variables. Here, we give some general examples and postpone details to the following chapters. Continuing with polyatomic intermolecular potentials, consider a molecular system AB breaking up into fragments A and B. The potential energies are functions of the nuclear position vectors. The molecular energy (electrons plus fixed nuclei) is $E^{AB}(\mathbf{Q}) = E_{el}^{AB}(\mathbf{Q}) + E_{nn}^{AB}(\mathbf{Q})$, where the first term is the sum of electron kinetic energies plus electron–electron repulsion and electron–nuclear attraction, and the second term is the nuclear–nuclear repulsion energy. The nuclear position variables are collected in $\mathbf{Q} = (\vec{R},\mathbf{q})$, with $\vec{R}$ the relative position vector

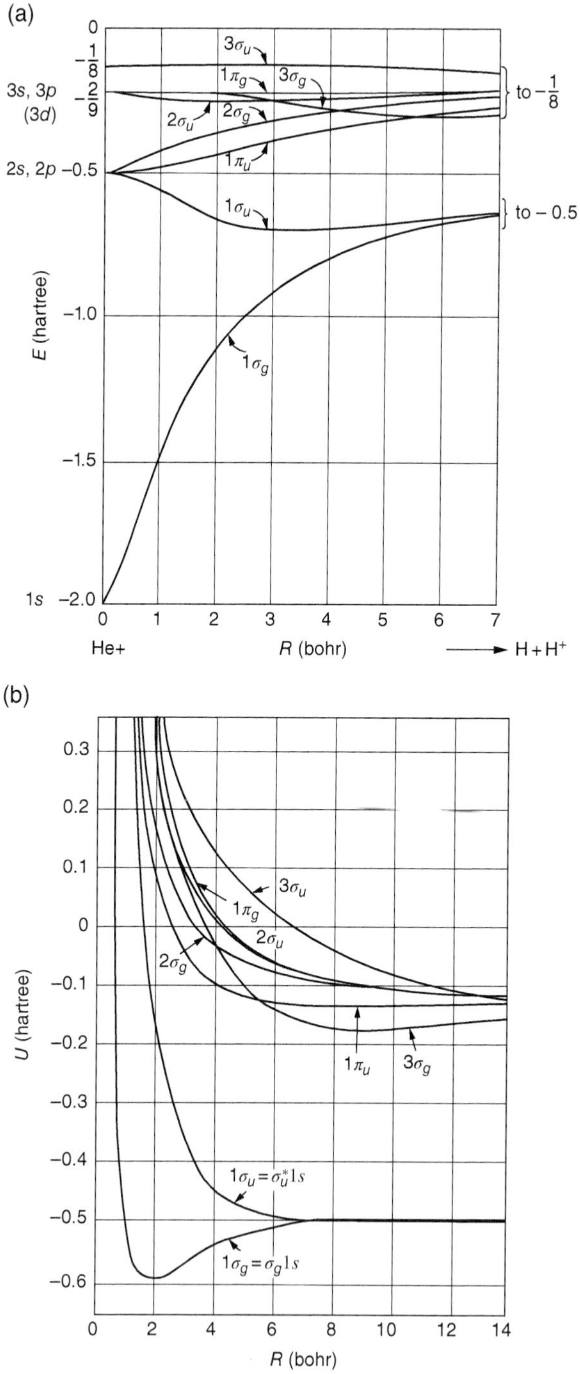

Figure 1.2 (a) Electronic orbital energy $\varepsilon_\mu(R)$ for ground and excited states of an electron in the H_2^+ molecular ion versus the distance R between the two protons, labeled by symmetry symbols. The limits at $R = 0$ and at infinite R are the electronic energies for He^+ and $H + H^+$, respectively. *Source:* adapted from Slater [20]. (b) Total potential energy $E_\mu(R) = U(R)$ of ground and excited states of H_2^+ versus interatomic distance R. *Source:* adapted from Slater [20].

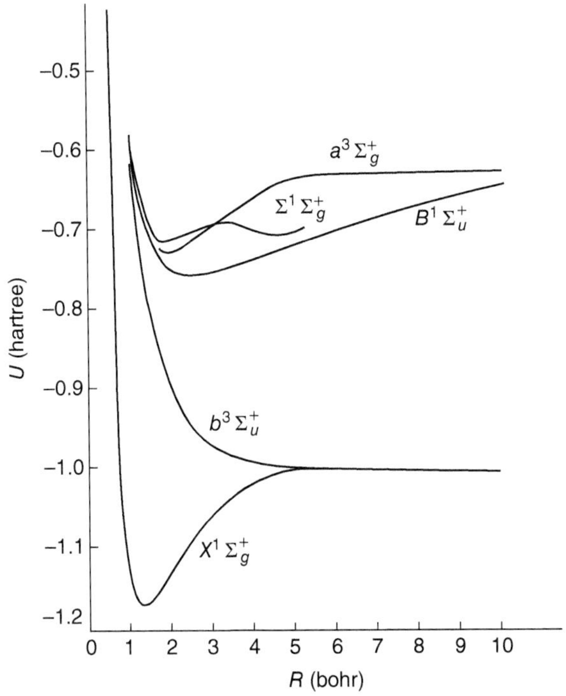

Figure 1.3 Ground and excited total potential energies for several states of H_2 versus the internuclear distance *R*. The limiting electronic energies for infinite *R* are ground and excited H + H energies. *Source:* from Kolos and Wolniewicz [21]. Reproduced with permission of AIP Publishing.

between the centers of mass of fragments *A* and *B*, and with $\mathbf{q} = (\mathbf{q}^A, \mathbf{q}^B)$ their internal and orientation coordinates. We can work in a reference frame with its *z*-axis passing through the centers of mass of *A* and *B*, so that the interaction potential energy is given by

$$V_{AB}(R, \mathbf{q}) = E^{AB}(R, \mathbf{q}) - E^A(\mathbf{q}^A) - E^B(\mathbf{q}^B)$$

where *R* is the distance between centers of mass, which becomes infinitely large as the pair breaks up.

Specific examples can be given for polyatomic intramolecular and intermolecular potentials. Starting with intramolecular potentials, we describe cases where only two nuclear coordinates or degrees of freedom Q_1 and Q_2 are changing, which leads to potential energy surfaces (PESs) $U(Q_1, Q_2)$ in a three-dimensional depiction. The energy can alternatively be represented by contours of equal energy, or equipotential isocontours, in the (Q_1, Q_2) plane.

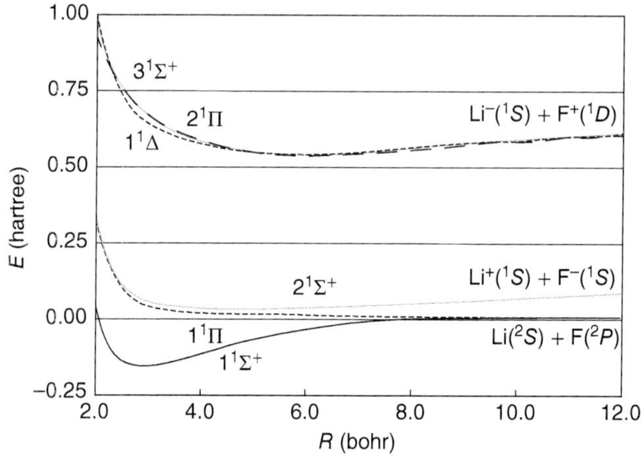

Figure 1.4 Potential energies E versus internuclear distance R for ground and excited states of LiF. The electronic ground state and five electronically excited states have been obtained with a many-electron treatment including electronic correlation energies. *Source:* from Yagi and Takatsuka [23]. Reproduced with permission of AIP Publishing.

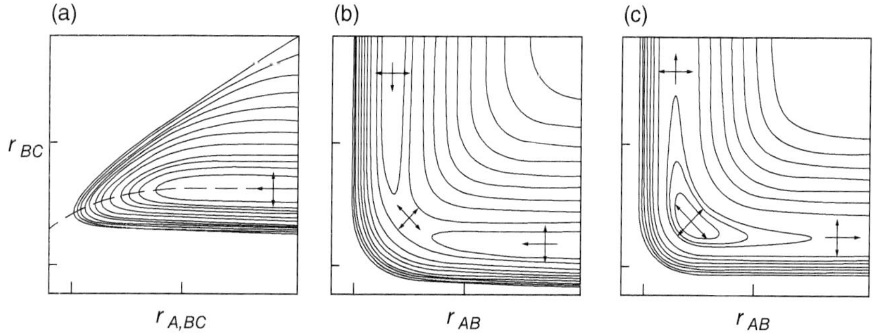

Figure 1.5 Typical PESs versus atomic positions for a triatomic *ABC* system with a fixed bond angle between *AB* and *BC*. (a) Isocontours for the interaction energy of a noble gas atom *A* interacting with a diatomic *BC*; (b) isocontours for the reaction $A + BC \rightarrow AB + C$, showing the location of an activation energy barrier; (c) isocontours for a stable compound *ABC* in a potential well, connected to dissociation valleys. *Source:* from Figure 2.3 of Smith [7]. Reproduced with permission of Elsevier.

Figure 1.5 shows typical energy isocontours of PESs: (i) with a single valley, as for He + H_2, versus the center of mass relative and internal coordinates $R = r_{A,\,BC}$ and r_{BC}; (ii) showing a rearrangement, e.g. $H + H_2 \rightarrow H_2 + H$, with a saddle point between two valleys in a PES versus bond distances; and (iii) a well for a stable species such as H_2O connected by a saddle point to valleys for dissociations [9].

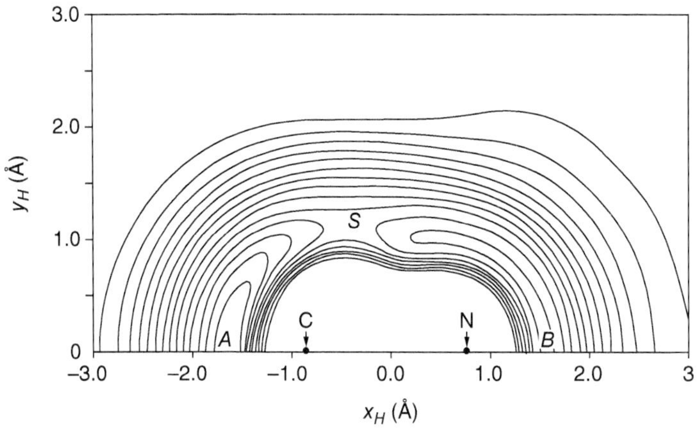

Figure 1.6 HCN PES isocontours for the HCN ⟶ HNC isomerization, with H changing position on a *x*–*y* plane and C and N fixed on the *x*-axis. The locations of minima are shown as *A* and *B*, for the two isomers, and a saddle point as *S*. *Source:* from Hirst [10]. Reproduced with permission of Taylor & Francis.

Figure 1.6 is an alternative description and shows the HCN ⟶ HNC isomerization by displaying the interaction energy of the hydrogen atom with the CN fragment as the atom is displaced between minima near C and N.

Plots of potential energies versus torsion angles for macromolecular systems also offer examples of surfaces with several minima. For example Figure 1.7 shows several minima of the potential energy of a polypeptide chain versus the consecutive dihedral angles ψ and φ. In all these cases, figures are only showing the lowest PES. The PESs of electronically excited states would have different shapes.

Figures 1.8 and 1.9 give specific examples of dissociation in HCN ⟶ H + CN and ⟶ N + CH, versus bond distances, and of atomic rearrangement from reactants to products in $F + H_2 \longrightarrow FH + H$.

More than one PES may be involved in electron transfer or in electronically excited systems, involving two position variables Q_1 and Q_2, or more. Two potential energy functions $V_g(Q_1, Q_2)$ and $V_e(Q_1, Q_2)$ for ground and excited electronic states, drawn in a 3D picture of V versus (Q_1, Q_2) with isocontours of constant energy, typically show intersections where couplings due to atomic displacements can be large [26]. Depending on how the electronic states couple through those displacements, the surfaces can intersect along a line forming a seam, or at a point giving a conical intersection. Seams are features of interaction potentials involving electron transfer in collisions between ions and neutral molecules, such as found in collisions of $H^+ + H_2$ [27] and of $H_2^+ + Ar$ [28]. Conical intersections appear in electron exchange between molecules [29], in isomerization [30], and in photoinduced chemistry [31].

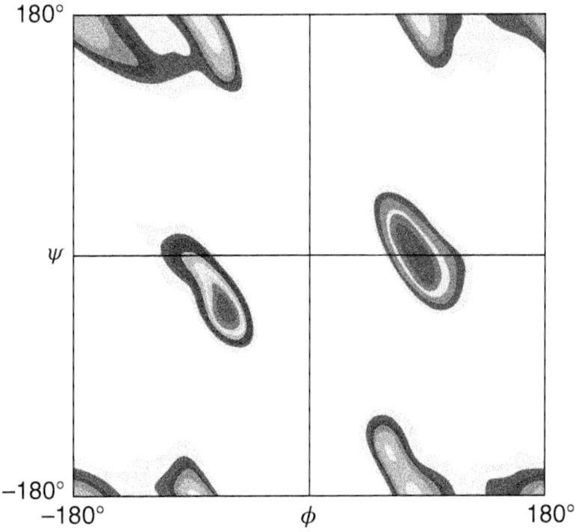

Figure 1.7 Peptide link (a glycyl residue of a polypeptide chain) isocontours of constant energy versus the torsion angles phi around the C–N bond and psi around the C–C bond (a Ramachandran diagram). Ovals lead to energy minima. *Source:* from Hovmoeller [24]. Reproduced with permission of John Wiley & Sons.

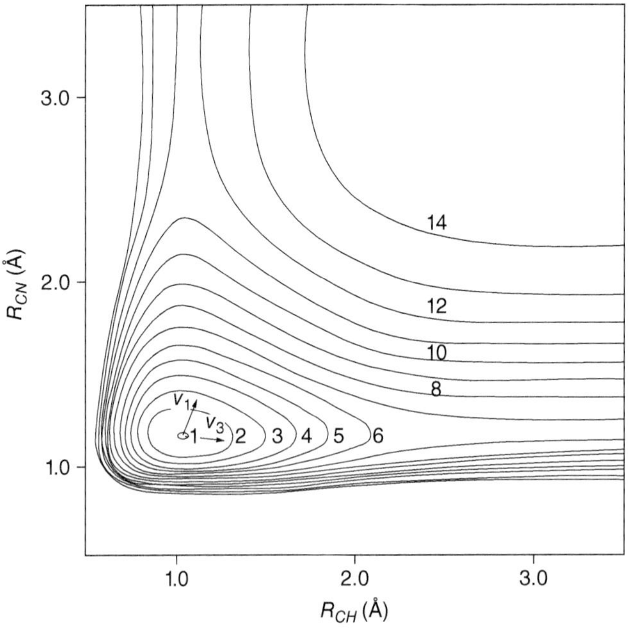

Figure 1.8 HCN PES versus bond distances C–H and C–N for the dissociation HCN → H + CN. Contour 1 is for the energy − 13.75 eV, and other contours are drawn at intervals of 1 eV. *Source:* from Hirst [10]. Reproduced with permission of Taylor & Francis.

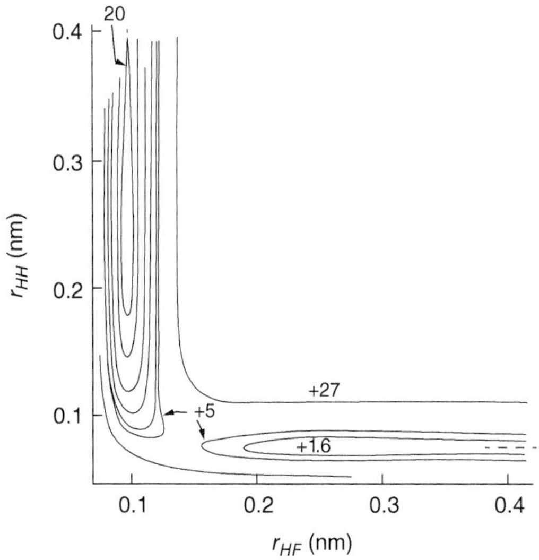

Figure 1.9 $F + H_2 \rightarrow FH + H$ PES for the collinear FHH configuration and contour energies given in kcal mol^{-1} (with 1.0 kcal mol^{-1} = 4.184 kJ mol^{-1}). *Source:* from Bender et al. [25]. Reproduced with permission of AIP Publishing.

Figure 1.10 shows the intersection of two potentials for collinear $Ar(^1S) + H_2^+$ $(^2\Sigma)$ and Ar^+ (^2P)+ H_2 $(^1\Sigma)$ and higher potentials versus the H–H distance when Ar is far removed, and also potentials for ArH^+ and ArH as a H is far away. The potentials for H_2^+ and H_2 versus the H–H distance are seen to cross, but when Ar is brought in this becomes a seam between two PESs, and their coupling leads to a linear avoided crossing of the PESs versus the Ar–H distances. A similar situation arises for $H^+ + H_2 \rightarrow H_2 + H^+$, or $\rightarrow H_2^+ + H$, involving two PESs, versus bond distances, but here the lowest one shows an energy well for H_3^+. The new feature in both cases is that the PESs have an avoided crossing or "seam" running along the entrance valley, corresponding to electron transfer as an approaching atom moves in.

Figure 1.11 shows two PESs interacting to give conical intersections in the isomerization of the C_2H_2 ethylene molecule. The picture at the top corresponds to ground and excited electronic states for isomerization into HC_2H_3, ethylidene, while the one at the bottom shows the ground and excited PESs for the pyramidalization intermediate $C_2H_2^*$ as obtained in [32]. Here, the variables are g and h, themselves functions of all the atomic positions, with g measuring the strength of the coupling of the states by the atomic displacements, and h giving the magnitude of energy splitting between the two PESs [31].

Similar treatments and images apply to interactions of atoms and molecules with an extended system, such as a solid surface or a cavity in a liquid. Figure 1.12 shows the PES of He approaching the surface LiF(001) of the solid, with a face-center-cubic (or fcc) structure for a lattice of F^- and Li^+ ions. The surface is perpendicular to a z-axis with the origin of a surface plane

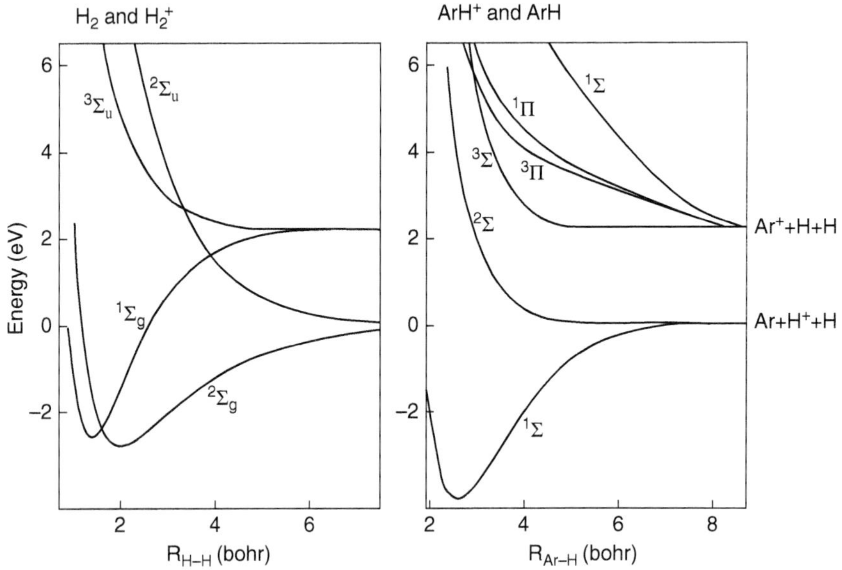

Figure 1.10 Potential energies for collinear ArH_2^+ showing their change with atom–atom distances, while the third atom is far away, with energies given relative to Ar + H^+ + H. The intersection in H–H curves leads to the appearance of a seam between potential energy surfaces as the Ar atom is brought in. *Source:* from Baer and Beswick [28]. Reproduced with permission of American Physical Society.

(x, y) at a F^- ion location, a cell cube side of 2.84 Å and a Li^+ at the center of the face. The isocontours reflect the periodicity of the surface and show deep wells on the x–z plane above each ion, for two chosen y values corresponding to lines along the negative and positive ions. A similar figure can be drawn for an atom approaching a molecule adsorbed on a solid surface.

All these PESs display special points where forces are null, defined by

$$\frac{\partial E}{\partial Q_j} = 0 \rightarrow \left\{Q_j^0\right\} = \mathbf{Q^0}$$

for at least one set $\{Q_j^0\}$, which locates maxima, minima, or saddle points; these can be distinguished by calculating the second derivatives $\partial^2 E/(\partial Q_j \partial Q_k)$ at the special points. The dynamics of motion on PESs are closely related to their topography as defined by the geometry of the special points. For example, minima are connected by paths of steepest ascent and descent along which one finds activation barrier energies, which in turn account for reaction rate magnitudes.

A different conceptual approach is needed to describe the energetics of reacting molecules in a medium such as a liquid solvent or a solid matrix, and in particular electron transfer in liquid solutions [34, 35]. Sets $\left\{\vec{r}_n\right\}$ of many atomic

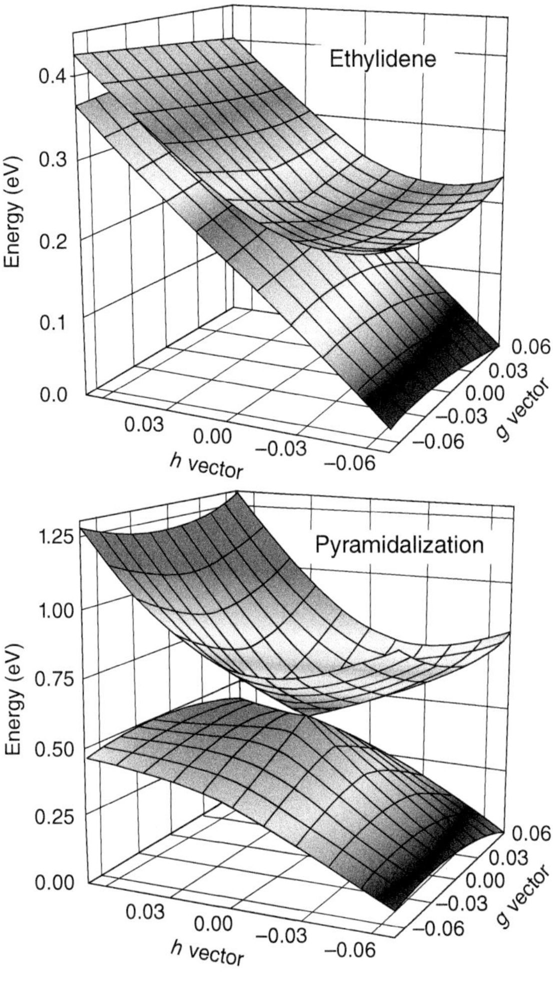

Figure 1.11 Conical intersections in the isomerization of the C_2H_2 ethylene molecule. The picture at the top corresponds to ground and excited electronic states for isomerization into HC_2H_3, ethylidene, while the one at the bottom shows the ground and excited PESs for the pyramidalization intermediate $C_2H_2^*$ as obtained in [32]. *Source:* from Ben-Nun and Martinez [32]. Reproduced with permission of Elsevier.

position variables are involved and one must instead work with collective structure variables $\boldsymbol{Y}\left(\{\vec{r}_n\}\right)$ which account for many atom rearrangements for given thermodynamical constrains such as temperature T, density ρ, and pressure p. Potential energies $V_\Gamma\left(\{\vec{r}_n\}\right), \Gamma = g, e,$ must be replaced by thermodynamical internal energies $U_\Gamma(\boldsymbol{Y}, T, \rho)$ or Gibbs free energies $G_\Gamma(\boldsymbol{Y}, T, p)$.

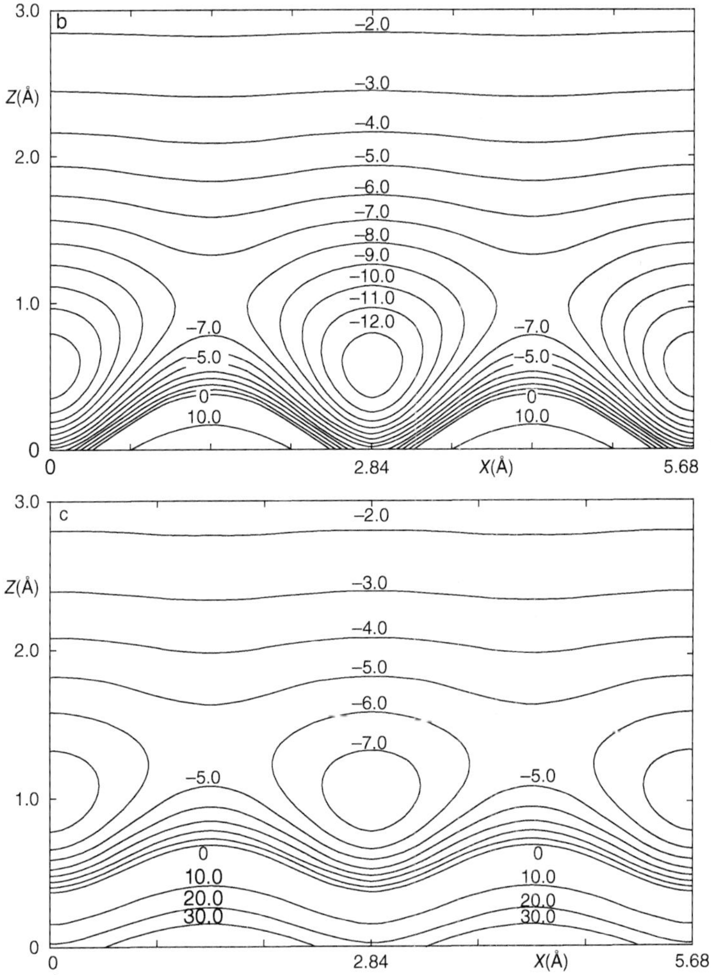

Figure 1.12 He approaching the surface LiF(001) of the fcc solid. The surface is perpendicular to a *z*-axis with the origin of a surface plane (*x*, *y*) at a F^- ion location, a cell cube side of 2.84 Å and a Li^+ at the center of the face. The isocontours in meV values reflect the periodicity of the surface and shows deep wells on the *x*–*z* plane above each ion, for two chosen *y* values of 0.00 Å (upper panel) and 1.42 Å (lower panel) corresponding to lines along the ions. *Source:* from Wolken [33]. Reproduced with permission of Springer Nature.

Figure 1.13 shows sketches of two Gibbs free energy surfaces for ground and excited electronic states $G_g(\mathbf{Y}, T, p)$ and $G_e(\mathbf{Y}, T, p)$ of reactants donor–acceptor pair D–A undergoing charge transfer into products pair D^+–A^-, such as a Fe^{2+}/Fe^{3+} mixed valence compound, in a liquid solution, with avoided conical

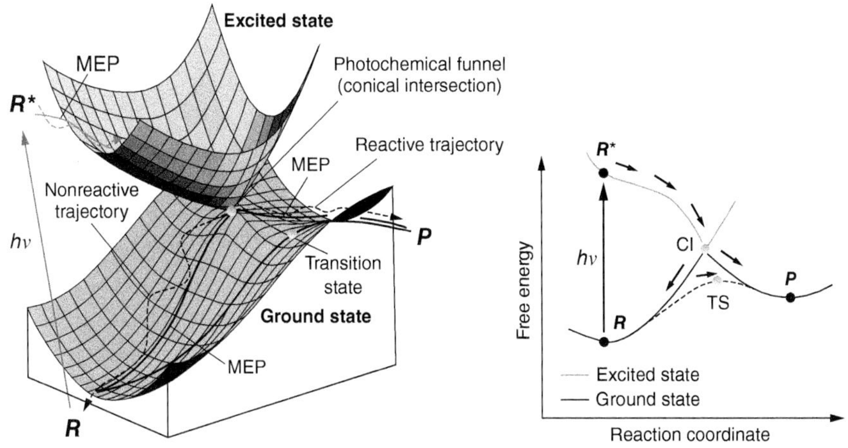

Figure 1.13 Gibbs free energy surfaces (FESs) for ground and excited electronic states in a system containing reactants ***R*** donor and acceptor compounds *D* and *A* in a liquid solution and the products ***P*** electron transfer ions D^+ and A^-. (a) Contours of FESs versus collective variables Y_1 and Y_2 chosen to be average liquid cavity radii around fragments *D* and *A*, showing minimum energy paths. (b) Free energy along the Y_s reaction path coordinate through a transition state between the two minima. *Source:* from Schapiro et al. [36]. Reproduced with permission of Royal Society of Chemistry.

intersections similar to the ones mentioned above now shown instead for free energies. The variables chosen here are average liquid cavity radii $\bar{R}_D = Y_1$ and $\bar{R}_A = Y_2$ around donor and acceptor fragments. The ground surface has two wells, for ground electronic states of the D–A and D^+–A^- arrangements, connected by a path Y_s at the surface through a saddle point. The reaction energy is ΔG_R and the splitting between electronically adiabatic free energy surfaces is given by $2G_R^0$. The upper surface displays a conical shape with a single minimum for the average radii of cavities around the two fragments interacting when the electronic state is excited.

1.4 Long-Range Interactions and Electrical Properties of Molecules

1.4.1 Electric Dipole of Molecules

Molecules usually have permanent electric dipole vectors $\vec{D}$, except when they have specific symmetries, such as inversion symmetry, which make the dipole null. Dipole physical units are Cm in the SI system, or the Debye, with $1\ \mathrm{D} = 10^{-18}$ cgs-esu $= 3.335 \times 10^{-30}$ Cm. Molecular dipoles are anisotropic and

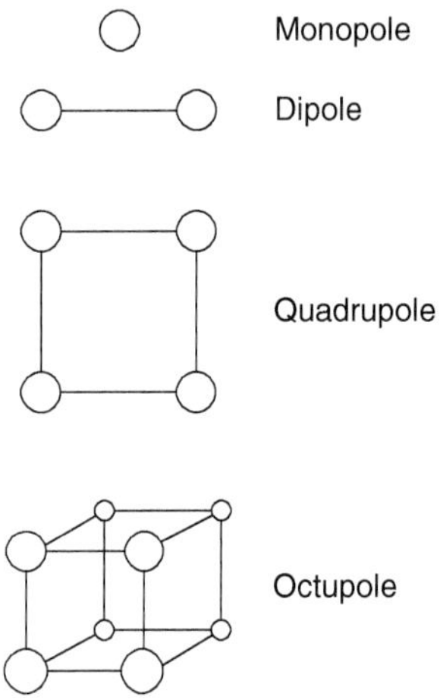

Figure 1.14 Point charge models of electrical multipoles.

can be estimated by adding the dipoles of its atom–atom bonds [16]. Examples are HF, HCl, HI, H_2O, CO_2, O_3, C_6H_6, and CH_4.

More generally, multipoles can be constructed as point charge models, with the monopole or 2^0 – pole given by a single charge, a dipole or 2^1 – pole given by two opposite charges at a distance L, and so on, as shown in Figure 1.14.

1.4.2 Electric Polarizabilities of Molecules

A molecule in a static electric field of components $\mathcal{E}_\xi$, $\xi = x, y, z$, acquires an induced dipole $D'_\eta = \sum_\xi \alpha_{\eta\xi}\mathcal{E}_\xi$ along a direction η proportional to the field strength, where $\alpha_{\eta\xi}$ is a dipolar polarizability tensor. To simplify here, we consider that both field and dipole are along the z-direction with $D' = \alpha\mathcal{E}$, where α is the zz-dipolar polarizability. For large fields, an expansion in the field strength gives also hyperpolarizabilities, with $D' = \alpha\mathcal{E} + \beta\mathcal{E}^2/2 + \ldots$ involving higher polarizabilities.

Dipoles and polarizabilities can be obtained from quantum mechanical perturbation theory as detailed in following chapters. Second-order perturbation theory shows that the polarizability is inversely proportional to molecular excitation energies. Quantum mechanical perturbation theory gives, in terms of unperturbed energies $E_n^{(0)}$ and molecular wavefunctions $\Phi_n^{(0)}$, static polarizabilities

$$\alpha = 2\sum\nolimits_{n\neq 0}\frac{|D_{n0}|^2}{E_n^{(0)} - E_0^{(0)}}$$

where $D_{n0} = \left\langle \Phi_n^{(0)} \,|\, \hat{D} \,|\, \Phi_0^{(0)} \right\rangle$ is a matrix element of the dipole operator. This shows that electronic polarizabilities tend to be larger for molecules with small electronic excitation energies, such as small highest occupied/lowest unoccupied (HOMO/LUMO) molecular orbital excitation energy. This gives larger polarizabilities for larger molecules with delocalized electrons.

The polarizability volume is given by $\alpha_v = \alpha/(4\pi\varepsilon_0)$, where $\varepsilon_0 = 8.854 \times 10^{-12}\,\mathrm{J^{-1}\,C^2m^{-1}} = 1/(4\pi)$ cgs-esu is the vacuum electrical permittivity. The value of α_v can be estimated for a molecule of size L using an average excitation energy $\Delta E \approx e^2/(4\pi\varepsilon_0 L)$, with e the electron charge in the SI units and a transition dipole $D_{n0} \approx eL$, giving $\alpha_v \approx L^3$.

The response of a molecule to an oscillating electric field with frequency ω is given by the dynamical polarizability $\alpha(\omega)$, obtained from second-order time-dependent perturbation theory in the form

$$\alpha(\omega) = (2/\hbar)\sum\nolimits_{n\neq 0}|D_{n0}|^2\omega_{n0}/\left(\omega_{n0}{}^2-\omega^2\right)$$

with $\hbar\omega_{n0} = E_n^{(0)} - E_0^{(0)}$. As ω increases, this function goes through infinite values at the transition frequencies and shows regions of nearly constant values in between for each type of motion: rotational, vibrational, and electronic, corresponding to the range of frequencies for radio, microwave, infrared, visible, and UV waves. A more detailed treatment accounts for the lifetime of excited states and gives finite polarizability values at transition frequencies, as shown in the following chapter.

1.4.3 Interaction Potentials from Multipoles

Long-range potential energies for two interacting molecules are obtained from their charge, dipole, quadrupole, etc., interactions both permanent or induced. Interaction between a 2^m-pole of species A and a 2^n-pole of species B can be called a $(2^m)_A$–$(2^n)_B$ interaction.

Interaction potential energies between multipoles can be obtained from charge models for molecules of size L when the distance between centers of mass satisfies $R \gg L$, with each one containing a collection of charges at given positions, by expanding the molecule–molecule interaction potential energy, written as a sum of charge pair interactions, in powers of L/R.

One finds that for the monopole–monopole interaction of charges C_A and C_B at distance R,

$$\left(2^0\right)_A - \left(2^0\right)_B : V(R) = \frac{C_A C_B}{4\pi\varepsilon_0 R}$$

For a charge–dipole interaction, with the dipole of B consisting of charges C_B and $-C_B$ at distance L, oriented along angles $\omega_B = (\theta_B, \varphi_B)$,

$$\left(2^0\right)_A - \left(2^1\right)_B : V(R) = f_B(\omega_B)\frac{C_A D_B}{4\pi\varepsilon_0 R^2}$$

with $D_B = C_B L$. The dipole–dipole interaction energy is

$$\left(2^1\right)_A - \left(2^1\right)_B : V(R) = f_{AB}(\omega_A, \omega_B)\frac{D_A D_B}{4\pi\varepsilon_0 R^3}$$

In general, the $(2^m)_A - (2^n)_B$ interaction between multipoles of magnitude $\mathcal{M}_A^{(m)}$ and $\mathcal{M}_B^{(n)}$ is of the form

$$(2^m)_A - (2^n)_B : V_{AB}^{(m,n)}(R) = f_{AB}(\omega_A, \omega_B)\frac{\mathcal{M}_A^{(m)}\mathcal{M}_B^{(n)}}{4\pi\varepsilon_0 R^{m+n+1}}$$

Induction energies are obtained from the interaction of an induced multipole with a permanent multipole. For the (induced dipole)–dipole interaction, this involves the induced dipole $D'_A = \alpha_A \mathcal{E}_A^{(B)}$ and the electric field $\mathcal{E}_A^{(B)} = -D_B/(4\pi\varepsilon_0 R^3)$ created at A by dipole B. The interaction energy is built increasing the field by increments $d\mathcal{E}_A^{(B)}$ as the polarized species is brought closer to the permanent multipole so that in increments $dV_{D'D} = -D'_A\, d\mathcal{E}_A^{(B)}$, which after integration over field values gives, using the polarization volume of A,

$$V_{D'D}(R) = -\frac{\alpha_A\left(\mathcal{E}_A^{(B)}\right)^2}{2} = -\frac{\alpha_{v,A}\, D_B{}^2}{2\pi\varepsilon_0 R^6}$$

Dispersion energies originate in (induced multipole)–(induced multipole) interactions, when charge fluctuations create transient dipole. For the interaction between transient dipoles $\tilde{D}_A = \alpha_A \tilde{\mathcal{E}}_A$ and $\tilde{D}_B = \alpha_B \tilde{\mathcal{E}}_B$, where each field is created by the transient dipole at the other molecule, the energy builds up to

$$V_{\tilde{D}\tilde{D}}(R) = -\frac{\alpha_A\left(\tilde{\mathcal{E}}_A^{(B)}\right)^2}{2} - \frac{\alpha_B\left(\tilde{\mathcal{E}}_B^{(A)}\right)^2}{2} = -\frac{C_{\tilde{D}\tilde{D}}}{R^6}$$

which is the van der Waals attractive long-range interaction, with $C_{\tilde{D}\tilde{D}} > 0$. This coefficient can be obtained from a quantum mechanical treatment and can be approximated in terms of the polarizabilities and ionization potentials of species A and B, as shown in following chapters.

1.5 Thermodynamic Averages and Intermolecular Forces

1.5.1 Properties and Free Energies

The total interaction energy of a macroscopic system such as a gas or condensed matter (liquid, solid, surface, etc.) at a temperature T must be obtained by doing statistical averages for given statistical distributions of the system components and their physical properties. The thermodynamical internal energy or the Gibbs free energy can be obtained from a classical treatment with probability distributions of the Maxwell–Boltzmann type, or from quantal statistical averages using thermal distributions of quantum states [37–39].

The classical statistical average of a many-atom property $F(Q, P_Q)$, where Q is the collection of all internal molecular position variables and center-of-mass locations, and the P_Q are their related momenta, is obtained from the Maxwell–Boltzmann distribution $w_{cl}(Q,P_Q;T,V,\mathcal{N})$, for a system at temperature T and with $\mathcal{N}=N/V$ molecules per unit volume in a volume V, in the form

$$\bar{F}_{T,V,\mathcal{N}} = \int dQ \int dP_Q F(Q,P_Q) w_{cl}(Q,P_Q;T,V,\mathcal{N})$$

In particular, the average energy $\bar{E}_{T,V,\mathcal{N}} = \mathcal{U}(\mathcal{N},V,T)$ is the internal energy. It is usually possible to obtain correct thermal averages of potential energies for a collection of molecules with a classical procedure, provided the system is not electronically excited.

In a quantal description, one first obtains the expectation value of the property, now given by the operator $\hat{F}(\hat{Q},\hat{P}_Q)$, a function of position and momentum operators, in each stationary quantum state $\Phi_n(Q)$ with energy $E_n(V,\mathcal{N})$ and the quantal statistical probability $w_n(T,V,\mathcal{N})$. The statistical average follows from

$$\bar{F}_{T,V,\mathcal{N}} = \sum\nolimits_n \langle F\rangle_n w_n(T,V,\mathcal{N})$$

$$\langle F\rangle_n = \langle \Phi_n | \hat{F} | \Phi_n\rangle = \int dQ\, \Phi_n(Q) F\left(Q, -\frac{i\partial}{\partial Q}\right)\Phi_n(Q)$$

When the operator is the energy of the system, the above procedure leads to expressions for the free energy and internal energy. For a closed system of $\mathcal{N}$ molecules per unit volume, at given temperature and with volume V, the Helmholtz free energy $\mathcal{A}(\mathcal{N},V,T)$ is obtained from the thermal partition function $Q_{th}(\mathcal{N},V,T) = \sum\nolimits_n \exp(-E_n/k_B T)$ by means of [37]

$$\mathcal{A} = -k_B T \ln[Q_{th}(\mathcal{N},V,T)]$$

from which the internal energy follows as $\mathcal{U}(\mathcal{N},V,T) = \mathcal{A} - T(\partial\mathcal{A}/\partial T)_V$. Changing variables from volume in $\mathcal{A}$ to pressure p gives the Gibbs free energy $\mathcal{G}(\mathcal{N},p,T)$.

1.5.2 Polarization in Condensed Matter

A molecular fluid subject to an external electric field $\mathcal{E}$ acquires a polarization per unit volume $\mathcal{P} = \chi_{el}\varepsilon_0\mathcal{E}$, where χ_{el} is the electrical susceptibility and ε_0 is the vacuum permittivity. This polarization can be obtained from an average molecular polarizability $\bar{\alpha}$ per unit volume, which depends on the temperature T and from the density (number of molecules per unit volume $\mathcal{N}$) of the fluid as an statistical average $\mathcal{P} = \mathcal{N}\bar{\alpha}\mathcal{E}_{loc}$, where $\mathcal{E}_{loc}$ is an average local electric field at the position of the molecule and depends on intermolecular forces. It is approximately given by the Lorentz formula $\mathcal{E}_{loc} = \mathcal{E} + \mathcal{P}/(3\varepsilon_0)$. For polar molecules one must also obtain the average dipole per unit volume, related to the permanent molecular dipole. This gives $(\bar{D})_T = D^2\mathcal{E}_{loc}/(3k_B T)$ for the thermal average of the permanent molecular dipole D, to which one must add the induced dipole. The average permanent dipole is the limit, for small fields $\mathcal{E}$, of a more general Langevin formula for the average [16].

In a dielectric (a fluid which does not conduct electricity), the presence of polarization alters the Coulomb interaction between charges, by changing the electric permittivity from ε_0 to $\varepsilon = \varepsilon_0\varepsilon_r$. The relative permittivity ε_r is given by the Debye equation

$$\frac{(\varepsilon_r - 1)}{(\varepsilon_r + 2)} = \frac{\rho \mathcal{P}_m}{M}$$

$$\mathcal{P}_m = \frac{N_A}{2\varepsilon_0}\left(\bar{\alpha} + \frac{D^2}{3k_B T}\right)$$

where N_A is the Avogadro number, ρ is the mass density, M the molar mass, and $\mathcal{P}_m$ is the molar polarization. This formula can be used also for dynamical pertmitivities when the field oscillates with a constant frequency, and also gives the electrical susceptibility insofar as $\chi_{el} = \varepsilon_r - 1$. When the permanent dipole is null, the equation gives a way to measure the polarizability, extracting it from optical measurements of the refractive index $n_r(\omega) = c/c_{med}(\omega) = \varepsilon_r(\omega)^{1/2}$, where c is the speed of light in vacuum and c_{med} in the fluid medium. The average polarization $\mathcal{P}_m(\omega)$ depends on the intermolecular forces in the fluid and provides information about them.

1.5.3 Pair Distributions and Potential of Mean-Force

The potential energy V_N of a fluid containing N molecules can frequently be expressed as a sum over interactions v of pairs (jk) of molecules so that

$$V_N = \sum\nolimits_{1 \le j < k \le N} v_{jk}$$

in which case the internal energy of the fluid can be obtained from a pair correlation function $g_2(R, \Omega_j, \Omega_k)$ for molecules j and k at a distance R and with orientations given by angles Ω_j and Ω_k. Introducing a thermal average $\bar{g}_2(R)$ over a distribution of orientations $(\Omega_j, \Omega_k; T)$, the average total potential energy $\overline{E_{pot}} = \mathcal{U}^{(pot)}(\mathcal{N}, V, T)$ is a function of the temperature and density which can be given in terms of the pair correlation function $\bar{g}_2(R; T, \mathcal{N})$ by [37]

$$\mathcal{U}^{(pot)}(\mathcal{N}, V, T) = \frac{N^2}{2V}\int d^3R\, v(R)\bar{g}_2(R; T, \mathcal{N})$$

It is also useful to introduce a statistical pair potential of mean force $w_2(R; T, \mathcal{N})$ by means of $\bar{g}_2(R; T, \mathcal{N}) = \exp[-w_2(R; T, \mathcal{N})/(k_B T)]$. It can be interpreted as the interaction between two molecules at a distance R when a thermal average is done over the remaining $N-2$ molecules so that the forces between them are both direct and also indirect through the medium.

For the dipole–dipole interaction of a pair at fixed distance R, the orientation average gives the Keesom formula,

$$\overline{[v_{DD}(R)]_T} = \bar{v}_{DD}{}^{(2)}(R) = -\frac{(f^2)_{sph}\left(\mathcal{M}_A^{(m)}\right)^2\left(\mathcal{M}_B^{(n)}\right)^2}{(4\pi\varepsilon_0 R^3)^2 k_B T}$$

which involves an orientation factor f squared after the leading thermal average term disappears. The original dipole–dipole interaction going as R^{-3} has become an average interaction going as R^{-6} as a result of orientational averaging [1].

Using a sum of pair potentials also helps to derive an equation of state. The pair potential can be written as a sum of a short-range hard sphere repulsion term $v^{(SR)}(R) = 0$, $R \geq R_0$, plus a long-range attraction term $v^{(LR)}(R) = -C_6/R^6$. The first one entails that molecular motions occur outside an excluded volume per pair $2b = 4\pi R_0^3/3$, and the second term means that the total attraction energy is $\mathcal{U}^{(attr)} = aN^2/V$ with $a = -2\pi\int_{R_0}^{\infty} dR\, R^2 v^{(LR)}(R)$ [37]. The equation of state is found then to be the well-known van der Waals equation, $(p + aN^2/V^2)(V - Nb) = Nk_BT$.

1.6 Molecular Dynamics and Intermolecular Forces

1.6.1 Collisional Cross Sections

Molecular species A and B can be prepared in molecular beams with given relative velocities and in specific internal states. Their collision in a crossed beam region leads generally into new species C and D emerging from the collision region at solid angles Ω_P where they are detected. Measurements can be stated in terms of differential collisional cross sections $I_{\beta\alpha}^{(L)}(\Omega_P, v_{rel})$ from reactants $A + B$ moving with relative velocity v_{rel} in a reactant channel α to products $C + D$ in channel β, and with species $R = C$ or D detected at angles Ω_P [8, 40, 41].

Cross sections can be calculated from a knowledge of the interaction potential energy between reactant and product species, in terms of the incoming reactant flux $J_\alpha^{(0)}$, as the number of incoming reactants A relative to B per unit area and unit time, and the rate $\dot{N}_{\beta\alpha}$ of the number of transitions from reactants to products per unit time. For an increment $d\Omega_P$ in detection angle, the rate increment is $d\dot{N}_{\beta\alpha} = J_\alpha^{(0)} d\sigma_{\beta\alpha}^{(L)}$ with $d\sigma_{\beta\alpha}^{(L)} = I_{\beta\alpha}^{(L)} d\Omega_P$ an increment of cross sectional area. The integral cross section is obtained from

$$\sigma_{\beta\alpha}^{(L)}(v_{rel}) = \int d\Omega_P I_{\beta\alpha}^{(L)}(\Omega_P, v_{rel}).$$

Calculations of these cross sections as functions of Ω_P, v_{rel} and the channel states, α, starting with the interaction potential energies and the structures of reactants and products, can be compared to experimental measurements to provide very detailed information on potential energy functions and parameters.

These calculations can be very demanding of computational resources. In some cases, when transitions among rotational, vibrational, and electronic states of the molecules have been averaged, calculations can be done with a classical mechanics treatment using equations of motion for classical positions and momentum variables $Q(t)$, $P_Q(t)$ of the constituent atoms. More accurately, the state-to-state cross sections must be obtained from quantal treatments which may involve time-dependent wavepackets or time-independent (stationary) wavefunctions. The stationary wavefunction Φ_E for a total energy $E = E_{rel} + E_{int}^{(\alpha)}$ is a function of nuclear an electronic position variables. Its asymptotic form as the distance R between the products C and D becomes large can be written as $\Phi_E \approx \Phi^{(0)} + \Phi^{(sc)}$ for $R \to \infty$, with $\Phi^{(0)}$ giving the incoming flux $J_\alpha^{(0)}$ and the scattered wave $\Phi^{(sc)}$ giving scattering amplitudes $f_{\beta\alpha}(\Omega_P, v_{rel})$ from which the differential cross section $I_{\beta\alpha}^{(L)}(\Omega_P, v_{rel}) = (p_\beta / p_\alpha) \left| f_{\beta\alpha}(\Omega_P, v_{rel}) \right|^2$ can be calculated [40]. The wavefunction Φ_E satisfies the time-independent Schroedinger equation with a Hamiltonian operator which contains the potential energy of the interacting reactant and product species, and therefore is shaped by the intermolecular forces.

A related subject is transport rates of molecular momentum (viscosity), kinetic energy (thermal conductivity), and mass (diffusion) in real gases and liquids. They are linked to intermolecular forces through collision integrals which can be calculated in many realistic situations, particularly when thermal averages allow use of classical mechanics and deflection functions [1, 8, 33].

1.6.2 Spectroscopy of van der Waals Complexes and of Condensed Matter

The spectroscopy or diffraction pattern of a molecule interacting with light of frequencies extending from radiowaves to UV waves provides extensive information on molecular parameters such as rotational and vibrational constants and oscillator strengths for electronic transitions [16]. Calculations of absorption–emission light shapes and diffraction intensities for given molecular structures, starting from their electronic structure, can be compared with results of measurements to obtain details of potential energy surfaces for atoms near equilibrium positions. A similar procedure can be followed for molecular complexes where molecules are held together by intermolecular forces to form van der Waals molecular pairs, but in this case, the theoretical treatment is complicated by the appearance of floppy motions and quantal tunneling effects, due to small energy variations among minima, maxima, and barriers in the intermolecular potentials.

The treatment of large amplitude motions in van der Waals complexes can be done separating the nuclear position variables into a set of position values $\{u_j\} = \boldsymbol{u}$ which give small displacements from equilibrium positions for rigid structural regions of the complex, and a remaining set $\{q_k\} = \boldsymbol{q}$ of position variables which can range over large distances. The complex dynamics and related spectra are treated starting from a Hamiltonian containing the potential energy $V(\boldsymbol{u}, \boldsymbol{q})$ and kinetic energy terms K_u, K_q, and K_{uq}. Terms containing $\boldsymbol{u}$ are expanded in its powers and treated as done for standard molecular spectra, while K_q and K_{uq} can be expressed in terms of a mass tensor $\boldsymbol{G} = [G_{kk'}(\boldsymbol{q},\boldsymbol{q}')]$ and generalized momenta $\boldsymbol{p} = -i\hbar G^{-\frac{1}{2}}(\partial/\partial\boldsymbol{q})G^{\frac{1}{2}}$ conjugate to $\boldsymbol{q}$, with $G = \det[\boldsymbol{G}]$. Examples of applications can be found for van der Waals complexes like $ArCH_4$ and ArC_6H_6 and hydrogen bonded complexes like $(H_2O)_2$ and $(HCl)_2$ [42, 43]. This approach is also applicable to the study of transient (or collision-induced) complexes detected through light absorption–emission in a collision region. The subject of van der Waals complexes continues to be quite active and has been recently reviewed in [44].

The mentioned treatment can be generalized to include scattering boundary conditions applicable to photodissociation of van der Waals complexes, which can be considered to be half-collisions and can be described as mentioned above with channel amplitudes for collisional phenomena [45].

Intermolecular forces participate in molecular spectra in condensed phases such as real gases, liquids, and molecular solids. They shape absorption–emission lines, Raman spectra, and electron transfer spectra through solvation effects. These spectra can be calculated and compared with experimental measurements to obtain information about the functional form and parameters of involved interaction potential energies [46–48]. Some of these aspects are considered in Chapter 7 on interactions of two many-atom systems.

1.7 Experimental Determination and Applications of Interaction Potential Energies

Knowledge on interaction potential energies can be extracted from experiments measuring properties that depend in a known way on molecular interactions. This requires introduction of realistic theoretical models and the ability to calculate measurable properties, for a range of choices on the form of the potential energies and their parameters. Experimental methods can be classified in three categories: (i) thermodynamic; (ii) spectroscopic and diffraction; and (iii) beam methods. Of these the last one provides the most detailed results when cross sections can be obtained as functions of deflection angles and relative velocities, and for state-to-state collisional transitions, insofar it deals with isolated systems of one or two molecules, but it is limited by the necessity to produce and detect beams of interacting species in known internal states. Spectroscopic methods are even more accurate, but they mostly provide information around equilibrium conformations. This is changing with the recent introduction of time-resolved spectroscopy methods. Following are lists of some relevant measurable properties.

1.7.1 Thermodynamics Properties

- Virial coefficients in equations of state of real gases versus temperature T and pressure p.
- Heat capacities and cohesive energies of solids and liquids versus temperature T and pressure p.
- Compressibilities, sound absorption, and dispersion in solids and liquids versus temperature T and pressure p.
- Adhesive energy of solid surfaces versus temperature T and pressure p.
- Viscosity, diffusion, and heat conduction transport coefficients of real gases versus temperature T and pressure p.
- Relaxation rates of temperature or pressure of real gases near equilibrium temperature T and pressure p.

1.7.2 Spectroscopy and Diffraction Properties

- Molecular spectra of weakly bound (van der Waals) atomic and molecular complexes formed in beams.
- Collision-induced spectra in gases versus light wavelengths at given temperature T and pressure p.
- Line shapes in condensed matter versus light wavelengths at given temperature T and pressure p.
- Relaxation times of quantum state populations in condensed matter versus temperature T and pressure p.
- Electron, neutron, and light diffraction intensities in condensed matter versus temperature T and pressure p.
- Time-resolved spectral intensities versus light wavelengths in pulsed laser (pumping–probing) experiments.

1.7.3 Molecular Beam and Energy Deposition Properties

- Cross sections for elastic, inelastic, and reactive collisions in crossed molecular beams.
- Energy loss cross sections for beams traversing fluids or solids.

- Photodissociation of atomic and molecular complexes and of clusters formed in beams.
- Cross sections for scattering of atoms and molecules by solid surfaces and by adsorbates.

1.7.4 Applications of Intermolecular Forces

Intermolecular forces are essential in many applications of molecular and materials properties to technologies contributing to social needs. Some properties (and their applications) are as follows:

- Storage of hydrogen in solids (fuel cells)
- Storage and transport of ions in solids (batteries)
- Synthesis of thermally stable and conducting surfaces for solar devices (solar panels)
- Delivery of compounds through biological cell membranes (pharmacology)
- Catalysis and photocatalysis in electrochemical cells (sustainable fuel production)
- Atmospheric reactions (environmental sciences)
- Efficient fuel combustion (transportation and energy)
- Solvation and lubricants (machinery)

References

1 Hirschfelder, J.O., Curtis, C.F., and Byron Bird, R. (1954). *Molecular Theory of Gases and Liquids*. New York: Wiley.
2 Hirschfelder, J.O. (ed.) (1967). *Intermolecular Forces* (Adv. Chem. Phys. vol. 12). New York: Wiley.
3 Margenau, H. and Kestner, N.R. (1971). *Intermolecular Forces*, 2e. Oxford, England: Pergamon Press.
4 Dainton, F.S. (ed.) (1965). *Faraday Society Discussions* (Intermolecular Forces vol. 40). London, England: Royal Society of Chemistry.
5 Lawley, K.P. (ed.) (1980). *Potential Energy Surfaces* (Adv. Chem. Phys. vol. 42). New York: Wiley.
6 Pullman, B. (ed.) (1982). *Intermolecular Forces*. Dordrecht, Holland: Reidel Publishing Company.
7 Smith, I.W.M. (1980). *Kinetics and Dynamics of Elementary Gas Reactions*. London, Great Britain: Butterworth.
8 Maitland, G.C., Rigby, M., Brain Smith, E., and Wakeham, W.A. (1981). *Intermolecular Forces*. Oxford, England: Oxford University Press.
9 Murrell, J.N., Carter, S., Farantos, S.C. et al. (1984). *Molecular Potential Energy Functions*. New York: Wiley.

10 Hirst, D.M. (1985). *Potential Energy Surfaces*. London, England: Taylor & Francis.
11 Stone, A.J. (2013). *The Theory of Intermolecular Forces*, 2e. Oxford, England: Oxford University Press.
12 Israelachvili, J. (2011). *Intermolecular and Surface Forces*, 3e. New York: Elsevier.
13 Kaplan, I.G. (2006). *Intermolecular Interactions*, 81. New York: Wiley.
14 Born, M. and Oppenheimer, J.R. (1927). Zur Quantentheorie der Molekeln. *Ann. Phys.* 84: 457.
15 Born, M. and Huang, K. (1954). *Dynamical Theory of Crystal Lattices, Appendix VIII*. Oxford, England: Oxford University Press.
16 Atkins, P.W. and Friedman, R.S. (1997). *Molecular Quantum Mechanics*. Oxford, England: Oxford University Press.
17 Schatz, G.C. and Ratner, M. (1993). *Quantum Mechanics in Chemistry*. Englewood, New Jersey: Prentice-Hall.
18 Cohen-Tanoudji, C., Diu, B., and Laloe, F. (1977). *Quantum Mechanics*, vol. 1 and 2. New York: Wiley.
19 Messiah, A. (1962). *Quantum Mechanics*, vol. 1 and 2. Amsterdam: North-Holland.
20 Slater, J.C. (1963). *Quantum Theory of Molecules and Solids*, vol. 1. New York: McGraw-Hill.
21 Kolos, W. and Wolniewicz, L. (1965). Potential-energy curves for the $X^1\Sigma_g^+$, $b^3\Sigma_u^+$, and $C^1\Pi_u$ states of the hydrogen molecule. *J. Chem. Phys.* 43: 2429.
22 Levine, I.N. (2000). *Quantum Chemistry*, 5e. New Jersey: Prentice-Hall.
23 Yagi, K. and Takatsuka, K. (2005). Nonadiabatic chemical dynamics in an intense laser field: Electronic wave packet coupled with classical nuclear motions. *J. Chem. Phys.* 123: 224103.
24 Hovmoeller, T. (2002). Conformations of amino acids in proteins. *Acta Cryst* D58: 768.
25 Bender, C.F., Pearson, P.K., O'Neil, S.V., and Schaefer, H.F. III (1972). Potential energy surface including electron correlation for the chemical $F+H_2 \rightarrow FH+H$ I. Preliminary surface. *J. Chem. Phys.* 56: 4626.
26 Baer, M. (2006). *Beyond Born-Oppenheimer: Electronic Nonadiabatic Coupling Terms and Conical Intersections*. New York: Wiley.
27 Preston, R.K. and Tully, J.C. (1971). Effects of surface crossing in chemical reactions: the H3+ system. *J. Chem. Phys.* 54: 4297.
28 Baer, M. and Beswick, J.A. (1979). Electronic transitions in the ion-molecule reaction $Ar^+ + H_2$ to $Ar + H_2^+$ and $ArH^+ + H$. *Phys. Rev. A* 19: 1559.
29 Worth, G.A. and Cederbaum, L.S. (2004). Beyond Born-Oppenheimer: molecular dynamics through a conical intersection. *Annu. Rev. Phys. Chem.* 55: 127.
30 Levine, B.G. and Martinez, T.J. (2007). Isomerization through conical intersections. *Annu. Rev. Phys. Chem.* 58: 613.

31 Domcke, W. and Yarkony, D.R. (2012). Role of conical intersections in molecular spectroscopy and photoinduced chemical dynamics. *Annu. Rev. Phys. Chem.* 63: 325.
32 Ben-Nun, M. and Martinez, T.J. (2000). Photodynamics of ethylene: ab initio studies of conical intersections. *Chem. Phys.* 259: 237.
33 Wolken, G.J. (1976). Scattering of atoms and molecules from solid surfaces. In: *Dynamics of Molecular Collisions Part A* (ed. W.H. Miller), 211. New York: Plenum Press.
34 Marcus, R. (1964). Chemical and electrochemical electron-transfer theory. *Annu. Rev. Phys. Chem.* 15: 155.
35 Hush, N.S. (1961). Adiabatic theory of outer sphere electron-transfer reactions in solution. *Trans. Faraday Soc.* 57: 557.
36 Schapiro, I., Melaccio, F., Laricheva, E.N., and Olivucci, M. (2011). Using the computer to understand the chemistry of conical intersections. *Photochem. Photobiol. Sci.* 10: 867.
37 McQuarrie, D.A. (1973). *Statistical Mechanics*. New York: Harper & Row.
38 Gray, C.G. and Gubbins, K.E. (1984). *The Theory of Molecular Fluids. I. Fundamentals*. Oxford, England: Clarendon Press.
39 Kittel, C. (2005). *Introduction to Solid State Physics*, 8e. Hoboken, NJ: Wiley.
40 Child, M.S. (1974). *Molecular Collision Theory*. New York: Academic Press.
41 Lawley, K.P. (ed.) (1975). *Molecular Scattering* (Adv. Chem. Phys. vol. 30). New York: Wiley.
42 Wormer, P.E.S. and van der Avoird, A. (2000). Intermolecular potentials, internal motions, and spectra of van der Waals and hydrogen-bonded complexes. *Chem. Rev.* 100: 4109.
43 Beswick, J.A. and Jortner, J. (1981). Intramolecular dynamics of van der Waals molecules. In: *Photoselective Chemistry, Part 1* (Adv. Chem. Phys. vol. 47) (eds. J. Jortner, R.D. Levine and S.A. Rice), 363. New York: Wiley.
44 Hobza, P. and Rezak, J. (eds.) (2016). *Non-Covalent Interactions*. Special Issue of Chemical Review, vol. 116. Washington DC: American Chemical Society.
45 Lawley, K.P. (ed.) (1985). *Photodissociation and Photoionization* (Adv. Chem. Phys. vol. 60). New York: Wiley.
46 Mukamel, S. (1995). *Principles of Nonlinear Optical Spectroscopy*. Oxford, England: Oxford University Press.
47 May, V. and Kuhn, O. (2000). *Charge and Energy Transfer Dynamics in Molecular Systems*. Berlin: Wiley-VCH.
48 Nitzan, A. (2006). *Chemical Dynamics in Condensed Phases*. Oxford, England: Oxford University Press.

2

Molecular Properties

CONTENTS

2.1 Electric Multipoles of Molecules

2.1.1 Potential Energy of a Distribution of Charges

The size of a molecule is of the order of a nanometer, while the wavelength of visible light is about 500 nm, and larger for infrared or radio waves. Therefore, the amplitude of an externally applied electric field can be assumed to be constant over space except in two situations: (i) for large atomic systems such as clusters, polymers, surfaces, or proteins; or (ii) for inhomogeneous fields created by a

Molecular Interactions: Concepts and Methods, First Edition David A. Micha.

neighbor molecule. These cases can be included in a general treatment by letting the electric field $\mathcal{E}(\vec{r},t)$ be a function of position $\vec{r}$ and time t. We ignore here the magnetic field associated to it because the interaction energy of a molecule with a magnetic field is much smaller than the interaction with the electric field, and will concentrate on electric multipoles and polarizabilities. Magnetic field effects can be considered as perturbations involving the magnetic properties of the molecules, such as their electronic spin and orbital magnetic moments.

We consider a molecule at rest with its center of mass located at the origin of a coordinate system $\mathcal{S}$. To start with we treat all charged particles as classical ones with well-defined positions and will later on allow for their quantal nature. The presence of electric charges in space creates an inhomogeneous electric potential $\phi(\vec{r})$ (also called $U(\vec{r})$ in some books). If a molecule with its own distribution of charges $\{C_I\}$ at locations $\{\vec{r}_I\}$ (with $I = a$ indicating a nucleus and $I = i$ an electron) enters this region of space, the potential energy function of particle positions is given by $V_{MF} = \sum_I C_I \phi(\vec{r}_I)$, due to the molecule-field interaction. This energy can be expressed in terms of the electric potential and its space derivatives around the origin of coordinates. In effect, using Cartesian coordinates $\xi = x, y, z$, and expanding in a series around $\xi = 0$,

$$\phi(\vec{r}) = \phi(\vec{0}) + \sum_{\xi} \xi \left(\frac{\partial \phi}{\partial \xi}\right)_0 + \frac{1}{2}\sum_{\xi,\eta} \xi\eta \left(\frac{\partial^2 \phi}{\partial \xi \partial \eta}\right) + \cdots$$

$$V_{MF} = \phi(0)\left(\sum_I C_I\right) + \sum_{\xi} \left(\frac{\partial \phi}{\partial \xi}\right)_0 \left(\sum_I \xi_I C_I\right) + \frac{1}{2}\sum_{\xi,\eta} \left(\frac{\partial^2 \phi}{\partial \xi \partial \eta}\right)_0 \left(\sum_I C_I \xi_I \eta_I\right) + \cdots$$

where summations over I display in order from the left: total charge $C = \sum_I C_I$, total dipole Cartesian component $D_\xi = \sum_I \xi_I C_I$, total quadrupole, and so on. We use the relation $C_e = e/(4\pi\varepsilon_0)^{1/2}$ between the electron charge C_e and its symbol e in esu units, with ε_0 the vacuum permittivity.

These expansions can be rewritten in terms of the electric field vector component $\mathcal{E}_\xi = -\partial\phi/\partial\xi$, and its space derivatives, and are valid whether the electric field is internal to a molecular system, or externally applied. Simplifications arise when the electric field is external and the molecule is not large so that the field can be assumed to be constant over its extent. Then the quadrupolar and higher terms can be omitted and a quantal treatment of the molecular properties can be derived from a quantized Hamiltonian operator $\hat{H}_{MF} = \phi(0)\left(\sum_I C_I\right) - \sum_\xi \mathcal{E}_\xi \left(\sum_I \hat{\xi}_I C_I\right)$, where operators are shown with carats [1–4].

2.1.2 Cartesian Multipoles

The distribution of nuclear and electronic charges in a molecule can be described in terms of electrical multipoles [1, 2, 5]. These can be constructed from classical mechanics and then converted into quantal operators.

Let particle I (a nucleus or electron) have charge C_I and be located at $\vec{r}_I$ with components $\xi_I = x_I, y_I, z_I$. Then the dipole vector component along direction ξ is given by

$$D_\xi = \sum_I C_I \xi_I$$

and its units are *Coulomb* × *meter* (or *Cm*) in the SI system or $e.a_0$ in atomic units; another usual unit is 1 *Debye* = 1 $D = 10^{-18}$ *esu*. The relation among them is 1 *au*(*dip*) = 2.5418 $D = 8.478 \times 10^{-30}$ Cm. The quadrupole tensor component along (ξ, η) is defined by

$$Q_{\xi\eta} = \sum_I C_I \left(3\,\xi_I \eta_I - r_I^2 \delta_{\xi\eta}\right)$$

where $\delta_{\xi\eta}$ equals one for $\xi = \eta$ and zero otherwise, and $r_I^2 = x_I^2 + y_I^2 + z_I^2$, constructed so that the trace $\sum_\xi Q_{\xi\xi} = 0$. An alternative definition of the quadrupole in the literature is $\Theta_{\xi\eta} = Q_{\xi\eta}/2$.

In general, the Cartesian components $\xi, \eta, \ldots \zeta$ of the n-th electric multipole are given by

$$M^{(n)}_{\xi,\eta,\ldots\zeta} = \frac{(-1)^n}{n!} \sum_I C_I r_I^{2n+1} \left[\frac{\partial}{\partial \zeta} \cdots \frac{\partial}{\partial \eta} \frac{\partial}{\partial \xi} \left(\frac{1}{r}\right)\right]_I$$

which is traceless when adding over a pair of identical indices, as follows from the relation $\nabla^2(1/r) = \sum_\xi \partial^2 r^{-1}/\partial \xi^2 = 0$.

2.1.3 Spherical Multipoles

The analytical form of electric multipoles changes as the molecule is rotated. To describe that change, it is more convenient to work with spherical components of the multipole, and the corresponding spherical components of the electric potential gradients. Components of rotational number $l \geq 0$ and projection $-l \geq m \geq l$ are given by

$$Q_{lm} = \left(\frac{4\pi}{2l+1}\right)^{1/2} \sum_I C_I r_I^l Y_{lm}(\vartheta_I, \varphi_I)$$

written in terms of the spherical harmonic function Y_{lm} of position angles, with well-known transformation properties under rotation. From $Y_{00}(\vartheta_I, \varphi_I) = 1/\sqrt{4\pi}$ one finds that Q_{00} is the total charge. Dipole and quadrupole multipoles follow from $l = 1$ and $l = 2$, respectively, and are related to Cartesian components, so that, for example $Q_{11} = \left(D_x + iD_y\right)/\sqrt{2} = Q^*_{1,-1}$ and $Q_{10} = D_z$. The potential energy of interaction takes the form $V_{MF} = \sum_{lm}(-1)^m Q_{lm} F_{l,-m}$, where the coefficients are related to the electric field and its derivatives. See [4, 6, 7] for details.

2.1.4 Charge Distributions for an Extended System

Here one introduces the density of charge per unit volume $c(\vec{r}) = \sum_I C_I\, \delta(\vec{r} - \vec{r}_I)$, as a sum of Dirac delta functions over all the point charges in the system, electrons (with $I = i$) and nuclei (with $I = a$). Alternatively, the sum is over valence electrons and ionic cores. One can analyze the distribution in terms of Cartesian or spherical components and relate them to the previous results. For example the dipole ξ-components of the extended system are $D_\xi = \int d^3r\, \xi\, c(\vec{r})$ and other multipoles can be obtained from higher moments of the charge density with respect to position components. However, such multipole description is not practical for an extended system, but possibly only for its molecular fragments. Instead, it is more general to work with the total charge density.

When particles are treated within quantum mechanics, the charge density becomes an operator $\hat{c}(\vec{r}) = \sum_I C_I\, \delta\left(\vec{r} - \hat{\vec{r}}_I\right)$, and its average values must be obtained from expectation values over quantal states. These are functions $\Phi_k(\boldsymbol{X}, \boldsymbol{Q})$ of all the electronic variables (positions and spins) $\boldsymbol{X}$ and all nuclear (or ionic core) positions $\boldsymbol{Q}$. Provided electronic and nuclear motions are not very strongly coupled, as it happens for low electronically excited states, the state functions are given in the Born–Oppenheimer approximation by products $\Phi_k(\boldsymbol{X}, \boldsymbol{Q}) = \Phi_J^{(nu)}(\boldsymbol{Q})\Phi_\alpha^{(el)}(\boldsymbol{X}; \boldsymbol{Q})$ of a many-electron wavefunction for fixed nuclei and a nuclear motions wavefunction describing vibrations and rotations of the molecule. The electronic functions $\Phi_\alpha^{(el)}(\boldsymbol{X}; \boldsymbol{Q})$ must be antisymmetric with respect to electron exchanges and can be constructed as superpositions of electronic determinantal wavefunctions. The nuclear wavefunctions $\Phi_J^{(nu)}(\boldsymbol{Q})$ depend on variable distances between nuclei and on molecular orientation angles and describe vibrational and rotational motions for each given electronic state.

It is convenient to separate nuclear and electronic charge densities and to use the Born–Oppenheimer approximation where electronic states are generated for fixed nuclear positions $\boldsymbol{Q}$, with the nuclei treated to begin with as classical particles so that $\hat{c}(\vec{r}) = \hat{c}^{(nu)}(\vec{r}) + \hat{c}^{(el)}(\vec{r})$ with $\hat{c}^{(nu)}(\vec{r}) = \sum_a C_a\, \delta\left(\vec{r} - \hat{\vec{r}}_a\right)$ and $\hat{c}^{(el)}(\vec{r}) = C_e \sum_i \delta\left(\vec{r} - \hat{\vec{r}}_i\right)$. This gives for an electronic state $\Phi_\alpha^{(el)}(\boldsymbol{X}; \boldsymbol{Q}) = \langle \boldsymbol{X} \mid \Phi_\alpha^{(el)}(\boldsymbol{Q})\rangle$, using the Dirac bra-ket notation and integrating over electronic variables, the quantal expectation value for the averaged electronic density $c_\alpha^{(el)}(\vec{r}; \boldsymbol{Q}) = \left\langle \Phi_\alpha^{(el)}(\boldsymbol{Q}) \mid \hat{c}^{(el)}(\vec{r}) \mid \Phi_\alpha^{(el)}(\boldsymbol{Q})\right\rangle$, and a total charge density $c_\alpha(\vec{r}, \boldsymbol{Q}) = c^{(nu)}(\boldsymbol{Q}) + c_\alpha^{(el)}(\vec{r}; \boldsymbol{Q})$.

This can be treated by decomposition into physically or mathematically convenient terms, introducing a variety of approaches: (a) A decomposition into atomic and bond densities (or atomic charge- and bond-order sum) suitable for large molecules; (b) decomposition into fragments densities for molecules in liquid solutions and at surfaces; (c) plane-wave decompositions for extended systems with many atoms; and (d) decompositions into finite elements for continuum models. These are described in some detail as follows:

a) The charge density operator can be represented by a matrix in a basis set of atomic orbitals, and matrix elements can be classified as atomic- or bond-like. If an electronic structure can be accurately described by a basis set of atomic orbitals $\chi_\nu^{(n)}$ centered at nucleus n and with atomic quantum numbers ν, then a related basis set of localized orthonormal orbitals $\varphi_\nu^{(n)}$ can be introduced using the overlap matrix $\mathbf{\Delta}$ with elements $\Delta_{\mu,\nu}^{(m,n)} = \left\langle \chi_\mu^{(m)} | \chi_\nu^{(n)} \right\rangle$ and the symmetric transformation orbital $\varphi_\nu^{(n)} = \sum_{m,\mu} | \chi_\mu^{(m)} \rangle \left(\mathbf{\Delta}^{-1/2} \right)_{\mu\nu}^{(m,n)}$, which is large at location n and contains small contributions of nearby atomic orbitals. It provides a partial completeness relation $\hat{I} = \sum_{m\mu} | \varphi_\mu^{(m)} \rangle \langle \varphi_\mu^{(m)} |$. With the identity $\hat{I}$ operating to right and left of the charge density operator, this becomes a sum over single atom and atom-pair terms,

$$\hat{c}^{(el)}(\vec{r}) = \sum_{m,\mu} | \varphi_\mu^{(m)} \rangle \left| \varphi_\mu^{(m)}(\vec{r}) \right|^2 \langle \varphi_\mu^{(m)} | + \sum_{m\mu \neq n\nu} | \varphi_\mu^{(m)} \rangle \varphi_\mu^{(m)}(\vec{r})^* \varphi_\nu^{(n)}(\vec{r}) \langle \varphi_\nu^{(n)} |.$$

The expectation value $c_\alpha^{(el)}(\vec{r})$ for a given electronic state α can be interpreted as an atomic charge- and bond-order sum, and terms can be given physical meaning [8].

b) Molecular fragments are identified by a subset of nuclear positions, usually chosen with well-known structure, each with its own multipoles and polarizations constructed to reproduce the fragments density, and the total charge density can be given for the combined fragments in the form of a sum over all the distributed multipoles and polarizations [4]. In each fragment, the multipoles and polarizations are located at chosen points $\vec{a}$ and are expressed in terms of localized functions. These are conveniently written using Gaussian exponentials, which can be readily translated between locations, of the form

$$L_{\vec{a}lm}(\vec{r}) = r_a^l \left[\frac{4\pi}{(2l+1)} \right]^{1/2} Y_{lm}(\vartheta_a, \varphi_a) \exp\left(-\zeta r_a^2\right)$$

where $(r_a, \vartheta_a, \varphi_a)$ are angular variables for $\vec{r} - \vec{a}$.

c) The density can be expanded in plane waves for a collection of vectors selected to form a grid in a reciprocal space. This is convenient when the charge densities are smoothly varying, which excludes ion core electrons. It is usually done enclosing the molecule in a volume $\Omega = L_1 L_2 L_3$ with periodic boundary conditions which restrict the reciprocal vectors to a denumerable set $\{\vec{q}_n\}$ with $\boldsymbol{n} = (n_1, n_2, n_3)$ a triplet of integers. The density operator is Fourier transformed from space to reciprocal space using that $\langle \vec{q}_n | \vec{q}_{n'} \rangle = \Omega^{-1} \int_\Omega d^3 r \exp(-i\vec{q}_n . \vec{r}) \exp(i\vec{q}_{n'} . \vec{r}) = \delta_{nn'}$, as

$$\hat{c}(\vec{r}) = \sum_n \exp(i\vec{q}_n . \vec{r}) \hat{c}_n, \quad \hat{c}_n = \int_\Omega \frac{d^3 r}{\Omega} \exp(-i\vec{q}_n . \vec{r}) \hat{c}(\vec{r})$$

and the operator coefficients are used in interactions. This expansion is common in treatments of solid-state properties and is particularly useful for polymers, surfaces, and solids with periodic atomic structure [9].

d) The density can be numerically represented by a decomposition of a continuous distribution of charges into three-dimensional finite elements. Each element is a solid tetrahedron $P_3 = a_0 + a_1 x + a_2 y + a_3 z$ with coefficients fixed by density values at the four corners. A set of tetrahedra can cover a general distribution of the density and can be used in the treatment of two interaction densities. Finite elements have been introduced in treatments of optical properties of atomic clusters [10] and are convenient also for biomolecular distributions.

Molecular interactions can be described in terms of these density decompositions. They are all approximations, and their accuracy must be tested by systematic improvement of the decomposition procedures.

2.2 Energy of a Molecule in an Electric Field

2.2.1 Quantal Perturbation Treatment

The energy of a molecule in an electric field must be obtained in quantum mechanics as the expectation value of the energy operator for the system. The starting point is the quantal Hamiltonian for a set of charged particles in an external electromagnetic field (e.m.) described by the vector and scalar potentials $\vec{A}(\vec{r},t)$ and $\phi(\vec{r},t)$ at each point in space. The electric field follows from $\vec{\mathcal{E}}(\vec{r},t) = -\partial \vec{A} / \partial t - \partial \phi / \partial \vec{r}$. For a homogeneous external field without external charges, $\phi = 0$ and the divergence $\nabla . \vec{A} = 0$, which simplify the molecular internal charge-field interaction as $\hat{H}_{MF} = -\sum_I C_I \vec{A} . \vec{p}_I / m_I$ to lowest order in the field strength, with $\vec{p}_I$ the charged particle momentum. The Hamiltonian can be further simplified when the wavelengths of the e.m. field is long compared to molecular dimensions so that only a dipolar coupling to an external electric field is needed [6].

This gives a Hamiltonian operator $\hat{H} = \hat{H}_M + \hat{H}_{MF}$ with two terms, corresponding to the energy operator of the molecule in free space plus the operator form of the molecule-field interaction energy $\hat{H}_{MF} = -\sum_{\xi}\hat{D}_{\xi}\mathcal{E}_{\xi}(t)$ for an externally applied electric field of components $\mathcal{E}_{\xi}$. This field can vary with position for a large molecule, but is otherwise constant over the molecular extension.

When the maximum applied external electric field amplitude $\mathcal{E}_{ext}$ is small compared to internal electric fields in the molecule, of magnitude $\mathcal{E}_{int}$, one can use the parameter $\lambda = \mathcal{E}_{ext}/\mathcal{E}_{int}$ as the perturbation expansion variable, with $\hat{H}_M = \hat{H}^{(0)}$ and $\hat{H}_{MF} = \lambda\hat{H}^{(1)}$. Perturbation solutions for the wavefunction can be obtained using a basis set of unperturbed molecular states $\left\{\Phi_l^{(0)}\right\}$, solutions of the Hamiltonian equation $\hat{H}^{(0)}\Phi_l^{(0)} = E_l^{(0)}\Phi_l^{(0)}$ with the orthonormalization condition $\left\langle\Phi_l^{(0)} \mid \Phi_m^{(0)}\right\rangle = \delta_{lm}$ and assuming completeness, or $\sum_l \left|\Phi_l^{(0)}\right\rangle\left\langle\Phi_l^{(0)}\right| = \hat{I}$, the identity operator [6]. The molecular wavefunctions $\Phi_k(\boldsymbol{X}, \boldsymbol{Q}) = \langle \boldsymbol{X}, \boldsymbol{Q} \mid \Phi_k\rangle$ can be taken as Born–Oppenheimer (adiabatic) products $\Phi_k(\boldsymbol{X},\boldsymbol{Q}) = \Phi_J^{(nu)}(\boldsymbol{Q})\Phi_\alpha^{(el)}(\boldsymbol{X};\boldsymbol{Q})$, with the bracket notation here indicating integration over both electronic and nuclear position variables.

The perturbation treatment depends on whether the perturbing field is static or changing with time. In the first case, it involves solving for the time-independent Schroedinger equation, $\left[\hat{H}^{(0)} + \lambda\hat{H}^{(1)}(0)\right]\Phi_l = E_l\Phi_l$ to obtain stationary states of fixed energy, while for a time-dependent field, the wavefunction must satisfy the equation of motion $\left[\hat{H}^{(0)} + \lambda\hat{H}^{(1)}(t)\right]\Psi(t) = i\hbar\partial\Psi/\partial t$, with an initial condition $\Psi(0) = \Psi^{(in)}$ determined by the state of the molecule before it is perturbed.

Solutions can be generated using operator methods, or in more detail introducing an expansion in the basis set of unperturbed states. This expansion gives physically meaningful expressions involving matrix elements of the multipole or charge density operators. The ground and excited states must be generated and the treatment is accurate provided enough excitations can be considered.

A general and compact perturbation treatment can also be done introducing resolvent operators and a basis set partitioning method. This also facilitates treatment of perturbation of a degenerate energy eigenvalue, with several unperturbed states of this same energy [6, 11–13]. An operator-based treatment is needed when the molecule is very large and its relevant unperturbed states are not all known. In this case, it is yet possible to account for property values by expanding in chosen basis sets, or with an expansion in a set of operator amplitudes.

2.2.2 Static Polarizabilities

For a field constant in time, properties can be obtained from the eigenvalue solutions of the quantal Hamiltonian equation $\hat{H}\Phi_k = E_k\Phi_k$. The stationary molecular states Φ_k and their energies E_k for each state k depend on the magnitude of the applied field and are given by the products $\Phi_k(\boldsymbol{X},\boldsymbol{Q}) = \Phi_I^{(nu)}(\boldsymbol{Q})\Phi_\alpha^{(el)}(\boldsymbol{X};\boldsymbol{Q})$ with the electronic functions $\Phi_\alpha^{(el)}(\boldsymbol{X};\boldsymbol{Q})$ antisymmetric with respect to electron exchanges, and constructed as superpositions of electronic determinantal wavefunctions.

The response of a molecule to a time-independent electric field can be expressed in terms of static polarizabilities, which can be obtained from a quantum mechanical treatment using perturbation theory, insofar as external fields are usually small compared to internal ones [5, 6, 14]. Expanding states and their energies in powers of the field strength parameter λ one has

$$\left(\hat{H}^{(0)} + \lambda\hat{H}^{(1)}\right)\left(\sum\nolimits_{n\geq 0}\lambda^n \Phi_k^{(n)}\right) = \left(\sum\nolimits_{p\geq 0}\lambda^p E_k^{(p)}\right)\left(\sum\nolimits_{q\geq 0}\lambda^q \Phi_k^{(q)}\right)$$

and equating coefficients of each power one finds for $n = p + q \geq 1$ that, after multiplication by $\Phi_k^{(0)*}$ and integration over its variables,

$$E_k^{(1)} = \left\langle \Phi_k^{(0)} \mid \hat{H}^{(1)} \mid \Phi_k^{(0)} \right\rangle,\ \Phi_k^{(1)} = \sum\nolimits_{l\neq k}\left\langle \Phi_k^{(0)} \mid \hat{H}^{(1)} \mid \Phi_l^{(0)} \right\rangle \Phi_l^{(0)} / \left(E_k^{(0)} - E_l^{(0)}\right)$$

for first-order corrections, and that to second order

$$E_k^{(2)} = \sum\nolimits_{l\neq k}\left|\left\langle \Phi_k^{(0)} \mid \hat{H}^{(1)} \mid \Phi_l^{(0)} \right\rangle\right|^2 / \left(E_k^{(0)} - E_l^{(0)}\right)$$

with similar expressions for higher-order perturbation corrections. Replacing $\lambda\hat{H}^{(1)} = -\sum_\xi \mathcal{E}_\xi \hat{D}_\xi$ in these equations one obtains the field dependence of the energy E_k at each order.

The total average dipole component $\langle D_\xi \rangle$ equals the permanent molecular dipole plus a dipole induced by the field, and this contains as the leading terms the molecular dipole polarizability plus higher polarizabilities. It is obtained as the derivative of the total energy with respect to the corresponding electric field component so that for a chosen state $k = a$ it is $\langle D_\xi \rangle_a = -(\partial E_a / \partial \mathcal{E}_\xi)$. This can be expanded in powers around $\mathcal{E}_\xi = 0$ to obtain

$$\langle D_\xi \rangle_a = D_{\xi,a}^{(0)} + \sum\nolimits_\eta \alpha_{\xi\eta,a}\mathcal{E}_\eta + \frac{1}{2}\sum\nolimits_{\eta,\zeta}\beta_{\xi\eta\zeta,a}\,\mathcal{E}_\eta\mathcal{E}_\zeta + \cdots$$

$$\alpha_{\xi\eta,a} = -\left(\frac{\partial^2 E_a}{\partial\mathcal{E}_\xi\partial\mathcal{E}_\eta}\right)_0$$

$$\beta_{\xi\eta\zeta,a} = -\left(\frac{\partial^3 E_a}{\partial\mathcal{E}_\xi\partial\mathcal{E}_\eta\partial\mathcal{E}_\zeta}\right)_0$$

where $\alpha_{\xi\eta}$ is a dipolar polarizability of the molecule and $\beta_{\xi\eta\zeta}$ is a dipolar hyperpolarizability. These tensor components are usually related by molecular symmetry and depend on the molecular orientation with respect to a reference frame. The polarizability is found from the second-order correction to the energy as

$$\alpha_{\xi\eta,a} = \sum\nolimits_{l \neq a} \frac{\left\langle \Phi_a^{(0)} \mid \hat{D}_\xi \mid \Phi_l^{(0)} \right\rangle \left\langle \Phi_l^{(0)} \mid \hat{D}_\eta \mid \Phi_a^{(0)} \right\rangle + c.c.}{E_l^{(0)} - E_a^{(0)}}$$

which is positive valued when a is the ground state, of lowest energy. Here *c.c.* stands for complex conjugate of the preceding form. The hyperpolarizability requires calculation of the third-order correction to the energy and can be found from recursion relations.

The static polarizability is related to the transition dipole for each state-to-state transition, and the largest contributions to the polarizability come from large transition matrix elements and small excitation energies. They can be calculated separating nuclear and electronic dipole operators and using the Born–Oppenheimer factorization of states.

The $\langle Q_{\xi\eta} \rangle_a$ Cartesian components of the total quadrupole are similarly obtained differentiating the total energy with respect to the field gradient tensor components $\mathcal{E}_{\xi\eta} = -\partial^2 \phi / \partial\xi\partial\eta$, to obtain the permanent molecular quadrupole plus its quadrupolar polarization components.

An alternative description is based on the expansion of the molecule-field interaction in terms of the spherical components of the dipole operator and of the electric field. Derivatives of the energy with respect to the spherical components of the electric field give the total spherical components of the dipole, as sums of a permanent value plus a value derived from the polarizability and contributions from hyperpolarizabilities [5]. Using spherical components, it is possible to give compact expressions for higher multipoles [4].

Molecular multipoles can be calculated as quantal averages so that the total molecular n-pole average is $\left\langle M_{\xi,\eta,\ldots\zeta}^{(n)} \right\rangle_a = \left\langle \Phi_a \mid \hat{M}_{\xi,\eta,\ldots\zeta}^{(n)} \mid \Phi_a \right\rangle$, a function of the applied electric field. Permanent n-pole and polarizabilities follow from the perturbation expansion keeping up to second-order terms.

2.3 Dynamical Polarizabilities

2.3.1 General Perturbation

The response of a molecule to a time-dependent electromagnetic field can instead be expressed in terms of dynamical polarizabilities, which follow from a quantal treatment and time-dependent perturbation theory [6]. As mentioned, the Hamiltonian can be simplified when the wavelengths of the e.m. field is long compared to molecular dimensions so that only a dipolar coupling to an external electric field is needed [6]. The associated Schroedinger equation of motion must be considered to obtain the time-dependent wavefunction $\Psi(t)$ of the molecule plus field, or more generally the time-evolution operator describing a general perturbation. It will be done here in particular for an external electric field which is initially null at times $t<0$ and grows to a given oscillating field of constant amplitude and is homogeneous over the molecular region. The perturbation expansion at time t leads to expressions involving integrals over earlier time t' and integral kernels which are functions of t and t', with Fourier transforms providing response functions. The treatment can also be conveniently done introducing time-delayed propagators [13].

The Hamiltonian is $\hat{H}=\hat{H}_M+\hat{H}_{MF}(t)$, with $\hat{H}_M$ a time-independent Hamiltonian for the isolated molecule in its stationary states, and $\hat{H}_{MF}(t)=-\hat{D}_z\mathcal{E}_z(t)$. As before, when the maximum value $\mathcal{E}_0$ of a time-dependent applied external electric field amplitude is small compared to internal electric fields in the molecule, of magnitude $\mathcal{E}_{int}$, one can use the parameter $\lambda=\mathcal{E}_0/\mathcal{E}_{int}$ as the perturbation expansion variable, with $\hat{H}_M=\hat{H}^{(0)}$ and $\hat{H}_{MF}(t)=\lambda\hat{H}^{(1)}(t)$, so that the time-dependent equation of motion for the state is

$$\left[\hat{H}^{(0)}+\lambda\hat{H}^{(1)}(t)\right]\Psi(t)=i\hbar\partial\Psi/\partial t$$

and it can be solved with the wavefunction expansion $\Psi(t)=\sum_{n\geq 0}\lambda^n\Psi^{(n)}(t)$ when the molecule at times $t<0$ is in a steady state $\Phi_a\exp(-iE_at/\hbar)$. After a short interval when the transient effects of excitation have disappeared, this can be taken to be the unperturbed state at times $t\geq 0$ with $\Psi_a^{(0)}(0)=\Phi_a$.

The perturbative treatment for dynamical polarizabilities can be done for a time-independent unperturbed Hamiltonian $\hat{H}^{(0)}$ using the time evolution operator $\hat{U}^{(0)}(t)=\exp\left(-i\hat{H}^{(0)}t/\hbar\right)$ for the unperturbed problem, which satisfies the equation

$$\hat{H}^{(0)}\hat{U}^{(0)}(t)=i\hbar\partial\hat{U}^{(0)}/\partial t,\quad \hat{U}^{(0)}(0)=\hat{I}$$

This operator is unitary, and its inverse satisfies $\hat{U}^{(0)}(t)^{-1}=\hat{U}^{(0)}(t)^{\dagger}$, its adjoint. Writing $\Psi(t)=\hat{U}^{(0)}(t)\Psi^{(I)}(t)$ and solving the full equation of motion for the interaction picture states $\Psi^{(I)}(t)$, which are constant in the absence of a perturbation so that $\Psi^{(I,\,0)}(t)=\Psi^{(I)}(0)$ and have the initial value $\Psi^{(I)}(0)=\Psi(0)$, gives the integral equation

$$\Psi^{(I)}(t)=\Psi^{(I)}(0)-\frac{i}{\hbar}\lambda\int_0^t dt'\,\hat{U}^{(0)}(t)^{\dagger}\hat{H}^{(1)}(t')\hat{U}^{(0)}(t')\Psi^{(I)}(t')$$

A solution by iteration of this integral equation provides the desired expansion in powers of λ. The first term to the right is the unperturbed solution of the time-dependent Schroedinger equation. The solution to first order is

$$\Psi^{(I,1)}(t) = -\frac{i}{\hbar}\int_0^t dt'\, \hat{U}^{(0)}(t)^\dagger \hat{H}^{(1)}(t')\hat{U}^{(0)}(t')\Psi^{(I)}(0)$$

and one has in the interaction picture to n-th order the recursive relation

$$\Psi^{(I,n)}(t) = -\frac{i}{\hbar}\int_0^t dt'\, \hat{U}^{(0)}(t)^\dagger \hat{H}^{(1)}(t')\hat{U}^{(0)}(t')\Psi^{(I,n-1)}(t')$$

Replacing results in the average total dipole, one finds the results for permanent and induced dipoles. Writing for the dipole the expansion $\langle D_z(t)\rangle_a = D_{z,a}^{(0)}(t) + \lambda D_{z,a}^{(1)}(t) + \lambda^2 D_{z,a}^{(2)}(t)/2 + \ldots$ one finds that $D_{z,a}^{(0)}(t) = \left\langle \Psi_a^{(0)}(t) \,|\, \hat{D}_z \,|\, \Psi_a^{(0)}(t)\right\rangle$, and that

$$D_{z,a}^{(1)}(t) = \left\langle \Psi_a^{(0)}(t) \,|\, \hat{D}_z \,|\, \Psi_a^{(1)}(t)\right\rangle + \left\langle \Psi_a^{(1)}(t) \,|\, \hat{D}_z \,|\, \Psi_a^{(0)}(t)\right\rangle$$

This first correction gives the induced dipole (or polarization) $P_{z,a}(t) = \lambda D_{z,a}^{(1)}(t)$ as

$$P_{z,a}(t) = \lambda\left\langle \Psi_a^{(0)}(t) \,|\, \hat{D}_z \hat{U}^{(0)}(t) \,|\, \Psi_a^{(I,1)}(t)\right\rangle + \lambda\left\langle \Psi_a^{(I,1)}(t) \,|\, \hat{U}^{(0)}(t)^\dagger \hat{D}_z \,|\, \Psi_a^{(0)}(t)\right\rangle$$

and can be expressed in terms of the integral over time as

$$P_{z,a}(t) = -\frac{i}{\hbar}\int_0^t dt' \left\langle \Psi_a^{(0)}(0) \,|\, \left[\hat{D}_z(t)\hat{D}_z(t') - \hat{D}_z(t')\hat{D}_z(t)\right] \,|\, \Psi_a^{(0)}(0)\right\rangle \mathcal{E}_z(t')$$

where the time-dependent dipole operator $\hat{D}_z(t) = \hat{U}^{(0)}(t)^\dagger \hat{D}_z \hat{U}^{(0)}(t)$ appears multiplying at two different time, in a quantal average giving a dipole-dipole time-correlation function. The induced linear polarization is found to be dependent on earlier values of the applied field and contains a delayed response to its application. This treatment is valid for a general perturbation and takes a special form for a harmonic perturbation.

In practice, calculations are usually done introducing the basis set of unperturbed states $|\Phi_k\rangle$ with energies E_k, to expand time-dependent states. In a compact Dirac notation with kets $|\Phi_k\rangle = |\,k\rangle$, bras $\langle k\,|$, and brackets indicating integration over state variables,

$$\Psi(t) = \sum\nolimits_k |\,k\rangle \exp(-iE_k t/\hbar)\, c_k(t)$$

$$i\hbar\frac{dc_k}{dt} = \sum\nolimits_l \left(\langle k \,|\, \hat{H}(t) \,|\, l\rangle - E_k\delta_{kl}\right) \exp(i\omega_{kl}t) c_l(t)$$

where $\omega_{kl} = (E_k - E_l)/\hbar$ is a transition frequency, and the coupled equations must be solved with the initial conditions $c_k(0) = \delta_{ka}$ for each initial state a. Expanding in powers of the field strength with $c_k(t) = \sum_{n \geq 0} \lambda^n c_k^{(n)}(t)$ it follows that $c_k^{(0)}(t) = c_k^{(0)}(0) = \delta_{ka}$ and therefore $c_k^{(n)}(0) = 0$ for $n \geq 1$. The changes of higher-order terms for $n \geq 1$ satisfy the iterative equations

$$i\hbar \frac{d\, c_k^{(n)}}{dt} = \sum_l \left\langle k \,|\, \hat{H}^{(1)}(t) \,|\, l \right\rangle \exp(i\omega_{kl} t)\, c_l^{(n-1)}(t)$$

which gives $c_k^{(1)}(t) = \int_0^t dt' \left\langle k \,|\, \hat{H}^{(1)}(t') \,|\, a \right\rangle \exp(i\omega_{ka} t')$, linear in the field and from an integral to be obtained for specific fields.

The average dipole takes values obtained from the coefficient. To the lowest order, $D_{z,a}^{(0)}(t) = \langle a \,|\, \hat{D}_z \,|\, a \rangle$ which is usually null by molecular symmetry of the state a. To the next order

$$D_{z,a}^{(1)}(t) = \sum_k \langle a \,|\, \hat{D}_z \,|\, k \rangle c_k^{(1)}(t) \exp(-i\omega_{ka} t) + c_k^{(1)}(t)^* \langle k \,|\, \hat{D}_z \,|\, a \rangle \exp(i\omega_{ka} t)$$

is linear in the field strength and provides the polarizability. The induced dipole or polarization is given by

$$P_{z,a}(t) = \sum_k \int_0^t dt' \, \langle a \,|\, \hat{D}_z \,|\, k \rangle \langle k \,|\, \hat{D}_z \,|\, a \rangle \mathcal{E}_z(t') \exp[i\omega_{ka}(t' - t)] + c.c.$$

which can be integrated for a given time-dependent field.

More generally, the polarization component $P_{\xi,\, a}(t)$ arises in response to an applied field $\mathcal{E}_\eta(t')$ present from time $t = 0$ at all times $0 \leq t' \leq t$, and can be expressed in terms of delayed response functions for phenomena both linear and nonlinear in the applied field, as [15]

$$P_{\xi,a}(t) = \sum_\eta \int_0^t dt_1 \, \alpha_{\xi\eta,a}^{(1)}(t, t_1) \mathcal{E}_\eta(t_1) + \sum_{\eta,\zeta} \int_0^t dt_1 \int_0^{t_1} dt_2 \, \alpha_{\xi\eta\zeta,a}^{(2)}(t, t_1, t_2) \mathcal{E}_\eta(t_1) \mathcal{E}_\zeta(t_2) + \cdots$$

written here in general using tensor components of response functions. To simplify, consider only the linear response in the first term, usually sufficient for studies of intermolecular interactions. For a time-independent unperturbed state a, it is a function of the difference $t' = t - t_1$ and it is convenient to introduce the retarded susceptibility $\alpha_{\xi\eta,a}^{(+)}(t') = \theta(t')\alpha_{\xi\eta,a}^{(1)}(t', 0)$, where $\theta(t')$ is the step function null at negative times. Furthermore, the lower integration limit can be replaced by $-\infty$ insofar as the electric field is null for negative times, and changing the integration variable from t_1 to t' gives the useful expression for the linear term,

$$P_{\xi,a}(t) = \sum_\eta \int_{-\infty}^{\infty} dt' \alpha_{\xi\eta,a}^{(+)}(t') \mathcal{E}_\eta(t - t')$$

which is a convolution form and has a Fourier transform from time to frequency, giving $\tilde{P}_{\xi,a}(\omega) = \sum_\eta \tilde{\alpha}^{(+)}_{\xi\eta,a}(\omega)\tilde{\mathcal{E}}_\eta(\omega)$ in terms of the transforms of the three expressions, using the definitions

$$\tilde{f}(\omega) = \int_{-\infty}^{\infty} dt \exp(i\omega t) f(t), f(t) = \int_{-\infty}^{\infty} \frac{d\omega}{2\pi} \exp(-i\omega t)\tilde{f}(\omega)$$

for a function $f(t)$. The dynamical polarizability $\tilde{\alpha}^{(+)}_{\xi\eta,a}(\omega)$ can be obtained from a perturbation calculation of the average polarization. The retarded susceptibility is in general

$$\alpha^{(+)}_{\xi\eta,a}(t) = -\frac{i}{\hbar}\left\langle \Psi_a^{(0)}(0) \mid \left[\hat{D}_\xi(t)\hat{D}_\eta(0) - \hat{D}_\eta(0)\hat{D}_\xi(t)\right] \mid \Psi_a^{(0)}(0)\right\rangle \theta(t)$$

where the time-dependent dipole density operator $\hat{D}_\xi(t) = \hat{U}^{(0)}(t)^\dagger \hat{D}_\xi \hat{U}^{(0)}(t)$ appears multiplying at two different time, in a dipole–dipole time-correlation function.

2.3.2 Periodic Perturbation Field

So far the field is a general function of time and may be a superposition of harmonic terms or a light pulse. When the applied field is harmonic of form $\mathcal{E}_z(t;\omega_L) = \mathcal{E}_0 \cos(\omega_L t)$ for positive times, the linear polarization acquires, after a transient interval, the same frequency but has two components, one in phase and one out of phase with the applied field. It can be written as

$$P_{z,a}(t;\omega_L) = \alpha^{(c)}_{zz,a}(\omega_L)\,\mathcal{E}_0 \cos(\omega_L t) + \alpha^{(s)}_{zz,a}(\omega_L)\,\mathcal{E}_0 \sin(\omega_L t)$$

$$\alpha^{(c)}_{zz,a}(\omega_L) = \int_0^\infty d\tau \alpha^{(+)}_{zz,a}(\tau)\cos(\omega_L \tau)$$

$$\alpha^{(s)}_{zz,a}(\omega_L) = \int_0^\infty d\tau \alpha^{(+)}_{zz,a}(\tau)\sin(\omega_L \tau)$$

The dynamical polarizabilities $\alpha^{(l)}_{zz,a}(\omega_L)$, $l = c, s$, can be derived from a perturbation treatment.

We start a perturbation treatment with an electric field $\mathcal{E}_\xi(t;\omega_L) = \mathcal{E}_0 \cos(\omega_L t + \varphi) f_r(t)$ oscillating with the harmonic light frequency ω_L and turned on by a rising function $f_r(t)$ null at times $t < 0$ and tending to 1.0 for long times t of interest, with a rise time parameter τ much smaller than measuring times of interest. To begin with, we assume that the field is along the z-axis and that the molecule is isotropic in state a. The polarization is then along the z-axis and the total dipole $\langle D_z(t)\rangle_a$ is the sum of a permanent dipole plus terms induced by the field. Once transient polarizations have disappeared, and the system has reached a steady state for the given light frequency, it is of the form

$\langle D_z(t;\omega_L)\rangle_a = D_{z,a}{}^{(0)} + P_{z,a}(t;\omega_L)$, the sum of the permanent dipole plus a polarization changing with the same frequency and given in general by [15]

$$P_{z,a}(t;\omega_L) = \int_0^t dt_1 \chi^{(1)}_{zz,a}(t,t_1;\omega_L)\mathcal{E}_z(t_1;\omega_L) + \int_0^t dt_1 \int_0^{t_1} dt_2 \chi^{(2)}_{zzz,a}(t,t_1,t_2;\omega_L)\mathcal{E}_z(t_1;\omega_L)\mathcal{E}_z(t_2;\omega_L) + \cdots$$

which displays a dipolar polarizability in the term linear with the field and a first hyperpolarizability in the quadratic term, as functions of the field frequency. To obtain explicit expressions for them, we perform the integrations over time.

One can introduce a specific electric field along the z-direction in the form $\mathcal{E}_z(t;\omega_L) = \mathcal{E}_0 \cos(\omega_L t)[1-\exp(-t/\tau)]$ and integrate the previous expression for $c_k^{(1)}(t)$ which give for $t \gg \tau > 0$

$$\lambda c_k^{(1)}(t) = \langle k | \hat{D}_z | a \rangle \frac{\mathcal{E}_0}{2\hbar} \left[\frac{e^{i(\omega_{ka}+\omega_L)t} - 1}{\omega_{ka}+\omega_L} + \frac{e^{i(\omega_{ka}-\omega_L)t} - 1}{\omega_{ka}-\omega_L} \right]$$

Replacing this in the form of the average dipole, it appears as a permanent dipole $D_{z,a}^{(0)}$ plus an induced dipole or polarization $P_{z,a}(t;\omega_L)$, and the result allows identification of the dynamical polarizability $\alpha_{zz}(\omega_L)_a$, in $\langle D_z(t)\rangle_a = D_{z,a}^{(0)} + P_{z,a}(t;\omega_L)$, as

$$P_{z,a}(t;\omega_L) = \alpha_{zz}(\omega_L)_a \mathcal{E}_0 \cos(\omega_L t)$$

$$\alpha_{zz}(\omega_L)_a = \frac{2}{\hbar} \sum_{k \neq a} \frac{\omega_{ka} \left| \langle k | \hat{D}_z | a \rangle \right|^2}{\omega_{ka}^2 - \omega_L{}^2}$$

This shows the polarizability as a function of the field frequency ω_L with infinite values as this goes through each transition frequency.

More accurately, a solution for the coefficient $c_a(t)$ to all orders in the coupling of molecule and field reveals that it must decay exponentially over time as state a makes spontaneous light-induced transitions to all states of lower energy. This gives its magnitude as a decay function $|c_a(t)| = \exp(-\gamma_a t/2)$ where γ_a is a decay rate. To account for this decay, it is convenient to return to the iterative coupled equations for $c_k^{(1)}(t)$ and replace $c_k^{(0)}(t) = \delta_{ka}$ to the right with $c_k^{(d)}(t) = \delta_{ka}\exp(-\gamma_a t/2)$ [16]. It is then found that to second order in the molecule-field coupling

$$\gamma_a = \sum_{k<a} (2\pi/\hbar^2) \left| \langle a | \hat{H}_{MF} | k \rangle \right|^2 \rho_L(\omega_{ka})$$

where $k < a$ signifies adding over states of lower energy, and ρ_L is a density of light per unit frequency. With the same time-dependent coefficient changes used in the dipole average, one finds its previous stationary term replaced by a decaying dipole $D_{z,a}^{(d)}(t) = \langle a | \hat{D}_z | a \rangle \exp(-\gamma_a t)$, and that the transition frequency is replaced by $\omega_{ka} + i\gamma_{ka}/2$ with $\gamma_{ka} = \gamma_k + \gamma_a$ the sum of initial and final decay rates, in the first-order coefficients. With this replacement, the dynamical polarizabilities becomes

$$\alpha_{zz}(\omega_L)_a = \frac{2}{\hbar} \sum_{k \neq a} \frac{\omega_{ka} |\langle k | \hat{D}_z | a \rangle|^2}{(\omega_{ka} - \omega_L)^2 + \gamma_{ka}^2/4}$$

which avoids the function's singularity at the transition frequencies and gives instead peaks near resonance $\omega_L = \omega_{ka}$ excitations.

The present treatment can be readily generalized to the case where the field and dipole components are not in the same direction. For a general orientation of the field, the dynamical polarizabilities become tensors and the total dipole is

$$\langle D_\xi(t) \rangle_a = D_{\xi,a}^{(0)} + P_{\xi,a}(t; \omega_L)$$

$$P_{\xi,a}(t; \omega_L) = \sum_\eta \alpha_{\xi\eta,a}(\omega_L) \mathcal{E}_{0\eta} \cos(\omega_L t)$$

$$\alpha_{\xi\eta,a}(\omega_L) = \frac{1}{\hbar} \sum_{k \neq a} \frac{\omega_{ka} \langle a | \hat{D}_\xi | k \rangle \langle k | \hat{D}_\eta | a \rangle}{(\omega_{ka} - \omega_L)^2 + \gamma_{ka}^2/4}$$

with similar expressions for the quadrupolar and higher dynamical polarizability tensors, replacing the dipole operator with a higher electrical multipole.

More generally, the decay rates originate in the interactions of the molecule with its medium, with light, and with other molecules, and the rates can be obtained extending the expansion basis set to include states with a continuum of energies, which describes the photons and the molecules in its medium. Solving the larger set of coupled equations for the expansion coefficients, the interactions are found to give in addition an energy shift ΔE_{ka} and new transition frequencies $\omega'_{ka} = \omega_{ka} + \Delta E_{ka}/\hbar$ [6].

An alternative treatment of decay can be developed quantizing the electromagnetic field, which introduces photons instead of the field and can describe the interaction of the molecule with photons, including stimulated absorption-emission and also spontaneous emission [14, 17].

2.4 Susceptibility of an Extended Molecule

A large molecule, an aggregate of molecules, a polymer, or a solid surface are examples of systems where there are very large numbers of states which are usually unknown and which would make summations over states an unpractical approach. Furthermore, the applied electric field may vary over the lengths of the molecular system, and it is necessary to generalize the treatment to include a position- and time-dependent electric potential $\phi(\vec{r},t)$.

For an extended system of charges of electrons and nuclei (or ionic cores) at operator locations $\hat{\vec{r}}_I$, the total charge density operator per unit volume at space location $\vec{r}$ is $\hat{c}(\vec{r}) = \sum_I C_I \delta(\vec{r} - \hat{\vec{r}}_I)$. The dipole operator is $\hat{\vec{D}} = \int dr^3 \hat{\vec{D}}(\vec{r})$ with dipole density $\hat{\vec{D}}(\vec{r}) = \vec{r}\,\hat{c}(\vec{r})$, and higher multipoles can be similarly defined and used to expand the charge density operator as a series containing multipolar polarizations. However, many such multipoles may be needed for an extended system, and the treatment is best done avoiding expansions in multipoles. Instead, it is better to work with the average, or expectation value, of the total charge density, and to return to the potential energy expression given by charges interacting with an electric potential, rewriting $V_{MF} = \sum_I C_I \phi(\vec{r}_I, t)$ as $\hat{V}_{MF} = \int dr^3 \hat{c}(\vec{r}) \phi(\vec{r},t)$ and working directly with the charge density operator [15, 18].

For a time-independent applied electric potential, the average charge density in state a is a sum $c_a(\vec{r}) = c_a^{(0)}(\vec{r}) + c'_a(\vec{r})$ of a permanent value plus an induced charge density, and the latter can be derived theoretically using a perturbation treatment with resolvent operators, which can be expressed in any suitable basis set instead of having to use unknown unperturbed states.

For a time-dependent electric potential $\phi(\vec{r},t)$, the average charge density in state a is a sum $c_a(\vec{r},t) = c_a^{(0)}(\vec{r}) + c'_a(\vec{r},t)$ of a permanent value plus an induced charge density changing over time, with the latter obtained from the density operators $\hat{c}(\vec{r},t)$, and response functions involving the time correlation of these densities at two times.

It is convenient to proceed within a formal operator approach to obtain expressions for a physical system in the form of response functions and to calculate these without expanding in the usually unknown states for an extended system. The average induced charge density $c'_a(\vec{r},t)$ arises in response to an applied electric potential $\phi(\vec{r},t')$ present from time $t = 0$ at all times $0 \le t' \le t$, and can be expressed in terms of delayed response functions for phenomena both linear and nonlinear in the applied field, as [15]

$$\begin{aligned} c'_a(\vec{r},t) = &\int_0^t dt_1 \int d^3 r_1 \chi_a^{(1)}(\vec{r},t;\vec{r}_1,t_1)\phi(\vec{r}_1,t_1) \\ &+ \int_0^t dt_1 \int_0^{t_1} dt_2 \int d^3 r_1 \int d^3 r_2 \chi_a^{(2)}(\vec{r},t;\vec{r}_1,t_1;\vec{r}_2,t_2)\phi(\vec{r}_1,t_1)\phi(\vec{r}_2,t_2) \\ &+ \cdots \end{aligned}$$

written here using response functions of time and positions. To simplify, consider only the linear response in the first term, usually sufficient for studies of intermolecular interactions. For a time-independent unperturbed state a, it is a function of the difference $t' = t - t_1$, and it is convenient to introduce the retarded susceptibility $\chi_a^{(+)}(\vec{r},\vec{r}_1; t') = \theta(t')\chi_a^{(1)}(\vec{r},t';\vec{r}_1,0)$ where $\theta(t')$ is the step function null for negative arguments. Furthermore, the lower integration limit can be replaced by $-\infty$ insofar as the electric field is null for negative times, and changing the integration variable from t_1 to t' gives the useful expression for the linear term,

$$c'_a(\vec{r},t) = \int_{-\infty}^{\infty} dt' \int d^3r_1 \chi_a^{(+)}(\vec{r},\vec{r}_1;t')\phi(\vec{r}_1,t-t')$$

which is a convolution form and has a Fourier transform from time to frequency giving $\tilde{c}'_a(\vec{r},\omega) = \sum_\eta \int d^3r_1 \tilde{\chi}_a^{(+)}(\vec{r},\vec{r}_1;\omega)\tilde{\phi}(\vec{r}_1,\omega)$ in terms of the transforms of the three expressions.

The dynamical susceptibility $\tilde{\chi}_a^{(+)}(\omega)$ can be obtained from a perturbation calculation of the average polarization. Proceeding as before for the dipolar response, now the charge polarization is found to be

$$c'_a(\vec{r},t) = -\frac{i}{\hbar}\int_0^t dt' \int d^3r' \left\langle \Psi_a^{(0)}(0) \mid \left[\hat{c}(\vec{r},t)\hat{c}(\vec{r}',t') - \hat{c}(\vec{r}',t')\hat{c}(\vec{r},t)\right] \mid \Psi_a^{(0)}(0)\right\rangle \phi(\vec{r}',t')$$

where the time-dependent density operator $\hat{c}(\vec{r},t) = \hat{U}^{(0)}(t)^\dagger \hat{c}(\vec{r})\hat{U}^{(0)}(t)$ appears multiplying at two different time. The bracket is in fact a function of only the difference $\tau = t - t'$ because the time evolution is done by a time-independent Hamiltonian, and this provides a result with a real valued retarded response function

$$\chi_{cc,a}^{(+)}(\vec{r},\vec{r}',\tau) = -\frac{i}{\hbar}\left\langle \Psi_a^{(0)}(0) \mid \left[\hat{c}(\vec{r},\tau)\hat{c}(\vec{r}',0) - \hat{c}(\vec{r}',0)\hat{c}(\vec{r},\tau)\right] \mid \Psi_a^{(0)}(0)\right\rangle$$

for $\tau > 0$ and $\chi_{cc}^{(+)}(\tau) = 0$ for $\tau < 0$. The lower integral limit can be replaced with $-\infty$, and the integration variable can be changed to τ giving

$$c'_a(\vec{r},t) = \int_{-\infty}^{\infty} d\tau \int d^3r' \chi_{cc,a}^{(+)}(\vec{r},\vec{r}',\tau)\phi(\vec{r}',t-\tau)$$

and the two sides can be Fourier transformed into $\tilde{c}'_a(\omega) = \tilde{\chi}_{cc,a}^{(+)}(\omega)\tilde{\phi}(\omega)$, which are complex valued functions of frequency.

The response function $\chi_{cc,a}^{(+)}(t)$ can be calculated in a variety of ways, using expansions of operators in a known basis set, or generating numerically a solution over time. It can also be constructed semiempirically to incorporate known features of the response. The response functions can be calculated introducing a convenient basis set $\{|\mu\rangle\}$ of states to expand operators and the reference unperturbed state $\Psi_a^{(0)}$, in which case they can be expressed in terms of the amplitudes $\langle \mu \mid \hat{c}(\vec{r},t) \mid \mu' \rangle = c_{\mu\mu'}(\vec{r},t)$. The response can alternatively be obtained from states

$|\Upsilon_a(\vec{r},t)\rangle = \hat{c}(\vec{r},t)\,|\Psi_a^{(0)}(0)\rangle$, which satisfy an equation of motion generated by $\hat{U}^{(0)}(t)$. This can be accomplished propagating the states numerically in time, or obtaining them from a variational procedure [19].

The connection between the previous polarizabilities $\alpha_{\xi\eta,a}^{(+)}$ and the present susceptibilities $\chi_{cc,a}^{(+)}$ can be obtained writing the polarization components as $P_{\xi,a}(t) = \int d^3r\,\xi c_a'(\vec{r},t)$ and the electric potential as $\phi(\vec{r}',t-t') = -\sum_{\eta'}\eta'\mathcal{E}_{\eta'}(\vec{r}',t-t')$, and using the equation given above for $c_a'(\vec{r},t)$ to find that

$$\alpha_{\xi\eta,a}^{(+)}(\tau) = \int d^3r \int d^3r'\,\xi\eta'\chi_{cc,a}^{(+)}(\vec{r},\vec{r}',\tau)$$

which are moment integrals of the susceptibility with respect to positions. Similar expressions can be obtained for higher multipolar polarizabilities from higher moment integrals. In an extended system, the integrals over space variables can be calculated with coarse-grained numerical procedures that can be adapted to the physical distribution of electronic densities and localized polarizabilities in atomic or bond components.

2.5 Changes of Reference Frame

So far the reference frame has been chosen with its origin at the center of mass of the molecule and with axes attached to the molecule. It is also convenient to choose the origin of coordinates to be at the center of mass of the nuclei (CMN) to simplify the treatment of electronic transitions for fixed nuclei. This is sufficient for most applications, but for higher accuracy it is necessary to include the masses of electrons with a so-called mass-polarization correction. The axes attached to the molecule form a body-fixed (BF) set, and the molecule is described in the $\mathcal{S}_{CMN-BF}$ frame. Interacting molecules are, however, rotated and displaced in a common laboratory space, and we need to know what happens to multipoles of a molecule when its BF frame has been translated or rotated with respect to a common frame $\mathcal{S}_L$ [5, 20]. An alternative description of changes moves a molecule from one initial position to another, both described in the same $\mathcal{S}_L$ frame, and gives different relations between multipoles insofar as it is equivalent to keeping the molecule fixed and changing the reference frame with the reverse movements.

Let $\mathcal{S}_L = \mathcal{S}$ and $\mathcal{S}_{CMN-BF} = \mathcal{S}'$ have the same origin and orientation of axes to begin with, and consider first a translation of $\mathcal{S}'$ by a displacement vector $\vec{L}$

while keeping the axes parallel, so that the origin position $\vec{R}'_{CMN} = \vec{R}_{CMN} + \vec{L}$, and charge positions relate instead by $\vec{r}'_I = \vec{r}_I - \vec{L}$. This gives for the average dipole component ξ in the BF frame,

$$\langle \hat{D}_\xi \rangle_{S'} = \langle \hat{D}_\xi \rangle_S - CL_\xi$$

which shows it has changed due to the translation, if the total charge $C = \sum_I c_I$ is different from zero. For the quadrupole $Q_{\xi\eta} = \sum_I C_I (3\xi_I \eta_I - r_I^2 \delta_{\xi\eta})$ one finds

$$\langle \hat{Q}_{\xi\eta} \rangle_{S'} = \langle \hat{Q}_{\xi\eta} \rangle_S + C(3L_\xi L_\eta - L^2 \delta_{\xi\eta}) + 2 \langle \vec{D} \rangle_S . \vec{L} \, \delta_{\xi\eta}$$

which changes with frame translation if either total charge or total dipole is not zero. The conversions between frames when the molecule is moved, instead of moving the frame, follow by changing $\vec{L}$ into $-\vec{L}$ in the above equations. Changes of the n-th multipole involve the n-th power of the L_ξ components, and therefore its changes can be large, and they must be considered together with all others.

A rotation of the BF while keeping the origins at the same location can be done moving the frame $\mathcal{S}'$ by the set of Euler angles $\Omega = (\alpha, \beta, \gamma)$ with respect to $\mathcal{S}$, by angle α around the z-axis of (a right-handed) $\mathcal{S}$, followed by a rotation $-\beta$ around an intermediate y''-axis, and after this a γ rotation around the final z'-axis [20]. For a column 3×1 matrix of elements (x, y, z) transformed into a new column with (x', y', z'), the 3×3 transformation matrix $A(\alpha,\beta,\gamma) = \boldsymbol{R}_{z'}(\gamma)\boldsymbol{R}_{y''}(\beta)\boldsymbol{R}_z(\alpha)$, a product of three axial rotation matrices, with components $A_{\xi\eta}$, can be used to transform dipole and higher multipole components and also polarizations. From $D_{\xi'} = \sum_\eta A_{\xi'\eta}(\alpha,\beta,\gamma) D_\eta$ and $\mathcal{E}_{\xi'} = \sum_\eta A_{\xi'\eta}(\alpha,\beta,\gamma)\mathcal{E}_\eta$ for the electric field, using the inverse of the matrix $\boldsymbol{A}$, one finds for the polarizability tensor components in the rotated frame,

$$\alpha_{\xi'\eta'} = \sum_{\xi,\eta} A_{\xi'\eta} \alpha_{\eta\zeta} (\boldsymbol{A}^{-1})_{\zeta\eta'}$$

which is a function of the Euler angles. The permanent quadrupole has the same angle dependence. If instead one rotates the molecule keeping the same reference frame, then the transformations of multipoles and polarizabilities are similar but involve the inverse rotation and Euler angles. The usual notation in this case is $(\alpha, \beta, \gamma) = (\varphi, \vartheta, \chi)$ for the orientation of the main axes of a rotated rigid body.

The spherical components of multipoles are simply transformed under a rotation operation $\hat{R}(\alpha,\beta,\gamma)$ by means of the Wigner rotational matrices $D^{(l)}_{m'm}(\alpha,\beta,\gamma)$, so that the transformed multipole is [4, 7]

$$\hat{R}(\alpha,\beta,\gamma) Q_{lm} = \sum_{m'} Q_{lm'} D^{(l)}_{m'm}(\alpha,\beta,\gamma)$$

The effect on a spherical multipole Q_{lm} of a translation by $\vec{L}$ is more complicated but can also be given in a compact way as a sum of multipoles of order $0 \leq l' \leq l$ with coefficients obtained from solid spherical harmonic functions of the translation components [4].

2.6 Multipole Integrals from Symmetry

Molecular symmetry can be used to identify which components of the average multipole tensors and polarization tensors must vanish, or are interrelated. For finite systems, this is done introducing point symmetry groups and symmetry adapted wavefunctions, and the symmetry of multipole and polarization components [5, 21]. An extension to include infinite systems with translational symmetry, such as crystalline solids, regular polymers, and surface lattices, is done with space groups or crystallographic groups [22]. Symmetry arguments can also be used to determine which transitions between molecular states will be forbidden, leading to selection rules. Here, we only summarize the extensive group theory for properties of molecules and solids.

Molecular symmetry is introduced identifying movements (or operations) with respect to geometric elements that leave the nuclear framework with a structure equivalent to the original one. The elements are the point of inversion (named i), the axis of rotation by an angle $2\pi/n$ (or C_n), the planes of reflection (σ), and the combination of a rotation axis and a perpendicular plane of reflection (improper axis S_n).

The set of symmetry movements G for a finite atomic system that leave unchanged at least one point of the structure is collected in a point group $\mathcal{G}$ of transformations. This contains h symmetry movements (the group order). A physically meaningful symmetry–conserving transformation moves not only the nuclear structure but also the electronic charge distribution bound to it. In the spirit of the Born–Oppenheimer treatment, it is convenient to focus on the symmetry of electronic states $\Phi(X; \boldsymbol{Q})$ for fixed nuclear positions. When the location of the nuclei uniquely determines the electronic state, as is usually the case, it is sufficient to inspect only the atomic structure to identify symmetry elements. States of molecules with open electronic shells or nonzero spin multiplicity may require additional symmetry considerations. The movement or operation G is defined by changes of position variables of all electrons and nuclei so that using Cartesian coordinates $(x,y,z) \xrightarrow{G} (x',y',z')_G$.

These movements generate changes in the electronic state functions, represented by symmetry operators $\hat{O}_G$ for each movement given by $\Phi(G\boldsymbol{X}; G\boldsymbol{Q}) = \hat{O}_G \Phi(\boldsymbol{X}; \boldsymbol{Q})$, which defines the symmetry operators in the space of states. These form a point group of operators isomorphic with the original point group of movements in real space. The theory of point groups and space groups and their matrix representations provide powerful and general methods for analyzing multipole and polarization components.

For each element G in a group of symmetry movements, the equation for stationary states $\hat{H}^{(el)}\Phi(\boldsymbol{X};\boldsymbol{Q}) = E(\boldsymbol{Q})\Phi(\boldsymbol{X};\boldsymbol{Q})$ is also satisfied by $\Phi(G\boldsymbol{X}; G\boldsymbol{Q})$, which means that the electronic Hamiltonian satisfies $\hat{O}_G\hat{H}^{(el)} = \hat{H}^{(el)}\hat{O}_G$, and $\hat{O}_G\Phi(\boldsymbol{X};\boldsymbol{Q})$ can be expressed as a combination of stationary eigenstates of the Hamiltonian. Stationary states Φ_μ with the same eigenenergy $E_r(\boldsymbol{Q})$ of degeneracy l_r form a basis set for an irreducible representation (or irrep) $\Gamma^{(r)}$ of the symmetry operator by means of matrices of the same order as the energy degeneracy. A generated set of irreducible matrix representations $\Gamma^{(r)}_{\mu\nu}(G)$ can be used to construct symmetry-adapted electronic states $\Phi^{(r)}_\lambda$ with projection operators so they undergo prescribed changes for given movements [5]. The projection operators can be simply constructed from the characters $\chi^{(r)}(G) = \sum_\mu \Gamma^{(r)}_{\mu\mu}(G)$ as

$$\hat{P}^{(r)} = \frac{l_r}{h}\sum\nolimits_G \chi^{(r)}(G)^* \hat{O}_G$$

and project the adapted state $\Phi^{(r)}_\lambda = \hat{P}^{(r)}\Phi_\lambda$ from a given state Φ_λ. Products of irreps generated by basis set products $\Phi^{(r)}_\lambda\Phi^{(s)}_\mu$ can be decomposed into sums shown schematically as $\Gamma^{(r)} \otimes \Gamma^{(s)} = \sum_p a^{(p)}_{rs}\Gamma^{(p)}$ with coefficients derivable from the related characters [5, 22]. This is done using Tables of characters and the orthogonality properties of the characters, $\sum_G \chi^{(r)}(G)^* \chi^{(s)}(G) = h\delta_{rs}$.

Symmetry considerations allow determination of what permanent multipole or polarization components must be null, and if a transition matrix elements between states must vanish indicating a selection rule. The integrals involved in quantum brackets are numbers $I = \int d\boldsymbol{X}d\boldsymbol{Q}\, f(\boldsymbol{X}, \boldsymbol{Q})$ clearly invariant under symmetry operations. Therefore, $f(G\boldsymbol{X}, G\boldsymbol{Q})$ must give the same integral value or this must be zero. Consequently, nonzero integrals must involve fully invariant integrands, and these can be identified using symmetry. The identification of null tensor components is frequently obvious from simple symmetry considerations, especially when the electronic state is unchanged by symmetry movements.

A homonuclear molecule such as H_2 has a center of inversion and as a results all three components of its average dipole in its ground electronic state, which is nondegenerate and invariant, must be null because they are identical to their opposite when the inversion movement is applied. The average dipole in the H_2O molecule must be located along the z-axis through O and the midpoint between H's because the perpendicular components are equal to their opposite upon a 180° rotation around the z-axis, and so on, for other molecules.

More complicated symmetries require the mathematical group treatment and this relies on the basic result that Dirac brackets involving the product of two electronic functions are null if they belong to different irreducible representations. In more detail, a theorem states that the direct product of two irreps contains the totally symmetric representation only if they are identical, and then only once. This can be applied as follows: using Tables of characters of point groups and of direct product decompositions [5].

The permanent value $\left\langle \Phi_\lambda^{(r)} \mid \mathcal{M}_{\alpha\beta}^{(n)} \mid \Phi_\lambda^{(r)} \right\rangle$ of a multipole in a state belonging to the r representation is found to likely differ from zero only if the multipole components change under symmetry operations so that the components span a representation that contains the same irreducible representations found in $\Gamma^{(r)} \otimes \Gamma^{(r)} = \sum_p a_{rr}^{(p)} \Gamma^{(p)}$. Otherwise, the integral in the bracket expression does vanish. Similarly, the matrix elements of the transition multipole $\left\langle \Phi_\lambda^{(r)} \mid \mathcal{M}_{\alpha\beta}^{(n)} \mid \Phi_\mu^{(s)} \right\rangle$ vanish if the multipole transformation gives a representation which does not contain one of the irreducible representations in $\Gamma^{(r)} \otimes \Gamma^{(s)}$.

As an example, consider again the H_2O molecule with symmetry movements E (none), C_2, $\sigma(xz)$, and $\sigma(yz)$ giving the point group symmetry $\mathcal{C}_{2v}$ and the fully symmetric irrep designated as A_1. From the group characters in Table 2.1 for $\mathcal{C}_{2v}$, and knowing that the ground state function is of type A_1 one concludes that the z-component of the permanent dipole, which also belongs to A_1, is different from zero because the product of ground states belongs to $\Gamma^{(A_1)} \otimes \Gamma^{(A_1)} = \Gamma^{(A_1)}$, the same symmetry as the z-component. But the x- and y-components have symmetry B_1 and B_2, which are absent, and vanishing expectation value integrals. Similar arguments lead to the conclusion that the x^2, y^2, and z^2 components of the permanent quadrupole are not zero, but that the xy, yz, and xz components must vanish. With regard to transition multipoles appearing in the polarization, one finds that the x-, y-, and z-components of the dipole may only induce excitations to states $\Phi_\lambda^{(r)}$ of the same symmetry B_1, B_2, and A_1, respectively, because these are the ones equal to the product of irreps $\Gamma^{(A_1)} \otimes \Gamma^{(r)} = \Gamma^{(r)}$ of the involved states. Similar considerations apply to other molecules with different point group symmetries.

Table 2.1 Group characters for the point group symmetry $\mathcal{C}_{2v}$.

C_{2v} *or* $2mm$	E	C_2	$\sigma(xz)$	$\sigma(yz)$	Basis set
A_1	1	1	1	1	z; x^2, y^2, z^2
A_2	1	1	−1	−1	R_z; xy
B_1	1	−1	1	−1	x, R_y; xz
B_2	1	−1	−1	1	y, R_x; yz

The treatment can also be done introducing the spherical components of multipoles. If the set $\{Q_{lm}\}$ with $-l \geq m \geq l$ spans a representation $\Gamma^{(l)}$, this can be decomposed using characters into irreps to consider the values of permanent and induced components. For example in H_2O, the irrep decomposition for the dipole set $\{Q_{1m}\}$ is $A_1 + B_1 + B_2$, and as before one finds that there is one permanent dipole (from Q_{10}) and that dipolar excitations from the ground state can occur into states of symmetry B_1 and B_2. The decomposition shows the same symmetries as the rotational basis functions in the Table 2.1.

Conditions under which transition matrix elements of multipole operators are null constitute selection rules which exclude many transitions. In addition to the ones relating to molecular symmetry, there are selection rules derived from the rotational and spin dynamics of involved states. For fixed nuclear positions, if electronic initial and final transition states have angular momentum quantum numbers (L_i, M_i) and (L_f, M_f), respectively, then a spherical multipole component with an angular distribution given by quantum numbers (l, m) will only lead to allowed transitions insofar as $|L_i - l| \leq L_f \leq L_i + l$ and $M_f = M_i + m$. Also, the multipole operators do not depend on electronic spin variables and cannot lead to transitions which change spin quantum numbers (S, M_S), so that we have also the selection rules $S_f = S_i$ and $M_{Sf} = M_{Si}$ [6].

2.7 Approximations and Bounds for Polarizabilities

2.7.1 Physical Models

Useful parametrized models of response functions, valid over a homogeneous region of space, have been introduced in the literature, such as the Debye model of dielectric relaxation with a decay time τ_D, where the dipolar time correlation is parametrized as $\alpha^{(+)}_{zz,a}(t) = \overline{(D_z^2)}_a \exp(-|t|/\tau_D)$, and one finds

$$\tilde{\alpha}^{(+)}_{zz,a}(\omega) = \overline{(D_z^2)}_a \frac{(1 - i\omega\tau_D)}{(1 + \omega^2 \tau_D{}^2)}$$

Other physical choices for the dipole time correlation lead to dynamical susceptibilities with Lorentzian or Gaussian distributions of frequencies. For a distribution of charges undergoing an oscillation of frequency ω_R, and a dipole correlation relaxing over time τ_R, the choice $\alpha^{(+)}_{zz,a}(t) = \overline{(D_z^2)}_a \exp(-|t|/\tau_R)$ $\cos(\omega_R t)$ leads to a Lorentzian susceptibility as above but with ω replaced by $\omega - \omega_R$. Another instance involves random changes over time in the charge distribution and is given by a Gaussian dipole correlation function like $\alpha^{(+)}_{zz,a}(\tau) = \overline{(D_z^2)}_a \exp(-t^2/\tau_G{}^2)$ from which a Gaussian distribution of frequencies is found for the dynamical susceptibility [23].

2.7.2 Closure Approximation and Sum Rules

The summation over states in the polarizability extends over molecular states $\lambda = (I, \alpha)$ composed of an electronic state α and rotational-vibrational states I for each state α. Fixing the nuclear positions $\mathbf{Q}$ and concentrating on the electronic polarizability, this can be expressed in terms of the electronic oscillator strength for electrons with mass m_e and charge C_e, defined as the dimensionless value $f_{\xi,\kappa\lambda}^{(el)}(\mathbf{Q}) = [2m_e/(e^2\hbar)]\omega_{\kappa\alpha}(\mathbf{Q})\left|\langle\kappa\,|\,D_\xi^{(el)}\,|\,\lambda\rangle_{\mathbf{Q}}\right|^2$, involving elements of the electronic dipole between electronic states κ and α and excitation energies $\hbar\omega_{\kappa\lambda}(\mathbf{Q}) = E_\kappa^{(0)}(\mathbf{Q}) - E_\lambda^{(0)}(\mathbf{Q}) = \Delta_{\kappa\lambda}$.

The static polarizability for a field only along the z-axis is

$$\alpha_{zz,\lambda} = 2\sum\nolimits_{\kappa\neq\lambda} \frac{\left|\left\langle \Phi_\lambda^{(0)}\,|\,\hat{D}_z\,|\,\Phi_\kappa^{(0)}\right\rangle\right|^2}{\Delta_{\kappa\lambda}}$$

and for a general field the orientation average is $\bar{\alpha}_\lambda = (1/3)\left(\alpha_{xx,\lambda} + \alpha_{yy,\lambda} + \alpha_{zz,\lambda}\right)$ given by

$$\bar{\alpha}_\lambda = \frac{e^2\hbar^2}{m_e}\sum\nolimits_{\kappa\neq\lambda} \frac{f_{\kappa\lambda}^{(el)}}{\Delta_{\kappa\lambda}{}^2}$$

where the averaged oscillator strength $f_{\kappa\lambda}^{(el)} = \left[2m_e/\left(3e^2\hbar^2\right)\right]\,\Delta_{\kappa\lambda}\left|\langle\kappa\,|\vec{D}^{(el)}\,|\,\lambda\rangle_{\mathbf{Q}}\right|^2$ contains the electronic dipole vector. A compact expression is obtained defining a weighted excitation energy Δ by means of

$$\sum\nolimits_{\kappa\neq\lambda}\left|\left\langle\kappa\,|\vec{D}^{(el)}\,|\,\lambda\right\rangle_{\mathbf{Q}}\right|^2/\Delta_{\kappa\lambda} = \sum\nolimits_{\kappa\neq\lambda}\left|\left\langle\kappa\,|\vec{D}^{(el)}\,|\,\lambda\right\rangle_{\mathbf{Q}}\right|^2/\Delta$$

where the numerator to the right, using the closure (or completeness) relation $\sum_{\kappa\neq\lambda}|\,\kappa\rangle\langle\kappa\,| = \hat{I} - |\,\lambda\rangle\langle\lambda\,|$ of the electronic basis set, is simply the dipole standard deviation value $(\Delta D)^2 = \left\langle\lambda\,|\left[\vec{D}^{(el)}\right]^2|\,\lambda\right\rangle_{\mathbf{Q}} - \left(\left\langle\lambda\,|\vec{D}^{(el)}\,|\,\lambda\right\rangle_{\mathbf{Q}}\right)^2$. This gives $\bar{\alpha}_\lambda = (2/3)(\Delta D)^2/\Delta$, which can be estimated taking $(\Delta D)^2$ to be of the order of the square of electron charge times the average atomic displacement $\delta\,|\,\mathbf{Q}|$ in the electronic state λ, and taking Δ to be of the order of the vertical ionization energy for fixed $\mathbf{Q}$. Alternatively, one finds $\bar{\alpha}_\lambda = \left(e^2\hbar^2/m_e\right)\Delta^{-2}\sum_{\kappa\neq\lambda} f_{\kappa\lambda}^{(el)}$ and from the sum rule for the oscillator strength [5] which gives $\sum_{\kappa\neq\alpha} f_{\kappa\alpha}^{(el)} = N_{el}^*$, equal to the effective total number of independent electrons partaking in excitation, one finds that $\bar{\alpha}_\lambda = (e^2\hbar^2/m_e)\Delta^{-2}N_{el}^*$, which provides another rough estimate.

For the dynamical polarizability, the denominator is instead $\Delta_{\kappa\lambda}{}^2 = (\omega_{k\lambda} - \omega_L)^2 + \gamma_{k\lambda}{}^2/4$ and provided the decay rate is negligible compared to excitation frequencies, it agrees with the static value when $\omega_L \to 0$. In the limit for $\omega_L \to \infty$ one finds $\bar{\alpha}_\lambda \approx (e^2\hbar^2/m_e)\sum_{\kappa\neq\lambda} f_{\kappa\lambda}^{(el)}/\omega_L{}^2 = (e^2\hbar^2/m_e)N_{el}^*/\omega_L{}^2$ approaching zero. These two limits can be incorporated into a generic form for the dynamical polarizability extended to complex values $\omega + iu$ of the frequency [24], which appear, for example in the intermolecular dispersion energy [25, 26]. The interpolation form $\alpha(iu) = a/(b^2 + u^2)$ can be rewritten using the static polarizability and the polarizability limit for $\omega_L \to \infty$ as $\alpha(iu) = \alpha^{(0)}/(1 + \alpha^{(0)}u^2/N_{el}^*)$.

More general sum rules can be applied to the calculation of dynamical polarizabilities of the form $\alpha(iu) = \sum_{n=1}^{N} a_n/(b_n^2 + u^2)$, for example starting from the sum $G_\lambda^{(k)}(\omega) = \sum_{\kappa\neq\lambda}(\Delta_{\kappa\lambda} - \omega)^k \langle\lambda|\hat{A}|\kappa\rangle_Q \langle\kappa|\hat{B}|\lambda\rangle_Q$, which agrees with the dipolar polarizability for $k = -2$ and when the operators are dipoles [25, 27]. These sums can be evaluated with convenient basis sets instead of a set containing usually unknown excited states of the unperturbed molecule.

2.7.3 Upper and Lower Bounds

Static polarizabilities can be obtained from time-independent, second-order perturbation theory of the energy of a molecule in an external field. This can be treated using the basis set completeness relation and projection operators, to provide upper and lower bounds to the energy and from this to get bounds to the polarizability [28, 29].

A similar treatment can be provided for upper and lower bounds to the dynamical polarizability extended to imaginary values of the complex frequency, $\beta(u) = \alpha(iu)$, as it appears in formulas for the intermolecular dispersion forces [30]. This has also been done using Pade approximants and Gaussian grid integration [31].

The charge C, dipole D, and mean polarizability $\bar{\alpha}$ of Li^+, Ne, H_2, HF, CO_2, H_2O and other selected species, are given in SI units in Table 2.2. Electric quadrupole moments are second-order tensors with components $Q_{\xi\eta}$, and the main quadrupole component $Q = \max(Q_{\xi\xi})$ take values (in units of $10^{-40}Cm^2$) of 0.237, 0.0, 2.12, 8.7, −14.3, and 9.087, respectively for the mentioned species. Multipole and polarizability values, however, depend on the chosen origin of coordinates for the charge and polarization distribution, which are assumed here to refer to the molecular center of mass.

Table 2.2 Charge *C*, electric dipole *D*, and mean polarizability $\bar{\alpha}$ of selected atoms and molecules in SI units.

SI units	$C/(10^{-19}\ C)$	$D/(10^{-30} Cm)$	$\bar{\alpha}/(10^{-40}\ J^{-2}\ C^2 m^2)$
Li^+	1.602	0.0	0.189
Na^+	1.602	0.0	0.994
Ne	0.0	0.0	0.437
H_2	0.0	0.0	0.911
N_2	0.0	0.0	1.97
HF	0.0	6.37	0.57
CO	0.0	0.390	2.20
CO_2	0.0	0.0	2.93
H_2O	0.0	6.17	1.65
CH_4	0.0	0.0	2.89

References

1 Jackson, J.D. (1975). *Classical Electrodynamics*. New York: Wiley.
2 Landau, L.D. and Lifshitz, E. (1975). *Classical Theory of Fields*, 4e. Oxford, England: Pergamon Press.
3 Hirschfelder, J.O., Curtis, C.F., and Bird, R.B. (1954). *Molecular Theory of Gases and Liquids*. New York: Wiley.
4 Stone, A.J. (2013). *The Theory of Intermolecular Forces*, 2e. Oxford, England: Oxford University Press.
5 Atkins, P.W. and Friedman, R.S. (1997). *Molecular Quantum Mechanics*. Oxford, England: Oxford University Press.
6 Cohen-Tanoudji, C., Diu, B., and Laloe, F. (1977). *Quantum Mechanics*, vol. 2. New York: Wiley-Interscience.
7 Zare, R.N. (1988). *Angular Momentum*. New York: Wiley.
8 McWeeny, R. (1989). *Methods of Molecular Quantum Mechanics*, 2e. San Diego, CA: Academic Press.
9 Martin, R.M. (2004). *Electronic Structure: Basic Theory and Practical Methods*. Cambridge, England: Cambridge University Press.
10 Kelly, K.L., Coronado, E., Zhao, L.-L., and Schatz, G.C. (2003). Optical properties of metal nanoparticles: influence of size, shape, and dielectric environment. *J. Phys. Chem. B* 107: 668–677.
11 Messiah, A. (1962). *Quantum Mechanics*, vol. 2. Amsterdam: North-Holland.
12 Lowdin, P.O. (1962). Studies in perturbation theory. IV. Projection operator formalism. *J. Math. Phys.* 3: 969.

13 Micha, D.A. (1974). *Effective Hamiltonian Methods for Molecular Collisions* (Adv. Quantum Chem.), 231. New York: Academic Press.
14 Loudon, R. (1973). *Quantum Theory of Light*. Oxford, England: Oxford University Press.
15 Mukamel, S. (1995). *Principles of Nonlinear Optical Spectroscopy*. Oxford, England: Oxford University Press.
16 Davydov, D.S. (1965). *Quantum Mechanics*. Reading, MA: Pergamon Press.
17 Cohen-Tannoudji, C., Dupont-Roc, J., and Grynberg, G. (1992). *Atom-Photon Interactions*. New York: Wiley.
18 May, V. and Kuhn, O. (2000). *Charge and Energy Transfer Dynamics in Molecular Systems*. Berlin: Wiley-VCH.
19 Karplus, M. and Kolker, H.J. (1963). Variation-perturbation approach to the interaction of radiation with atoms and molecules. *J. Chem. Phys.* 39: 1493.
20 Arfken, G. (1970). *Mathematical Methods for Physicists*, 2e. New York: Academic Press.
21 Cotton, F.A. (1990). *Chemical Applications of Group Theory*, 3e. New York: Wiley.
22 Tinkham, M. (1964). *Group Theory and Quantum Mechanics*. New York: Mc-Graw-Hill.
23 McQuarrie, D.A. (1973). *Statistical Mechanics*. New York: Harper & Row.
24 Mavroyannis, C. and Stephen, M.J. (1962). Dispersion forces. *Mol. Phys.* 5: 629.
25 Dalgarno, A. and Davison, W.D. (1966). The calculation of Van der Waals interactions. *Adv. At. Mol. Phys.* 2: 1.
26 Kramer, H.L. and Herschbach, D.R. (1970). Combination rules for van der Waals force constants. *J. Chem. Phys.* 53: 2792.
27 Hirschfelder, J.O., Brown, W.B., and Epstein, S.T. (1964). Recent developments in perturbation theory. *Adv. Quantum Chem.* 1: 255.
28 Lowdin, P.O. (1966). Calculation of upper and lower bounds of energy eigenvalues in perturbation theory by means of partitioning techniques. In: *Perturbation Theory and its Applications in Quantum Mechanics* (ed. C.H. Wilcox), 255–294. New York: Wiley Publication.
29 Lindner, P. and Lowdin, P.O. (1968). Upper and lower bounds in second order perturbation theory and the Unsold approximation. *Int. J. Quantum Chem.* 2 (S2): 161.
30 Goscinski, O. (1968). Upper and lower bounds to polarizabilities and Van de Waals forces I. General theory. *Int. J. Quantum Chem.* 2: 761.
31 Langhoff, P., Gordon, R.G., and Karplus, M. (1971). Comparison of dispersion force bounding methods with applications to anisotropic interactions. *J. Chem. Phys.* 55: 2126.

3

Quantitative Treatment of Intermolecular Forces

CONTENTS

Molecular Interactions: Concepts and Methods, First Edition. David A. Micha.

3.1 Long Range Interaction Energies from Perturbation Theory

3.1.1 Interactions in the Ground Electronic States

Basic concepts of intermolecular forces can be extracted from models of electronic structure chosen here to simplify the mathematical treatment [1, 2]. The main simplification derives from the Born–Oppenheimer separation of electronic and nuclear motions whereby we consider, as a first good approximation, how electrons behave while keeping fixed the nuclear positions. This amounts to treating nuclei as point charges while describing electrons by means of their electronic wavefunctions or in terms of their electronic charge densities. A reference frame, called here the CMN-SF frame, is chosen for two interacting species A and B, with its center at the position of the center-of-mass of the nuclei and with axes orientations fixed in the space common to A and B. Interaction potential energies between A and B become functions of the relative position vector $\vec{R}$ of the center of mass of B relative to that of A, and of the set of internal nuclear position coordinates $\boldsymbol{Q}_{int} = (\boldsymbol{Q}_A, \boldsymbol{Q}_B)$ including internal position variables of A and B.

We consider here a system of two molecules A and B with fixed nuclei and electronic particles labeled by I and J, respectively, and with charge distributions as obtained from their electronic wavefunctions $\Psi_\kappa^{(AB)}$ for a state κ. The Hamiltonian operator $\hat{H}_A$ of species A, written in the coordinate representation in terms of the position and momentum operators $\hat{\vec{r}}_a$ and $\hat{\vec{p}}_a = -i\hbar\nabla_a$ for nuclei $a = 1$ to $N_{nu}^{(A)}$ and operators $\hat{\vec{r}}_i$ and $\hat{\vec{p}}_i = -i\hbar\nabla_i$ for electrons $i = 1$ to $N_{el}^{(A)}$, is given for fixed nuclei by the sum of the kinetic energy of electrons plus the Coulomb potential energies of interaction between nuclei and electrons and among electrons, as

$$\hat{H}_A = \hat{K}_A^{(e)} + \hat{V}_A^{(e,n)} + \hat{V}_A^{(e,e)}$$

$$\hat{K}_A^{(e)} = -\frac{\hbar^2}{2m_e}\sum\nolimits_i \nabla_i^2, \quad \hat{V}_A^{(e,n)} = \frac{1}{4\pi\varepsilon_0}\sum\nolimits_{a,i} \frac{C_a C_e}{\left|\hat{\vec{r}}_a - \hat{\vec{r}}_i\right|},$$

$$\hat{V}_A^{(e,e)} = \frac{1}{4\pi\varepsilon_0}\sum\nolimits_{i<j} \frac{C_e^2}{\left|\hat{\vec{r}}_j - \hat{\vec{r}}_i\right|}$$

where C_a and C_e are the electric charges of nucleus a and the electron, in the SI units.

The electrostatic energy operator must be added to the Hamiltonian operators of A and B to describe their interaction, given in general for $R > R_{min}$ by

$$\hat{H}_{AB} = \hat{H}_A + \hat{H}_B + \hat{H}_{AB}^{(int)}\left(\vec{R}\right)$$

$$\hat{H}_{AB}^{(int)}\left(\vec{R}\right) = (4\pi\varepsilon_0)^{-1}\sum\nolimits_I\sum\nolimits_J \frac{C_I C_J}{\left|\vec{R} - \hat{\vec{r}}_{IA} + \hat{\vec{r}}_{JB}\right|}$$

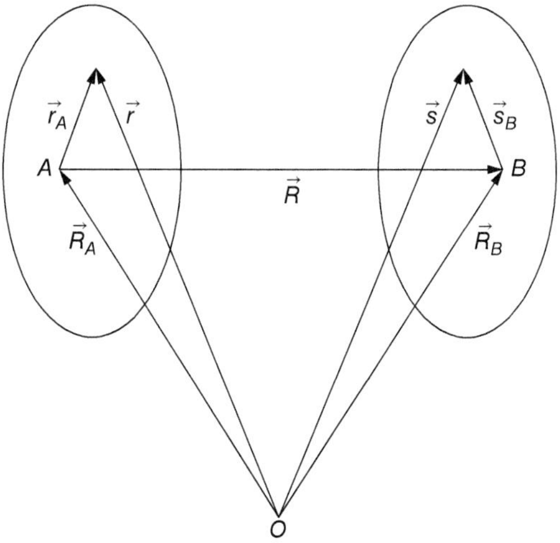

Figure 3.1 Charges and center-of-mass position vectors for species *A* and *B*.

Charges I of A are located at $\vec{r}_I = \vec{R}_A + \vec{r}_{IA}$ and charges J of B are at $\vec{s}_J = \vec{R}_B + \vec{s}_{JB}$, with $\vec{R}_A$ and $\vec{R}_B$ the center-of-mass of nuclei in A and B. The definitions of charge and CM position vectors are shown in Figure 3.1. Each of the Hamiltonians $\hat{H}_S$, $S = A, B$, contain the kinetic energy and Coulomb interactions of all particles in species S. The interaction or coupling term between A and B is written assuming that there are no interchanges of particles (electrons or nuclei) between them.

The form of the coupling can be written in different ways depending on whether the two interacting systems contain only few atoms, or they are extended many-atom systems. For many atoms, it is advantageous to introduce the charge density operators per unit volume, containing particles $I = a, i$ for nuclei and electrons in species A and $J = b, j$ for B. The charge density operators are $\hat{c}_A(\vec{r}) = \sum_I C_I \delta(\vec{r} - \vec{r}_I)$ in terms of the Dirac delta function, and $\hat{c}_B(\vec{s}) = \sum_J C_J \delta(\vec{s} - \vec{s}_J)$, and the Coulomb interaction energy is given by

$$\hat{H}_{AB}^{(int)}\left(\vec{R}\right) = (4\pi\varepsilon_0)^{-1} \int d^3r \int d^3s \; \hat{c}_A\left(\vec{r}\right) \left|\vec{r} - \vec{s}\right|^{-1} \hat{c}_B\left(\vec{s}\right)$$

States and energies of the whole system satisfy the eigenvalue equation $\hat{H}_{AB}\Psi_\kappa^{(AB)} = E_\kappa^{(AB)}\Psi_\kappa^{(AB)}$, with energies and states dependent on the intermolecular distance R. This can be solved with a perturbation expansion to a convenient order, assuming that the interaction is small, introducing the internal electronic states of A and B, and the zeroth order electronic states $\left|\Psi_\kappa^{(0)}\right\rangle = |j_A\rangle|k_B\rangle = |j_A k_B\rangle$ of the unperturbed (non-interacting) pair AB with zeroth-order energies $E_\kappa^{(0)} = E_j^{(A)} + E_k^{(B)}$. Here, it is assumed that insofar as A and B are electronically bound states, their wavefunctions of electronic coordinates fall exponentially with distance and their overlap is negligible at large R so that electronic exchange and charge transfer between A and B can be neglected. It is then sufficient to start with an interaction energy Hamiltonian written in terms of the sets of unperturbed states of A and B assumed to form complete systems, as

$$\lambda\hat{H}_{AB}^{(1)}\left(\vec{R}\right) = \sum\nolimits_{j,k}\sum\nolimits_{j'k'}|j_A k_B\rangle\left\langle j_A k_B\,|\,\hat{H}_{AB}^{(int)}\,|\,j'_A k'_B\right\rangle_R \langle j'_A k'_B|$$

The matrix element $\left\langle j_A k_B\,|\,\hat{H}_{AB}^{(int)}\,|\,j'_A k'_B\right\rangle$ can be re-expressed as a sum of products of factors relating separately to A and B and has the meaning of an interaction where a transition in A from state j_A to j'_A is coupled to a transition in B from state k_B to k'_B.

For two molecules A and B enclosed by spheres of radius a_A and a_B, the interaction potential energy is small at distances larger than a R_{min} and a suitable expansion parameter is $\lambda = (a_A + a_B)/R_{min}$. Expanding the energy and wavefunction in powers of λ the Hamiltonian equation is

$$\left(\hat{H}^{(0)} + \lambda\hat{H}^{(1)}\right)\left(\sum\nolimits_{n\geq 0}\lambda^n\Psi_\kappa^{(n)}\right) = \left(\sum\nolimits_{p\geq 0}\lambda^p E_\kappa^{(p)}\right)\left(\sum\nolimits_{q\geq 0}\lambda^q\Psi_\kappa^{(q)}\right)$$

where $\hat{H}^{(0)} = \hat{H}_A + \hat{H}_B$ and $\lambda\hat{H}^{(1)} = \hat{H}_{AB}^{(1)}\left(\vec{R}\right)$. Equating on both sides the factors multiplying λ^p, a recursion procedure follows as

$$\left(\hat{H}^{(0)} - E_\kappa^{(0)}\right)\Psi_\kappa^{(0)} = 0$$

$$\left(\hat{H}^{(0)} - E_\kappa^{(0)}\right)\Psi_\kappa^{(1)} + \hat{H}^{(1)}\Psi_\kappa^{(0)} = E_\kappa^{(1)}\Psi_\kappa^{(0)}$$

$$\left(\hat{H}^{(0)} - E_\kappa^{(0)}\right)\Psi_\kappa^{(p)} + \hat{H}^{(1)}\Psi_\kappa^{(p-1)} = E_\kappa^{(p)}\Psi_\kappa^{(0)} + \sum\nolimits_{q=1}^{p-1}E_\kappa^{(q)}\Psi_\kappa^{(p-q)}$$

for $p \geq 2$, and with the energy to p-th order given by the recursion relation

$$E_\kappa^{(p)} = \left\langle\Psi_\kappa^{(0)}\,|\,\hat{H}^{(1)}\,|\,\Psi_\kappa^{(p-1)}\right\rangle - \sum\nolimits_{q=1}^{p-1}E_\kappa^{(q)}\left\langle\Psi_\kappa^{(0)}\,|\,\Psi_\kappa^{(p-q)}\right\rangle$$

where the normalization $\left\langle\Psi_\kappa^{(0)}\,|\,\Psi_\kappa^{(0)}\right\rangle = 1$ has been chosen. Higher-order wavefunction terms can alternatively be constructed so that $\left\langle\Psi_\kappa^{(0)}\,|\,\Psi_\kappa^{(p-q)}\right\rangle = 0$ for $p - q \geq 1$, which imposes the intermediate normalization $\left\langle\Psi_\kappa^{(0)}\,|\,\Psi_\kappa\right\rangle = 1$. Additional relations can be used to show that knowledge of $\Psi_\kappa^{(p)}$ is sufficient to obtain the $E_\kappa^{(2p+1)}$ energy [3]. In particular, $E_\kappa^{(1)} = \left\langle\Psi_\kappa^{(0)}\,|\,\hat{H}^{(1)}\,|\,\Psi_\kappa^{(0)}\right\rangle$, $E_\kappa^{(2)} = \left\langle\Psi_\kappa^{(0)}\,|\,\hat{H}^{(1)}\,|\,\Psi_\kappa^{(1)}\right\rangle$, and $E_\kappa^{(3)} = \left\langle\Psi_\kappa^{(1)}\,|\left(\hat{H}^{(1)} - E_\kappa^{(1)}\right)|\,\Psi_\kappa^{(1)}\right\rangle$.

We consider first perturbation energies when both A and B are separated by a distance R and in their ground electronic state $(0_A, 0_B)$. The first-order energy is

$$E_{0,0}^{(1)}(R) = \left\langle 0_A, 0_B \mid \hat{H}^{(1)} \mid 0_A, 0_B \right\rangle$$

with A in its ground state interacting with B also in its ground state. This is the classical electrostatic interaction energy $E_{0,0}^{(els)}(R) = \lambda E_{0,0}^{(1)}(R) = \left\langle 0_A, 0_B \mid \hat{H}_{AB}^{(int)} \mid 0_A, 0_B \right\rangle$.

To second-order

$$E_{0,0}^{(2)}(R) = \sum_{(j,k)\neq(0,0)} \frac{\left|\left\langle 0_A, 0_B \mid \hat{H}^{(1)} \mid j_A, k_B \right\rangle\right|^2}{E_{0,0}^{(0)} - E_{j,k}^{(0)}}$$

and the double sum over j and k, which must not be both ground states, can be separated into two expressions: one where only j or k is the ground state, called an induction potential energy; and a second double sum where both j and k differ from the ground state, called a dispersion potential energy. The induction energy is therefore

$$E_{0,0}^{(ind)}(R) = -\sum_{j\neq 0} \frac{\left|\left\langle 0_A, 0_B \mid \hat{H}_{AB}^{(int)} \mid j_A, 0_B \right\rangle\right|^2}{E_j^{(A)} - E_0^{(A)}} - \sum_{k\neq 0} \frac{\left|\left\langle 0_A, 0_B \mid \hat{H}_{AB}^{(int)} \mid 0_A, k_B \right\rangle\right|^2}{E_k^{(B)} - E_0^{(B)}}$$

and has the meaning of B in its ground state interacting with A in a transition state for the first term, and the reverse in the second term. For the dispersion energy one has a double sum where A in a state transition interacts with B also in a state transition, in accordance with

$$E_{0,0}^{(dsp)}(R) = -\sum_{j\neq 0}\sum_{k\neq 0} \frac{\left|\left\langle 0_A, 0_B \mid \hat{H}_{AB}^{(int)} \mid j_A, k_B \right\rangle\right|^2}{E_j^{(A)} - E_0^{(A)} + E_k^{(B)} - E_0^{(B)}}$$

This can be made more specific by introducing details of the AB pair interaction. Here, the denominators have been changed to show them as always positive, with the minus sign in front of expressions indicating that these are negative quantities corresponding to attraction between A and B when these are in their ground states. This is not necessarily the case for the perturbation energies of their excited states.

3.1.2 Interactions in Excited Electronic States and in Resonance

The perturbation treatment requires reconsideration when the reference state, ground, or excited, has the same, so-called degenerate, energy as other states. In this case, a useful procedure is to lift the degeneracy by including the part of $\hat{H}_{AB}^{(int)}$ within the subspace of degenerate functions in the Hamiltonian operator to zeroth order and solving for new eigenstates and eigenenergies within the subspace of the previous degenerate states with the expectation that the new energy values will not be degenerate. The perturbation treatment can then proceed as before with the new zeroth order states and energies [3, 4].

When the two species are the same and one of them is in an excited electronic state A^*, the interactions in $A + A^*$ lead to quantum coherence of its states with those of $A^* + A$ and consequent resonance energy transfer. The coherent states replacing $|j_A, k_B\rangle$ are now $|j,k\rangle_{AA^*}^{(\pm)} = 2^{-1/2}(|j_A,k_{A^*}\rangle \pm |k_A,j_{A^*}\rangle)$ with zeroth order energy $E_K^{(0)} = E_j^{(A)} + E_k^{(A^*)}$ and the first-order resonance interaction energy is for $R > R_{min}$,

$$E_{AA^*}^{(\pm)} = 2^{-1}\Big(\Big\langle j_A,k_{A^*} \,|\, \hat{H}_{AB}^{(int)} \,|\, j_A,k_{A^*}\Big\rangle + \Big\langle k_A,j_{A^*} \,|\, \hat{H}_{AB}^{(int)} \,|\, k_A,j_{A^*}\Big\rangle$$
$$\pm\, 2\Big\langle j_A,k_{A^*} \,|\, \hat{H}_{AB}^{(int)} \,|\, k_A,j_{A^*}\Big\rangle\Big)$$

so that the first two terms are again electrostatic interaction energies from the charge distributions of A and A^* in states j or k, but the last term is an addition involving transitions between states j and k for each species. This involves an energy of interaction between two electronic transition densities. The last term contains state transfer integrals which obey different selection rules and lead to longer range interactions, absent when there is no state coherence.

This subject is closely related to the theory of molecular excitons [5] and intermolecular electronic energy transfer [6].

3.2 Long Range Interaction Energies from Permanent and Induced Multipoles

3.2.1 Molecular Electrostatic Potentials

We model the interaction of two molecules A and B with their center of mass (or CM) located at positions $\vec{R}_A$ and $\vec{R}_B$ and with relative position $\vec{R} = \vec{R}_B - \vec{R}_A$. We need the Coulomb, or electrostatic potential, $\phi_{els}(\vec{s})$ created by the charges in A at a field position $\vec{s}$ near B where its charges may be found.

The interaction energy of charges in A with those in B must be separately treated depending on whether the distance between A and B is large compared to their size, or not. In the first case, we deal with long-range (LR) interactions, and we can express the energy as an expansion in inverse powers of R and in terms of the quantal electronic states of the isolated compounds. The short-range interactions must, however, be treated differently starting from the electronic structure of the whole pair and its quantal states, allowing for electronic rearrangement. Indicating the electrostatic, or Coulomb, interaction of the two sets of charges as $H_{AB}^{(int)}\left(\vec{R}\right)$, its long-range form can be written as $H_{AB}^{(LR)}\left(\vec{R}\right) = f_d(R) H_{AB}^{(int)}\left(\vec{R}\right)$, with $f_d(R)$ a function changing from zero at short distances into 1.0 at long distances with a transition around a minimum distance R_{min} and a transition region of width $a_A + a_B$, the sum of radii of spheres enclosing the charge distributions of A and B. The total Hamiltonian is then $\hat{H}_{AB} = \hat{H}_A + \hat{H}_B + H_{AB}^{(int)}\left(\vec{R}\right)[1 - f_d(R)] + H_{AB}^{(LR)}\left(\vec{R}\right)$, and one can concentrate on the LR part to begin with.

We consider here situations where the two charge distributions do not overlap, with $R > R_{min}$, a minimum distance, large, and with $r_I \ll R$ and $s_J \ll R$. We work in the CMN-SF frame, take the origin of coordinates at the CM of A and expand the inverse distance $\left|\vec{s} - \vec{r}\right|^{-1} = \left|\vec{R} + \vec{s}_B - \vec{r}_A\right|^{-1}$ between generic field and charge locations in a power series for small r_A/R', where $\vec{R}' = \vec{R} + \vec{s}_B$, with components $R'_\xi = (X', Y', Z')$. This is done using that $\left(\partial |\vec{R}' - \vec{r}_A| / \partial \xi_A\right)_{\xi_A = 0} = R'_\xi / R'$ $= -\partial |\vec{R}' - \vec{r}_A| / \partial R'_\xi$ which leads to

$$\frac{1}{|\vec{R}' - \vec{r}_A|} = \left(1 - \sum\nolimits_\xi \xi_A \frac{\partial}{\partial R'_\xi} + \frac{1}{2} \sum\nolimits_{\xi,\eta} \xi_A \eta_A \frac{\partial^2}{\partial R'_\xi \partial R'_\eta} + \ldots\right) \frac{1}{R'}$$

and an expression for $\phi_{els}\left(\vec{R}'\right) = (4\pi\varepsilon_0)^{-1} \sum\nolimits_I C_I \left|\vec{R}' - \vec{r}_{IA}\right|^{-1}$ like

$$\phi_{els}\left(\vec{R}'\right) = \frac{1}{4\pi\varepsilon_0}\left(C^{(A)} - \sum\nolimits_\xi D_\xi^{(A)} \frac{\partial}{\partial R'_\xi} + \sum\nolimits_{\xi,\eta} \frac{1}{6} Q_{\xi\eta}^{(A)} \frac{\partial^2}{\partial R'_\xi \partial R'_\eta} + \ldots\right) \frac{1}{R'}$$

written in terms of total charge, dipole, and quadrupole components

$$C = \sum\nolimits_I C_I, \; D_\xi = \sum\nolimits_I C_I \, \xi_I$$
$$Q_{\xi\eta} = \sum\nolimits_I C_I \left(3\,\xi_I \eta_I - \delta_{\xi\eta} r_I^2\right)$$

for compound A. These can also be given in terms of spherical multipole functions of spherical coordinates, which are useful in the description of two interacting few-atom systems.

This is next illustrated in a simple case where we describe the interaction of compound A with charges 1 and 2, interacting with compound B containing charges 3 and 4. For two charges C_1 and C_2 at locations z_1 and z_2 on a z-axis along direction $\vec{n}_z$, the Coulomb electrostatic potential at field location $\vec{s} = \vec{R}$, the CM of B, with components (X, Y, Z) is

$$\phi_{els}\left(\vec{R}\right) = C_1/(4\pi\varepsilon_0 R_1) + C_2/(4\pi\varepsilon_0 R_2)$$

where $R_j = |\vec{R} - \vec{r}_j|$ and $\vec{r}_j = z_j\vec{n}_z$, $j = 1, 2$.

Expanding R_j^{-1} in a power series for small r_j/R, and using that $(\partial R_1/\partial z_1)_{z_1=0} = Z/R = -\partial R/\partial Z$, we have

$$\frac{1}{R_1} = \left(1 + z_1\frac{\partial}{\partial z_1} + \frac{1}{2}z_1^2\frac{\partial^2}{\partial z_1^2} + \ldots\right)\frac{1}{R_1} = \left(1 - z_1\frac{\partial}{\partial Z} + \frac{1}{2}z_1^2\frac{\partial^2}{\partial Z^2} + \ldots\right)\frac{1}{R}$$

and similarly for charge 2; further using that $Z/R = \cos(\Theta)$ we find

$$\phi_{els}\left(\vec{R}\right) = \phi_{els}^{(0)}\left(\vec{R}\right) + \phi_{els}^{(1)}\left(\vec{R}\right) + \phi_{els}^{(2)}\left(\vec{R}\right)$$

$$\phi_{els}^{(0)}\left(\vec{R}\right) = \frac{C}{4\pi\varepsilon_0 R}$$

$$\phi_{els}^{(1)}\left(\vec{R}\right) = -\frac{D\cos(\Theta)}{4\pi\varepsilon_0 R^2}$$

$$\phi_{els}^{(2)}\left(\vec{R}\right) = \frac{(Q/2)\{[3\cos^2(\Theta) - 1]/2\}}{4\pi\varepsilon_0 R^3}$$

where we have introduced the total charge $C = C_1 + C_2$ (a 2^0-pole), dipole $D = C_1z_1 + C_2z_2$ (a 2^1-pole), and quadrupole $Q = 2\left(C_1z_1^2 + C_2z_2^2\right)$ (a 2^2-pole) of the system. This shows the electrostatic potentials created by charge, dipole, and quadrupole distributions. Higher-order terms include octupoles, and higher 2^n-poles. The electric potential of a 2^n-pole is found to vary with large R as $R^{-(n+1)}$.

The electrostatic field vector $\vec{\mathcal{E}} = -\partial\phi_{els}/\partial\vec{R}$ has spherical components (for the above charges on the z-axis) given by

$$\mathcal{E}_R = -\frac{\partial\phi_{els}}{\partial R},\ \mathcal{E}_\Theta = -\frac{1}{R}\frac{\partial\phi_{els}}{\partial\Theta},\ \mathcal{E}_\Phi = 0$$

3.2.2 The Interaction Potential Energy at Large Distances

Continuing with the simple model, we apply the previous expansion to the electrostatic potential due to the A charges, and find the interaction Coulomb energy of a system B of charges in that potential [7, 8]. We consider the system B with two charges 3 and 4 located at positions $\vec{R}_3 = \vec{R} + \vec{r}_3$ and $\vec{R}_4 = \vec{R} + \vec{r}_4$ in the potential of charges 1 and 2 (system A). The interaction energy function is now

$$H_{AB}^{(int)} = C_3 \phi_{els}^{(A)}\left(\vec{R}_3\right) + C_4 \phi_{els}^{(A)}\left(\vec{R}_4\right)$$

This can be further expanded now for small r_j/R, $j = 3, 4$, which brings in the multipoles of B, and we introduce $C_B = C_3 + C_4$ and $\vec{D}_B = C_3\vec{r}_3 + C_4\vec{r}_4$, and a similar notation for the A multipoles. Using a vector notation and noticing that the interaction of a 2^m-pole of A with a 2^n-pole of B gives a term with an $R^{-(m+n+1)}$ dependence, we find for $R > R_{min}$

$$H_{AB}^{(int)}\left(\vec{R}\right) = (4\pi\varepsilon_0)^{-1} \sum_{m,n} F_{m,n}^{(AB)} R^{-(m+n+1)}$$

$$F_{0,0}^{(AB)} = C_A\, C_B$$

$$F_{0,1}^{(AB)} = -C_A \vec{D}_B . \vec{n},\; F_{1,0}^{(AB)} = -C_B \vec{D}_A . \vec{n}$$

$$F_{1,1}^{(AB)} = \vec{D}_A . \vec{D}_B - 3\left(\vec{D}_A . \vec{n}\right)\left(\vec{D}_B . \vec{n}\right)$$

where $\vec{n} = \vec{R}/R$ is a unit vector pointing from A to B. Additional terms contain higher-order multipoles and can be derived by inspection. These expressions have the same form in a quantal treatment where the electrostatic energy and dipoles are operators. The electrostatic potential energy of a quantal state of the A–B pair of compounds is obtained from the quantal expectation value of this potential energy operator for the given state.

A general treatment requires an expansion of the inverse of $\vec{R}' = \vec{R} + \vec{s}_B$ for small s_B/R, which brings in all the multipoles of B, multiplying those of A. The electrostatic energy operator

$$H_{AB}^{(int)} = \sum_{J \in B} C_J^{(B)} \phi_{els}^{(A)}\left(\vec{R} + \vec{s}_{JB}\right)$$

can be expanded now for small s_{JB}/R as

$$H_{AB}^{(LR)} = \left(C^{(B)} + \sum_{\xi} D_{\xi}^{(B)} \frac{\partial}{\partial R_{\xi}} + \frac{1}{6} \sum_{\xi,\eta} Q_{\xi\eta}^{(B)} \frac{\partial^2}{\partial R_{\xi} \partial R_{\eta}} + \ldots\right) \phi_{els}^{(A)}\left(\vec{R}\right) f_d(R)$$

and written for large distances in a very general form introducing Cartesian tensors $\boldsymbol{t}$ constructed from unit vectors $n_{\xi} = R_{\xi}/R$ from

$$\frac{\partial}{\partial R_{\xi}} \frac{1}{R} = -\frac{R_{\xi}}{R^3} = -\frac{n_{\xi}}{R^2} = -\frac{t_{\xi}}{R^2}$$

$$\frac{\partial^2}{\partial R_{\xi} \partial R_{\eta}} \frac{1}{R} = \frac{3 n_{\xi} n_{\eta} - \delta_{\xi\eta}}{R^3} = \frac{t_{\xi\eta}}{R^3}$$

and $t_{\xi\eta\zeta} = -15 n_\xi n_\eta n_\zeta + 3(n_\xi \delta_{\eta\zeta} + n_\eta \delta_{\xi\zeta} + n_\zeta \delta_{\xi\eta})$. In general, the tensor with n indices is

$$t^{(n)}_{\xi\eta\ldots\chi} = R^{n+1} \partial_\xi \partial_\eta \cdots \partial_\chi R^{-1}$$

for higher derivatives, with the notation $\partial_\xi = \partial / \partial R_\xi$ [9]. These tensors depend on the orientation angles of the relative position vector $\vec{R} = \vec{R}_B - \vec{R}_A$ and their sign changes by $(-1)^n$ if this direction is reversed. The electrostatic potential is then

$$\phi^{(A)}_{els}\left(\vec{R}\right) = \frac{1}{4\pi\varepsilon_0}\left(\frac{C^{(A)}}{R} + \sum\nolimits_\xi D^{(A)}_\xi \frac{t_\xi}{R^2} + \sum\nolimits_{\xi,\eta} \frac{1}{6} Q^{(A)}_{\xi\eta} \frac{t_{\xi\eta}}{R^3} + \ldots\right)$$

which is in turn differentiated n times and multiplied times the 2^n-th multipole of B to obtain the interaction energy operator as [9, 10]

$$\begin{aligned}\hat{H}_{AB}{}^{(LR)} = (4\pi\varepsilon_0)^{-1} \Bigg[& \frac{C^{(A)} C^{(B)}}{R} + \sum\nolimits_\xi \left(C^{(A)} \frac{t_\xi}{R^2} \hat{D}^{(B)}_\xi - \hat{D}^{(A)}_\xi \frac{t_\xi}{R^2} C^{(B)}\right) \\ & + \frac{1}{6}\sum\nolimits_{\xi,\eta}\left(C^{(A)} \frac{t_{\xi\eta}}{R^3} \hat{Q}^{(B)}_{\xi\eta} + \hat{Q}^{(A)}_{\xi\eta} \frac{t_{\xi\eta}}{R^3} C^{(B)}\right) \\ & - \sum\nolimits_{\xi,\eta} \hat{D}^{(A)}_\xi \frac{t_{\xi\eta}}{R^3} \hat{D}^{(B)}_\eta \\ & - \frac{1}{6}\sum\nolimits_{\xi,\eta,\zeta}\left(\hat{D}^{(A)}_\xi \frac{t_{\xi\eta\zeta}}{R^4} \hat{Q}^{(B)}_{\eta\zeta} + \hat{Q}^{(A)}_{\xi\eta} \frac{t_{\xi\eta\zeta}}{R^4} \hat{D}^{(B)}_\zeta\right) + \cdots \Bigg] f_d(R)\end{aligned}$$

Alternatively, introducing multipolar tensor factors $T^{(m,n)}_{\mu,\nu}(R) = f_d(R) t^{(m,n)}_{\mu,\nu}$ dependent on orientation angles for each pair of interacting multipoles, the interaction Hamiltonian is compactly given by

$$\hat{H}^{(LR)}_{AB} = (4\pi\varepsilon_0)^{-1} \sum\nolimits_{m\mu} \sum\nolimits_{n\nu} \hat{M}^{(m)}_\mu(A) \frac{T^{(m,n)}_{\mu,\nu}(R)}{R^{m+n+1}} \hat{M}^{(n)}_\nu(B)$$

where μ designates the components of the 2^m multipole, with three values for the dipole, six for the quadrupole, and so on.

In a reference frame with the Z-axis along the intermolecular relative position, the perturbation energies depend on R and are given as before in terms of the integrals $\left\langle \Psi^{(0)}_\kappa \mid \hat{H}^{(1)} \mid \Psi^{(0)}_\lambda \right\rangle$. Here, $\hat{H}^{(1)}$ contains a sum of $R^{-(m+n+1)}$ terms, and the energy term $E^{(p)}_\kappa(R)$ is itself an expansion in powers of R^{-1}, of the form

$$E^{(p)}_\kappa(R) = \sum\nolimits_{n \geq n^{(p)}} C^{(p)}_n R^{-n}$$

where $n^{(p)}$ is the lowest power of R^{-1} which appears to order p in the energy expansion, and increases as p increases.

For the ground state of the pair, the energy varies as the relative distance decreases from infinity, and the intermolecular potential energy evolving from the ground state $g = (0_A, 0_B)$ of the pair is

$$V_g^{(AB)}(R) = E_g^{(AB)}(R) - E_g^{(AB)}(\infty)$$

where the asymptotic term is simply $E_g^{(AB)}(\infty) = E_0^{(A)} + E_0^{(B)}$, the sum of energies of the isolated molecules in their ground state. This intermolecular potential energy is therefore a double perturbation expansion of form

$$V_g^{(AB)}(R) = \sum\nolimits_{p \geq 1} \lambda^p \sum\nolimits_{n \geq n^{(p)}} C_n^{(p)} R^{-n}$$

and the lower limits $n^{(p)}$ to the summations must be found by inspection of the asymptotic expansion series in powers of R^{-1}. The coefficient of R^{-n} must be obtained from all orders in λ relevant to a desired accuracy.

An extension of the perturbation treatment is needed when the states of A or B are degenerate [3, 4, 11], as is frequently the case for excited electronic states.

The expansions in inverse powers of R can be expected to converge only asymptotically, particularly when applied to atoms or compounds with open electronic shells, as has been noticed in a survey of several examples [12]. The terms A_n/R^n making the series expansion of $V_g^{(AB)}(R)$ are frequently such that the convergence criterion based on the quotient of two adjacent terms gives $|A_{n+1}/(A_n R)| \rightarrow \infty$ for $n \rightarrow \infty$ indicating series divergence. However, the sum of the first N terms in the expansion usually gives a good approximation to the exact potential energy so that $| V_g^{(AB)}(R) - \sum_{n=0}^{N} A_n/R^n | \rightarrow 0$ as $N \rightarrow \infty$ for R larger than a critical value dependent on N, indicating an asymptotic or semiconvergent series. This justifies using a sum of inverse power terms to calculate the potential energy for large distances.

3.2.3 Electrostatic, Induction, and Dispersion Forces

We consider here the ground states of the pair. To first order, the total energy is in general the expectation value $\lambda E_\kappa^{(1)} = \langle j_A k_B | \hat{H}_{AB}^{(int)} | j_A k_B \rangle$, a sum where each term contains the expectation values of the $(2^m)_A$ and $(2^n)_B$ multipoles of A and B. In the ground state $(j_A, k_B) = (0_A, 0_B)$. Therefore, each term is of the form

$$V_{m,n}^{(els)}(R) = (4\pi\varepsilon_0)^{-1} C_{m+n+1}^{(els)} R^{-(m+n+1)}$$

where $C_{m+n+1}^{(els)}$ is a constant obtained from the permanent multipoles in the ground states.

To second order, the potential energy operator appears in two factors, between the ground and excited states of the (A, B) pair, and again back from excited states to the ground state. Depending on the nature of the excitation, we can distinguish induction and dispersion contributions in the second-order energies.

In the induction energy terms, the first factor involves the transition from (*A*, *B*) to (A, B^*) in which B^* is excited and acquires distortion multipoles. This factor is an interaction between the $(2^m)_A$ permanent multipole of *A* and the $\left(2^{n'}\right)_{B^*}$ induced multipole of B^*. The second factor in the induction energy involves an induced multipole $\left(2^{n''}\right)_{B^*}$ and a permanent $(2^p)_A$ and represents the interaction leading back to the ground state. Therefore, the interaction is of the form $C^{(ind)}_{m+p+n'+n''+2} R^{-(m+p+n'+n''+2)}$. However, selection rules for allowed transitions into and out of states of B^* usually requires that $n' = n''$, so that the induction potential energy is usually

$$V^{(ind)}_{m,p,n'}(R) = -(4\pi\varepsilon_0)^{-2} C^{(ind)}_{m+p+2n'+2} R^{-(m+p+2n'+2)}$$

A similar term arises from $B \rightarrow A^* \rightarrow B$, with the induced multipole $\left(2^{m'}\right)_{A^*}$ interacting with $(2^n)_B$ up and $(2^q)_B$ down in energy. The coefficient here is positive as can be seen from the second-order correction to the ground state energy. The leading induction term is typically one with $m = p$ and $n = q$. For ion-neutral interactions, it usually is a (permanent monopole)–(induced dipole) interaction going as R^{-4}.

In the dispersion interaction terms, we have temporary transitions from (A, B) to (A^*, B^*) and back to (A, B), in nonenergy conserving transient events allowed by quantum mechanics. This can be interpreted as a fluctuation of charges in *A* which creates a distortion multipole $\left(2^{m'}\right)_{A^*}$; this induces fluctuations in *B*, which changes into B^* with multipole $\left(2^{n'}\right)_{B^*}$ as both compounds are excited. The same interacting multipoles bring the system back to (A, B). The interaction potential is therefore of the form

$$V^{(dsp)}_{m',n'}(R) = -(4\pi\varepsilon_0)^{-2} C^{(dsp)}_{2(m'+n'+1)} R^{-2(m'+n'+1)}$$

with the leading term coming from (induced dipole)–(induced dipole) interactions going as R^{-6}. The three types of interaction processes, electrostatic, induction, and dispersion, are diagrammed in Figure 3.2.

Going to higher powers in the perturbation expansion in λ of the intermolecular potential energy gives additional terms combining permanent and induced multipole interactions. The total potential energy can be written as $V(R) = V^{(els)}(R) + V^{(ind)}(R) + V^{(dsp)}(R) + V^{(pol)}(R)$, an asymptotic expansion in powers of R^{-1} with a varying radius of convergence, where a polarization term from higher-order perturbations in λ combines induction and dispersion interactions.

When the two species are the same, and one is electronically excited, there is in addition a resonance potential energy $V^{(res)}(R)$ resulting from the quantal

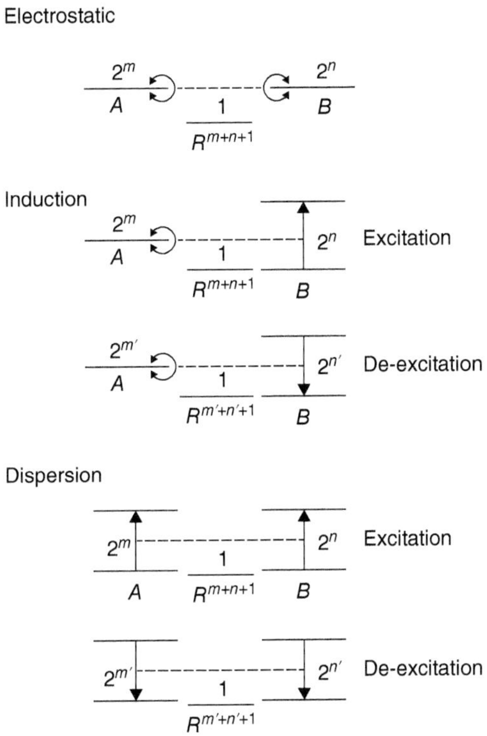

Figure 3.2 Interactions (dashed lines) of electronic transitions (arrows) in electrostatic, induction, and dispersion phenomena between A and B.

coherence of states in $A + A^*$ and $A^* + A$. This leads to additional interactions between transition multipoles with the same behavior as the electrostatic term, and in the case of the (transition dipole)–(transition dipole) interaction, it gives to first-order perturbation $V^{(res)}(R) = (4\pi\varepsilon_0)^{-1}C^{(res)}\, R^{-3}$, of longer range than induction or dispersion terms.

3.2.4 Interacting Atoms and Molecules from Spherical Components of Multipoles

The interaction energies involving atoms and small molecules can be conveniently treated using a few spherical multipoles per compound, with each component containing charges I given by

$$Q_{lm} = \left(\frac{4\pi}{2l+1}\right)^{1/2} \sum\nolimits_I C_I r_I^l Y_{lm}(\vartheta_I, \varphi_I)$$

for $l \geq 0$ and $-l \geq m \geq l$, written in terms of the spherical harmonic function Y_{lm} of position angles, with well-known transformation properties under rotation.

The electrostatic potential $\phi_{els}^{(A)}\left(\vec{R}\right) = (4\pi\varepsilon_0)^{-1}\sum_I C_I \left|\vec{R}-\vec{r}_{IA}\right|^{-1}$ can be expanded in the spherical harmonics for $\vec{R}$ and $\vec{r}_{IA}$ as

$$\phi_{els}^{(A)}\left(\vec{R}\right) = (4\pi\varepsilon_0)^{-1}\sum_I C_I \sum_l \frac{1}{R}\left(\frac{r_{IA}}{R}\right)^l \frac{4\pi}{2l+1}\sum_m Y_{lm}(\vartheta_{AI},\varphi_{AI})Y_{lm}(\vartheta_R,\varphi_R)^*$$

which gives in terms of multipoles

$$\phi_{els}^{(A)}\left(\vec{R}\right) = (4\pi\varepsilon_0)^{-1}\sum_{l,m}\left(\frac{4\pi}{2l+1}\right)^{\frac{1}{2}}\frac{1}{R^l}Q_{lm}\left(\{\vec{r}_{IA}\}\right)Y_{lm}(\vartheta_R,\varphi_R)^*$$

This potential acts on the charge multipoles of B to give a pair interaction energy with a relatively simple form provided the local reference frames $\mathcal{S}_A$ and $\mathcal{S}_B$ are parallel and have their z-axes parallel to the intermolecular position vector $\vec{R}$ [13], as in

$$\hat{H}_{AB}^{(LR)} = (4\pi\varepsilon_0)^{-1}\sum_{l_a\geq 0}\sum_{l_b\geq 0}\hat{F}_{AB}^{(l_a,l_b)}R^{-(l_a+l_b+1)}f_d(R)$$

$$\hat{F}_{AB}^{(l_a,l_b)} = \sum_{\mu=-\lambda}^{\mu=\lambda}\frac{(-1)^\lambda(l_a+l_b)!}{[(l_a-\mu)!(l_a+\mu)!(l_b-\mu)!(l_b+\mu)!]^{1/2}}Q_{l_a,\mu}\left(\{\vec{r}_{IA}\}\right)Q_{l_b,-\mu}\left(\{\vec{r}_{JB}\}\right)$$

where λ is the smallest between l_a and l_b. The multipoles here can be readily rotated in each frame using the Wigner rotational matrices $D_{m'm}^{(l)}(\alpha,\beta,\gamma)$, functions of Euler angles for each rotation [14, 15].

Electrostatic, induction, and dispersion energies follow from perturbation expansions containing integrals of multipoles over electronic states of A and B. For atoms, integrals of the multipoles of A between electronic states with given angular momentum quantum numbers, $|\nu_A L_A M_A\rangle$ and $|\nu_A' L_A' M_A'\rangle$, are zero except if the selection rules $M_A' = M_A + \mu$ and $|L_A' - L_A| \leq l_a \leq L_A' + L_A$ apply, and similarly for B. These rules substantially decrease the number of excited states needed in calculations of the interaction energies.

3.2.5 Interactions from Charge Densities and their Fourier Components

It is possible to avoid expansions of interaction energies in terms of multipoles, by instead using charge densities and their Fourier transforms [16, 17]. The charge density operator of all the charges I in species A distributed over space can be re-expressed in terms of its Fourier components $Q_A\left(\vec{k}\right)$ as in

$$\hat{c}_A\left(\vec{r}\right) = \sum_I C_I\delta\left(\vec{r}-\hat{\vec{r}}_{IA}\right) = \int\frac{d^3k}{(2\pi)^3}\exp\left(-i\vec{k}.\vec{r}\right)\hat{Q}_A\left(\vec{k}\right)$$

with the inverse $\hat{Q}_A\left(\vec{k}\right) = \int d^3r \exp\left(i\vec{k}.\vec{r}\right)\hat{c}_A\left(\vec{r}\right) = \sum_I C_I \exp\left(i\vec{k}.\vec{\hat{r}}_{IA}\right)$ and similarly for species B. The interaction energy operator

$$\hat{H}_{AB}^{(int)}\left(\vec{R}\right) = (4\pi\varepsilon_0)^{-1}\int d^3r \int d^3s\, \hat{c}_A\left(\vec{r}\right)\left|\vec{r}-\vec{s}\right|^{-1}\hat{c}_B\left(\vec{s}\right)$$

can be also re-expressed in terms of Fourier components using that $\left|\vec{r}\right|^{-1} = \int\left[d^3k/(2\pi)^3\right]\exp\left(-i\vec{k}.\vec{r}\right)k^{-2}$, to obtain the result

$$\hat{H}_{AB}^{(int)}\left(\vec{R}\right) = (4\pi\varepsilon_0)^{-1}\int \frac{d^3k}{2\pi^2k^2}\exp\left(i\vec{k}.\vec{R}\right)\hat{Q}_A\left(-\vec{k}\right)\hat{Q}_B\left(\vec{k}\right)$$

to be incorporated into the first- and second-order perturbation expansions given above for the interaction energies. Here, we concentrate on the dispersion component to second order, which contains denominators like $E_j^{(A)} - E_0^{(A)} + E_k^{(B)} - E_0^{(B)} = a + b$, and use the Casimir–Polder integral (for positive a, b)

$$\frac{2}{\pi}\int_0^{\infty}\frac{du\, a\, b}{(a^2+u^2)(b^2+u^2)} = \frac{1}{a+b}$$

to write for the dispersion energy in Section 3.1.1, with $\omega = i\upsilon$

$$E_{0,0}^{(dsp)}\left(\vec{R}\right) = -(4\pi\varepsilon_0)^{-2}\int\frac{d^3k}{2\pi^2k^2}\int\frac{d^3k'}{2\pi^2k'^2}\exp\left[i\left(\vec{k}-\vec{k}'\right).\vec{R}\right]$$
$$\frac{\hbar}{2\pi}\int_0^{\infty} d\upsilon\, \chi_A\left(\vec{k},\vec{k}';i\upsilon\right)\chi_B\left(-\vec{k},-\vec{k}';i\upsilon\right)$$
$$\chi_A\left(\vec{k},\vec{k}';i\upsilon\right) = \frac{2}{\hbar}\sum_j\frac{\omega_j}{\omega_j^2+\upsilon^2}\left\langle 0\right|\hat{Q}_A\left(-\vec{k}\right)\left|j\right\rangle\left\langle j\right|\hat{Q}_A\left(\vec{k}'\right)\left|0\right\rangle$$

with $\hbar\omega_j = E_j^{(A)} - E_0^{(A)}$, and similarly for B. This gives the dispersion energy as a sum over grid points in $\vec{k}$ and $\vec{k}'$ spaces and avoids an expansion in inverse powers of R. The factors χ_S are dynamical susceptibilities evaluated at imaginary-valued frequencies for each species S.

Further, using the plane wave decomposition into spherical waves,

$$\exp\left(i\vec{K}.\vec{R}\right) = \sum_{l=0}^{\infty}\sum_{m=-l}^{l} i^l 4\pi\, Y_l^m(\vartheta_K,\varphi_K)^* Y_l^m(\vartheta_R,\varphi_R) j_l(KR)$$

where the spherical Bessel function $j_l(x)$ satisfies asymptotic conditions $j_l(x) \approx A_l x^{-l-1}\sin(x)$ for $x \to \infty$ and $j_l(x) \approx B_l x^l$ for $x \to 0$, it is possible to provide a detailed analysis of the two limits for large and small R [17]. It is found that for large R one recovers the inverse R multipole expansion previously generated, and that now the present treatment gives finite values also for small R, so that the dispersion energy for all R can be written as

$$V_{m',n'}^{(dsp)}(R) = -(4\pi\varepsilon_0)^{-2} f_{2(m'+n'+1)}^{(dsp)}(R) C_{2(m'+n'+1)}^{(dsp)} R^{-2(m'+n'+1)}$$

with $f(R) \approx 1$ for $R \to \infty$ and $f(R) \approx 0$ for $R \to 0$ [17]. A detailed analysis here shows that for $R \to 0$ one finds $V_{m,n}^{(dsp)}(R) \approx -AR^{2|m-n|} + BR^{2|m-n|+2}$. Therefore, using the Fourier transform of charge distributions provides a rigorous procedure for avoiding the divergences of the inverse R expansions. For the (induced dipole)–(induced dipole), interaction, with $m = 1$ and $n = 1$, it gives $V_{1,1}^{(dsp)}(R) \approx -A + BR^2$ when $R \to 0$.

3.3 Atom–Atom, Atom–Molecule, and Molecule–Molecule Long-Range Interactions

3.3.1 Example of Li^++Ne

The ground electronic state of Li^+ in the atomic LS coupling scheme, which neglects spin-orbit coupling, is $1s^2\ {}^1S$, and that of Ne is $1s^2 2s^2 2p^6\ {}^1S$. Their states are labeled as ${}^{2S+1}L(M_L, M_S)$. For a charged Li^+ and neutral Ne, and since their ground state parity $(-1)^L = 1$ is even, their permanent dipole is zero by the parity selection rule, and they have quadrupoles and higher multipoles. Locating the atoms along the z-axis at a distance R, with Li^+ as A and Ne as B, the interaction Hamiltonian is $\hat{H}_{LR}^{(AB)} = (4\pi\varepsilon_0)^{-1}\left(\hat{F}_{01}^{(AB)} R^{-2} + \hat{F}_{11}^{(AB)} R^{-3}\right)$, plus higher-order terms in the inverse distance variable. In their ground state $(0_A, 0_B)$, the first-order energy perturbation expression gives $V^{(els)}(R) = 0$.

Second-order perturbation terms involve excited atomic states. For Li^+ they are $1sns\ {}^1S$, $1snp\ {}^1P$, $1snd\ {}^1D$, ..., with $n \geq 2$; for Ne, and ignoring core electrons, they are $2p^5 3p\ {}^1S$, ..., $2p^5 3s\ {}^1P$, $2p^5 3p\ {}^1P$, $2p^5 4s\ {}^1P$, $2p^5 3d\ {}^1P$, ..., $2p^5 3p\ {}^1D$, $2p^5 3d\ {}^1D$, ..., all of which are coupled to the ground states by multipole transitions allowed by selection rules. All the excitations are to total spin singlet states, with $S = 0$. Considering that a 2^m multipole has parity $(-1)^m$ and angular momentum quantum numbers $(L, M) = (m, \mu)$, it follows that transitions from initial to final states must satisfy the parity rule $(-1)^{L_i} = (-1)^{m+L_f}$ and the angular momentum rules $|L_i - m| \leq L_f \leq L_i + m$ and $M_{Lf} = M_{Li} + \mu$. The rules can be imposed going from Cartesian components $\xi = x, y, z$ to spherical ones $= 0, \pm 1$, and using the dipole spherical components $D_0 = D_z$ and $D_{\pm 1} = (D_x \pm D_y)/\sqrt{2}$.

For the induction potential energy we find, with $\hat{F}_{01}^{(AB)} = C^{(A)} \vec{n}_z . \vec{D}^{(B)} = C^{(A)} \hat{D}_0^{(B)}$,

$$V^{(ind)} = -(4\pi\varepsilon_0)^{-2} \left(C^{(A)}\right)^2 \alpha^{(B)}(0) / (2R^4)$$

$$\alpha^{(B)}(0) = 2\sum_{k \neq 0} \left| \left\langle 0_B \,|\, \hat{D}_0^{(B)} \,|\, k_B \right\rangle \right|^2 / \left(E_k^{(B)} - E_0^{(B)}\right)$$

where the second line gives the static dipolar polarizability of B, and the excited states are $k_B = n_B\,{}^1P(M_L = 0)$. The dispersion potential energy follows from $\hat{F}_{11}^{(AB)} = \hat{D}_{+}^{(A)}\hat{D}_{-}^{(B)} + \hat{D}_{-}^{(A)}\hat{D}_{+}^{(B)} - 2\hat{D}_{0}^{(A)}\hat{D}_{0}^{(B)}$, and its matrix elements between $\langle 0_A 0_B|$ and $|j_A k_B\rangle$. Noticing that selection rules restrict the three terms to be nonzero only one at a time, the dispersion energy is given by three $\sigma = \pm, 0$ components like

$$V_{\sigma}^{(dsp)} = -\frac{1}{(4\pi\varepsilon_0)^2 R^6}\sum\nolimits_{j\neq 0}\sum\nolimits_{k\neq 0}\frac{\left|\left(\hat{D}_{\sigma}^{(A)}\right)_{0j}\left(\hat{D}_{-\sigma}^{(B)}\right)_{0k}\right|^2}{\left|\epsilon_{j0}^{(A)} + \epsilon_{k0}^{(B)}\right|} = -\frac{C_6^{(dsp)}}{(4\pi\varepsilon_0)^2 R^6}$$

where $\left(\hat{D}_{\sigma}^{(A)}\right)_{0j} = \langle 0_A \,|\, \hat{D}_{\sigma}^{(A)} \,|\, j_A\rangle$ and the $\epsilon_{j0}^{(A)} = E_j^{(A)} - E_0^{(A)}$ are excitation energies, as $V^{(dsp)} = V_{+}^{(dsp)} + V_{-}^{(dsp)} + 4\,V_{0}^{(dsp)}$. Here, the excited states for A mediated by $\hat{D}_{0}^{(A)}$ and $\hat{D}_{\pm}^{(A)}$ are instead $j_A = n_A\,{}^1P(M_L = 0, \pm 1)$ and similarly for B.

For a pair of neutral atoms such as He+Ne, the induction energy disappears and only the dispersion terms remain. The (induced dipole)–(induced quadrupole) dispersion energy goes as R^{-8}.

3.3.2 Interaction of Oriented Molecular Multipoles

Interaction energies depend on the Euler angles $\Gamma_A = (\alpha, \beta, \gamma)_A$ giving the orientation of multipoles of A, and similarly for B, situated in a common space fixed reference frame for the A–B pair. Their transformation from the local body-fixed (BF) frames of A and B to their common space-fixed (SF) frame can be generally done in terms of matrix representations of the rotations operators for each multipole.

A rotation operator $\hat{R}(\Gamma_A)$ applied to the Cartesian coordinates ξ, η in the SF frame generates a 3×3 matrix $\boldsymbol{A}(\Gamma_A)$ giving the new components $\xi' = \sum_\eta A_{\xi'\eta}(\Gamma_A)\eta$ in the BF frame of A, with a similar expression for the Cartesian coordinates of species B. The dipole components are given by $\hat{D}_{\xi'}^{(A)} = \sum_\eta A_{\xi'\eta}(\Gamma_A)\hat{D}_{\eta}^{(A)}$ with a similar expression for $\hat{D}_{\xi'}^{(B)}$ in terms of its Euler angles Γ_B. This is generalized to other multipoles, which contain two or more matrix factors $\boldsymbol{A}(\Gamma_A)$ for quadrupoles or higher multipoles. An alternative compact derivation can be based on the spherical components $\hat{Q}_{lm}^{(A)}$ of species A, which is transformed from the BF to the SF frame by means of the Wigner rotational matrices $D_{m'm}^{(l)}(\Gamma_A)$ as shown in Section 2.5.

Induction potentials $V^{(ind)}$ depend in these cases on the multipole orientation angles through the static polarizability $\alpha^{(B)}(0;\Gamma_B)$, and dispersion potentials $V_{\sigma}^{(dsp)}$ depend on the orientations of the two species, through $C_6^{(dsp)}(\Gamma_A,\Gamma_B)$.

3.3.3 Example of Li^++HF

The treatment for Li^++HF goes along similar lines, with the same electronic states for Li^+. The main differences are that now HF is a heteropolar molecule with a permanent electric dipole, and with electronically excited states which have only a good angular momentum quantum number Λ along the diatomic axis so that applicable selection rules for HF are different. The ground state of HF is a $^1\Sigma$, for $\Lambda = 0$, and dipolar excitations can occur to $\Lambda = 0, \pm 1$. Only spin singlet states are allowed. The interaction must be treated in a space fixed (SF) XYZ reference frame, while the selection rules for the diatomic are applicable in a body fixed (BF) $X'Y'Z'$ frame attached to the diatomic.

The interaction potentials depend on the relative orientation of the diatomic. Letting the diatomic be the compound B with its center of mass (CM) at the origin of coordinates in the SF frame, the BF frame with the same origin is located by the Euler angles (α, β, γ) of rotation from SF to BF. Choosing the three atoms to be on the SF XZ-plane, the relative position vector from A to the CM of B along the SF Z-axis, and the diatomic nuclei on the BF Z'-axis, the Euler angles are $(0, \beta, 0)$. The transformation of multipole components is done with the 3×3 rotation matrix $\boldsymbol{A}(\alpha, \beta, \gamma)$ and in this case, this simply gives for the dipole in the SF frame $D_\xi^{(B)} = \sum_{\eta'} D_{\eta'}^{(B)} A_{\eta'\xi}(0,\beta,0) = \sum_{\eta'} D_{\eta'}^{(B)} [\boldsymbol{R}_Y(\beta)]_{\eta'\xi}$, a rotation around the $Y = Y'$ axis, in terms of dipole components in the BF frame. These give the operator relations $\hat{D}_X^{(B)} = \cos(\beta)\hat{D}_{X'}^{(B)} + \sin(\beta)\hat{D}_{Z'}^{(B)}$, $\hat{D}_Z^{(B)} = -\sin(\beta)\hat{D}_{X'}^{(B)} + \cos(\beta)\hat{D}_{Z'}^{(B)}$, and $\hat{D}_Y^{(B)} = \hat{D}_{Y'}^{(B)}$. Similar transformations apply for the quadrupoles and involve two rotational matrix factors.

To obtain the potential energies, we need the matrix elements of permanent and transition dipole operators. In the BF frame, the diatomic electronic states have axial symmetry and are labeled by the axial angular momentum quantum numbers. Matrix elements satisfy selection rules so that for the permanent components we have $D_{Z'}^{(B)} = \left\langle {}^1\Sigma \,|\, \hat{D}_{Z'}^{(B)} \,|\, {}^1\Sigma \right\rangle = D^{(B)}$, $D_{X'}^{(B)} = D_{Y'}^{(B)} = 0$. Transition dipole matrix elements of the form $\left\langle {}^1\Sigma \,|\, \hat{D}_{Z'}^{(B)} \,|\, {}^1|\Lambda| \right\rangle = 0$ except for $\Lambda = 0, \pm 1$, with parallel (for $\Lambda = 0$) and perpendicular components, respectively. This leads to parallel and perpendicular polarizabilities appearing in the induction and dispersion energies. Taking the angle between the diatomic dipole vector and the position vector from B to A to be $\theta = \pi - \beta$, the atom–diatom (A–D) potentials are, with $P_2(u) = (3u^2 - 1)/2$ [18]

$$V_{A-D}^{(els)} = (4\pi\varepsilon_0)^{-1} C^{(A)} D^{(B)} R^{-2} \cos(\theta)$$

$$V_{A-D}^{(ind)} = -\frac{\left(C^{(A)}\right)^2 \bar{\alpha}^{(B)}}{(4\pi\varepsilon_0)^2 2R^4}\left[1 + \frac{2}{3}\left(\frac{\alpha_\parallel - \alpha_\perp}{\alpha}\right)^{(B)} P_2(\cos\theta)\right] - \frac{\alpha^{(A)}\left(D^{(B)}\right)^2}{(4\pi\varepsilon_0)^2 R^6} P_2(\cos\theta)$$

with $\bar{\alpha}^{(B)} = \left(\alpha_{\|} + 2\alpha_{\perp}\right)^{(B)}/3$, an average static polarizability for the HF diatomic, while the dispersion potential becomes

$$V_{A-D}^{(dsp)} = -\frac{C_{6\|}^{(dsp)} + 2\,C_{6\perp}^{(dsp)}}{(4\pi\varepsilon_0)^2 R^6}\left[1 + \frac{1}{3}\left(\frac{\alpha_{\|} - \alpha_{\perp}}{\alpha}\right)^{(B)} P_2\left(\cos\theta\right)\right]$$

in terms of parallel and perpendicular dispersion coefficients involving corresponding transition dipoles of HF. Additional terms must appear if interactions involving permanent and transient quadrupoles were included, containing higher powers of R^{-1} and of $\cos(\theta)$. The interactions of a neutral atom with a heteropolar diatomic are obtained letting $C^{(A)} = 0$ above, while the interaction of a charged atom with a homopolar molecule follows from $D^{(B)} = 0$.

The charge C, dipole D, and mean polarizability $\bar{\alpha}$ of Li^+, Ne, H_2, HF, H_2O, and other selected species have been given in SI units in Table 2.2 of Chapter 2, and can be used to calculate electrostatic, induction, and dispersion energies, however keeping in mind that multipole and polarizability values depend on the chosen origin of coordinates for the charge and polarization distribution, which refer here to the molecular center of mass.

3.4 Calculation of Dispersion Energies

3.4.1 Dispersion Energies from Molecular Polarizabilities

A dispersion potential energy for A and B in their ground states, of form

$$V_{\sigma}^{(dsp)} = -\frac{1}{(4\pi\varepsilon_0)^2 R^6}\sum\nolimits_{j\neq 0}\sum\nolimits_{k\neq 0}\frac{\left|\left(\hat{D}_{\sigma}^{(A)}\right)_{0j}\left(\hat{D}_{-\sigma}^{(B)}\right)_{0k}\right|^2}{\left|\epsilon_{j0}^{(A)} + \epsilon_{k0}^{(B)}\right|}$$

can be conveniently rewritten as an integral with factors relating separately to species A and B, which allows calculations of the pair A–B dispersion potential from the polarizabilities of the components. This is done following [19] using again the Casimir–Polder integral

$$\frac{2}{\pi}\int_0^{\infty}\frac{du\,a\,b}{(a^2 + u^2)(b^2 + u^2)} = \frac{1}{a + b}$$

valid for $a > 0, b > 0$. The right-hand side denominator is like the denominator in the dispersion potential, and choosing $a = \epsilon_{j0}^{(A)}, b = \epsilon_{k0}^{(B)}$, the integral form leads to

$$V_{\sigma}^{(dsp)} = -\frac{1}{(4\pi\varepsilon_0)^2 R^6}\frac{1}{2\pi}\int_0^{\infty} du\,\alpha_{\sigma,0}^{(A)}\left(\frac{iu}{\hbar}\right)\alpha_{-\sigma,0}^{(B)}\left(\frac{iu}{\hbar}\right)$$

where

$$\alpha_{\sigma,0}^{(A)}\left(\frac{iu}{\hbar}\right) = 2\sum_{j\neq 0}\frac{\epsilon_{j0}^{(A)}\left|\left(\hat{D}_{\sigma}^{(A)}\right)_{0j}\right|^{2}}{\left(\epsilon_{j0}^{(A)}\right)^{2}+u^{2}}$$

is a positive valued function decreasing and vanishing as u goes from zero to $u \to \infty$. It follows that the dispersion potential energy can be constructed from the pair polarizabilities evaluated at imaginary-value frequencies $\omega = iu/\hbar$. The dispersion energy written for Cartesian or spherical dipoles can also be re-expressed as shown, and the procedure can also be applied to dispersion interactions involving quadrupoles or higher multipoles of A or B, introducing the corresponding multipolar polarizabilities.

An alternative expression for the orientation averaged polarizability is

$$\bar{\alpha}_{0}^{(A)}\left(\frac{iu}{\hbar}\right) = \left(\frac{c_e^2\hbar^2}{m_e}\right)\sum_{j\neq 0}\frac{f_{j0}^{(A,el)}}{\left[\left(\epsilon_{j0}^{(A)}\right)^{2}+u^{2}\right]} = \beta_0^{(A)}(u)$$

where the averaged electronic oscillator strength is $f_{j0}^{(A,el)} = \left[2m_e/\left(3c_e^2\hbar^2\right)\right]\epsilon_{j0}^{(A)}\left|\left\langle j\left|\vec{D}_{el}^{(A)}\right|0\right\rangle_{Q}\right|^{2}$ from the dipole vector at fixed nuclear positions.

3.4.2 Combination Rules

Approximate expressions for the dynamical polarizabilities translate into useful approximations for the dispersion coefficients. They also lead to combination rules which provide values of the coefficient for the pair A–B in terms of the coefficients of A–A and B–B.

Provided the optically active electrons in compounds A and B, typically their valence electrons, can be described as independent on the average, the well-known Kuhn–Thomas sum rule $\sum_j f_{j0}^{(A,el)} = N_{el}^{(A)^*}$, the number of active electrons of A, and similarly for B, gives for $u \to \infty$ the asymptotic value $u^2\beta_0^{(A)} \approx \left(c_e^2\hbar^2/m_e\right)N_{el}^{(A)^*}$. This together with the static value $\beta_0^{(A)}(0) = \bar{\alpha}_0^{(A)}(0)$ at zero frequency gives an interpolation function

$$\beta_0^{(A)}(u) = \bar{\alpha}_0^{(A)}(0)\left[1+\left(\frac{u}{\bar{u}^{(A)}}\right)^{2}\right]^{-1}$$

with a new parameter satisfying $\bar{\alpha}_0^{(A)}(0)\left(\bar{u}^{(A)}\right)^2 = \left(c_e^2\hbar^2/m_e\right)N_{el}^{(A)^*}$, that can be used to evaluate the integral over u [20]. It gives the approximate dispersion coefficient

$$\tilde{C}_6^{(AB)} = \frac{3}{2}\frac{\bar{\alpha}_0^{(A)}(0)\,\bar{\alpha}_0^{(B)}(0)}{\left(\bar{\alpha}_0^{(A)}(0)/N_{el}^{(A)*}\right)^{1/2} + \left(\bar{\alpha}_0^{(B)}(0)/N_{el}^{(B)*}\right)^{1/2}}$$

which can be calculated from static polarizabilities and the known number of valence electrons for A and B.

A slightly modified version of this procedure provides useful combination rules [21, 22]. The parameter $\bar{u}^{(A)}$ appears in $\tilde{C}_6^{(AA)}$ and it can be extracted as $\bar{u}^{(A)} = (4/3)\left(\tilde{C}_6^{(AA)}/\bar{\alpha}_0^{(A)}(0)^2\right)$ which involves two measurable quantities. Its values are given in [22] for atoms and small molecules and are (in au's) for He, Ar, Na, H_2, and CH_4 equal to 1.02, 0.706, 0.0774, 0.610, and 0.649, respectively, with larger values of 1.71 and 1.401 for He^+ and Na^+, all relating to electronic excitation energies. Using the form of $\bar{u}^{(A)}$ in $\beta_0^{(A)}(u)$, and performing the integration over u for the A–B pair gives the combination rule

$$\tilde{C}_6^{(AB)} = \frac{2\,C_6^{(AA)}\,C_6^{(BB)}}{\left(\alpha^{(B)}/\alpha^{(A)}\right)C_6^{(AA)} + \left(\alpha^{(A)}/\alpha^{(B)}\right)C_6^{(BB)}}$$

which allows calculation of the dispersion coefficient for two different interacting compounds A and B, given the coefficients for the two identical pairs A–A and B–B, and has been found to be accurate within a few percent for many combinations of atoms [22].

3.4.3 Upper and Lower Bounds

The second-order perturbation energies providing induction and dispersion interactions require sums over excited states, most of which are usually not known. Approximate expressions can however be obtained that involve only a few excited states, and that provide upper and lower bounds to the second-order energies. Here, we follow an operator-based treatment [23, 24] which is also closely related to a method of sum over states [13].

Writing the unperturbed, zeroth order states of pair A–B as a product of states of A and B like $|k\rangle_{AB} = |p\rangle_A |q\rangle_B$, the second-order energy for the ground state $|0\rangle_{AB} = |0\rangle_A |0\rangle_B$ can be written in terms of the resolvent operator

$$\hat{R}_0 = -\sum\nolimits_{k\geq 1} \frac{|k\rangle\langle k|}{E_k^{(0)} - E_0^{(0)}}$$

where unperturbed pair energies have been ordered in the sequence $E_0^{(0)} \leq E_1^{(0)} \leq E_2^{(0)} \leq$ and so on, and all denominators are positive valued giving a negative valued operator. The second-order energy is now $E_0^{(2)} = \left\langle 0|\hat{H}^{(1)}\hat{R}_0\hat{H}^{(1)}|0\right\rangle_{AB}$

Replacing in the resolvent $E_k^{(0)} - E_0^{(0)}$ with the smaller $E_1^{(0)} - E_0^{(0)}$ in all the quotients, so they all become larger, and using the completeness relation $\sum_{k \ge 0} | k\rangle\langle k | = \hat{I}$, the identity operator, and the complementary projector $\hat{P} = \hat{I} - |0\rangle\langle 0|$ it follows that

$$0 \ge \hat{R}_0 \ge -\frac{\hat{P}}{\left(E_1^{(0)} - E_0^{(0)}\right)}$$

which gives the energy bounds

$$0 \ge E_0^{(2)} \ge -\frac{\left\langle 0 | \hat{H}^{(1)} \hat{P} \hat{H}^{(1)} | 0 \right\rangle_{AB}}{\left(E_1^{(0)} - E_0^{(0)}\right)}$$

This can be calculated from a knowledge of only the $|0\rangle_{AB} = |0\rangle_A |0\rangle_B$ state of the pair and the integrals in $\left\langle 0 | \hat{H}^{(1)} | 0 \right\rangle$ and $\left\langle 0 | \hat{H}^{(1)} \hat{H}^{(1)} | 0 \right\rangle$, and has a numerator containing the squared standard deviation of $\hat{H}^{(1)}$ in the state $|0\rangle_{AB}$.

For the dispersion energy of two hydrogen atoms in their ground state $|0\rangle = |1s_0\rangle$, the dipole–dipole interaction Hamiltonian is $\hat{H}^{(1)} = (4\pi\varepsilon_0)^{-1} C_e^2 (x_A x_B + y_A y_B - 2 z_A z_B) R^{-3}$ and its averages are $\left\langle 0 | \hat{H}^{(1)} | 0 \right\rangle = 0$ and $\left\langle 0 | \hat{H}^{(1)} \hat{H}^{(1)} | 0 \right\rangle = (4\pi\varepsilon_0)^{-2} C_e^4 R^{-6} 2 \langle r^2 \rangle_{1s}^2 / 3$, after consideration of spatial symmetry and selection rules. This together with $\langle r^2 \rangle_{1s} = 3a_0^2$ and $E_1^{(0)} - E_0^{(0)} = C\left(E_{2p0} - E_{1s0}\right) = 3C_e^2/(4a_0)$ gives a dispersion energy lower bound of $V^{(dsp)}(R) C - 8.0\left(C_e^2/a_0\right)\left(a_0^6/R^6\right)$, with C_e and a_0 the electron charge and the Bohr radius both equal to 1.0 in atomic units, where $4\pi\varepsilon_0 = 1.0$. This lower bound contains a numerical factor of the correct magnitude and not far from the actual value of nearly 6.50 au. The next term in the multipole expansion of $\hat{H}^{(1)}$ comes from the dipole–quadrupole interaction and the dispersion energy can be similarly calculated to find in $V^{(dsp)}$ a factor $\left(1 + 22.5 a_0^2/R^2\right)$ [1]. This indicates that asymptotic convergence for large R must start around $R \approx 5.0 a_0$.

Better bounds can be found if one knows ground and excited states up to $k = N$, in which case one can define a partial resolvent and write

$$\hat{R}_0^{(N)} = -\sum_{k=1}^{N} \frac{|k\rangle\langle k|}{E_k^{(0)} - E_0^{(0)}} \ge \hat{R}_0 \ge \hat{R}_0^{(N)} - \sum_{k=N+1}^{\infty} \frac{|k\rangle\langle k|}{E_{N+1}^{(0)} - E_0^{(0)}}$$

which gives the better bounds

$$\left\langle 0 \,|\, \hat{H}^{(1)} \hat{R}_0^{(N)} \hat{H}^{(1)} \,|\, 0 \right\rangle \geq E_0^{(2)}$$
$$\geq \left\langle 0 \,|\, \hat{H}^{(1)} \hat{R}_0^{(N)} \hat{H}^{(1)} \,|\, 0 \right\rangle - \left\langle 0 \,|\, \hat{H}^{(1)} \left(\hat{I} - \hat{\mathcal{P}}^{(N)} \right) \hat{H}^{(1)} \,|\, 0 \right\rangle_{AB} / \left(E_{N+1}^{(0)} - E_0^{(0)} \right)$$

with $\hat{\mathcal{P}}^{(N)} = \sum_{k=1}^{N} | k \rangle \langle k |$ a projection operator on the known subspace of N states excluding the ground state $k = 0$. A calculation can now proceed involving integrals of states only within this subspace. Induction and dispersion energies can be obtained as done above, by letting the states k contain only one ground state of the A–B pair, to generate the induction interaction of one permanent multipole and a transient one, or by excluding the two ground states, to generate the dispersion interaction of two transient multipoles.

Calculations can be done quite generally introducing a new basis set suitable for expansion of the $\hat{R}_0^{(N)}$ and $\hat{\mathcal{P}}^{(N)}$ operators [23, 24]. Using a notation with the $1 \times N$ row matrix of chosen states $|\boldsymbol{f}\rangle = [|f_1\rangle, \ldots, |f_N\rangle\,]$, and constructing the projector $\hat{\mathcal{P}}^{(N)} = |\boldsymbol{f}\rangle(\langle \boldsymbol{f}|\boldsymbol{f}\rangle)^{-1}\langle \boldsymbol{f}|$ where $\langle \boldsymbol{f}|$ is a $N \times 1$ column matrix, a positive-valued operator $\hat{A}$ is found to satisfy the inequality $\langle g \,|\, \hat{A} \,|\, g \rangle \geq \left\langle g \,|\, \hat{A}^{1/2} \hat{\mathcal{P}}^{(N)} \hat{A}^{1/2} \,|\, g \right\rangle$. This can be used with $|g\rangle = \hat{H}^{(1)} \,|\, 0 \rangle$, $\hat{A} = -\hat{R}_0$, and $\left| \left(-\hat{R}_0\right)^{1/2} \boldsymbol{f} \right\rangle = |\boldsymbol{h}\rangle$, all orthogonal to $|0\rangle$, to obtain the upper bound

$$E_0^{(2)} \leq \left\langle 0 \,|\, \hat{H}^{(1)} \,|\, \boldsymbol{h} \right\rangle \left\langle \boldsymbol{h} \,|\, \left(E_0^{(0)} - \hat{H}^{(0)} \right) |\, \boldsymbol{h} \right\rangle^{-1} \left\langle \boldsymbol{h} \,|\, \hat{H}^{(1)} \,|\, 0 \right\rangle$$

Using instead $\hat{A} = \hat{R}_0 - \hat{P}\left(E_0^{(0)} - E_1^{(0)} \right)^{-1}$ and $|\hat{A}^{\frac{1}{2}} \boldsymbol{f}\rangle = \left(\hat{H}^{(0)} - E_1^{(0)} \right) |\, \boldsymbol{k}\rangle$, with $|\boldsymbol{k}\rangle$ in the orthogonal complement space of $\hat{P}$, one obtains a lower bound from [23]

$$\langle g \,|\, \hat{A} \,|\, g \rangle = E_0^{(2)} - \frac{\left\langle 0 \,|\, \hat{H}^{(1)} \hat{P} \hat{H}^{(1)} \,|\, 0 \right\rangle}{E_0^{(0)} - E_1^{(0)}}$$
$$\geq \left\langle 0 \,|\, \hat{H}^{(1)} \left(\hat{H}^{(0)} - E_1^{(0)} \right) |\, \boldsymbol{k} \rangle \left(\left\langle \boldsymbol{k} \,|\, \left(\hat{H}^{(0)} - E_0^{(0)} \right) \right. \right.$$
$$\times \left. \left(\hat{H}^{(0)} - E_1^{(0)} \right) |\, \boldsymbol{k}\rangle \right)^{-1} \left\langle \boldsymbol{k} \,|\, \left(\hat{H}^{(0)} - E_1^{(0)} \right) \hat{H}^{(1)} \,|\, 0 \right\rangle$$

The calculation of bounds to the induction energy in the pair A–B in its ground state can be done selecting the functions in $|\boldsymbol{h}\rangle$ and $|\boldsymbol{k}\rangle$ to generate the interaction of a permanent dipole with a transition dipole, with $|0\rangle = |0_A, 0_B\rangle$, $|h_j\rangle = |0_A, h_{Bj}\rangle$ or $|h_k\rangle = |h_{Ak}, 0_B\rangle$, the $|h_{Ak}\rangle$ orthogonal to $|0_A\rangle$, and similarly for B and for the $|k_j\rangle$. The dispersion energy calculations must involve states $|h_j\rangle = |h_{Aj}, h_{Bj}\rangle$ where both components are orthogonal to the ground states of A and B.

The sets $|\boldsymbol{h}\rangle$ and $|\boldsymbol{k}\rangle$ can be also chosen to be a single variational state $|\chi\rangle = |\chi_A, \chi_B\rangle$ which can be varied to maximize or minimize bounds as needed.

These procedures allow the introduction of any convenient basis set, which need not be made of unperturbed eigenfunctions. In an application to the dispersion energy of two hydrogen atoms, it is found that using just a few hydrogenic states for each atom is not accurate because of the neglect of the hydrogen atom continuum. Better results are obtained with a new basis set of functions $\langle \vec{r} \,|\, nlm \rangle$ of the electronic coordinates constructed from Laguerre radial functions, and sequentially adding new basis functions [23]. In this way upper and lower bounds to the dispersion energy for the ground state of H + H can be made to converge to the highly accurate value $C_6 = 6.4990267$ au using only seven basis functions.

Dispersion interaction energies can also be obtained doing Gaussian quadratures for the integrals over imaginary frequencies appearing in the Casimir–Polder expressions, using Pade approximants for the polarizabilities versus frequency, or choosing the expansion states in $|\boldsymbol{h}\rangle$ or $|\boldsymbol{k}\rangle$ to generate energies as sums over states of terms containing oscillator strengths. Some of these procedures have been compared in the literature [4, 13, 25].

3.4.4 Variational Calculation of Perturbation Terms

A computationally convenient alternative to the calculation of perturbation corrections to the energy by means of expansions in basis sets is to relate the wavefunction equations for each perturbation order to variational functionals optimized to obtain perturbed functions and energies. This was done first for the dispersion interaction energy of H + H and He + He [26], and was later expanded within the perturbation theory of intermolecular forces [4, 13, 27].

The functional $\mathcal{J}[\tilde{\Phi}] = \left\langle \tilde{\Phi} \,|\, \left(\hat{H}^{(0)} + \lambda \hat{H}^{(1)} - \tilde{E}\right) \,|\, \tilde{\Phi} \right\rangle$ of a variational function $\tilde{\Phi}$ can be made stationary, with variations $\delta\mathcal{J} = 0$ giving an equation for $\tilde{\Phi}$ or its parameters, and then $\mathcal{J} = 0$ giving its energy $\tilde{E}$. These two can first be expanded in powers of λ to extract the functional at each order. To simplify here, one can assume that the exact solution to $\hat{H}^{(0)}\Phi^{(0)} = E^{(0)}\Phi^{(0)}$ is known and normalized, and let $\tilde{\Phi}^{(0)} = \Phi^{(0)}$ while solving for the higher orders so that $\left\langle \tilde{\Phi}^{(p)} \,|\, \Phi^{(0)} \right\rangle = 0$ for $p \geq 1$. This in particular gives to second order,

$$\mathcal{J}^{(2)}[\tilde{\Phi}] = \left\langle \tilde{\Phi}^{(1)} \,|\, \left(\hat{H}^{(0)} - \tilde{E}^{(0)}\right) \,|\, \tilde{\Phi}^{(1)} \right\rangle + \left\langle \tilde{\Phi}^{(0)} \,|\, \left(\hat{H}^{(1)} - \tilde{E}^{(1)}\right) \,|\, \tilde{\Phi}^{(1)} \right\rangle$$
$$+ \left\langle \tilde{\Phi}^{(1)} \,|\, \left(\hat{H}^{(1)} - \tilde{E}^{(1)}\right) \,|\, \tilde{\Phi}^{(0)} \right\rangle - \tilde{E}^{(2)}$$

with $\tilde{E}^{(1)} = \left\langle \tilde{\Phi}^{(0)} \,|\, \hat{H}^{(1)} \,|\, \tilde{\Phi}^{(0)} \right\rangle$ known. With accurate functions $\tilde{\Phi}^{(0)}(X_A, X_B)$ of electronic coordinates and choices of variational functions of the form

Table 3.1 Dispersion coefficients from variational wavefunctions.

Atom pair	$C_6^{(DD)}$(au)	$C_6^{(DQ)}$ (au)	$C_6^{(QD)}$ (au)
H + H	6.4990	124.399	124.399
H + He	2.817	13.14	28.64
He + He	1.459	7.05	7.05

$\tilde{\Phi}^{(1)}(X_A,X_B)=\tilde{\Phi}^{(0)}(X_A,X_B)f(X_A,X_B)H^{(AB)}(X_A,X_B)$ where $f(X_A, X_B)$ is a sum of factors with powers of electronic coordinates in A and B, it has been possible to obtain this way highly accurate values of dipole–dipole $C_6^{(DD)}$ and dipole–quadrupole $C_8^{(DQ)}$ dispersion energies for interacting atoms with one and two electrons, including H + H, H + He, and He + He [13]. Their collected values are shown in Table 3.1, in au's.

3.5 Electron Exchange and Penetration Effects at Reduced Distances

3.5.1 Quantitative Treatment with Electronic Density Functionals

As the interacting species A and B are brought closer with decreasing R, electronic clouds start to penetrate each other, and it is necessary to account for quantal electron exchange. Effects depend on whether the electronic shells of the species are closed, with full occupation of electron orbitals, or open due to partial occupations. For a species with closed electronic shells, a resulting physical effect is the Pauli exclusion of electrons from occupied orbitals, which amounts to a repulsive force, with electronic density moving away from the intermolecular region and into the outer regions. If one or both species have an open shell with partly occupied orbitals, electrons can move into the intermolecular region to form covalent bonds and create attraction forces. In addition there is a probability of electron transfer forming pairs $A^+ + B^-$ or $A^- + B^+$, which brings in electrostatic attraction forces. To obtain these forces as functions of the distance R one must properly antisymmetrize the total electronic wavefunctions with respect to electron exchange, and must allow for electronic redistribution. The electronic wavefunction $\Phi^{(AB)}(X; \mathbf{Q})$ of all electronic position and spin variables in $X = (X_A, X_B)$, for fixed nuclear (or ion core) positions $\mathbf{Q} = \left\{\vec{R}_a\right\}$, can be constructed to contain a set $\boldsymbol{\lambda} = \{\lambda_j\}$ of parameters chosen to be optimized. This is done requiring that the total energy functional $\mathcal{E}\left[\Phi^{(AB)}\right] = \left\langle\Phi^{(AB)} \mid \hat{H}_Q \mid \Phi^{(AB)}\right\rangle_Q / \left\langle\Phi^{(AB)} \mid \Phi^{(AB)}\right\rangle_Q = E^{(AB)}(\boldsymbol{\lambda}; \mathbf{Q})$ must have a

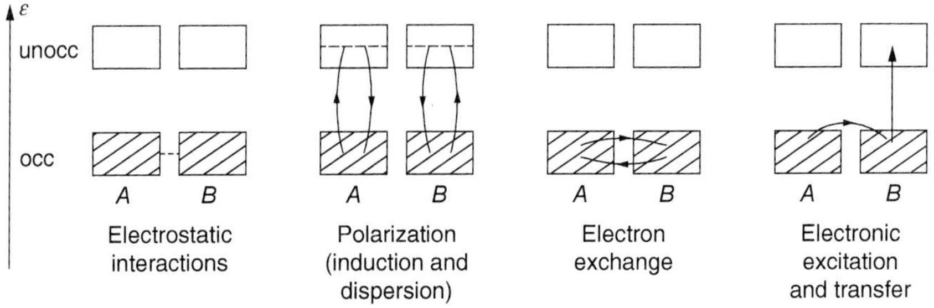

Figure 3.3 Schematic illustration of electrostatic, polarization (induction and dispersion), electron exchange, and electronic excitation phenomena in transitions involving closed (shown dashed) and open (shown empty) electronic shell interactions.

minimum as a function of the parameters, for the ground state or for each state of lowest energy with a given symmetry given by fixed $\boldsymbol{Q}$. The optimized state $\Phi^{(AB)}$ can be taken as the reference or zeroth order state to be further corrected by combining it with a basis set of orthonormal states to be used in a perturbation expansion.

A qualitative discussion can be based on electronic orbitals and their charge distributions. The several phenomena, appearing at short and long distances for closed–closed, closed–open, and open–open shell interactions are schematically illustrated in Figure 3.3.

For each species A or B, electronic energy levels ε of orbitals are found to be occupied as shown by the dashed bands in the figure, or to be unoccupied. Electrostatic interactions are always present and obtained from electronic charge densities, polarization (induction and dispersion) occurs at both long and short distances, electron exchange appears as the distance is reduced and orbitals start to overlap, and electronic excitation and charge transfer occur more likely at short distances due to electron correlations and the Pauli exclusion of electrons from occupied orbitals.

A great deal about interactions at intermediate and short distances can be understood and estimated from the relationship between energies and electronic densities established by the Hohenberg–Kohn theorem [28]. Given a set of nuclei a (or ion cores) at fixed positions $\vec{R}_a$, electrons are distributed among them and show unique spikes at their locations. This means that the external potential energy $V_{ext}(\vec{r})$ provided by the nuclei uniquely creates an electron number distribution $\rho(\vec{r})$ per unit volume. The theorem states that the reverse is also verified for the nondegenerate electronic ground state and lowest energy electronic state of each symmetry, and that there is a unique functional $\mathcal{E}[\rho(\vec{r})]$ which reaches a minimum when a trial density function is varied to reach the correct one. The density derives from a many-electron wavefunction $\Phi[X; \rho(\vec{r})]$ which is constrained to provide the density but may be unknown in detail. Letting the potential energy of repulsion among nuclei be $V_{NN}(\mathbf{Q})$ and the interaction energy between an electron and all the nuclei $v_{Ne}(\vec{r}; \mathbf{Q})$, the total energy is decomposed as

$$\mathcal{E}[\rho(\vec{r}); \mathbf{Q}] = V_{NN}(\mathbf{Q}) + \int d^3r\, \rho(\vec{r})\, v_{Ne}(\vec{r}; \mathbf{Q}) + F[\rho(\vec{r}); \mathbf{Q}]$$

where $F[\rho(\vec{r}); \mathbf{Q}] = \langle \Phi[\rho(\vec{r})] \mid (\hat{K}_{el} + \hat{H}_{ee}) \mid \Phi[\rho(\vec{r})] \rangle_{\mathbf{Q}}$ in terms of the electronic kinetic energy and electron–electron potential energy operators. Further separating in F the Coulomb electron–electron interaction energy $V_{Coul}[\rho(\vec{r})]$ the remainder is a functional containing kinetic energy, exchange and correlation effects, so that

$$F[\rho(\vec{r}); \mathbf{Q}] = V_{Coul}[\rho(\vec{r})] + G_{kin}[\rho(\vec{r}); \mathbf{Q}] + G_{xc}[\rho(\vec{r}); \mathbf{Q}]$$

with the last term describing a combined exchange-correlation energy that can further be decomposed into separate exchange and correlation terms as $G_{xc}[\rho(\vec{r}); \mathbf{Q}] = G_{exc}[\rho(\vec{r}); \mathbf{Q}] + G_{cor}[\rho(\vec{r}); \mathbf{Q}]$.

The electron–electron Coulomb energy functional follows from the Hamiltonian operator

$$\hat{H}^{(ee)} = \left(\frac{1}{2}\right) \int d^3r_1 \int d^3r_2 \frac{c_e^2}{|\vec{r}_1 - \vec{r}_2|} \sum_{m \neq n} \delta(\vec{r}_1 - \hat{\vec{r}}_m)\, \delta(\vec{r}_2 - \hat{\vec{r}}_n)$$

$$\sum_{m \neq n} \delta(\vec{r}_1 - \hat{\vec{r}}_m)\, \delta(\vec{r}_2 - \hat{\vec{r}}_n)$$
$$= \sum_{m,n} \delta(\vec{r}_1 - \hat{\vec{r}}_m)\, \delta(\vec{r}_2 - \hat{\vec{r}}_n) - \delta(\vec{r}_1 - \vec{r}_2) \sum_m \delta(\vec{r}_1 - \hat{\vec{r}}_m)$$

where the double sum in the last line includes $m = n$. This gives

$$\sum_{m \neq n} \delta(\vec{r}_1 - \hat{\vec{r}}_m)\, \delta(\vec{r}_2 - \hat{\vec{r}}_n) = \hat{\rho}^{(1)}(\vec{r}_1)\, \hat{\rho}^{(1)}(\vec{r}_2) - \delta(\vec{r}_1 - \vec{r}_2)\, \hat{\rho}^{(1)}(\vec{r}_1)$$

with $\hat{\rho}^{(1)}(\vec{r}) = \sum_{1 \le m \le N} \delta(\vec{r} - \hat{\vec{r}}_m)$, to be replaced in the double integral. The expectation value of $\hat{H}^{(ee)}$ for a given wavefunction then leads to the functional of the electronic number densities $\rho(\vec{r})$ at locations $\vec{r}$ and $\vec{r}'$,

$$V_{Coul}[\rho(\vec{r})] = \frac{1}{2} \int d^3r \int d^3r' \frac{c_e^2}{|\vec{r} - \vec{r}'|} [\rho(\vec{r})\rho(\vec{r}') - \rho(\vec{r})\delta(\vec{r} - \vec{r}')]$$

with $c_e^2 = e^2/(4\pi\varepsilon_0)$ in SI units or $c_e^2 = e^2$ in cgs statCoul units.

This provides a framework for discussion of the intermediate and short-distance interaction energies for species with closed or open electronic shells. The $G[\rho(\vec{r}); \mathbf{Q}]$ functionals have been extensively treated and constructed from information about the inhomogeneous electron gas [29, 30] and by parametrizations designed to reproduce the energetics of sets of molecules [31]. The so-called DFT approach has been widely applied to molecular and solid-state phenomena. Separating nuclear position variables as $\boldsymbol{Q} = \left(\vec{R}, \boldsymbol{Q}_A, \boldsymbol{Q}_B\right)$ and doing calculations for varying R, the most recent functionals give quite acceptable results for ground electronic state equilibrium distances R_e and dissociation energies $D_e = E_0(\infty) - E_0(R_e)$ in many applications involving intermediate and short distances.

However, in applications of intermolecular forces to the A–B pair, it is necessary that as $R \to \infty$, one must find $\rho^{(AB)}\left(\vec{r}; \vec{R}, \boldsymbol{Q}_A, \boldsymbol{Q}_B\right) \approx \rho^{(A)}\left(\vec{r}; \boldsymbol{Q}_A\right) + \rho^{(B)}\left(\vec{r}; \boldsymbol{Q}_B\right)$, and also that the functional must satisfy the limiting form

$$F_{AB}\left[\rho(\vec{r}); \vec{R}, \boldsymbol{Q}_A, \boldsymbol{Q}_B\right] \approx F_A\left[\rho^{(A)}(\vec{r}); \boldsymbol{Q}_A\right] + F_B\left[\rho^{(B)}(\vec{r}); \boldsymbol{Q}_B\right]$$

This is a demanding restriction which is not obeyed in the available functionals and is the subject of active research at this time.

For the intermediate and short ranges (or reduced range, RR), two useful computational procedures have been devised for a system with N_e electrons. One is suitable for electronic rearrangement and bonding [32], with electrons of spin states $\eta = \alpha$, β described by independent (Kohn–Sham, or KS) spin-orbitals $\psi_j(\vec{r}, \zeta) = \psi_j^{(\eta)}(\vec{r})\eta(\zeta)$, chosen as the trial functions to be varied in the functionals, in terms of which the electron density for each spin is

$$\rho^{(\eta)}(\vec{r}) = \sum_{j=1}^{N_e^{(\eta)}} \left|\psi_j^{(\eta)}(\vec{r})\right|^2$$

and the total density is $\rho(\vec{r}) = \rho^{(\alpha)}(\vec{r}) + \rho^{(\beta)}(\vec{r})$. The other (Gordon–Kim, or GK) [33] is suitable for the interaction of two species where the densities of A and B do not change much over distances R, so that $\rho^{(AB)}\left(\vec{r}; \vec{R}, \boldsymbol{Q}_A, \boldsymbol{Q}_B\right) = \rho^{(A)}\left(\vec{r}; \boldsymbol{Q}_A\right) + \rho^{(B)}\left(\vec{r}; \boldsymbol{Q}_B\right)$ is a good approximation at all R. These are complementary approaches which can be applied to species with closed or open electronic shells.

For a pair of species both of which have closed shells, such as Ne + Ne, the total density can be simply chosen as $\rho^{(AB)}(\vec{r}) = \rho^{(A)}(\vec{r}) + \rho^{(B)}(\vec{r})$. Each species density is obtained from its electronic structure and the functional $F^{(AB)}\left[\rho(\vec{r})\right]$ does not need variation to obtain an optimal minimum. Results are acceptable for the short and intermediate ranges and provide a good estimate of equilibrium distance and dissociation energy. It must however be extended to include the long-range polarization terms, and this has been done to some extent in following work, by allowing displacement and distortion of the densities [34]. Applications with some improvements in the functionals have included diatomic and polyatomic species for which the dispersion energy terms have been added at large distances [35–37] Calculation of interaction energies at very short ranges also requires adding distortion densities for very short distances [38].

For a pair of species, such as Ne + Na, where one has closed shells and the other has open shells, it is necessary to allow for a partial redistribution of the electronic density of the open shells and to work with spin densities to allow for the unpaired electrons. If these are in species B, a possible choice of a variational density is $\rho^{(AB)}(\vec{r}) = \rho^{(A)}(\vec{r}) + \rho^{(B^*)}(\vec{r})$ where the second term can better be treated within the KS formalism by choosing $\rho^{(B,\eta)}(\vec{r}) = \left|\sum_j c_{j\eta}\,\varphi_j^{(B)}(\vec{r})\right|^2$, a sum including orbitals at species B with electron spin η, forming hybrid atomic orbitals, or more general distortion orbitals. If B is a molecule, coefficients of molecular orbitals can be optimized by minimizing its total energy functional.

For two species with open shells, such as Na + Na, the KS choice is suitable for intermediate and short distances, and the functional must be minimized with respect to variations of KS spin-orbitals $\varphi_{j\eta}^{(AB)}$ with occupation numbers $\nu_{j\eta}$ that extend over the two species, as $\rho^{(AB)}(\vec{r}) = \sum_{j\eta}\nu_{j\eta}\left|\varphi_{j\eta}^{(AB)}(\vec{r})\right|^2$ with $\varphi_{j\eta}^{(AB)}(\vec{r}) = \sum_\mu c_{\mu,j\eta}\chi_\mu(\vec{r})$ a linear combination of atomic orbitals χ_μ centered at the nuclei. This is best done in the case of chemical bonding by introducing spin densities to account for open shells of separated species [39].

The procedure for pairs of atoms was implemented in early work using a density $\rho^{(AB)}(\vec{r}) = \rho_{HF}^{(A)}(\vec{r}) + \rho_{HF}^{(B)}(\vec{r})$ with the atomic densities from Hartree–Fock calculations, and with a simple Thomas–Fermi–Dirac functional for $F\left[\rho(\vec{r});R\right]$ [40]. Results for the short-range interaction energy of many (Z = 2–36) atom pairs could be well fitted at short distances with the Born–Mayer function $V(R) = A\exp(-bR)$, and it was found that reasonable accuracy, of the order of 1%, could be obtained for different atoms A and B from the combination rule $V^{(AB)} = (V^{(AA)}V^{(BB)})^{1/2}$. This approach was substantially extended with the GK procedure to include both short and intermediate distances and to obtain the location and depth of van der Waals potential minima. Using Hartree–Fock atomic densities, and kinetic energy, exchange, and correlation energy functionals for the inhomogeneous electron gas provided by Thomas–Fermi, Dirac, and a local energy density expansion, respectively [33], remarkably good results could be obtained by comparison with ab initio calculations and experiment. This is shown in Figure 3.4 for Ar + Ar, and in Table 3.2 where GK values for R_m and ϵ_m are compared to experimental ones. For Ne–Ne it gives $R_m = 2.99$ Å = 299 pm and $\epsilon_m = 56 \times 10^{-16} erg = 56 \times 10^{-23}$ J for the minimum location and well energy, compared with experimental values of $R_m = 303$ pm and $\epsilon_m = 63 \times 10^{-23}$ J.

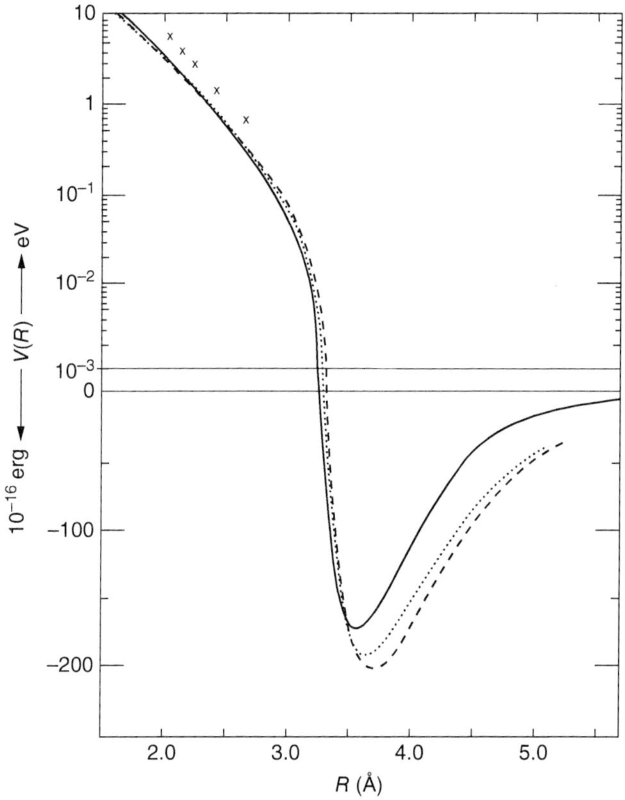

Figure 3.4 The Ar–Ar potential energy versus interatomic distance: full line GK calculations [33]; dashed line from gas experiments, and dotted line from beam scattering experiments. *Source:* from Gordon and Kim [33].

Table 3.2 Calculated and measured values of potential energy parameters (from [33]) in SI units.

Atom pair	**R_m (pm) calculated**	**R_m (pm) measured**	**ϵ_m ($\times 10^{-23}$ J) calculated**	**ϵ_m ($\times 10^{-23}$ J) measured**
Ne–Ne	299	303	56	63
Ar–Ar	363	370	175	195
Ne–Ar	342	338–364	81	84–115

Source: Reproduced with permission of AIP Publishing.

Similar accuracy is obtained for other pairs of atoms. The treatment was later improved with a more accurate functional correcting for the electronic self-energy and was applied to molecular species with similar encouraging results, also for interaction potential anisotropies [35].

At present, there is no satisfactory density functional treatment of intermolecular forces to cover all ranges with a consistent formulation, and it seems best to construct the exchange-correlation potential energy in the form

$$V_{xc}(R) = G_{xc}\left[\rho^{(AB)}(\vec{r});\mathbf{Q}\right] - G_{xc}\left[\rho^{(A)}(\vec{r});\mathbf{Q}_A\right] - G_{xc}\left[\rho^{(B)}(\vec{r});\mathbf{Q}_B\right]$$
$$= V_{xc}^{(nr)}(R)[1 - f_d(R;R_d,w_d)] + f_d(R;R_d,w_d)\,V_{xc}^{(pol)}(R)$$

where $V_{xc}^{(nr)}$ describes the exchange-correlation of electrons at short and intermediate ranges, or near-range (nr), as modeled by a density functional, and $V_{xc}^{(pol)}$ includes long-range induction and dispersion energies, with a transition between the two provided by a damping function f_d going smoothly from 1.0 at large R to zero at small R, with a drop around a damping location R_d of width w_d. At large distances, $V_{xc}^{(pol)}(R)$ must correctly go into an exchange-correlation expression constructed from properties of the separate species A and B. The $V_{xc}^{(nr)}(R)$ extended to large distances contains some of the long-range polarization interaction, and conversely $V_{xc}^{(pol)}(R)$ at short distances contains some electronic exchange-correlation. Therefore, it is necessary to dampen their magnitudes as shown to avoid doubly counting of exchange-correlation. Various choices have been proposed and tested for functions such as this, which damp with a function $f_d^{(n)}(R)$ each long-range energy varying as R^{-n}, containing values of the parameters suggested by electronic structure treatments or chosen for computational convenience to fall more rapidly than R^n as distances decrease [41, 42]. The damping function must be chosen so that potential energy surfaces or derived forces are smooth over all distances.

Alternatively, the near-range term may contain also the long-range electrostatic and induction energies when these appear in the DFT treatment, with the remaining polarization term equal to a calculated dispersion energy. Ongoing efforts to provide useful treatments at all distances have been presented in several recent reviews [43–46].

3.5.2 Electronic Rearrangement and Polarization

A deeper and more general understanding of interaction energies requires treatment of the electronic states of many-atom systems. It is easier to follow the related mathematical treatment of electronic rearrangement in interactions, by focusing on simple systems such as homonuclear and heteronuclear diatomics, and results for $H + H^+$, $H + H$, $H + He$, and $He + He$. The building components are atomic orbitals (AOs) $\chi_\mu(\vec{r})$, with $\mu = (Nnlm)$ for nucleus $N = a$, b and atomic quantum numbers n, l, m. The treatment can conveniently be done in a body-fixed (BF) reference frame where the z-axis points from nucleus a to b. Wavefunctions depend then only on the distance R, and the Hamiltonian has axial symmetry around the z-axis, as well as reflection symmetry through a plane containing the z-axis.

For an electron in the field of two ion cores at positions $\vec{R}_a$ and $\vec{R}_b$ with ion charges Z_aC_e and Z_bC_e, a simple molecular orbital written as a linear combination of atomic orbitals (a MO-LCAO) is

$$\varphi_j(\vec{r}) = c_{aj}\chi_a(\vec{r}) + c_{bj}\chi_b(\vec{r})$$

to be normalized to $\langle \varphi \mid \varphi \rangle = 1$ and with two variational parameters $c_{\mu j}(R)$ which can be obtained minimizing the electronic energy $\epsilon^{(el)}$ for fixed nuclei, obtained from the Hamiltonian for an electron (1), $\hat{h}^{(el)}(1) = -(\hbar^2/2m_e)\left(\partial^2/\partial\vec{r}_1^{\,2}\right)$ $-Z_a c_e^{\,2}/r_{a1} - Z_b c_e^{\,2}/r_{b1}$. This leads for each $j = 1, 2$ to the matrix equation

$$\sum\nolimits_{\mu=a,b} c_\mu \left(H_{\mu\nu} - \epsilon^{(el)} S_{\mu\nu}\right) = 0$$

with $\nu = a, b$, $S_{\mu\nu} = \langle \chi_\mu \mid \chi_\nu \rangle$ the overlap integral, and $H_{\mu\nu} = \left\langle \chi_\mu \mid \hat{h}^{(el)} \mid \chi_\nu \right\rangle$ an integral for the electronic Hamiltonian matrix element. The AOs satisfy certain combination rules: (i) They combine only when they have the same spatial symmetry, since otherwise $S_{\mu\nu} = H_{\mu\nu} = 0$; (ii) they combine more strongly (with comparable coefficient values) if $H_{aa} \cong H_{bb}$ as seen from $(c_a/c_b)^2 = (H_{bb} - \epsilon^{(el)}S_{bb})/(H_{aa} - \epsilon^{(el)}S_{aa})$; (iii) they combine more strongly when there is a larger overlap between atomic charge distributions at a and b, which increases the magnitudes of coupling terms H_{ab} and S_{ab}.

Solving for the coefficients requires that the determinant of the matrix they multiply must be zero, or $(\alpha - \epsilon)(\beta - \epsilon) - (\gamma - \epsilon S) = 0$ in the notation $\alpha = H_{aa}$, $\beta = H_{bb}$, $\gamma = H_{ab}$, $\epsilon = \epsilon^{(el)}$, $S = S_{ab}$, and with $S_{aa} = S_{bb} = 1$. This gives energy roots [47]

$$\epsilon_\pm = -A \pm \left(A^2 - B\right)^{1/2},$$

$$A = \frac{[-(\alpha+\beta) + 2\gamma S]}{(1-S^2)}, \quad B = \frac{(\alpha\beta - \gamma^2)}{(1-S^2)}$$

and energy eigenvalues $\epsilon_1^{(el)} = \epsilon_- \le \epsilon_2^{(el)} = \epsilon_+$. The corresponding linear combination coefficients follow from

$$\left(\frac{c_b}{c_a}\right)_j = -\frac{\gamma - \epsilon_j^{(el)} S}{\beta - E_j^{(el)}} = -\frac{\alpha - \epsilon_j^{(el)}}{\gamma - E_j^{(el)} S}$$

and the normalization condition $c_a^2 + c_b^2 + 2c_a c_b S = 1$ for each j. Furthermore, the MOs can be labeled by the axial rotation quantum number $m_z = 0, \pm 1, \pm 2, \ldots$, as being of symmetry $\lambda = |m_z|$ and type $\sigma, \pi, \delta, \ldots$. The coefficients change with R and in the limit $R = 0$ the MO becomes a united atom AO of nuclear charge number $Z_a + Z_b$ and atomic quantum numbers (nlm), with $m = m_z$. Furthermore, the energy functions of R for MOs of the same symmetry must not cross as R changes, except by numerical coincidence, because any electronic coupling term $\hat{U}^{(el)}$ in the Hamiltonian must then lead to coupling integrals $\left\langle \varphi_j \,|\, \hat{U}^{(el)} \,|\, \varphi_{j'} \right\rangle$ different from zero and the energy roots of a new eigenvalue equation must be different, indicating their repulsion. All this leads to a MO energy correlation diagram as R changes from separated atoms at large values to the united atom, as shown in Figure 3.5.

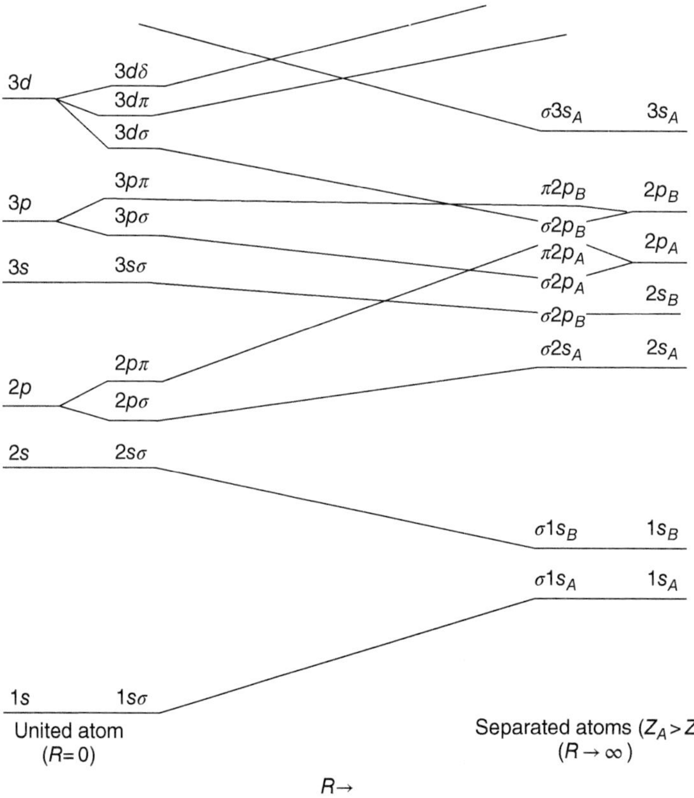

Figure 3.5 Energy-level correlation diagram (not to scale) for the one-electron heteronuclear diatomic molecules. *Source:* Figure 4.1, page 96 of Bransden and McDowell [48]. Reproduced with permission of Oxford University Press.

In H + H$^+$, or for two identical ion cores, the molecular reflection symmetry across a plane perpendicular to the internuclear axis leads to $c_a = \pm c_b$ for real valued MOs, and the normalized LCAOs are $\varphi_{g,u}(\vec{r}) = [\chi_a(\vec{r}) \pm \chi_b(\vec{r})]/[2(1 \pm S_{ab})]^{1/2}$, with the gerade and ungerade forms for the + and − signs. The nlm hydrogenic orbital for an electron at position $\vec{r}$ at a nucleus a with nuclear charge Z_a is $\chi_{a,nlm}(\vec{r}) = \langle \vec{r} \,|\, anlm \rangle = R_{nl}(r_a) Y_{lm}(\vartheta_a, \varphi_a)$. For the 1$s$ orbital $\chi_{a,1s0}(\vec{r}) = \langle \vec{r} \,|\, a1s \rangle = (Z_a/a_0)^{3/2} 2 \exp(-2Z_a r_a/a_0).(2\pi^{1/2})^{-1}$ with a_0 the Bohr radius, of energy $E_{1s} = -Z_a^2 e^2/(2a_0)$, when using Gaussian units [49]. Here, $S_{ab} = S$ decreases exponentially for large R as $\exp(-R/a_0)$. An energy correlation diagram from the separated atoms to the united atom for homonuclear diatomics is shown in Figure 3.6.

The total energy is the sum of nuclear repulsion plus electronic energy, $E_g(R) = c_e^2/(4\pi\varepsilon_0 R) + E_g^{(el)}$, with

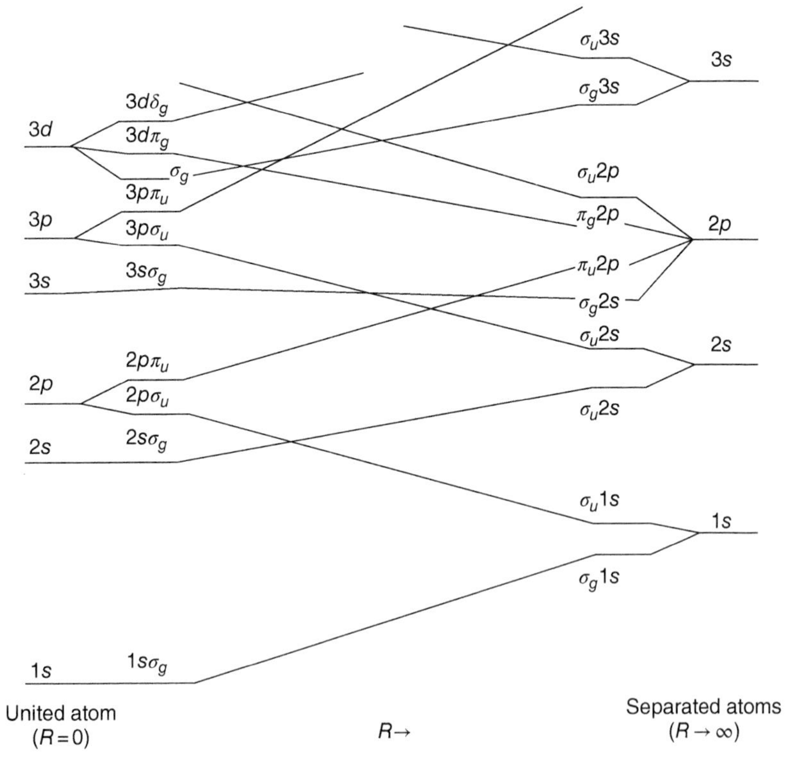

Figure 3.6 Energy-level correlation diagram (not to scale) for the one-electron homonuclear diatomic molecules. *Source:* Figure 4.3, page 98 of Bransden and McDowell [48]. Reproduced with permission of Oxford University Press.

$$E_g^{(el)}(R) = \epsilon_{1s} - \frac{j+k}{1+S}$$

$$j(R) = \frac{C_e^2}{4\pi\varepsilon_0}\int d^3r \frac{\chi_a(\vec{r})^2}{|\vec{r}-\vec{R}_b|}, \quad k(R) = \frac{C_e^2}{4\pi\varepsilon_0}\int d^3r \frac{\chi_a(\vec{r})\chi_b(\vec{r})}{|\vec{r}-\vec{R}_a|}$$

where j is called the Coulomb integral and describes the attraction between the electron in orbital a and the nucleus at b, a term expected from classical electrostatic. The integral k, however, is purely quantal and accounts for the interaction of the overlap of charges centered at a and b with the nucleus at a, a quantal penetration effect. For large distances, one has $j(R) \approx C_e^2/(4\pi\varepsilon_0 R)$ plus a term exponentially decreasing as $\exp(-2R/a_0)$, while $k(R) \approx 0$ exponentially as $\exp(-R/a_0)$, so that the results from pure electrostatic are recovered at large R.

The choice of atomic orbitals with a nuclear charge number $Z_a = 1.0$ gives a potential energy $V_0(R) = E_0(R) - E_{1s}$ repulsive at short distances and attractive at large ones with a minimum at $R_e = 2.50$ au and a dissociation energy $D_e = E_0(\infty) - E_0(R_e) = 0.065$ au, to be compared with the accurate values of $D_e = 0.1026$ au = 2.79 eV at $R_e = 2.00$ au. This is qualitatively correct and can be improved minimizing the energy functional $\mathcal{E}[\varphi]$ with respect to parameters in a trial function.

A flexible trial function must describe the polarization of atomic orbitals, and the effective nuclear charge at short distance keeping in mind that as the internuclear distance decreases the charge changes from $Z_a = 1.0$ au at large distances to the united atom limit at $R = 0$ of value $Z_a = 2.0$ au. This can be done allowing for variable charge numbers Z in exponents to get the energy optimized. Polarization of H by H^+ can be treated adding to the $1s0$ orbital a portion of the $2pz$ orbital, pointing from one nucleus to the other, to form a hybrid orbital $\chi_a(\vec{r}) = \chi_{a,1s0}(\vec{r}) + \lambda\chi_{a,2pz}(\vec{r})$, with

$$\chi_{a,2pz}(\vec{r}) = \left(\frac{Z_a}{2a_0}\right)^{3/2}\left(\frac{Z_a r_a}{a_0}\right)\exp\left(-\frac{Z_a r_a}{2a_0}\right)\left(2\pi^{1/2}\right)^{-1}\cos(\vartheta_a)$$

and λ treated as a variational parameter which depends on R. Varying parameters for each R one finds optimal values $Z_{1s}(R)$, $Z_{2p}(R)$, and $\lambda(R)$ [50] and a minimum at $R_e = 2.01$ au with $D_e = 0.1004$ au = 2.73 eV when $Z_{1s} = 1.246$, $Z_{2p} = 2.965$, and $\lambda = 0.138$, a much better result, showing the importance of introducing effective nuclear (or ion core) charges and orbital hybridization as functions of a decreasing distance.

A connection can be made with the interaction energy at large distances. Keeping the variation parameter λ but letting $Z_{1s} = Z_{2p} = 1.0$ at large distances, the minimum of $\mathcal{E}(\lambda)$ provides the analytical form of the parameter and the energy, which is found to agree with the long-range charge-induced dipole interaction energy going as R^{-4} with a transition dipole and a polarizability obtained from the integral $\langle 1s0 \,|\, z \,|\, 2pz\rangle$. The long-range dependence on R^{-1}, however, is altered in this homonuclear system when there is electronic excitation, by the quantal coherence between the two equivalent states of the electron at large distances, which leads to a R^{-3} resonance interaction in first order of perturbation theory.

The MO constructed for the ground state of H + H^+ can be labeled by its symmetry as the $1\sigma_g$ MO φ_g with energy $E_0(R) = E_g(R)$, and describes how the electronic distribution extends into the internuclear region. An excited $1\sigma_u$ with the MO $\varphi_u(\vec{r}) = [\chi_a(\vec{r}) - \chi_b(\vec{r})]/[2(1-S_{ab})]^{1/2}$ has the energy

$$E_u(R) = \frac{C_e^2}{4\pi\varepsilon_0 R} + \epsilon_{1s} - \frac{j-k}{1-S}$$

and describes an electron distribution that has instead moved from inside the internuclear region to the outside. When more than one electron is present outside closed subshells, their correlation leads to interactions between electronic configurations made up of both φ_g and φ_u MOs. In general for a homonuclear diatomic one can locate an electron at each nucleus with the MO combinations that reconstruct AOs, $\chi_{a,b}(\vec{r}) = (1 + S_{ab})^{1/2}\varphi_g(\vec{r}) \pm (1 - S_{ab})^{1/2}\varphi_u(\vec{r})$.

3.5.3 Treatments of Electronic Exchange and Charge Transfer

Turning to H + H, with two interacting electrons, the new features as distances decrease are electron exchange and correlation changing with R. Given the AOs $\chi_a(\vec{r})$ and $\chi_b(\vec{r})$, two electron states $\Phi(\boldsymbol{x}_1, \boldsymbol{x}_2)$ depend on position and the spin variable ζ in each $\boldsymbol{x}_j = (\vec{r}_j, \zeta_j)$, and must be constructed to satisfy antisymmetry with respect to electron exchange. This can be done applying the antisymmetry projection operator $\hat{\mathcal{A}} = 2^{-1}(\hat{I} - \hat{\mathcal{P}}_{12})$ written in terms of the permutation operator $\hat{\mathcal{P}}_{12}$ of electrons 1 and 2, that is idempotent insofar $\hat{\mathcal{A}}^2 = \hat{\mathcal{A}}$. A valence bond (VB) function with the electrons in spin up and spin down states $\alpha(\zeta)$ and $\beta(\zeta)$ coupled to form a singlet state (of total spin quantum numbers $S = 0$, $M_S = 0$) $\Theta_{0,0}(\zeta_1, \zeta_2) = [\alpha(\zeta_1)\beta(\zeta_2) - \alpha(\zeta_2)\beta(\zeta_1)]/2^{1/2}$ is, normalized and with a short form for arguments,

$$\begin{aligned}\Phi^{(cov)}(\boldsymbol{x}_1,\boldsymbol{x}_2) &= C_{cov}\hat{\mathcal{A}}\chi_a(1)\chi_b(2)\Theta_{0,0}(1,2)\\ &= \frac{\chi_a(1)\chi_b(2) + \chi_a(2)\chi_b(1)}{(2 + 2S_{ab}^2)^{1/2}}\Theta_{0,0}(1,2)\end{aligned}$$

which describes the covalent bonding of the atoms. This form can be used in calculations with the AOs chosen as 1s orbitals or as hybrids, to form the $H_2(^1\Sigma)$ state. The electronic Hamiltonian can be decomposed as $\hat{H}^{(el)} = \hat{H}_a(1) + \hat{H}_b(2) + \hat{H}_{ab}(1,2)$ with $\hat{H}_a(1)\chi_a(1) = \epsilon_a\chi_a(1)$ involving the atomic Hamiltonian and its 1*s* AO, and $\hat{H}_{ab}(1,2)$ for the interatomic Coulomb interactions. The corresponding electronic energy is given by the variational function for this state, as

$$E^{(cov)} = 2\epsilon_a + \frac{Q_{ab} + A_{ab}}{1 + \Delta_{ab}}$$

where $\Delta_{ab} = S_{ab}^2$, $Q_{ab} = \int d(1)d(2)\chi_a(1)\chi_b(2)\hat{H}_{ab}(1,2)\chi_a(1)\chi_b(2)$ is an interatomic Coulomb integral, and $A_{ab} = \int d(1)d(2)\chi_a(1)\chi_b(2)\hat{H}_{ab}(1,2)\chi_a(2)\chi_b(1)$ is an interatomic exchange integral containing the differential overlap $\chi_a(1)\chi_b(1)$. The A_{ab} integral has negative values for H + H and contributes most of the binding energy.

Numerical results show a repulsive potential energy at short distances and attraction at long distances with a minimum giving values for R_e and D_e. Using AOs with a variable orbital exponent corresponding physically to an effective nuclear charge number, which varies from 1.0 at large distances to 2.0 at the united atom limit, the so-called Heitler–London–Wang wavefunction gives $R_e = 0.744$ A and $D_e = 3.78$ eV with an effective charge number $\zeta = 1.166$, to be compared with the accurate values $R_e = 0.741$ A and $D_e = 4.75$ eV. The discrepancy is due to a shortcoming of the wavefunction at short distances, where the potential energy is found to be too repulsive because electron transfer at short distances has been ignored. Results are much improved if one adds an ionic state describing $H^+ + H^-$,

$$\Phi^{(ion)}(\boldsymbol{x}_1, \boldsymbol{x}_2) = \frac{\chi_a(1)\chi_a(2) + \chi_b(1)\chi_b(2)}{\left(2 + 2\,S_{ab}^2\right)^{1/2}}\Theta_{0,0}(1,2)$$

to form a covalent-ionic extended VB function given by

$$\Phi^{(EVB)}(\boldsymbol{x}_1, \boldsymbol{x}_2) = A_{cov}\Phi^{(cov)}(\boldsymbol{x}_1, \boldsymbol{x}_2) + A_{ion}\Phi^{(ion)}(\boldsymbol{x}_1, \boldsymbol{x}_2)$$

This covalent-ionic trial function can be used to minimize the electronic energy by varying the parameters A_{cov}, A_{ion} for each distance *R*, and provides more accurate values $D_e = 4.03$ eV for $\zeta = 1.19$ even when using only 1*s* atomic orbitals. Introducing hybrid AOs which combine 1*s*0 and 2*pz* atomic orbitals further allows treatment of induction and dispersion forces which then appear at large distances and are given in terms of the transition dipole integrals $\langle 1s0 \,|\, z \,|\, 2pz \rangle$.

Extensions to many-electron homonuclear or heteronuclear diatomics and to polyatomics can be more readily implemented using an alternative description based on MO-LCAO molecular orbitals, which for a diatomic is $\varphi(\vec{r}) = c_a\chi_a(\vec{r}) + c_b\chi_b(\vec{r})$ such as we found for $H + H^+$. Antisymmetry is incorporated through the formation of Slater determinants, and a linear combination of determinants, or configuration interaction (CI), can be formed to construct a wavefunction of high accuracy. The procedure is to identify each element in a

$N \times N$ matrix $[A_{kj}]$ with the j-th one-electron spin-orbital $\psi_j(\boldsymbol{x}_k)$, a function of the k-th electron position and spin variables, so that the variable index identifies the k-th row, and the function index identifies the j-th column. The determinant of the matrix is then $\det[A_{kj}] = \det[\psi_j(\boldsymbol{x}_k)]$ and a determinant formed with a set of N spin-orbitals $\psi_{j_1}, \psi_{j_2}, \ldots, \psi_{j_N}$ can simply be written as $|\psi_{j_1}\psi_{j_2}, \ldots, \psi_{j_N}|$ with an implicit understanding of the electron variables, designating a chosen electronic configuration.

Early work on few-electron diatomics, including also discussions of physical aspects important in the treatment of intermolecular forces, can be found in Refs. [1, 10]. The coefficients of AOs in MOs, the coefficients in the linear combinations of configurations, and the exponents of AOs corresponding to effective nuclear charges have all been used as variational parameters to obtain accurate results. An alternative to variation of AO exponents is to introduce linear combinations of AOs with several fixed Z-values (called double-Z, triple-Z, ..., basis sets), chosen to allow for electronic shell splittings and for variations of effective nuclear charges with distance R [47, 51, 52].

For H + H the electronic Hamiltonian is $\hat{H}^{(el)}(1,2) = \hat{h}^{(el)}(1) + \hat{h}^{(el)}(1) + c_e^2 r_{12}^{-1}$ and a single determinant trial wavefunction $\Phi_{RHF}(1, 2) = |1\sigma_g\alpha(1)1\sigma_g\beta(2)|$, in a compact notation for determinantal wavefunctions, generates a restricted Hartree–Fock state where the MO can be written as a LCAOs. This is acceptable at short and intermediate distances, but incorrect at large distance. It is found already for H_2 in its determinantal state $|1\sigma_g\alpha 1\sigma_g\beta|$, that a single determinant gives the wrong asymptotic behavior, insofar it describes at large distances a mixture of H + H and $H^+ + H^-$ states.

To avoid this, a useful procedure is to start from the VB states and re-express its AOs in terms of MOs, in which case a linear combination of configurations, or configuration interaction (CI), is usually found [47, 51]. This leads to the appearance of two electronic configurations, $|1\sigma_g\alpha 1\sigma_g\beta|$ and $|1\sigma_u\alpha 1\sigma_u\beta|$. A linear combination of these two in a configuration interaction (CI) wavefunction

$$\Phi^{(CI)}(\boldsymbol{x}_1,\boldsymbol{x}_2) = A_G\,|1\sigma_g\alpha(1)1\sigma_g\beta(2)| + A_U\,|1\sigma_u\alpha(1)1\sigma_u\beta(2)|$$

with coefficients to be varied, leads to exactly the same states and results obtained from the covalent-ionic treatment [47]. Induction and dispersion interactions can be described by introducing AO hybrids, or alternatively by adding configurations constructed with an atomic basis set containing polarization atomic orbitals.

The two-electron heteronuclear diatomic He + H^+ can be treated using MOs constructed with He 1s and H 1s AOs, with the simplest wavefunction of form $\Phi_{RHF}(1, 2) = |1\sigma\alpha(1)1\sigma\beta(2)|$. Long-range induction and dispersion interactions can again be incorporated by means of basis sets of AOs containing at least the 2pz type in hybrid AOs, or by means of CIs. An accurate treatment here involves several configurations, capable of describing the electronic structures going to both $He(1s^2)$ and $He^+(1s) + H(1s)$ at large distances, and structures that go to the united atom with configuration $1s^2$ for a nuclear charge number $Z = 3$ for short distances. Other configurations are also relevant to the description of this system at intermediate distances [48].

The four-electron system He+He is an example of two closed shell species which can also be described by a CI. The leading configuration involves MOs as linear combinations of $1s$ orbitals at each He, and a single determinant $\Psi_{RHF}(1,\ 2,\ 3,\ 4) = |1\sigma_g\alpha(1)1\sigma_g\beta(2)1\sigma_u\alpha(3)1\sigma_u\beta(4)|$ suffices to qualitatively describe the short-range interaction and to go to the correct asymptotic limit $He(1s^2) + He(1s^2)$ at large distances. However, to obtain an accurate description at all distances, it is necessary to work with a larger basis set of AOs and at least include atomic polarization to generate induction and dispersion interactions as R decreases. The united atom limit is $1s^2 2p_0{}^2$.

The accurate treatment of two interacting many-electron species at all distances is very challenging. Theoretical methods and computational algorithms are being developed to account for many related properties describing systems as supermolecules. In addition to CI treatments, many-electron perturbation theory of the Moeller–Plesset type, with proper accounting of linked diagrams to assure electronic size consistency [52], coupled cluster theory [53], and symmetry adapted perturbation theory are important methods available and being developed to do *ab initio* calculations of intermolecular forces [54].

For systems with two or more electrons, interacting species may form states of higher total spin. For example in H + H, the $H_2(^3\Sigma)$ is formed from the spin states $\Theta_{S,M_s}(\zeta_1,\zeta_2)$ with $S = 1$, $M_S = 0, +1, -1$ which are given by $\Theta_{1,0}(\zeta_1,\ \zeta_2) = [\alpha(\zeta_1)\beta(\zeta_2) + \alpha(\zeta_2)\beta(\zeta_1)]/2^{1/2}$, $\Theta_{1,1}(\zeta_1,\ \zeta_2) = \alpha(\zeta_1)\alpha(\zeta_2)$, and $\Theta_{1,-1}(\zeta_1,\ \zeta_2) = \beta(\zeta_1)\beta(\zeta_2)$, all of which are even under electron permutation and must be matched by odd permutation functions of positions. A simple state of valence-bond type is $C[\chi_a(1)\chi_b(2) - \chi_a(1)\chi_b(2)]\Theta_{1,0}(1, 2)$ which can be re-expressed in terms of MOs $\varphi_{g,u}(\vec{r})$ as done for the bonding $H_2(^1\Sigma)$. The electronic energy expression for the new triplet state, however, contains the VB atomic exchange integral with the opposite sign and its energy is purely repulsive at all distances,

$$E^{(rep)} = 2\epsilon_a + \frac{Q_{ab} - A_{ab}}{1 - \Delta_{ab}}$$

and goes to the $H(1s)$ ground states at large distances.

In a MO description, the state $H_2(^3\Sigma)$ is given by the determinantal wavefunction $|1\sigma_g\alpha 1\sigma_u\beta|$ and this goes in the united atom limit to the higher atomic configuration $1s^1 2p_0{}^1$, which alternatively explains its repulsive energy at short and intermediate distances. This $H_2(^3\Sigma)$ state does also have a shallow well at large distances as the result of attraction due to long-range polarization, which can be described with hybrid AOs in a VB treatment, or with a CI in a MO treatment.

3.6 Spin-orbit Couplings and Retardation Effects

A more detailed treatment of molecular interactions requires consideration of quantum relativistic phenomena. The imposition of relativistic invariance of quantum mechanical and electromagnetism equations with respect to the Lorentz transformation means that one must consider coupled electric $\vec{\mathcal{E}}$ and magnetic $\vec{\mathcal{H}}$ fields in accordance with the Maxwell equations, and the Dirac equation for the electron in an electromagnetic field [55]. However, for the purpose of calculating interaction energies it is sufficient to consider the limit of small electron velocities v relative to the speed of light c, and to obtain results by mean of perturbation theory with the small expansion parameter v/c. Two main consequences are the appearance of new Hamiltonian terms involving the electron spin, and retardation effects in the very long range interaction of species.

Molecules with heavy atoms of large atomic number Z have their electronic structure affected by relativistic phenomena associated to large electron speeds with a fraction of the speed of light. In these cases, electron states must be obtained from a relativistic equation. For states of a single electron described by the Dirac equation, states are given by spinors with four spatial components, with two classified as strong and two as weak ones insofar they are smaller to order $(v/c)^2$. Molecular interactions can be accurately treated with a formalism which keeps only the strong components and gives rise to the electron intrinsic angular momentum, or spin, $\vec{s}$ with spin-up and spin-down electronic states [11]. This spin creates a magnetic dipole $\vec{\mu}_s$ which interacts with a magnetic field $\vec{\mathcal{H}}$ to add an energy $-\vec{\mu}_s.\vec{\mathcal{H}}$ to the Hamiltonian operator for the electron. An electron moving in the field of a nucleus also acquires a spin-orbit energy coupling $\hat{h}_{SO} = \xi(r)\vec{s}.\vec{l}$ dependent on the electron distance r to the nucleus and its momentum $\vec{p}$ through the orbital angular momentum $\vec{l} = \vec{r} \times \vec{p}$. The Hamiltonian contains in addition other terms dependent also on the electron momentum: a mass-polarization term and the so-called Darwin term involving the electronic density in contact with an electron-positron distribution at the nucleus, both of which are of the order $\alpha^2 \approx (1/137)^2$, with $\alpha = e^2/\hbar c$ the fine structure constant, compared to the nonrelativistic 1s orbital energy, and are therefore usually insignificant in calculations of intermolecular energies.

However, the spin-orbit coupling energy can make a difference for heavy atoms and in the interaction of systems at very large distances $R \geq 100\ a_0$. Relativistic effects in systems with two or more electrons can be treated with the Breit–Pauli equation, as given in appendix A of Ref. [10] and in [56]. The applications to intermolecular forces are reviewed in Ref. [57]. They are based on the Hamiltonian $\hat{H} = \hat{H}_{NRel} + \alpha^2 \hat{H}_{Rel}$ and a perturbation treatment with a basis set of zeroth-order state eigenfunctions of the nonrelativistic operator $\hat{H}_{NRel}$. The relativistic interaction operator decomposes as $\hat{H}_{Rel} = \hat{H}_{LL} + \hat{H}_{SS} + \hat{H}_{SL} + \hat{H}_{MP} + \hat{H}_{D}$, where in order the energy terms are coupling of magnetic dipoles created by orbiting electrons; coupling of magnetic dipoles from electron spins; spin-orbit couplings within each species and between spin in one and orbital angular momentum in the other; relativistic, or mass polarization, correction due to variation of the electron mass with its speed; and the Darwin-like contact term at the atomic nuclei. The calculation of interaction energies in the simpler case of atom–atom systems is done by constructing diatomic states of specific electronic angular momenta, depending on the strength of spin-orbit couplings [58], and proceeding to do perturbation theory for the relativistic interactions [59].

The other mentioned important relativistic effect on intermolecular forces appears at very large distances R between species A and B, so that photons of wavelength λ emitted during electronic density fluctuations in one must travel a large distance $R \approx \lambda$ to be absorbed by the other and are delayed by the finite speed of light. The first treatment of this effect by means of quantum electrodynamics showed that the previous R^{-6} dependence derived for the dipole–dipole dispersion interaction turns into R^{-7} due to retardation [19]. This can be derived in a perturbation treatment where radiation is quantized and interacting with A and B, using a Hamiltonian $\hat{H} = \hat{H}_A + \hat{H}_B + \hat{H}_L + \hat{H}_{AL} + \hat{H}_{BL}$ where $\hat{H}_L$ and $\hat{H}_{AL}$ are the Hamiltonians of quantized light and of its interaction with A. Perturbation terms must be added up to the fourth order (for photon emission and absorption at both A and B) to obtain a result valid for distances R going from very large to the previous large distances and showing the change in R^{-n} behavior [60, 61]. An alternative to the perturbation treatment is provided by response theory including density fluctuations of both electrons and light, that bypasses the need for high order perturbation theory and is convenient for the treatment of very long-range forces in large systems [62]. Retardation effects in van der Waals interactions for two and three atoms have also been considered within molecular quantum electrodynamics [63] in a treatment that includes both short-range and long-range interactions. This describes the interactions as resulting from the exchange of photons between two species and has also been extended to three body interactions.

3.7 Interactions in Three-Body and Many-Body Systems

3.7.1 Three-Body Systems

The long-range interactions of a system of three atoms A, B, and C was treated using perturbation theory and was developed for a Hamiltonian operator containing dipole–dipole interaction potentials between pairs of neutral atoms in

$S = 0$ states [64]. It was pointed out that the magnitude of a three-body dispersion interaction as obtained from third-order perturbation theory is of magnitude $I_{at}\alpha_{at}^3\left(R_{AB}^3R_{BC}^3R_{CA}^3\right)^{-1}$ where I_{at} and α_{at} are atomic ionization and dipolar polarization magnitudes, and R_{LM} is the distance between atoms L and M. The angular dependence of the energy on the interior angles of the triangle formed by the atoms is $3\cos(\gamma_A)\cos(\gamma_B)\cos(\gamma_C) + 1$, with γ_L the angle at the location of atom L. This factor changes sign depending on the shape of the triangle and indicates that the three-atom dispersion energy can be either positive or negative. Perturbation theory shows that the dispersion energy is pairwise additive to second order, and that a three-atom interaction term appears first in the energy function to third order. This can be expressed in terms of dipolar polarizabilities for each atom. However, when species contain permanent multipoles, induction energies appear, and it is found that pairwise additivity is lost even to second order [12].

More generally A, B, and C can be molecules, and their pair interactions may contain not only monopole and dipole operators but also higher order multipoles. It is convenient to introduce pair interactions in terms of charge distributions, and to give a more general treatment using the resolvent operator. The simplest case involves three species sufficiently separated so that their properties are not affected by electron exchange. This allows allocation of charges I of electrons and nuclei to each separate species $L = A, B, C$.

The Hamiltonian is now, with species labeled by indices L and M,

$$\hat{H}_{ABC} = \hat{H}_A + \hat{H}_B + \hat{H}_C + \hat{H}_{ABC}^{(int)}, \quad \hat{H}_{ABC}^{(int)}\left(\vec{R}_{AB},\vec{R}_{BC},\vec{R}_{CA}\right) = \sum\nolimits'_{L\neq M}\hat{H}_{LM}^{(int)}\left(\vec{R}_{LM}\right)$$

$$\hat{H}_{LM}^{(int)}\left(\vec{R}_{LM}\right) = (4\pi\varepsilon_0)^{-1}\int d^3r\int d^3s\,\hat{c}_L(\vec{r}_L)\left|\vec{R}_{LM}+\vec{r}_L-\vec{s}_M\right|^{-1}\hat{c}_M(\vec{s}_M)$$

where the interaction is a sum of pairs, and the charge density operators are functions of positions with respect to the center of mass of each species L or M. The equation for three-body states can be solved expressing the three-body resolvent operator in terms of two-body operators in a way similar to what has been done for three-body transition operators [65–67]. In a compact notation, arrangement $a = 0$ is $A + B + C$ with all species noninteracting, and arrangements $a = 1$, 2, and 3 refer to cases where A, B, or C are, respectively, noninteracting while the other two atoms interact. The zeroth-order Hamiltonian is $\hat{H}^{(0)} = \hat{H}_A + \hat{H}_B + \hat{H}_C$, and interaction Hamiltonians in each arrangement are $\hat{V}^{(0)} = 0$ and $\hat{V}_a$, so that the perturbation Hamiltonian is $\hat{H}_{ABC}^{(int)} = \lambda\hat{H}^{(1)} = \sum_a\hat{V}_a$.

Perturbation theory expressions give wavefunctions and energies to second and third order. For each of the three species starting in its ground state $|0_L\rangle$, the unperturbed state is $\left|\Psi_G^{(0)}\right\rangle = |0_A0_B0_C\rangle$ with energy $E_G^{(0)} = E_0^{(A)} + E_0^{(B)} + E_0^{(C)}$.

The first-order energy and state satisfy, for the given G state, $\lambda E^{(1)} = \left\langle \Psi^{(0)} \,|\, \sum_a \hat{V}_a \,|\, \Psi^{(0)} \right\rangle = \sum_a E_a^{(int)}$ and

$$\left(\hat{H}^{(0)} - E^{(0)}\right) \lambda \Psi^{(1)} = \sum_a \left(E_a^{(int)} - \hat{V}_a\right) \Psi^{(0)}$$

and second- and third-order energies are obtained from $E^{(2)} = \left\langle \Psi^{(0)} \,|\, \hat{H}^{(1)} \,|\, \Psi^{(1)} \right\rangle$, and $E^{(3)} = \left\langle \Psi^{(1)} \,|\, \left(\hat{H}^{(1)} - E^{(1)}\right) \,|\, \Psi^{(1)} \right\rangle$, and involve only $\Psi^{(1)}$. This satisfies an inhomogeneous differential equation that can be solved by inspection in terms of two-body solutions as in

$$\Psi^{(1)} = \sum_a \Psi_a^{(1)}$$

$$\left(\hat{H}^{(0)} - E^{(0)}\right) \lambda \Psi_a^{(1)} = \left(E_a^{(int)} - \hat{V}_a\right) \Psi^{(0)}$$

where each arrangement term in the wavefunction involves excited states of two species and the ground state of the third one so that $\Psi_1^{(1)}$ is a combination of states $|0_A j_B j_C\rangle$ and similarly for the other two arrangements. It follows from this and from the orthonormality of the states $|j_L\rangle$, $L = A, B, C$, that one can separate in each arrangement the induction interaction integrals, where one of the species remains in its state and the other is excited, from the dispersion interaction integrals where both are excited. This gives after analysis of all integrals of the type $\left\langle \Psi^{(0)} \,|\, \hat{V}_a \,|\, \Psi_b^{(1)} \right\rangle$, containing products with arrangement interactions $\hat{V}_a$ and $\hat{V}_b$ [1]

$$E^{(2)}(ABC) = E_{ind}^{(2)}(ABC) + E_{dsp}^{(2)}(ABC),$$

$$E_{dsp}^{(2)}(ABC) = E_{dsp}^{(2)}(AB) + E_{dsp}^{(2)}(BC) + E_{dsp}^{(2)}(CA)$$

and indicates that the dispersion interaction of the three species is additive to the second order, while the induction interaction is intrinsically a three-body function.

The analysis can be extended to third order where one has that

$$E^{(3)} = \left\langle \Psi^{(1)} \,|\, \sum_a \left(\hat{V}_a - E_a^{(int)}\right) \,|\, \Psi^{(1)} \right\rangle$$

contains 27 integrals of the type $\left\langle \Psi_b^{(1)} \,|\, \hat{V}_a \,|\, \Psi_c^{(1)} \right\rangle$ involving products with the three arrangement interactions $\hat{V}_1, \hat{V}_2$, and $\hat{V}_3$. When the arrangement interactions are expanded in the basis set $|j_A j_B j_C\rangle$, many of the integrals are zero and the remaining ones can be classified as giving three-body induction or dispersion energies. In particular, the dispersion energy arises from the interaction of density fluctuations of the three species. They undergo a sequence of transitions from the ground state for each pair first to excited states which transfer energy in a second interaction and next return to the ground states in a third interaction, all involving transition charge densities $\langle j_L \mid \hat{c}_L(\vec{r}_L) \mid j_L' \rangle$. A special case detailed in the literature considers the dipole–dipole interaction energy operator

$$\hat{H}_{LM}^{(dd)}\left(\vec{R}_{LM}\right) = -(4\pi\varepsilon_0)^{-1}\sum\nolimits_{\xi,\eta} \hat{D}_{\xi}^{(L)} \frac{t_{\xi\eta}}{R_{LM}^3} \hat{D}_{\eta}^{(M)}$$

which when entered into the third-order dispersion energy leads to [1]

$$E_{dsp}^{(3)}(ABC) = C_9^{(dsp)}(ABC)\frac{1 + 3\cos(\gamma_A)\cos(\gamma_B)\cos(\gamma_C)}{(4\pi\varepsilon_0)^3 R_{AB}^3 R_{BC}^3 R_{CA}^3}$$

with a factor $C_9^{(dsp)}(ABC)$ which can be expressed in terms of the dynamical polarizabilities of the three species, as [13, 68]

$$C_9^{(dsp)}(ABC) = \frac{3\hbar}{\pi}\int_0^{\infty} du\, \alpha^{(A)}(iu)\alpha^{(B)}(iu)\alpha^{(C)}(iu)$$

Some calculated values of $E_{dsp}^{(3)}$ for H and He systems, and discussions of the role of nonadditivity in condensed matter are given in the literature [2, 13, 69].

3.7.2 Many-Body Systems

A system composed of N species A, B, C, ..., N has a total interaction energy $E_{int}(AB \ldots N)$ which can be decomposed into two-body, three-body, and further energy terms, writing

$$E_{int}(AB\ldots N) = E_2(AB\ldots N) + E_3(AB\ldots N) + \ldots + E_N(AB\ldots N)$$

with $E_2(AB \ldots N)$ calculated as the sum of all two-body interaction energies, $E_3(AB \ldots N)$ calculated from all groupings of three-body energies excluding two-body ones, and so on. The last term is a single quantity obtained from the interaction of the N bodies when contributions from all smaller groupings have been subtracted.

The sum over many-body terms is not necessarily rapidly convergent. The convergence can be improved introducing statistical concepts such as the two-body potential of mean force for a collection of N identical bodies, which applies to the interaction of two bodies when the other $N-2$ bodies are described as a medium with a statistical spatial distribution for given temperature and average density. This has been done to obtain the energies of atomic clusters, and cohesive energy of condensed matter systems such as molecular liquids and crystals [2, 70, 71].

References

1 Margenau, H. and Kestner, N.R. (1971). *Intermolecular Forces*, 2e. Oxford, England: Pergamon Press.
2 Maitland, G.C., Rigby, M., Brain Smith, E., and Wakeham, W.A. (1981). *Intermolecular Forces*. Oxford, England: Oxford University Press.
3 Dalgarno, A. (1961). Stationary perturbation theory. In: *Quantum Theory*, vol. 1 (ed. D. Bates), 171. New York: Academic Press.
4 Hirschfelder, J.O., Byers Brown, W., and Epstein, S.T. (1964). Recent developments in perturbation theory. *Adv. Quantum Chem.* 1: 255.
5 Davydov, A.S. (1971). *Theory of Molecular Excitons*. New York: Plenum.
6 Foerster, T. (1948). Intermolecular energy migration and fluorescence. *Ann. Phys.* 2: 55.
7 Jackson, J.D. (1975). *Classical Electrodynamics*. New York: Wiley.
8 Landau, L.D. and Lifshitz, E. (1975). *Classical Theory of Fields*, 4e. Oxford, England: Pergamon Press.
9 Stone, A.J. (1996). *The Theory of Intermolecular Forces*. Oxford, England: Oxford University Press.
10 Hirschfelder, J.O., Curtis, C.F., and Bird, R.B. (1954). *Molecular Theory of Gases and Liquids*. New York: Wiley.
11 Cohen-Tanoudji, C., Diu, B., and Laloe, F. (1977). *Quantum Mechanics*, vol. 2. New York: Wiley-Interscience.
12 Kaplan, I.G. (2006). *Intermolecular Interactions*, 81. New York: Wiley.
13 Dalgarno, A. and Davison, W.D. (1966). The calculation of van der Waals interactions. *Adv. At. Mol. Phys.* 2: 1.
14 Zare, R.N. (1988). *Angular Momentum*. New York: Wiley.
15 Edmonds, A.R. (1960). *Angular Momentum in Quantum Mechanics*. Princeton NJ: Princeton University Press.
16 Linder, B. and Rabenold, D.A. (1972). Unified treatment of van der Waals forces between two molecules of arbitrary sizes and electron delocalization. *Adv. Quantum Chem.* 6: 203.
17 Koide, A. (1976). A new expansion for dispersion forces and its applications. *J. Phys. B: Atomic Mol. Phys.* 9: 3173.
18 Buckingham, A.D. (1967). Permanent and induced molecular moments and long-range intermolecular forces. *Adv. Chem. Phys.* 12: 107.
19 Casimir, H.B.G. and Polder, D. (1948). The influence of retardation on the London-van der Waals forces. *Phys. Rev.* 73: 360.
20 Mavroyannis, C. and Stephen, M.J. (1962). Dispersion forces. *Mol. Phys.* 5: 629.
21 Tang, K.T. (1969). Dynamical polarizabilities and van der Waals coefficients. *Phys. Rev.* 177: 108.
22 Kramer, H.L. and Herschbach, D.R. (1970). Combination rules for van der Waals force constants. *J. Chem. Phys.* 53: 2792.

23 Linder, P. and Lowdin, P.O. (1968). Upper and lower bounds in second order perturbation theory and the unsold approximation. *Int. J. Quantum Chem.* Symp. Series 2 S: 161.

24 Goscinski, O. (1968). Upper and lower bounds to polarizabilities and van der Waals forces. *Int. J. Quantum Chem.* 2: 761.

25 Langhoff, P., Gordon, R.G., and Karplus, M. (1971). Comparison of dispersion force bounding methods with applications to anisotropic interactions. *J. Chem. Phys.* 55: 2126.

26 Slater, J.C. and Kirkwood, J.G. (1931). van der Waals forces in gases. *Phys. Rev.* 37: 682.

27 Dalgarno, A. (1967). New methods for calculating long-range intermolecular forces. In: *Advances in Chemical Physics*, 143. New York: Wiley.

28 Hohenberg, P. and Kohn, W. (1964). Inhomogeneous electron gas. *Phys. Rev. B* 136: 864.

29 Lundqvist, S. and March, N.H. (eds.) (1983). *Theory of the Inhomogeneous Electron Gas*. New York: Plenum Press.

30 Trickey, S.B. (Special Editor) (1990). *Advances in Quantum Chemistry: Density Functional Theory of many-Fermion Systems*. San Diego, CA, USA: Academic Press.

31 Parr, R.G. and Yang, W. (1989). *Density Functional Theory of Atoms and Molecules*. Oxford, England: Oxford University Press.

32 Kohn, W. and Sham, L.J. (1965). Quantum density oscillations on an inhomogeneous electron gas. *Phys. Rev. A* 137: 1697.

33 Gordon, R.G. and Kim, Y.S. (1972). Theory for the forces between closed shell atoms and molecules. *J. Chem. Phys.* 56: 3122.

34 Kim, Y.S. and Gordon, R.G. (1974). Unified theory for the intermolecular forces between closed shell atoms and ions. *J. Chem. Phys.* 61 (1).

35 Parker, G.A., Snow, R.L., and Pack, R.T. (1976). Intermolecular potential surfaces from electron gas methods. I. Angle and distance dependence of the He + CO2 and Ar + CO2 interactions. *J. Chem. Phys.* 64: 1668.

36 Clugston, M.J. (1978). The calculation of intermolecular forces. A critical examination of the Gordon-Kim model. *Adv. Phys.* 27: 893.

37 Cohen, J.S. and Pack, R.T. (1974). Modified statistical method for intermolecular potentials: combining rules for higher van der Waals coefficients. *J. Chem. Phys.* 61: 2372.

38 Smith, F.T. (1972). Atomic distortion and the combining rule for repulsive potentials. *Phys. Rev. A* 5: 1708.

39 Gunnarsson, O. and Lundqvist, B.I. (1976). Exchange and correlation in atoms, molecules, and solids by the spin density functional formalism. *Phys. Rev. B* 13: 4274.

40 Gaydaenko, V.I. and Nikulin, V.K. (1970). Born-Mayer interaction potential for atoms with Z=2 to Z=36. *Chem. Phys. Lett.* 7: 360.

41 Knowles, P.J. and Meath, W.J. (1986). Non-expanded dispersion and induction energies, and damping functions for molecular interactions with application to HF-He. *Mol. Phys.* 59: 965.
42 Tang, K.T. and Toennies, J.P. (1984). An improved simple model for the van der Waals potential based on universal damping functions for the dispersion coefficients. *J. Chem. Phys.* 80: 3726.
43 Dobson, J.F. and Gould, T. (2012). Calculation of dispersion energies. *J. Phys. Condens. Matter* 24: 073201.
44 Berland, K., Cooper, V.R., Lee, K. et al. (2015). van der Waals forces in density functional theory: a review of the vdW-DF method. *Rep. Prog. Phys.* 78: 066501.
45 Grimme, S., Hansen, A., and Brandenburg, J.G.C. (2016). Dispersion corrected mean-field electronic structure methods. *Chem. Rev.* 116: 5105.
46 Zhao, Y. and Truhlar, D.G. (2008). The M06 suite of energy functionals: systematic testing of four M06 functionals and 12 other functionals. *Theor. Chem. Accounts* 120: 215.
47 Levine, I.N. (2000). *Quantum Chemistry*, 5ee. New Jersey, USA: Prentice-Hall Inc.
48 Bransden, B.H. and McDowell, M.R.C. (1992). *Charge Exchange in the Theory of Ion-Atom Collisions*. Oxford, England: Clarendon Press.
49 Atkins, P.W. and Friedman, R.S. (1997). *Molecular Quantum Mechanics*. Oxford, England: Oxford University Press.
50 Weinhold, F. (1971). Dickinson energy of H2+. *J. Chem. Phys.* 54: 530.
51 Szabo, A. and Ostlund, N.S. (1982). *Modern Quantum Chemistry*. New York: Macmillan.
52 Hehre, W.J., Radom, L., Schleyer, P.v., and Pople, J.A. (1986). *Ab initio Molecular Orbital Theory*. New York: Wiley.
53 Bartlett, R.J. (1981). Many-body perturbation theory and coupled cluster theory for electronic correlation in molecules. *Annu. Rev. Phys. Chem.* 32: 359.
54 Jeziorski, B., Moszynski, R., and Szalewicz, K. (1994). Perturbation theory approach to intermolecular potential energy surfaces of van der Waals complexes. *Chem. Rev.* 94: 1887.
55 Messiah, A. (1962). *Quantum Mechanics*, vol. 2. Amsterdam: North-Holland.
56 Bethe, H.A. and Salpeter, E.E. (1957). *Quantum Mechanics of One- and Two-electron Atoms*. Berlin: Springer-verlag.
57 Hirschfelder, J.O. and Meath, W.J. (1967). The nature of intermolecular forces. *Adv. Chem. Phys.* 12: 3.
58 Herzberg, G. (1950). *Molecular Spectra and Molecular Structure I. Spectra of Diatomic Molecules*, 2e. Princeton, NJ: Van Nostrand.
59 Chang, T.Y. (1967). Moderately long range intermolecular forces. *Rev. Mod. Phys.* 39: 911.
60 Power, E.A. (1967). Very long-range (retardation effect) intermolecular forces. In: *Intermolecular Forces* (Advan. Chem. Phys. vol. 12), 167. New York: Wiley.

61 Levin, F.S. and Micha, D.A. (eds.) (1993). *Long-Range Casimir Forces: Theory and Recent Experiments on Atomic Systems*. New York: Plenum Press.
62 McLachlan, A.D. (1963). Retardation dispersion forces between molecules. *Proc. Royal Soc. A* 271: 387.
63 Salam, A. (2011). Molecular quantum electrodynamics of radiation-induced intermolecular forces. *Adv. Quantum Chem.* 62: 1.
64 Axilrod, B.M. and Teller, E. (1943). Interaction of the Van der Waals type between three atoms. *J. Chem. Phys.* 11: 299.
65 Micha, D.A. (1972). Collision dynamics of three interacting atoms: the Faddeev equations. *J. Chem. Phys.* 57: 2184.
66 Micha, D.A. (1985). Rearrangement in molecular collisions: a many-body approach. In: *Theory of Chemical Reaction Dynamics*, vol. II, Chap. 3 (ed. M. Baer). Boca Raton, Florida: CRC Press.
67 Micha, D.A. (2017). Quantum partitioning methods for few-atom and many-atom dynamics. In: *Advances in Quantum Chemistry*, vol. 74, 107. London, England: Elsevier.
68 Salam, A. (2016). *Non-Relativistic Quantum Electrodynamics of the van der Waals Dispersion Interaction*. Cham, Switzerland: Springer.
69 Kihara, T. (1958). Intermolecular forces. *Adv. Chem. Phys.* 1: 267.
70 Kaplan, I.G. (1999). Role of electron correlation in nonadditive forces and ab initio model potentials for small metal clusters. In: *Advances in Quantum Chemistry*, vol. 31, 137. San Diego, CA: Academic Press.
71 McQuarrie, D.A. (1973). *Statistical Mechanics*. New York: Harper & Row.

4

Model Potential Functions

CONTENTS

4.1 Many-Atom Structures

The structure of a many-atom system with N_{at} atoms is defined by three Cartesian coordinates per atom and has $3N_{at}$ degrees of freedom. A rigid structure in vacuum has invariant properties under overall translations and rotations so that a description in a reference frame with its origin at the center of mass of the nuclei

Molecular Interactions: Concepts and Methods, First Edition. David A. Micha.

and its Cartesian axes fixed in space, a CMN-SF frame, decreases the required number of degrees of freedom. Overall, translation can eliminate three degrees of freedom, and in addition three (for a nonlinear system) or two (for a linear system) variables can also be eliminated with overall rotational invariance so that the rigid structure requires only $N_{df} = 3N_{at} - 6$ variables, or $N_{df} = 3N_{at} - 5$ for a linear system. They can be chosen as atom–atom distances, of which there are $N_{df}(N_{df} - 1)/2$. For a two atom system, this is the distance R between centers of mass, and for a nonlinear three atom system, A–B–C one can choose atom–atom distances R_{AB}, R_{BC}, R_{CA}, or sometimes two bond distances R_{AB}, R_{BC} and the common bond angle α_{ABC}. Generally, for a system with an interaction energy invariant under overall rotations and described by a single potential energy function (which must go to zero as any atom is removed to infinity) this potential energy is a $V(\{R_{IJ}\})$ function which depends on the interatomic distances between all atom pairs IJ. As distances are varied the energy generates a potential energy surface (PES) in a space of $N_{df}(N_{df} - 1)/2$ degrees of freedom, with peaks and valleys which can be pictured in three dimensions or as isocontours in two dimensions, by freezing all but two variables at a time.

In addition to this distances dependence, the potential energy may also depend on electron spin variables if the system has open electronic shells. It may also be multivalued for a given atomic structure if more than one PES is needed to account for electronic rearrangement due to electronic charge transfer or excitation. This requires consideration of couplings between PESs and is treated in a following chapter.

The potential can be decomposed in a variety of ways, some with more physical content than others. A number of criteria can be imposed on the form of $V(\{R_{IJ}\})$ to make it useful in the treatment of energetics and dynamics properties. They are the following:

a) Asymptotic consistency, meaning that as a fragment is removed to infinity, the total potential energy must become a sum of the potential energies for the parts.
b) Size consistency so that if the whole system is a collection of N identical units, the total potential energy must change linearly with a changing number N of units.
c) Proper spatial and spin symmetry.
d) Physical continuity of potential values and gradients with respect to distances, as distances change from short ranges to long ranges.

These PESs contain parameters to be fixed by means of calculated or measured procedures, and this requires:

a) An algebraic form allowing good fits as the structure varies, and practical for given computational resources.

b) Parameters which can be determined with a relatively small number of data points.
c) Flexible shape so that a fitted PES converges to an accurate one as the number of data points are increased.
d) Physically meaningful parameters providing insight as structure or composition are varied.

Model potential functions can be constructed with functional forms suggested by physical properties and parametrized to reproduce interaction energies known from measurements or from atomistic calculations. They can be classified in accordance with the atomic and electronic structure of the involved species. These can be atoms, molecules, clusters, solid surfaces, polymers, biomolecules, and complexes formed from them. Their interactions depend on whether they are in their ground electronic state or electronically excited, and whether they are stable during interactions or reactive. All these possibilities require a variety of functional forms, with parameters obtained from model calculations [1–6], or from experimental measurements. Experimental extraction of potential energy parameters from spectroscopy, molecular beam scattering, and molecular kinetics have been covered, for example in Refs. [7–9].

For three atoms A, B, C, we can use atomic orbitals $\chi_{a\lambda}, \chi_{b\mu}, \chi_{c\nu}$ centered at positions $\vec{R}_n$ of nuclei $n = a, b, c$, to construct molecular orbitals φ_j as linear combinations of atomic orbitals (or LCAOs). These compose many-electron determinantal wavefunctions D_K from which an optimized configuration interaction (CI) state $\Phi = \sum_K c_K D_K$ is obtained by minimizing the system's energy [10]. The total energy E can be calculated from a Hamiltonian given by a sum over a one-electron and a two-electron operator, as

$$E = \left\langle \Phi \mid \left(\hat{H}^{(1)} + \hat{H}^{(2)} \right) \mid \Phi \right\rangle$$

The expectation value involves integrals with one-electron operators between two atomic orbitals or two-electron operators with four atomic orbitals, with up to three centers for this system. The coefficients c_K of the CI expansion, however, depend on all atom locations. Collecting energy terms dependent on one, two, and three nuclear positions, one can write

$$E = E_A + E_B + E_C + E_{AB} + E_{BC} + E_{CA} + E_{ABC}$$

$$V = E - (E_A + E_B + E_C)$$

with one-center $E_J\left(\vec{R}_j\right)$, two-center $E_{IJ}\left(\vec{R}_i,\vec{R}_j\right)$, and three-center energies varying with nuclear positions. Hence, the potential energy V for three atoms can be written as a sum of two-center and three-center terms, $V = V^{(2)} + V^{(3)}$, each a sum of pairs and triplets.

Similarly for N atoms we have $V_N = V^{(2)} + V^{(3)} + \ldots + V^{(N)}$, with terms dependent on the positions of the N atoms. Here, it may happen that electrons are localized over few near-neighbors or next-to-near-neighbor centers, in which case many of the multicenter integrals can be neglected, and the CI coefficients and energies can be organized into sums for interacting molecular fragments. A two-fragment interaction term can be written as a sum of short-range and long-range interactions, and similarly for three- and four-fragment terms.

For a system in its electronic ground state, a useful decomposition is, labeling atoms in numerical fashion,

$$V(\{R_{lm}\}) = \sum\nolimits_{l<m} V^{(2)}(R_{lm}) + \sum\nolimits_{l<m<n} V^{(3)}(R_{lm}, R_{mn}, R_{nl}) + \ldots$$

from which a general force field can be generated. The two-atom terms can be parametrized as shown in the next section, while the three atom terms can be added as corrections which are dampened as interatomic distances increase beyond some boundary values R^0_{lm}, and have adjustable parameters at short distances. Labeling the pair $(lm) = \mu = 1, 2, 3, \ldots$, and with $R_{lm} = r_\mu$, one such function is

$$V^{(3)}(r_\mu, r_\nu, r_\pi) = \left\{\prod\nolimits_\mu \left(\frac{1}{2}\right)\left[1 - \tanh\gamma_\mu\left(r_\mu - r^0_\mu\right)\right]\right\}\left[b_0 + \sum\nolimits_\mu b_\mu\left(r_\mu - r^0_\mu\right)\right.$$
$$\left. + \sum\nolimits_{\mu<\nu} b_{\mu\nu}\left(r_\mu - r^0_\mu\right)\left(r_\nu - r^0_\nu\right), +, \ldots\right]$$

where the factor with a tanh goes to zero outside the boundary r^0_μ values, and the coefficients b can be chosen to shape the region within boundaries.

Alternative forms including also four-atom terms can be written in terms of bond distances, bond angles, and dihedral (torsion) angles, as well as distances between nonbonding atoms, and are the basis for treatments of the molecular dynamics of polymers and biomolecules.

4.2 Atom–Atom Potentials

4.2.1 Standard Models and Their Relations

To begin with, we consider the simplest case of two interacting atoms in their ground electronic states, classified by their electronic shell structures. For atoms they are (i) two closed shell atoms with a van der Waals interaction; (ii) two closed shell ions forming an ionic bond; (iii) an open shell ion and one closed shell ion in a transition complex; and (iv) two open shell atoms forming a chemical bond [11]. Potential energy functions $V(R)$ of the interatomic distance R can be given in terms of their values $V = 0$, V_m, V_i, ... at special locations R_0, R_m, R_i, ... at the zero-value or core, minimum, and inflection points, as shown in Figure 4.1, with magnitudes and functional forms dependent on their physical origin.

We deal here first with pairs where one or both atoms have a closed electronic shell and are not charged, so that they do not form a chemical bond, but may form a van der Waals complex with an interaction potential energy $V_{vdW}(R)$. The potential minimum energy $V_m = -\epsilon$ and location R_m are chosen here as units of energy and length, to introduce the reduced variables $x = R/R_m$ and $f(x) = V(R)/\epsilon$. Analytical forms usually employed are the Lennard Jones $(n,6)$, Exp-6(α), also called the Buckingham or Slater potential, and the Morse(a) potential. They all have three parameters, or only one in reduced variables. Their form and range of the remaining parameter are

$$f_{LJ}(x;n) = (n-6)^{-1}(6x^{-n} - nx^{-6}),\ 8 \le n \le 20$$
$$f_{E6}(x;\alpha) = 6\exp[-\alpha(x-1)] - \alpha x^{-6},\ 8 \le \alpha \le 20$$
$$f_M(x;a) = \exp[-2a(x-1)] - 2\exp[-a(x-1)],\ 4 \le a \le 8$$

From these, we can find the curvatures at the minimum to be, respectively,

$$\kappa = \left(\frac{d^2 f}{dx^2}\right)_{x=1} = 6n,\ = 6\alpha\frac{\alpha-7}{\alpha-6},\ = 2a^2$$

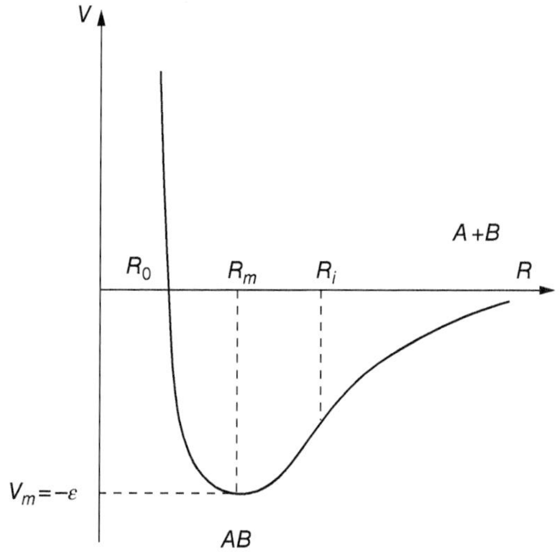

Figure 4.1 Special points R_0, R_m, R_i, in the potential energy function V versus the interatomic distance R between atoms A and B, for the core radius, minimum, and inflection distances.

and the location x_0 of the zero of the potential,

$$x_0 = \left(\frac{6}{n}\right)^{1/(n-6)}, \approx 1 - \frac{\ln(\alpha/6)}{\alpha - 6}, = 1 - \ln(2)/a$$

The location R_i and energy V_i of the inflection point can also be obtained from the root x_i of the second derivative $d^2f/dx^2 = 0$. The Exp-6 and Morse potentials are unphysical at short distances and show an inner inflection point as R decreases. Therefore, these potential functions must be limited to values of R larger than an ion core R_C, and may better be written as functions of $R - R_C$ with $V(R_C)$ kept much larger than relevant energies. Alternatively the Exp-6 potential can be modified multiplying its short range term times a R^{-n} factor with $n > 6$. The Morse potential is frequently used to treat chemical bond interactions and is parametrized with results from spectroscopy. An alternative is the extended Rydberg potential

$$f_{Ryd}(x) = -\left[1 + \sum\nolimits_k a_k (x-1)^k\right] \exp[-a_1(x-1)]$$

with $1 \le k \le 3$, which gives a potential energy $V_{Ryd} = -\epsilon$ at the minimum R_m and a value at the origin of $V_{Ryd}(0) = \epsilon[\sum_k a_k(-1)^{k+1} - 1] \exp(a_1) > 0$.

Other potential functions apply to interactions of atomic and molecular ions. For an interaction between a neutral species and an ionic one, it is necessary to add a charge-induced dipole term going as R^{-4}. The interaction of two closed shell charged ions such as in Na^+Cl^- can be described with the potential, $V_{ion}(R) = A \exp(-\alpha R) - Z_+Z_-e^2R^{-1}$, with $e = c_e/(4\pi\varepsilon_0)^{1/2}$ in SI units, and parameters extracted from rotational and vibrational spectroscopy. More accurately, the potential function must include the van der Waals dispersion term going as R^{-6}, the charge-induced dipole term going as R^{-4}, and the interaction between each distorted closed shell dipole with the other ion going as R^{-7}, as can be also found from third-order perturbation theory.

4.2.2 Combination Rules

These rules allow us to obtain parameters for a heteronuclear diatomic pair AB from the homonuclear pairs AA and BB. For the long range $-C_{AB}R^{-6}$ dispersion energy coefficients, we have [12]

$$C_{AB} = \frac{2C_{AA}C_{BB}}{[(\alpha_B/\alpha_A)C_{AA} + (\alpha_A/\alpha_B)C_{BB}]}$$

where α_A is the static polarizability of atom A. For the short range we can use the Born–Mayer potential

$$V_{AA}(R) = A_{AA}\exp(-b_{AA}R)$$

and the combination rule

$$V_{AB}(R) = [V_{AA}(R)V_{BB}(R)]^{1/2}$$

giving $A_{AB} = (A_{AA}A_{BB})^{1/2}$ and $b_{AB} = (b_{AA} + b_{BB})/2$. The parameter b_{AA} derives from the electronic charge distribution of the outer shell and the size of the atomic core of inner shells; if the former is given (in a hydrogenic description) by the effective nuclear charge ζ_A, it follows that $b_{AA} = b^0_{AA} + c\zeta_A$, with the first term given by the A core radius.

4.2.3 Very Short-Range Potentials

These are determined by the distortion of overlapping electron distributions, which happens due to the "Pauli exclusion" repulsion arising from the requirement that electronic wavefunctions must be antisymmetric. It leads to distortion multipoles which oppose each other and therefore lead to repulsion; their R-dependence relates to the overlap $\Delta^{(A)}$ of charge densities, which introduces a factor $\Delta^{(A)}_{dst}(R) = \Delta^{(A)}(R_{Ac})\exp[-\lambda_A(R - R_{Ac})]$ for distances larger than a core value R_{Ac} in the interaction function. If the highest distortion multipole at each species is an n-pole, then we have an interaction for AA of the form

$$V_{AA}(R) = V_{AA}(R_{Ac})\left(\frac{R}{R_{Ac}}\right)^{-N}\exp[-2\lambda_A(R - R_{Ac})]$$

with $N = 2n + 1$. More elaborate models for very short distances have also been introduced [13]. Interactions $V_{AB}(R)$ for heteronuclear systems follow from combination rules there.

4.2.4 Local Parametrization of Potentials

Local parametrizations can be done turning on and off potentials in certain regions, or constructing them in piecewise fashion. Local parametrizations are useful when semiempirical information is introduced for certain ranges of distances.

Short-range damping functions $g_s(R)$ are constructed to change around a damping location R_d and go from 1.0 at short distances to zero at infinite values of R, within a range R_d/λ. For example,

$$g_s(R; R_d, \lambda) = [1 - \exp(-\lambda)]\left[1 - \exp\left(-\lambda\frac{R_d - R}{R_d}\right)\right]^{-1}$$

can be used to select the short-range potential using $R_d = R_0$ and to dampen the long range past that point, while its complement $\bar{g}_s = 1 - g_s$ dampens the potential inside its inflection point using $R_d = R_i$. Choosing $R_d = R_0$, and $\lambda \ll R_0$ the decomposition

$$V(R) = g_s(R)V(R) + \bar{g}_s(R)V(R) = V^{(S)}(R) + V^{(L)}(R)$$

separates short-range and long-range components, with the second term usually small at all distances and suitable for perturbative treatments of thermodynamical and transport properties. This separation is shown in Figure 4.2.

Another use for these functions is in the switching from ionic to covalent potential energy functions as distance increases, with a crossing at a location $R_d = R_x$ for example in alkali–halide interactions, where one can use

$$V(R) = V_{ion}(R)g_s(R) + V_{cov}(R)\bar{g}_s(R)$$

The damping functions, however, have large derivatives around R_d that can affect the calculation of forces in studies of molecular dynamics, and their change with R must be locally balanced in the transition region before derivatives with respect to position variables are taken, to avoid artificial forces. Furthermore, if the damping function is combined with a long-range interaction going as R^{-n}, then care must be taken to compensate for, or exclude, the divergence of the interaction as $R \rightarrow 0$. This can be done, for example using instead

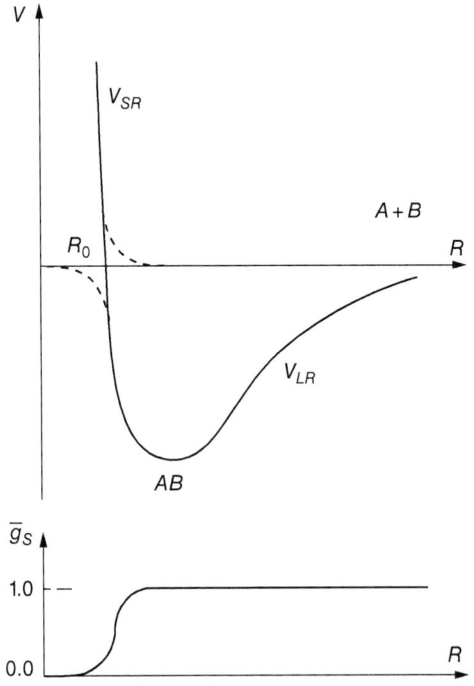

Figure 4.2 Separation of a short-range (or *SR*) potential energy component from intermediate and long-range (or *LR*) components, using a switching (or long-range damping) function $g_s(R)$ and its complement $\bar{g}_s(R)$.

$(R + \delta R)^{-n}$ with a small δR parameter, or multiplying the long-range interaction times a new damping function $f_d^{(n)}(R) = \bar{g}_s(R)$ given by [14]

$$f_d^{(n)}(R; a_n, b_n) = \left[1 - \exp\left(-a_n R - b_n R^2\right)\right]^n$$

which leads to a constant value at the origin for the potential energy containing R^{-n}, and has two parameters as needed to fix the transition location and its width. An alternative proposed form contains only the Born–Mayer parameter b and can be used if the origin is excluded [15].

Piecewise potentials can be put together from exponentials, Morse, and van der Waals potentials at short, intermediate, and long-range distances, respectively, and joining them with spline functions chosen to assure a smooth function and smooth derivative; this is the so-called Exponential–Spline–Morse–Spline–van der Waals (ESMSV) potential [16, 17]. The spline function between points x_j and x_{j+1} is

$$S(x) = \sum\nolimits_{k=0}^{3} a_k \left(\frac{x - x_j}{x_{j+1} - x_j}\right)^k$$

introduced with its bounds chosen to bracket the zero and inflection points and can be used to construct very accurate potential functions and forces over a wide range of distances. The four parameters are fixed by imposing continuity of the function and its derivative at the two ends of its range. The spline can also be used to smoothly join avoided crossing potentials, for example from an outer covalent potential to an inner ionic one as in NaCl.

4.3 Atom–Molecule and Molecule–Molecule Potentials

4.3.1 Dependences on Orientation Angles

For an atom A interacting with a diatomic $M = BC$ in a CMN-BF frame with its z-axis along the two centers of mass, let $\vec{R}$ be the position of A with respect to the CMN of BC, $\vec{r}$ the position vector from B to C, and θ the angle between the two directions, as shown in Figure 4.3, also for given interatomic distances R_{ac} and R_{ab} and a chosen angle at A. For a nondegenerate state, the potential energy is a unique function of only the three variables R, r, θ (after eliminating the rigid body three translational and three orientational degrees of freedom, a total of six variables).

The angular dependence can be treated differently depending on whether the interaction anisotropy is weak or strong. In the first case, it is expanded in Legendre polynomials $P_l(\cos\theta)$, and the coefficient functions for each term

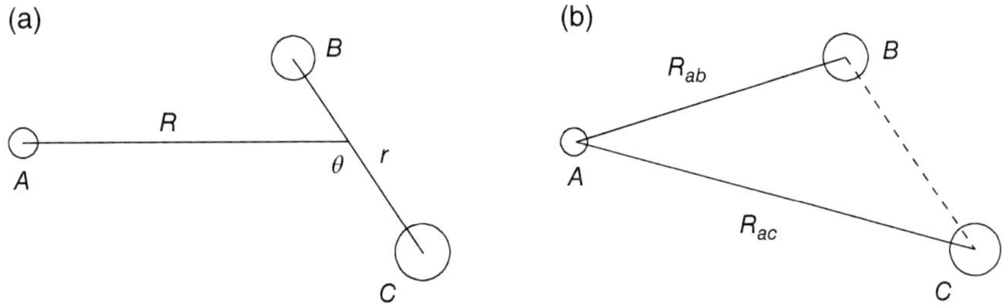

Figure 4.3 (a) Center-of-mass distance variables in the triatomic $A + BC$. (b) Variables for a description in terms of atom-pair potential energies, for a chosen distance between B and C.

are parametrized as given functions of R, and r, with parameters expanded around the equilibrium bond value of r,

$$V(R,r,\theta) = \sum\nolimits_l V_l(R,r)P_l(\cos\theta)$$

and the expansion functions can be further decomposed into short-range and long-range terms as

$$V_l(R,r) = V_l^{(S)}(R,r) + V_l^{(L)}(R,r)$$
$$V_l^{(S)}(R,r) = A_l(r)\exp[-b_l(r)R]$$
$$V_l^{(L)}(R,r) = v_{BC}(r)\delta_{l0} + \sum\nolimits_n C_{l,n}(r)f_d^{(n)}(R)R^{-n}$$

Only a few Legendre polynomials are needed for weak anisotropy and, if M is a homonuclear BB molecule, symmetry eliminates all odd polynomials from the expansion. The A, b, and C parameters can be expanded in powers of the displacement $\rho = r - r_e$ from the equilibrium bond length r_e. For each angle, it is possible to show potential energies as isocontours of V versus R and r. Isocontours of potential energy surfaces describing the atomic interactions are shown in Figure 4.4. The top panel is the same as Figure 1.5 in Chapter 1: Part (a), including also the internal diatomic potential, v_{BC} (r) which must be added to the long-range interaction for $l = 0$ (Figure 4.4). Part (b) shows isocontours for a reaction $A + BC \rightarrow AB + C$ with an activation barrier, such as $H + H_2 \rightarrow H_2 + H$, while Part (c) shows isocontours for a stable intermediate ABC such as H_2O with a fixed bond angle, and varying atom–atom distances R_{ab} and R_{bc}. The bottom panel is the same as Figure 1.6, and shows the potential energy of H approaching CN in HCN, in a (x,y) plane with the coordinate origin at the midpoint between C and N.

An alternative parametrization avoids the expansion in polynomials and instead introduces parameters dependent on r and θ, for each value of the latter.

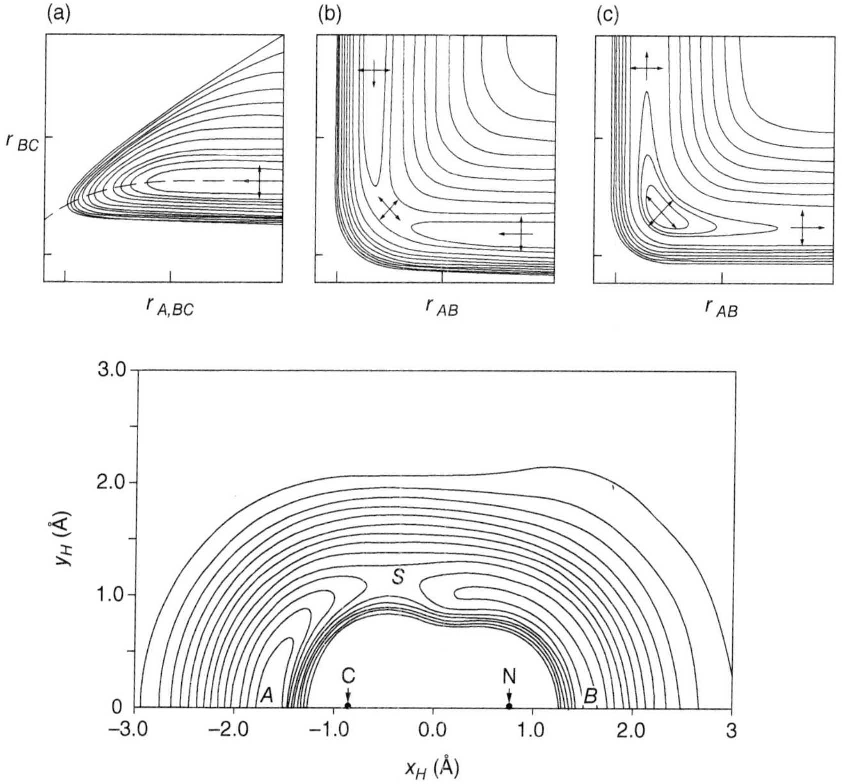

Figure 4.4 Isocontours for three-atom systems. Top panel: generic isocontours for (a) an inert gas A interacting with a diatomic BC, (b) the reaction $A + BC \rightarrow AB + C$ with an activation barrier, and (c) a stable intermediate ABC with a fixed bond angle, from [18]. Bottom panel: potential energy of H approaching CN in HCN, in a (x,y) plane with the coordinate origin at the midpoint between C and N, from [19]. The locations of minima are shown as A and B, for the two isomers, and a saddle point is shown as S.

For example the three atoms may be on a line or in a "T" conformation, with different parameters. Then we have

$$V(R,r,\theta) = V^{(S)}(R,r,\theta) + V^{(L)}(R,r,\theta)$$
$$V^{(S)}(R,r,\theta) = A(r,\theta)\exp[-b(r,\theta)R]$$
$$V^{(L)}(R,r,\theta) = v_{BC}(r) + \sum_n C_n(r,\theta) f_d^{(n)}(R) R^{-n}$$

This may be described as a method of variable parameters, where the atom–atom functional forms are used, but the parameters are allowed to change with internal coordinates and orientation of the molecules.

The case of a linear system with $\theta = 0$ is frequently described with a model in which the repulsion depends only on the distance from A to the nearest (B) atom, so that $V^{(S)}(R, r) = A\exp[-b(R-\gamma r)]$, $\gamma = m_C/(m_B + m_C)$, in terms of atomic masses. More generally, parameters can be expanded in Legendre polynomials $P_l(\cos\theta)$ to account for angle variations. For example results of extensive electronic configuration interaction calculations for $He + H_2$ have been used to parametrize its interaction as

$$V^{(S)}(R,r,\theta) = C\exp(-\alpha_0 R + \alpha_0 \rho R)[A(\theta) + B(\theta)\rho]$$

$$V^{(L)}(R,r,\theta) = -R^{-6}C_6(\theta)f_d(R)$$

with $\rho = r - r_e$ the bond length displacement from equilibrium, the expansions $A(\theta) = A^{(0)} + A^{(2)}P_2(\cos\theta)$, $B(\theta) = B^{(0)} + B^{(2)}P_2(\cos\theta)$, and $C_6(\theta) = C^{(0)} + C^{(2)}P_2(\cos\theta)$, and $f_d(R)$ damping the long range potential [20].

For diatom–diatom interactions and interactions involving polyatomic molecules we must specify the orientation of each species in a common reference frame, and either expand the interaction energy in a basis set of Wigner rotational functions [21], or let parameters depend on orientation angles. The interaction potential for a molecule A with internal degrees of freedom $\boldsymbol{Q}_A$, for example normal mode coordinates, and orientation given by the local $(\mathrm{CMN})_A$ Euler angles $\Gamma_A = (\alpha, \beta, \gamma)_A$ interacting with molecule B in the joint CMN-BF frame with the z-axis along $\vec{R}$, is a function $V^{(AB)}(R, \boldsymbol{Q}_A, \Gamma_A, \boldsymbol{Q}_B, \Gamma_B)$. This function can be expanded in a basis set of rotational functions of angles or it can be considered as a functional of parameters which depend on orientation angles.

Alternatively, the molecule–molecule interaction in a CMN-BF frame can be written as a function of the locations $\vec{R}_a$ of distributed charges c_a for molecule A and similarly for B,

$$V_{AB} = V_{AB}^{(SR)}(R) + \sum\nolimits_{a,b} \frac{c_a c_b}{4\pi\varepsilon_0 R_{ab}}$$

with the first term giving the short-range repulsive interaction between the CMN of A and B, and with the summation containing locations and values of charges chosen to reproduce molecular charge densities and their interaction and suitable only for large distances R_{ab}. This information can be extracted from an electronic structure calculation of the AB pair, which would provide a charge and bond order matrix from which charges and locations can be defined [22].

The set of charges at each molecule can alternatively be replaced by a set of distributed multipoles, which then give the total potential energy in terms of multipole–multipole interactions [5]. Figure 4.5 shows the TIP5P model for water [23], with four charges $a = 1$–4 representing hydrogens with positive point

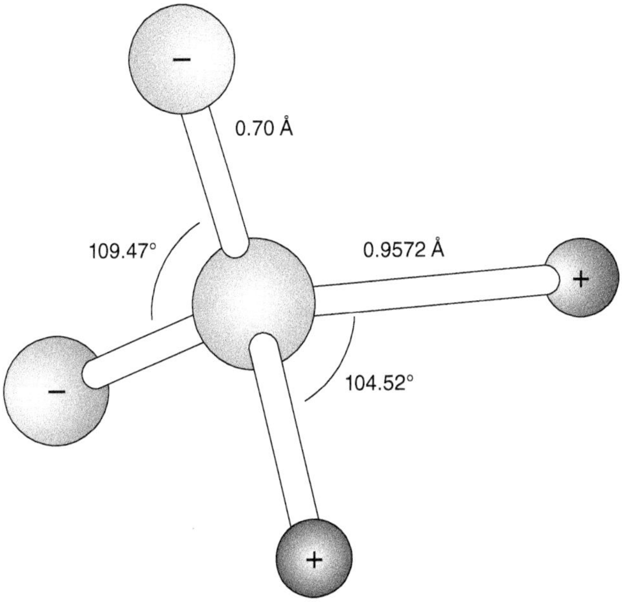

Figure 4.5 The H_2O monomer TIP5P charge model used for H_2O–H_2O interactions.

charges of 0.241e and lone pairs with balancing negative values, in addition to the oxygen atom which interacts with other oxygen atoms only through a pair potential.

The water–water interaction can then be constructed from an LJ(12,6) potential energy function between the two oxygen atoms with parameters $\sigma_0 = 3.12A$ for its core radius and $\varepsilon_{OO} = 0.16$ kcal mol^{-1} for its well depth, plus point charge interactions as shown in Figure 4.5 for the hydrogen atoms and lone pairs, and is of the form

$$V_{WW} = 4\varepsilon_{OO}\left[\left(\frac{\sigma_0}{R_{OO}}\right)^{12} - \left(\frac{\sigma_0}{R_{OO}}\right)^{6}\right] + \sum_{a,b}\frac{c_a c_b}{4\pi\varepsilon_0 R_{ab}}$$

which leads to a physically acceptable description of the local tetrahedral structure of liquid water, and other properties. This function is suitable only for large distances R_{ab} but can be modified introducing damping factors for the Coulomb a,b interactions, or adding bond dipoles and quadrupoles, to improve properties.

4.3.2 Potentials as Functionals of Variable Parameters

For an atom-diatomic A–D system, we can start from the atom–atom $V(R)$ potential expression with parameters ε, R_m, γ and make these dependent on internal coordinates and relative orientation,

$$V_{AD}(R,r,\theta) = V[R;\varepsilon(r,\theta),R_m(r,\theta),\gamma(r,\theta)]$$

with the expansion

$$\varepsilon(r,\theta) = \sum_l \varepsilon_l(r)P_l(\cos\theta)$$

and similarly for the other parameters for small anisotropies. For example with the Exp-6(α) parametrization, the short-range exponential parameter α can be made to vary with the BC distance and its orientation relative to A, and can be expanded in Legendre polynomials if the anisotropy is small. For interacting molecules A and B, this generalizes to $V_{AB}(R,\mathbf{Q}_A,\Gamma_A,\mathbf{Q}_B,\Gamma_B) = V[R;\ \varepsilon(\mathbf{Q}_A,\Gamma_A,\mathbf{Q}_B,\Gamma_B),\ R_m(\mathbf{Q}_A,\Gamma_A,\mathbf{Q}_B,\Gamma_B),\ \gamma(\mathbf{Q}_A,\Gamma_A,\mathbf{Q}_B,\Gamma_B)]$.

4.3.3 Hydrogen Bonding

Attraction between two species A and B mediated by a hydrogen atom plays an important role in aqueous solutions and in binding of nucleic acids such as DNA and RNA. A simple description involves covalent bonding to one of the species (like A) and noncovalent bonding to the other (as B), in a A—$H\cdots B$ structure. Given the electronic density distribution of this system, it is possible to construct models of its energetics based on distributed multipoles and polarizabilities, such as the interaction dipoles of the A—H bond and of the lone electron pair in :B, plus repulsion energies at short distances [5]. More accurately, a many-electron treatment mixes its electronic states in a quantum superposition of the states of A—$H\cdots B$ and $A\cdots H$—B. The potential for H moving between A and B may show a double well or a wide shallow well, and depends on the bond distances R_{AH}, R_{HB} and the bond angle α_{AHB}. For H covalently bonded to A, its interaction with B in A—$H\cdots B$ can be parametrized as the function of the distance between H and the nonbonding B

$$V_{AHB} = \varepsilon_{HB}\left[5\left(\frac{R_{HB}^{(m)}}{R_{HB}}\right)^{12} - 6\left(\frac{R_{HB}^{(m)}}{R_{HB}}\right)^{10}\right] f_{ang}(\cos\alpha_{AHB})$$

The overall attraction between organic molecules A and B mediated by H is of the order 10–25 kJ mol^{-1}.

4.3.4 Systems with Additive Anisotropic Pair-Interactions

It is sometimes accurate to use sums of atom-pair potentials, cutting off the long-range potentials at short distances to avoid over-binding there. For $A + BC$, we add model pair potentials (A,I) over each atom $I = B,C$ interacting with A, but let the potential depend on the orientation ω_i of the BC bond relative to the direction of AI, a function of θ in Figure 4.3 (a),

$$V\left(\vec{R}_a,\vec{R}_b,\vec{R}_c\right) = \sum\nolimits_{i=b,c} v_{AI}(R_{ai},\omega_i)$$

$$v_{AI}(R_{ai},\omega_i) = v_{AI}^{(S)}(R_{ai},\omega_i) + v_{AI}^{(L)}(R_{ai},\omega_i)f_{d,i}(R_{ai})$$

where $R_{ai} = |\vec{R}_a - \vec{R}_i|$, and the long-range term has been dampened at short interatomic distances.

4.3.5 Bond Rearrangements

In situations where bonds can be broken or formed, it is possible to obtain PESs from a Valence-Bond (VB) treatment starting from many-electron functions constructed with one orbital per active electron in the rearranging bonds. For a diatomic AB, such as H_2, with two active electrons in atomic orbitals $\chi_a(1)$ and $\chi_b(2)$ centered at the A and B nuclei, and a diatomic Hamiltonian written as $\hat{H}_{AB} = \hat{H}_A + \hat{H}_B + \hat{H}'_{AB}$, the interaction potential energy for singlet and triplet states ${}^{1,3}\Sigma$ can be written as [22]

$$V_{AB}\left({}^{1,3}\Sigma\right) = \frac{(Q_{AB} \pm A_{AB})}{(1 \pm \Delta_{AB})}$$

$$Q_{AB} = \int d(1)d(2)\chi_a(1)^*\chi_b(2)^*\hat{H}'_{AB}\chi_a(1)\chi_b(2)$$

$$A_{AB} = \int d(1)d(2)\chi_a(2)^*\chi_b(1)^*\hat{H}'_{AB}\chi_a(1)\chi_b(2)$$

where the + sign corresponds to the singlet, and the *VB* Coulomb integral Q_{AB} and exchange integral A_{AB} are shown in terms of the atomic orbitals while $\Delta_{AB} = S_{ab}^2$ is the square of the integral of the orbitals overlap. Reverse relations give expressions for the integrals in terms of the singlet and triplet energies, which are known from accurate calculations or from spectroscopic measurements. The singlet and triplet energy functions can be parametrized as Morse $f_M(x;a) = \exp[-2a(x-1)] - 2\exp[-a(x-1)]$ and anti-Morse $f_{AM}(x;b) = \exp[-2b(x-1)] + 2\exp[-b(x-1)]$ potential functions respectively, with adjustable parameters. The overlap integral follows from the form of atomic orbitals, which for $1s$ orbitals of the form $\chi_a(\vec{r}) = \left(\zeta_a^3/\pi a_0^3\right)^{1/2}\exp\left(-\zeta_a\left|\vec{r}-\vec{R}_a\right|/a_0\right)$ gives $S_{ab}(R_{ab}) = [1 + (\eta R_{ab}/a_0) + ((\eta R_{ab}/a_0)^2/3)]\exp(-\eta R_{ab}/a_0)$ and can be used as an adjustable function of the parameter η for each atom pair, to shape the PES.

For the three-atom system ABC with three active electrons described with three atomic orbitals, a London-like interaction potential energy function suitable for treatments of reactive atom-diatom systems undergoing a rearrangement such as $A + BC \rightarrow AB + C$ can be constructed from a generalized *VB* description [22, 24]. Two three-electron *VB* wavefunctions ${}^2\Phi_I$ and ${}^2\Phi_{II}$ can be formed to describe the doublet states corresponding to two independent bondings $A\text{—}B\cdots C$ and $A\cdots B\text{—}C$. The lowest PES is obtained from the Hamiltonian eigenvalues in this two-state description and can be given a simple form provided orbital overlaps are neglected. It is however made more flexible, so it can be applied to a variety of triatomics, by reintroducing overlap terms parametrized to reproduce desired features such as the location of a surface saddle point at an energy barrier for reaction, in which case it is the so-called London–Eyring–Polanyi–Sato (LEPS) function [3, 25, 26]

$$V_{ABC} = (1 + \Delta_{ABC})^{-1}\Big\{Q_{AB} + Q_{BC} + Q_{CA} - 2^{-1/2}\big[(A_{AB} - A_{BC})^2 + (A_{BC} - A_{CA})^2 + (A_{CA} - A_{AB})^2\big]^{1/2}\Big\}$$

and the integrals can be obtained from known singlet and triplet potential energies for each pair, while the triatomic overlap Δ_{ABC} can be used as an adjustable parameter to obtain a desired activation barrier in a rearrangement $A + BC \rightarrow AB + C$. In the Polanyi version [27], the term Δ_{ABC} is omitted and replaced with three distance dependent parameters $\Delta_{IJ} = S_{IJ}^2$ introduced in the expressions for the singlet and triplet energies of each IJ pair, from which the *VB* Coulomb and exchange integrals have been obtained, to allow for desired energy barrier and added attraction or repulsion between reactants or products. This gives

$$Q_{IJ} = \frac{[{}^1V_{IJ} + {}^3V_{IJ} + \Delta_{IJ}({}^1V_{IJ} - {}^3V_{IJ})]}{2}$$

in terms of singlet and triplet potential energy functions of R and an overlap integral, and with a similar expression for A_{IJ} replacing ${}^3V_{IJ}$ with $-{}^3V_{IJ}$. A generic isocontour for a reaction with an activation barrier is found in the top Figure 4.4a, while Figure 4.6 shows the case of the collinear $F + H_2 \rightarrow FH + H$ reaction with an activation barrier in the entrance channel. An analogous PES expression can be written for systems with four active electrons [22, 26] such as $H_2 + D_2$.

Spin-dependent pair potentials may instead be used to write the interaction of open shell atoms in terms of spin operators, which can be parametrized using pair interaction functions for specific multiplet states. These potentials can be derived from electronic structure states in a *VB* formulation [30]. For three monovalent atoms the pair potential operators are $\hat{V}_{IK}(R_{IJ}) = V_{IJ}^{(c)}(R_{IJ}) + V_{IJ}^{(s)}(R_{IJ})\hat{\vec{S}}_I.\hat{\vec{S}}_J$ where $\hat{\vec{S}}_I$ is the spin operator of atom I with two spin states. For three monovalent atoms, the spin operator $\hat{V}_{AB} + \hat{V}_{BC} + \hat{V}_{CA}$ can be diagonalized in a basis set of two doublet three-atom spin states to recover the LEPS expression.

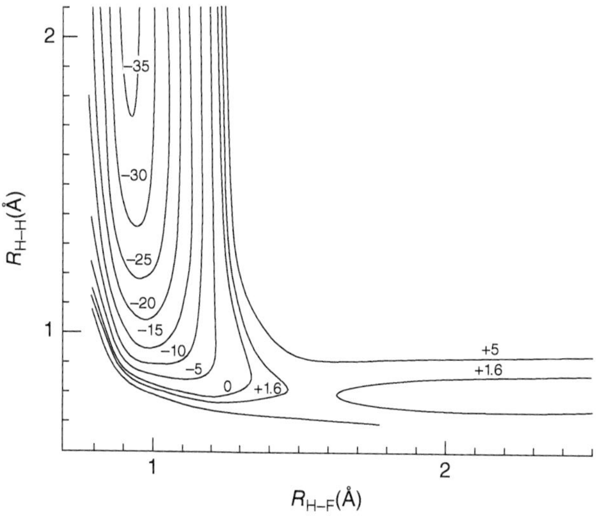

Figure 4.6 Energy isocontours for collinear FH_2 versus atom–atom distances, showing an early activation barrier as F approaches H_2. *Source:* from Bender et al. [28], almost identical to the figure from Bender et al. [29].

4.4 Interactions in Extended (Many-Atom) Systems

4.4.1 Interaction Energies in Crystals

We consider to begin with atoms or molecules forming a lattice with nuclear positions at rest, and surrounded by electrons, and a solid with a number density ρ_{sld}. The lattice energy is then a potential energy $U_{sld}^{(pot)}(\rho_{sld})$ for the system at a temperature $T = 0$ K. As the temperature is increased, one must add to this the kinetic energy of nuclear motion and the excitation energy of thermal electrons, $U_{sld}^{(kin)}(\rho_{sld},\beta)$ to obtain the total thermodynamical internal energy $U_{sld}(\rho_{sld}, \beta)$, with $\beta = 1/(k_B T)$ the inverse temperature in units of the Boltzmann constant k_B, that gives the cohesive energy at the given density and temperature.

In crystals made up of inert gas atoms, such as Ne(s), or of ions, such as NaCl(s), a good description of their crystal structure and cohesive energy can be extracted from models containing only the atom–atom or ion–ion interactions.

Crystals with covalent bonding, such as Si(s), must be treated as many-electron systems with electronic densities localized at ion cores and along chemical bonds.

In metals, such as Cu(s), some electrons are localized into ion cores and others are delocalized over the whole crystal with the exclusion of the ion cores, and treatments can be based on the concept of embedded atoms [31]. This is also helpful in the treatment of unsaturated hydrocarbon chains with delocalized electrons.

The cohesive energy of a molecular crystal with N_{at} atoms at rest in a sample volume Ω, such as crystals composed of identical noble gas atoms, can be written as a sum of pair potentials $v(R_{ij})$ between atoms i and j in a lattice, with suitable parameters which must depend on the atomic density $\rho_{sld} = N_{at}/\Omega$. The solid potential energy is then approximated by

$$V_{sld} = \frac{1}{2}\sum\nolimits_{i \neq j} v(R_{ij})$$

with pair potential parameters $v_m(\rho_{sld})$ for the well depth, and $l(\rho_{sld})$ for the nearest neighbors distance, at the minimum of energy between them, as functions of the solid density. It can be conveniently re-expressed in terms of a relative atom–atom distance $p_{ij} = R_{ij}/l$, so that the lattice sum for a long range R^{-n} interaction would be $(1/2)\sum_{i \neq j}(R_{ij})^{-n} = l^{-n}\alpha_n$, with a lattice sum $\alpha_n = (1/2)\sum_{i \neq j}(p_{ij})^{-n}$ characteristic of the lattice symmetry and independent of density. A starting point for calculation of the cohesive energy is a sum of pair potentials for the isolated pair, with $\rho_{sld} = 0$.

For the LJ(12,6) pair potential written in terms of its energy minimum $|v_m|$ and the location R_0 of its zero energy value, and with a close packed (*fcc*) lattice structure at equilibrium, the potential energy for identical noble gas atoms is

$$V_{latt} = \frac{1}{2} N_{at} 4 \, | \, v_m \, | \left[\left(\frac{R_0}{l} \right)^{12} \alpha_{12}(fcc) - \left(\frac{R_0}{l} \right)^{6} \alpha_6(fcc) \right]$$

The lattice summations have been evaluated and are $\alpha_{12}(fcc) = 12.13188$ and $\alpha_6(fcc) = 14.45392$, not far from the number 12 of nearest neighbors. The equilibrium distance l_{eq} is obtained setting the force $dV_{latt}/dl = 0$ to get $l_{eq}/R_0 = 1.09$, indicating that the internuclear distance has increased going from an isolated to an embedded atom pair. The measured values for Ne, Ar, Kr, and Xe are 1.14, 1.11, 1.10, and 1.09, in quite good agreement. The remaining discrepancy can be corrected by including in the cohesive energy the zero-point energy of vibrational motion of each atom as it oscillates around its equilibrium position [31].

Ionic crystals, such as NaCl(s) with the *fcc* lattice symmetry, contain short-range repulsions and long-range Coulomb electrostatic attractions. They are treated introducing a diatomic ion, like Na^+Cl^-, in a unit cell of a lattice, and adding over atom pair interactions. The solid cohesive energy is

$$V_{latt} = N_{diat} \left[v^{(SR)}(l) - \frac{Z_+ Z_- \, c_e^2}{4\pi\varepsilon_0\varepsilon_r l} \alpha_1(latt) \right]$$

where l is the distance between adjacent positive and negative ions, ε_r is the dielectric constant of the medium, the charge numbers Z are positive, and $\alpha_1 = \alpha_M$ is the Madelung lattice constant. It is obtained from two interpenetrating *fcc* lattices for Na^+ and Cl^-. Taking one of the negative ions as a reference center in the solid, one can write the constant as a sum of terms for its interaction with positive and negative ions, and the sum can be done partitioning the lattice around the reference ion into nearly neutral groups of ionic charges in cubes of increasing size until convergence. The result for the *fcc* lattice is 1.747558. For comparison, the value for the *bcc* lattice, such as in the CsCl(s) crystal, is 1.762670 [31]. The short-range term can be written as a Born–Mayer exponential repulsion $A\,\exp(-\alpha\, l)$ and the equilibrium value l_{eq} of the interionic distance can be found from the location of the zero-value force. Replacing this into the potential energy one finds the Born–Mayer equilibrium form

$$V_{latt} = -N_{diat}\frac{Z_+ Z_- c_e^2}{4\pi\varepsilon_0\varepsilon_r l_{eq}}\left(1-\frac{1}{\alpha l_{eq}}\right)\alpha_M(latt)$$

This shows that cohesion increases for more highly charged ions and for more steeply repulsive short-range interactions.

More accurately, the cohesive energy must also contain three- and many-atom terms such as $\sum_{l<m<n} V^{(LR)}(R_{lm}, R_{mn}, R_{nl})$, to account for nonadditivity of long-range interactions. The many-atom effects can be included in a variety of ways which depend on the nature of the structure. For solid crystalline Si, a useful functional form has been given by Tersoff [32], which incorporates a many-atom factor in the long-range interaction. The total energy can be written as

$$V(\{R_{ab}\}) = \sum_{l<m} f_c(R_{lm})[A\exp(-\lambda_1 R_{lm}) - B(\{R_{ab}\})\exp(-\lambda_2 R_{lm})]$$

The first term in the square parenthesis describes the repulsion energy between atoms l and m, and the second term gives a modified attraction energy where the function $B(\{R_{ab}\})$ depends on all the atom–atom distances needed to describe the environment of the (lm) pair. This attraction factor is constructed as a monotonically decreasing function of the number of bonds of l and m with other atoms. It also depends on the strength of the bonds to those two atoms competing with the (lm) bond, and on the angles between the (lm) bond and the other bonds. The cut-off function $f_c(R)$ is chosen to equal 1.0 at short distances and to smoothly fall to 0.0 at large distances. Results for various phases of solid structures can be fitted with suitable parameters this way.

The zero-point vibrational motion energy of the lattice atoms must be added even at the temperature of 0 K. If the temperature is increased, then parameters like the atom–atom distance in the expressions for V_{latt} depend on the density

ρ_{sld} and temperature in β, and become thermal potential energies $U_{latt}^{(pot)}$ to which one must add the thermal kinetic energy of vibrating atoms in a lattice, to obtain the internal energy U_{latt}, or cohesive energy, of the solid.

The solid dielectric properties alter the strength of the electrostatic and polarization interactions, decreasing them by the magnitude of the dielectric constant ε_r of the medium and affecting the values of the V_{latt} potential energy. A general treatment of dielectric effects can be based on the dynamical susceptibility of the solid, by considering the response of the whole system to an applied external electric field $\vec{\mathcal{E}}_{ext}$ [31]. The local electric field $\vec{\mathcal{E}}_{lcl} = \varepsilon_r \varepsilon_0 \vec{\mathcal{E}}_{ext}$ at the position of an atom in the solid can be constructed adding fields coming from the surfaces of the solid with an inner idealized cavity, plus the field of the atoms inside it, as

$$\vec{\mathcal{E}}_{lcl} = \vec{\mathcal{E}}_{ext} + \vec{\mathcal{E}}_{out} + \vec{\mathcal{E}}_{inn} + \vec{\mathcal{E}}_{atm}$$

where $\vec{\mathcal{E}}_{out}$ is the depolarization field from charges at the outer surface of the solid, $\vec{\mathcal{E}}_{inn}$ is the field from the charges at the inner surface of the cavity, and $\vec{\mathcal{E}}_{atm}$ is created by all the polarized atoms in the cavity. The sum $\vec{\mathcal{E}}_{ext} + \vec{\mathcal{E}}_{out} = \vec{\mathcal{E}}_{appl}$, also called the Maxwell field, depends on the shape of the outer surface and gives the total applied field. The field from the inner surface charges is the so-called Lorentz field which for a spherical surface can be obtained from the polarization vector $\vec{P} = \sum_j \rho_j \vec{d}_j = \sum_j \rho_j \alpha_j \vec{\mathcal{E}}_{lcl}$ of the sample, where ρ_j is the number of atoms of type j per unit volume, with atomic induced dipole $\vec{d}_j$ and polarizability α_j, and equals $\vec{\mathcal{E}}_{inn} = \vec{P}/(3\varepsilon_0)$ [31]. The last term $\vec{\mathcal{E}}_{atm}$ depends on the distribution of atoms in the cavity and can be seen by symmetry that it equals zero for a cubic lattice. This gives $\vec{\mathcal{E}}_{lcl} = \vec{\mathcal{E}}_{appl} + \vec{P}/(3\varepsilon_0)$ in SI units, with the dielectric susceptibility $\chi = \varepsilon_r - 1$ introduced by $\vec{P} = \varepsilon_0 \chi \vec{\mathcal{E}}_{appl}$. This assumes an isotropic medium or a cubic crystal structure. More accurately, one must use a susceptibility tensor $\chi_{\xi\eta}$ and extend the treatment using Cartesian components in $P_\xi = \varepsilon_0 \sum_\eta \chi_{\xi\eta} \mathcal{E}_{appl,\eta}$. Considering a single component and isotropy, with $P = \left(\sum_j \rho_j \alpha_j\right) \left[\mathcal{E}_{appl} + P/(3\varepsilon_0)\right]$, and solving for $\chi = P/\left(\varepsilon_0 \mathcal{E}_{appl}\right)$ one obtains the relation [31]

$$\frac{\varepsilon_r - 1}{\varepsilon_r + 2} = \frac{1}{3\varepsilon_0} \sum_j \rho_j \alpha_j$$

which provides the dielectric constant needed in the solid potential energy V_{latt}.

4.4.2 Interaction Energies in Liquids

The structure of a fluid made up of atoms or molecules can be described in terms of a probability distribution of its component particles, which depends on the average density ρ and temperature in β. To simplify, we first consider a liquid of N atoms and a probability $P^{(N)}\left(\vec{R}_1,\ldots,\vec{R}_N\right)d^3R_1\ldots d^3R_N$ for finding each atom j (=1 to N) in volume element d^3R_j. This allows us to define a reduced density for a set of n atoms irrespective of their label, as [33]

$$\rho^{(n)}\left(\vec{R}_1,\ldots,\vec{R}_N\right) = [N!/(N-n)!]P^{(n)}\left(\vec{R}_1,\ldots,\vec{R}_N\right)$$

where the factor to the right comes from integration of the whole probability over the position variables of N–n particles. In particular, when n = 1 we have for liquid volume Ω_{liq} that $\Omega_{liq}^{-1}\int d^3R_1\rho^{(1)}\left(\vec{R}_1\right) = N/\Omega_{liq} = \rho$, while $\rho^{(2)}\left(\vec{R}_1,\vec{R}_2\right)$ describes the distribution of atom pairs and $\rho^{(3)}$ gives the triplets distribution. Liquid structure correlation functions $g^{(n)}\left(\vec{R}_1,\ldots,\vec{R}_N\right)$ are defined by $\rho^{(n)} = \rho^n g^{(n)}$. In an isotropic liquid, these functions are invariant under reference frame rotations and depend only on relative distances, so that, for example the relative distribution of atoms at a distance $R = \left|\vec{R}_1 - \vec{R}_2\right|$ around a fixed one at position $\vec{R}_1 = \vec{0}$ averaged over orientations is $\bar{\rho}^{(2)}(R)/\rho = \rho\bar{g}^{(2)}(R)$, which leads to $g^{(2)}(\infty) \approx 1$.

Interaction energies in liquids and in liquid solutions can be obtained from thermal averages and are constructed from statistical pair distribution functions $\bar{g}^{(2)} = g(R;\rho,\beta)$, dependent on the liquid number density ρ and an inverse temperature $\beta = 1/(k_BT)$, with $\rho g(R;\rho,\beta)R^2dR$ counting the number of atoms found in a spherical shell of width dR. The simplest dependence, only on R, is for closed shell atoms, and this must be generalized to include variables for orientation angles of molecules, or for anisotropies of open shell atoms in covalent bonding. The function $g(R;\rho,\beta)$ can be measured with X-ray diffraction methods, or calculated using molecular dynamics.

For a closed thermodynamical system described by a statistical canonical ensemble with $N \gg 1$ atoms in a volume Ω, and density $\rho = N/\Omega$, with up to n atoms interacting with a potential energy $V_n\left(\vec{R}_1,\ldots,\vec{R}_N\right)$, the *n-ple* distribution function is given by [33]

$$g^{(n)}\left(\vec{R}_1,\ldots,\vec{R}_N;\rho,\beta\right) = \Omega^N Z_N(\Omega,\beta)^{-1}\int d^3R_{N+1}\ldots d^3R_N \exp\left[-\beta V_n\left(\vec{R}_1,\ldots,\vec{R}_N\right)\right]$$

where $Z_N = \int d^{3N}R \exp\left[-\beta V_n\left(\vec{R}_1,\ldots,\vec{R}_N\right)\right]$ is the configuration integral.

The pair distribution $g(R;\rho,\beta)$ is small for small R where repulsion takes place, followed by a large peak around the distance R_m for the pair potential energy minimum, and shows subsequent minima and maxima corresponding to atomic shells, as distance increases [33]. A pair distribution function is shown in Figure 4.7 obtained from a molecular dynamics simulation with the Lennard–Jones pair potential function with parameters ε and R_0, as a function of the reduced distance R/R_0 and for the reduced temperatures $T^* = k_B T/\varepsilon$ and densities $\rho^* = R_0{}^3\rho$, with R_0 the core radius.

The potential energy of interaction in the simplest case, when the density is low and many-atom effects can be ignored, is given as a sum of pair $v(R)$ potential energies as

$$U_{liq}^{(pot)}(\rho,\beta) = \frac{1}{2}\int d^3R_1 d^3R_2 v(R_{12})\bar{\rho}^{(2)}\left(\vec{R}_1,\vec{R}_2\right) = \frac{N}{2}\rho\int_0^\infty dR\, 4\pi R^2 v(R) g(R;\rho,\beta)$$

To this one must add the thermal kinetic energy $U_{liq}^{(kin)}(\rho,\beta) = 3Nk_BT/2 = 3N/(2\beta)$ to obtain the thermodynamical internal energy

$$U_{liq}(\rho,\beta) = \frac{3N}{2\beta} + \frac{N}{2}\rho\int_0^\infty dR\, 4\pi R^2 v(R) g(R;\rho,\beta)$$

or cohesive energy, at the given density and temperature. Different choices for the atom pair potential have been used in the calculations of correlation functions.

A perturbation theory for calculating $U_{liq}^{(pot)}(\rho,\beta)$ can be based on the assumption of additive pair potentials and separation of short-range and long-range contributions in $v(R) = v^{(S)}(R) + v^{(L)}(R)$, so that $V_N = V_N^{(S)} + V_N^{(L)}$, with the

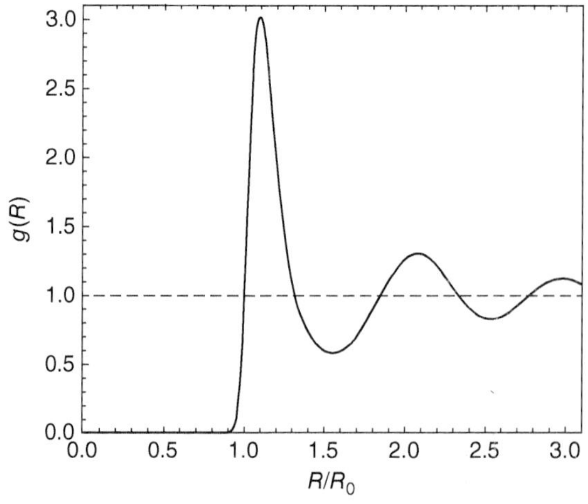

Figure 4.7 Radial pair distribution function $g(R;\rho,\beta)$ for an atomic fluid with a Lennard–Jones potential at reduced temperatures $T^* = k_BT/\varepsilon$ and densities $\rho^* = R_0^3\rho$, with R_0 the core radius, as shown. Calculated for $T^* = 0.71$ and $\rho^* = 0.844$. *Source:* from Wikipedia, 2009, public domain.

short-range term used to construct a reference potential energy $U_{liq}^{(0)}(\rho,\beta) = U_{liq}^{(S)}(\rho,\beta)$, and the long-range part treated as a perturbation to obtain a correction $U'_{liq}(\rho,\beta) = U_{liq}^{(L)}(\rho,\beta)$. The short-range summations over pairs can be usually limited to the average number z_{NN} of nearest neighbors, while the long-range part is treated as a small perturbation [33]. Restricting the nearest neighbors to a narrow distance around the maximum of the pair distribution function, which coincides with the minimum of the pair potential, provides a rough estimate of the short-range potential energy as $U_{liq}^{(S)}(\rho,\beta) = Nv(R_m)z_{NN}/2$.

Contributions from the *LR* interactions can be described by means of the Fourier components of the new function $h(\vec{R}) = g(\vec{R}) - 1$, given by the structure factor

$$h(\vec{k}) = \rho(2\pi)^{-3/2}\int d^3R \exp\left(i\vec{k}.\vec{R}\right)\left[g(\vec{R}) - 1\right]$$

from which the pair distribution function is obtained by the inverse Fourier transform, as

$$g(\vec{R}) = 1 + (2\pi)^{-3/2}\int d^3k \exp\left(-i\vec{k}.\vec{R}\right)h(\vec{k})/\rho$$

Replacing this in the integrals over R for $U_{liq}^{(L)}(\rho,\beta)$ we find the long-range potential energy of the liquid solution expressed in terms of the Fourier transform $\tilde{v}^{(L)}(\vec{k})$ of the long-range part of the pair interaction potentials, and of the structure factor, as

$$U_{liq}^{(L)} = \frac{N}{2}\left[\int d^3k\, \tilde{v}^{(L)}\left(-\vec{k}\right)h(\vec{k}) - \rho\int_0^\infty dR\, 4\pi R^2 v^{(L)}(R)\right]$$

with

$$\tilde{v}^{(L)}(\vec{k}) = (2\pi)^{-3/2}\int d^3R \exp\left(i\vec{k}.\vec{R}\right)v^{(L)}(R)$$

For the present potential and distribution functions dependent on only the radial variable, the transforms are

$$\tilde{v}^{(L)}(k) = (2\pi)^{-3/2}k^{-1}\int_0^\infty dR\, 4\pi R \sin(kR) v^{(L)}(R)$$

$$h(k) = \rho(2\pi)^{-3/2}k^{-1}\int_0^\infty dR\, 4\pi R \sin(kR)\,[g(R) - 1]$$

involving long-range electrostatic and polarization energy terms, cut-off at short distances, and dependent only on the magnitude k. The structure factor can be measured by means of X-ray diffraction techniques, in which case it would be available for calculation of the LR internal energy. It can also be generated from an expansion in powers of the density ρ to generate an equation of state for real gases and liquids. Alternatively, a perturbation theory can be developed for liquids, using the SR internal energy as a reference to zeroth order, and the LR interaction potential as a perturbation energy. The van der Waals equation of state for the pressure readily emerges from this treatment when the pair distribution function is approximated by a hard sphere expression [33].

The function $g(R;\rho,\beta)$ can also be used to introduce the concept of the potential of mean force $w^{(2)}(R;\rho,\beta) = -\beta^{-1}\ln[g(R;\rho,\beta)]$ which goes to zero at infinite distances and is shaped by the direct pair interaction $v(R)$, but in addition accounts for changes of the interaction of an atom at the origin with another at distance R, due to their indirect interactions through surrounding medium particles. The shape of the potential of mean force as R changes follows from its definition. It shows in particular a minimum at the location of the first maximum of the radial distribution, followed by decreasing oscillations as R increases through the medium. It can be obtained at low densities for real gases from the expansion in powers of the density, $g(R;\rho,\beta) = g_0(R;\beta) + \rho g_1(R;\beta) + \ldots$ and the Boltzmann thermal distribution for a dilute system which gives $g_0(R;\beta) = \exp[-\beta v(R)]$ [33].

More accurate treatments for $g^{(2)}\left(\vec{R}_1,\vec{R}_2\right)$ are derived from the triplets distribution $g^{(3)}\left(\vec{R}_1,\vec{R}_2,\vec{R}_3\right)$. To cut a many-molecule chain of equations, it is sometimes accurate to assume that the potential of mean force for triplets satisfies $w^{(3)}\left(\vec{R}_1,\vec{R}_2,\vec{R}_3\right) = w^{(2)}\left(\vec{R}_1,\vec{R}_2\right) + w^{(2)}\left(\vec{R}_2,\vec{R}_3\right) + w^{(2)}\left(\vec{R}_1,\vec{R}_3\right)$, a sum of pair interactions, even if this is not the case for the original triplet potentials. In this case $g^{(3)}\left(\vec{R}_1,\vec{R}_2,\vec{R}_3\right) = g^{(2)}\left(\vec{R}_1,\vec{R}_2\right)g^{(2)}\left(\vec{R}_2,\vec{R}_3\right)g^{(2)}\left(\vec{R}_1,\vec{R}_3\right)$ and all thermodynamical properties can be described in terms of pair distribution functions and pair potentials of mean force.

Generalizations are possible for a liquid of anisotropic molecules, and for liquid solutions. The pair distribution function of two polyatomic molecules A and B can be described in a space fixed (SF) frame in terms of their relative position $\vec{R}_{AB}$ and their Euler angles of orientation, $\Gamma_J = (\alpha,\beta,\gamma)_J$, $J = A, B$. It is frequently of interest to select certain orientations with respect to the intermolecular position, and to average over the remaining angles of the molecular orientations. For two identical linear molecules A_1 and A_2 with orientation angles $\Omega_j = (\vartheta,\varphi)_j$, $j = 1, 2$, the averaged pair distribution function can be expanded in spherical harmonics as

$$g(R_{12},\Omega_1,\Omega_2) = 4\pi\sum_{l,l',m} g_{l,l',m}(R_{12})\,Y_{l,m}(\Omega_1)Y_{l,-m}(\Omega_2)$$

and the coefficients can be related to partial potentials of mean force for molecules with fixed orientation. Expansions in terms of the Wigner rotational functions $D^{(l)}_{mm'}(\Gamma_j)$ [21] can instead be done for nonlinear molecules.

An alternative for interacting polyatomics is to represent each one with a set of potential centers a and b, possibly anisotropic, for A and B, respectively, and to introduce center-pair distribution functions so that, for example for interacting H_2O molecules one would define pair distributions g_{HH}, g_{OO}, and g_{HO} and use them to construct the internal energy of liquid water.

As in the case for solids, the long-range electrostatic and polarization interaction energies depend on the dielectric constant of the liquid. This can be obtained from the polarization $\vec{P}_{liq} = \sum_j \rho_j \vec{d}_j = \sum_j \rho_j \alpha_j \vec{\mathcal{E}}_{lcl}$ given by the sum of induced atomic dipoles per unit volume in an external electric field. In the simple case of a liquid of average density ρ with a single atomic species of induced dipole $\vec{d}\left(\vec{R}\right)$, the local polarization is $\vec{P}_{liq}\left(\vec{R}\right) = \vec{d}\left(\vec{R}\right)\rho\left(\vec{R}\right)$, where $\rho\left(\vec{R}\right) = \sum_j \delta\left(\vec{R}-\vec{R}_j\right)$ is the local density of atoms. Its thermal average is usually a smooth function over space within the liquid, which introduces the liquid pair distribution function in $\bar{\rho}\left(\vec{R}\right) = \rho g\left(\vec{R};\rho,\beta\right)$, giving

$$\overline{\vec{P}_{liq}}\left(\vec{R}\right) = \rho\,\alpha\left(\vec{R}\right) g\left(\vec{R};\rho,\beta\right)\vec{\mathcal{E}}_{lcl}\left(\vec{R}\right)$$

in terms of the local atomic polarizability $\alpha\left(\vec{R}\right)$. From this, one can obtain the dielectric constant ε_r of the liquid as done before for a solid, given by

$$\frac{\varepsilon_r - 1}{\varepsilon_r + 2} = \frac{1}{3\varepsilon_0}\int d^3R\rho\,\alpha\left(\vec{R}\right) g\left(\vec{R};\rho,\beta\right)$$

which however must be generalized for anisotropic species requiring a polarizability tensor, and for polar molecules with a permanent dipole [34].

4.5 Interaction Energies in a Liquid Solution and in Physisorption

4.5.1 Potential Energy of a Solute in a Liquid Solution

The interaction energy between a solute molecule M at location $\vec{R}_M$ and solvent molecules S_n at location $\vec{R}_n$ can be constructed from electronic structure calculations of the energy of M perturbed by surrounding solvent molecules plus the energies of interaction of the solvent molecules among themselves, obtained by considering the solution as organized into solute and solvent fragments [35].

The treatment is simpler when there are no chemical bondings between solute and solvent species, and one can use already constructed pair and triplet interaction energies. The liquid solution total interaction energy takes the form

$$V_{sol} = V_{MS}^{(Coul)} + V_{MS}^{(pol)} + V_{MS}^{(rep)} + V_{SS}^{(Coul)} + V_{SS}^{(pol)} + V_{SS}^{(rep)}$$

which includes Coulomb (or electrostatic), polarization (induction and dispersion), and repulsion (electron penetration and exchange) contributions between the molecule and solvent species through the pair potential energies $v_{MS}^{(int)}$, with $int = Coul, pol, rep$, and solvent–solvent interaction potential functions $v_{SS}^{(int)}$, plus possibly contributions from triplet interactions. The interactions among solvent species have been described in the previous section. Here we concentrate on the total interaction potential energy $V_{slt} = V_{MS}^{(Coul)} + V_{MS}^{(pol)} + V_{MS}^{(rep)}$ acting on the solute molecule M.

It helps to distinguish short-range (SR) and long-range (LR) interactions, with SR sums containing only nearby molecules, while sums of LR terms are averaged over large distances, are smoother, and can be expanded in a Fourier series or averaged over space. To proceed, it is convenient to separately treat short-range interactions in $V_{MS}^{(rep)}$ as differently from the other, long-range interactions. Short-range interactions between solute or solvent molecules are important only between nearest neighbors or next nearest neighbors. We assume to simplify first that the molecules are isotropic on average. The short-range (repulsion) interaction energy $V_{slt}^{(SR)} = V_{slt}^{(rep)}$ for the solute is a sum of short-range pair potentials for atoms in the solvent S, which can be written as

$$V_{slt}^{(SR)} = \sum\nolimits_{n \in S} v_{MS}^{(SR)}\left(|\vec{R}_M - \vec{R}_n|\right)$$

However, the SR potential extends only a short distance away from the origin and involve only the regions of pair distributions coming mostly from nearest neighbors, and next-nearest-neighbors shells to a smaller extent. Restricting the nearest-neighbors to a narrow distance around the maximum of the pair distribution functions, which coincide with the minima $R_{MS}^{(m)}$ of the pair potential, a rough estimate of the short-range potential energy near thermal equilibrium at the solute molecule M is

$$\left(V_{slt}^{(SR)}\right)_{eql} \approx v_{MS}^{(SR)}\left(R_{MS}^{(m)}\right) z_{MS}^{(NN)}$$

where $z_{MS}^{(NN)}$ is the number of nearest neighbor species S. The SR term can also be written as

$$V_{slt}^{(SR)} = \int d^3R\, v_{MS}^{(SR)}\left(|\vec{R}_M - \vec{R}|\right) \rho_S\left(\vec{R}\right)$$

where $\rho_S(\vec{R}) = \sum_n \delta(\vec{R} - \vec{R}_n)$ is the local density of solvent species. Furthermore, the solute potential energy can be statistically averaged over a thermal configuration distribution of the solvent, which contains the average density

$$\bar{\rho}_S(\vec{R}) = Z_N(\Omega,\beta)^{-1} \int d^{3N}R \exp\left[-\beta V_N\left(\vec{R}_1,\ldots,\vec{R}_N\right)\right] \sum_n \delta\left(\vec{R} - \vec{R}_n\right)$$

to obtain

$$\bar{V}_{slt}^{(SR)}\left(\vec{R}_M\right) = \rho_S \int d^3R\, v_{MS}^{(SR)}\left(|\vec{R}_M - \vec{R}|\right) g_{MS}\left(\vec{R};\rho_S,\beta\right)$$

in terms of the pair distribution function g_{MS} for the M and S species insofar $\bar{\rho}_S(\vec{R}) = \rho_S g_{MS}(\vec{R};\rho_S,\beta)$, with ρ_S the number of solvent molecules per unit volume, that can be obtained from a potential of mean force. More generally the pair potential function of R must be replaced by a function containing internal distances and orientation angle variables for the M and S molecules, and statistical averages can be performed over all internal variables.

The remaining long-range interactions due to electrostatic and polarization terms are again given by sums of long-range pair potentials suitably cut-off at short distances, $v_{MS}^{(LR)} = v_{MS}^{(Coul)} + v_{MS}^{(pol)}$, which contribute to an energy $V_{slt}^{(LR)}$. The long-range term becomes

$$V_{slt}^{(LR)} = \sum_n v_{Mn}^{(LR)}\left(\left|\vec{R}_M - \vec{R}_n\right|\right) = \int d^3R\, v_{MS}^{(LR)}\left(|\vec{R}_M - \vec{R}|\right)\rho_S\left(\vec{R}\right)$$

This integrand is usually a smooth function over space within the solid. It can be thermally averaged to obtain

$$\bar{V}_{slt}^{(LR)}\left(\vec{R}_M\right) = \rho_S \int d^3R\, v_{MS}^{(LR)}\left(|\vec{R}_M - \vec{R}|\right) g_{MS}\left(\vec{R};\rho,\beta\right)$$

Here, the radial integral is nearly zero for values smaller than the radius $R_{MS,0}$, where the LR potential is large but the pair distribution function vanishes. Beyond this radius the integrand is a smooth function of R which can be approximated by an analytical expression. It can be obtained for an attractive polarization potential of form $v_{MS}^{(LR)}(R) = -C_{MS}^{(s)}R^{-s}$, with $s = 6$ for a dispersion potential or $s = 4$ for induced polarization, approximating g_{MS} by its hard sphere form $g_{MS}(R) = 0$ for $R \le R_{MS}^{(c)}$ and $g_{MS}(R) = 1$ for $R > R_{MS}^{(c)}$. This follows from a potential of mean force w_{MS} strongly repulsive inside a cavity radius $R_{MS}^{(c)}$ of the order of the sum of van der Waals radii for M and S. With the origin of coordinates at the CMN of M and for an isotropic solvent, introducing the relative position vector $\vec{r} = \vec{R}_M - \vec{R}$ we find

$$\left(\bar{V}_{slt}^{(LR)}\right)_{pol} = \rho_S \int d^3r\, v_{MS}^{(LR)}(r)\, g_{MS}\left(\left|\vec{R}_M - \vec{r}\right|;\rho,\beta\right) \approx \rho_S \int_{R_{MS}^{(c)}}^{\infty} 4\pi r^2 dr\, v_{MS}^{(LR)}(r)$$

$$\approx \frac{4\pi\rho_S\, C_{MS}{}^{(s)}\left(R_{MS}^{(c)}\right)^{-s+3}}{(-s+3)}$$

giving a negative (attractive) solute potential energy which is larger for a smaller solute cavity radius. A similar approximation for the solvent density is obtained assuming that solvent atoms are packed with hard sphere radius $R_{SS}^{(c)}$ so that $\rho_S = \left[4\pi\left(R_{SS}^{(c)}\right)^3/3\right]^{-1}$, the inverse of the volume per solvent species.

A more accurate treatment can be derived using what is known about the pair distribution function and related structure factor, from measurements or from molecular dynamics calculations. Introducing the Fourier transform of the LR pair potential and its inverse transform,

$$\tilde{v}_{MS}^{(LR)}\left(\vec{k}\right) = (2\pi)^{-3/2} \int d^3R\, v_{MS}^{(LR)}\left(\vec{R}\right) \exp\left(i\vec{k}.\vec{R}\right),$$

$$v_{MS}^{(LR)}\left(\vec{R}\right) = (2\pi)^{-3/2} \int d^3k\, \tilde{v}_{MS}^{(LR)}\left(\vec{k}\right) \exp\left(-i\vec{k}.\vec{R}\right)$$

we find from the sum over pair potentials

$$V_{slt}^{(LR)} = (2\pi)^{-3/2} \int d^3k\, \tilde{v}_{MS}^{(LR)}\left(\vec{k}\right) \exp\left(-i\vec{k}.\vec{R}_M\right) \sum\nolimits_n \exp\left(i\vec{k}.\vec{R}_n\right)$$

Further, considering that the transform of the thermally averaged number density $\bar{\rho}_S\left(\vec{R}\right)$ of the solvent is $\tilde{\rho}_S\left(\vec{k}\right) = (2\pi)^{-3/2}\left[\sum\nolimits_n \exp\left(i\vec{k}.\vec{R}_n\right)\right]_{th}$ we find for the LR part,

$$\bar{V}_{slt}^{(LR)} = \int d^3k\, \tilde{v}_{MS}^{(LR)}\left(\vec{k}\right) \tilde{\rho}_S\left(\vec{k}\right) \exp\left(-i\vec{k}.\vec{R}_M\right)$$

A general way to evaluate this involves the introduction of the Fourier transform $h_{MS}\left(\vec{k}\right)$ of the M–S pair correlation function $h_{MS}\left(\vec{R}\right) = g_{MS}\left(\vec{R}\right) - 1$, that is a structure factor measurable by X-ray scattering. In detail,

$$h_{MS}\left(\vec{k}\right) = \rho_S (2\pi)^{-3/2} \int d^3R \exp\left(i\vec{k}.\vec{R}\right)\left[g_{MS}\left(\vec{R}\right) - 1\right]$$

from which the pair distribution function is obtained by the inverse Fourier transform, as

$$g_{MS}\left(\vec{R}\right) = 1 + (2\pi)^{-3/2}\int d^3k \exp\left(-i\vec{k}.\vec{R}\right) h_{MS}\left(\vec{k}\right)/\rho_S$$

Replacing this in the integrals over $\vec{R}$, we find the solute long-range potential energy expressed in terms of the Fourier transforms of the long-range part of the pair interaction potential and of the structure factors, as

$$\bar{V}_{slt}^{(LR)}\left(\vec{R}_M\right) = \int d^3k\, \tilde{v}_{MS}^{(LR)}\left(\vec{k}\right) h_{MS}\left(\vec{k}\right) \exp\left(-i\vec{k}\vec{R}_M\right) - \rho_S \int d^3R\, v_{MS}^{(LR)}\left(|\vec{R}_M - \vec{R}|\right)$$

Further simplification follows from observing that for the present isotropic potentials and distributions, the potential Fourier transforms and the present structure factors depend only on the wavevector magnitude k and are given by

$$\tilde{v}_{MS}^{(L)}(k) = (2\pi)^{-3/2} k^{-1} \int_0^\infty dR\, 4\pi R \sin(kR) v_{MS}^{(L)}(R)$$

$$h_{MS}(k) = \rho_S (2\pi)^{-3/2} k^{-1} \int_0^\infty dR\, 4\pi R \sin(kR)\, [g_{MS}(R) - 1]$$

which allows calculation of $V_{slt}^{(L)}$ from one-dimensional integrals over k. The structure factor $h_{MS}(k)$ can also be calculated from pair potentials by means of perturbation theory [33].

The extension to nonspherical, rotating, and vibrating M and S molecules requires reintroduction of internal orientation angles and atomic displacement variables in pair potentials and in pair distribution functions, and statistical averages over all internal variables of M and S components of the liquid solution.

The mean interaction $w_{MM}(R)$ between two solute molecules in the solution differs from the bare interaction pair potential $v_{MM}(R)$ due to medium effects and is a potential of mean force which depends on the dielectric constant of the solvent. A general treatment can be based on the susceptibility function of the solvent, by considering the response of the whole system to an applied external electric field $\vec{\mathcal{E}}_{ext}$ [36].

4.5.2 Potential Energies of Atoms and Molecules Adsorbed at Solid Surfaces

To begin with we consider an atom A physisorbed at position $\vec{R}_A$ on the surface of a solid made up of atoms or ions located at positions $\vec{R}_n$, $n = 1$ to N_{nu}, with all positions given in a reference frame with an origin at a chosen point $\vec{R}_O$ and a

z-axis perpendicular to the solid surface and pointing into a vacuum. The solid may contain several types of atoms, and may have a crystalline structure or be amorphous, such as a silicon solid with its surface dangling bonds saturated with hydrogen (Si(s):H), or titanium dioxide (TiO_2(s)). In the absence of chemical bonding between the adsorbate and the surface atoms, it is usually accurate to construct the total adsorption potential energy V_{ads} as a sum of pair potential energies $v_{An}\left(\vec{R}_A - \vec{R}_n\right)$, so that for fixed positions of atoms in the solid, $V_{ads}\left(\vec{R}_A\right) = \sum_n v_{An}\left(\vec{R}_A - \vec{R}_n\right)$. The treatment is then similar to the one for a solute in a solvent. However, for metal and semiconductor surfaces with delocalized electrons, the pair potentials v_{An} must be constructed to account for embedding effects as treated in Chapter 8, and are also functionals of the electronic densities at the surfaces. What follows here applies to nonbonding adsorbates and localized surface electrons.

Separating short range (*SR*) and long range (*LR*) terms in the pair potential one can write $V_{ads} = V_{ads}^{(SR)} + V_{ads}^{(LR)}$ to derive approximations. The short range potential can be obtained from the nearest neighbors (NTs) of the adsorbed atom, so that

$$V_{ads}^{(SR)}\left(\vec{R}_A\right) \approx \sum_{n \in NT} v_{An}^{(SR)}\left(\vec{R}_A - \vec{R}_n\right)$$

which can be evaluated for a known location of the adsorbate on the surface, and from the atomic structure of the surface and inner layers. This can be improved by adding next-NTs.

The long range term becomes

$$V_{ads}^{(LR)}\left(\vec{R}_A\right) = \sum_n v_{An}^{(LR)}\left(\vec{R}_A - \vec{R}_n\right)$$

which is usually a smooth function over space within the solid. For a solid containing atoms of type $J = B, C, \ldots$ it can be written in terms of the number densities $\rho_{sld}^{(J)}\left(\vec{R}\right) = \sum_{n \in J} \delta\left(\vec{R} - \vec{R}_n\right)$ of the solid as

$$V_{ads}^{(LR)}\left(\vec{R}_A\right) = \int d^3R \sum_J v_{AJ}^{(LR)}\left(\vec{R}_A - \vec{R}\right) \rho_{sld}^{(J)}\left(\vec{R}\right)$$

Introducing the Fourier transform of the *LR* pair potential and its inverse transform,

$$\tilde{v}_{AJ}^{(LR)}\left(\vec{k}\right) = (2\pi)^{-3/2} \int d^3R\, v_{AJ}^{(LR)}\left(\vec{R}\right) \exp\left(i\vec{k}.\vec{R}\right),$$

$$v_{AJ}^{(LR)}\left(\vec{R}\right) = (2\pi)^{-3/2} \int d^3k\, \tilde{v}_{AJ}^{(LR)}\left(\vec{k}\right) \exp\left(-i\vec{k}.\vec{R}\right)$$

we find

$$V_{ads}^{(LR)}\left(\vec{R}_A\right) = (2\pi)^{-3/2}\int d^3k \, \exp\left(-i\vec{k}.\vec{R}_A\right)\sum\nolimits_J \tilde{v}_{AJ}^{(LR)}\left(\vec{k}\right)\sum\nolimits_{n\in J}\exp\left(i\vec{k}.\vec{R}_n\right)$$

Further considering that the transform of the thermally averaged number density of the solid is $\tilde{\rho}_{sld}^{(J)}\left(\vec{k}\right) = (2\pi)^{-3/2}\left[\sum\nolimits_{n\in J}\exp\left(i\vec{k}.\vec{R}_n\right)\right]_{th}$ we get for the thermally averaged LR part,

$$\bar{V}_{ads}^{(LR)}\left(\vec{R}_A\right) = \int d^3k \, \exp\left(-i\vec{k}.\vec{R}_A\right)\sum\nolimits_J \tilde{v}_{AJ}^{(LR)}\left(\vec{k}\right)\tilde{\rho}_{sld}^{(J)}\left(\vec{k}\right)$$

Special cases follow when the atoms form a crystalline lattice, and when one can assume that on the average the solid has a constant density which vanishes past the surface.

The treatment simplifies when all atoms in the solid are the same, with $v_{AJ} = v_A$. The total energy operator for a solid in a volume Ω is then $V_{ads}\left(\vec{R}_A\right) = \int_\Omega d^3R \, v_A\left(\vec{R}_A - \vec{R}\right)\rho_{sld}\left(\vec{R}\right)$ and this can be statistically averaged over a thermal distribution of atom locations for a solid at temperature T giving a density $\bar{\rho}\left(\vec{R};T\right) = \rho_{sld}g\left(\vec{R};\rho_{sld},\beta\right)$ in terms of the pair distribution function of atoms in the solid relative to the origin of coordinates, and an average potential energy of adsorption

$$\bar{V}_{ads}\left(\vec{R}_A;T\right) = \rho_{sld}\int_\Omega d^3R \, v_A\left(\vec{R}_A - \vec{R}\right)g\left(\vec{R};\rho_{sld},\beta\right)$$

which can be further treated separating SR and LR terms. For the SR term one again has $V_{ads}^{(S)}\left(\vec{R}_A\right) \approx \sum\nolimits_{n\in NT} v_A^{(S)}\left(\vec{R}_A - \vec{R}_n\right)$, while for the LR we can estimate the potential energy on the adsorbate by assuming that the solid is crystalline, or homogeneous with a smooth surface at distance D from the adsorbed atom.

Consider the case where the solid surface is on a (xy)-plane with atoms labeled by integers $\boldsymbol{n} = (n_x, n_y, n_z)$ in a cubic crystalline lattice generated by translations of a unit cell with axes along orthogonal unit vectors $\vec{u}_x, \vec{u}_y, \vec{u}_z$ and edges of length a_x, a_y, a_z, with one atom per cell at locations $X_{\boldsymbol{n}} = X_0 + n_x a_x$ and $Y_{\boldsymbol{n}} = Y_0 + n_y a_y$ with $-\infty < n_\xi < \infty$. The locations along the z-direction are $Z_{\boldsymbol{n}} = Z_0 - n_z a_z$ with $0 \le n_z < \infty$. The density is now a lattice periodic function $\rho_{sld}\left(\vec{R}\right) = \rho_{latt}\left(\vec{R}\right) = \rho_{latt}\left(\vec{R} + \vec{t}_{\boldsymbol{n}}^{(s)}\right)$, with a surface periodic translation $\vec{t}_{\boldsymbol{n}}^{(s)} = n_x a_x \vec{u}_x + n_y a_y \vec{u}_y$. The condition that the density must be invariant under translations along x- and y-directions by steps a_x and a_y means that now an expansion in related wavevector components must only contain values

$k_x = 2\pi m_x / a_x = g_{m_x}^{(x)}$ and similarly for k_y, with integers $-\infty < m_\xi < \infty$ [31]. The density is given by

$$\rho_{latt}\left(\vec{R}\right) = \sum_{m} \exp\left(i\vec{g}_{m}^{(s)}.\vec{R}^{(s)}\right) \int_{-\infty}^{+\infty} dk_z \exp(ik_z Z) \tilde{\rho}_{latt}\left(\vec{g}_{m}^{(s)}, k_z\right)$$

where $\vec{g}_{m}^{(s)} = 2\pi\left(m_x \vec{u}_x / a_x + m_y \vec{u}_y / a_y\right)$ and $\vec{R}^{(s)} = x\vec{u}_x + y\vec{u}_y$ are surface vectors. The density $\tilde{\rho}_{latt}$ in the reciprocal space of $\vec{k}$ vectors can be extracted by taking its inverse transform and using it in the adsorbate expression

$$V_{ads}^{(L)}\left(\vec{R}_A\right) = \sum_{m} \exp\left(i\vec{g}_{m}^{(s)}.\vec{R}_A^{(s)}\right) \int dk_z \tilde{v}_A^{(L)}\left(\vec{g}_{m}^{(s)}, k_z\right) \tilde{\rho}_{latt}\left(\vec{g}_{m}^{(s)}, k_z\right) \exp(-ik_z Z_A)$$

with the lattice density obtained for $\vec{R}^{(s)} = \vec{t}_{n}^{(s)}$ as

$$\tilde{\rho}_{latt}\left(\vec{g}_{m}^{(s)}, k_z\right) = (2\pi)^{-3/2} \sum_{n} \exp\left(i\vec{g}_{m}^{(s)}.\vec{t}_{n}^{(s)}\right) \exp(ik_z Z_n) = (2\pi)^{-3/2} \sum_{n} \exp(ik_z Z_n)$$

The integral and summations over lattice points can be evaluated given the Fourier transform of the pair interaction potential.

For the solid treated as a homogeneous structure, and a pair interaction $v_A^{(L)}(R) = -C_A^{(s)} R^{-s}$ between the adsorbed atom and an atom within a solid element of volume, integration within the solid with a surface perpendicular to a z-axis can be done adding over its elementary rings of radius r and width dr parallel to the surface with volume $d\Omega_{rng} = 2\pi r dr dz$, and with the number of atoms in this ring volume being $dN_{rng} = \rho_{sld} d\Omega_{rng}$. Taking the adsorbate distance to that element of volume to be $R_A = \left(r^2 + z_A^2\right)^{1/2}$, with the z_A variable originating at A and growing into the solid, the thermally averaged LR potential energy is then

$$\bar{V}_A^{(s)}(D) = -2\pi\rho_{sld} C_A^{(s)} \int_0^\infty dr \int_D^\infty dz_A \frac{r}{\left(r^2 + z_A^2\right)^{s/2}} = -\frac{2\pi\rho_{sld}\, C_A^{(s)}}{[(s-2)(s-3)D^{s-3}]}$$

This gives, for the van der Waals dispersion potential with $s = 6$, the well-known D^{-3} dependence on the distance from the adsorbate to the surface.

This result must be modified when the adsorbate is surrounded by a medium (such as a liquid) outside the solid, to account for the dielectric constant of that medium [36]. Here, we can assume that the dielectric effects have been incorporated into the constant $C_A^{(s)}$.

4.6 Interaction Energies in Large Molecules and in Chemisorption

4.6.1 Interaction Energies Among Molecular Fragments

Large molecules such as polymers, proteins, and nucleic acids can be described as made up of atomic groups or of molecular fragments $\mathcal{J} = \mathcal{A}, \mathcal{B}, \mathcal{C}, \ldots$ usually selected on the basis of their chemical structure and bonding strength, interacting among themselves as described by their electronic density distributions. The assumption is that to zeroth order the total electronic wavefunction $\Phi^{(0)}$ is a superposition of products of fragment wavefunctions $\Phi_J^{(0)}$ for each fragment $\mathcal{J}$, antisymmetrized for the exchange of electron variables among all fragments. Each fragment has an electronic density $\rho_J^{(0)}(\vec{r})$ distributed among atomic centers, bonds, and localized electronic lobes. As a consequence the total electronic density $\rho^{(0)}(\vec{r})$ is a sum of intra-fragment and inter-fragment density terms, or $\rho^{(0)}(\vec{r}) \approx \sum_J \rho_J^{(0)}(\vec{r}) + \sum_{JK} \rho_{JK}^{(0)}(\vec{r})$, if one discards additional smaller terms involving more than two fragments. A large two-fragment term indicates that there is chemical bonding between those fragments.

Interfragment interactions contain short-range repulsion, electronic exchange, electrostatic (or Coulomb), polarization, and dispersion terms. For the calculation of electrostatic interaction energies, fragment electronic densities can be alternatively represented by a collection of weighted electronic point charge densities as in $\rho_J^{(0)}(\vec{r}) = \sum_{j \in J} w_j \delta\left(\vec{r} - \vec{R}_j\right)$ located at sites $\vec{R}_j$ of its atoms, bonds, and lobes. The total density $\rho^{(0)}(\vec{r})$ can be decomposed into terms labeled by j and k, and interfragment electrostatic interaction energies can be obtained from the interactions of sites j in $\mathcal{J}$ and k in $\mathcal{K}$ [5].

To improve on that zeroth order approximation when molecular fragments $\mathcal{J}$ and $\mathcal{K}$ display chemical bonding, it is necessary to account for the electronic rearrangement at the bonds in terms of more accurate molecular wavefunctions Φ_{JK} calculated for the bonded $\mathcal{JK}$ pair. An improved density is now of the form $\rho(\vec{r}) \approx \sum_J \rho_J(\vec{r}) + \sum_{JK} \rho_{JK}(\vec{r})$ and includes intrafragment $\Delta\rho_J(\vec{r}) = \rho_J(\vec{r}) - \rho_J^{(0)}(\vec{r})$ and interfragment $\Delta\rho_{JK}(\vec{r}) = \rho_{JK}(\vec{r}) - \rho_{JK}^{(0)}(\vec{r})$ density changes, which can be represented by additional electrical multipoles at the locations of point charges. Repulsion and exchange interaction energies can be extracted from quantal electronic structure calculations [1, 37], or from mixed (quantum mechanics)/(molecular mechanics) (QM/MM) treatments [38, 39]. Electrostatic induction and dispersion interaction energies can be obtained from a set of electrical multipoles and a set of polarizabilities for each fragment and their pair interactions. Improved fragment electronic density and polarizability distributions accounting for bonding can be calculated from more accurate wavefunctions Φ_{JK} [5, 35].

When intramolecular fragment–fragment interactions are thermally averaged by atomic motions and solvent effects, less than complete details are needed in the description of properties of fragments, and these can be treated in an average sense with continuum models, or with coarse graining methods. It also helps to distinguish short-range and long-range fragment interactions, with *SR* sums containing only nearby fragments, while sums of *LR* terms can be expanded in a Fourier series or simply averaged over space.

Given the electronic wavefunction Ψ_M of the large molecule M for fixed nuclear positions, possibly in a medium of solvent molecules with their charge and polarization densities, the liquid solution total potential energy takes the form

$$V_{M,sol} = \left\langle \Psi_M \mid \left(\hat{H}_M^{(0)} + \hat{H}_{MS}^{(Coul)} + \hat{H}_{MS}^{(pol)} + \hat{H}_{MS}^{(rep)} \right) \mid \Psi_M \right\rangle$$
$$+ V_{SS}^{(Coul)} + V_{SS}^{(pol)} + V_{SS}^{(rep)}$$

which includes Coulomb (or electrostatic), polarization (induction and dispersion), and repulsion (electron penetration and exchange) potential energy contributions between the molecule and solvent species through the Hamiltonian operators $\hat{H}_{MS}^{(int)}$, with $int = Coul, pol, rep$, and solvent–solvent interaction potential functions $V_{SS}^{(int)}$. The Hamiltonian operator $\hat{H}_M^{(0)}$ contains the electronic kinetic energies and all the molecular electronic and nuclear Coulomb interactions.

The wavefunction Ψ_M can be obtained from perturbation theory starting from a wavefunction $\Psi_M^{(0)}$ of the isolated M, and adding a perturbation term Ψ'_M due to the presence of solvent species. This additional term depends parametrically on the positions of solvent species and their electronic charge densities. The expectation values of the Hamiltonian operators for M, and the M–S interaction energies can be constructed from information about bonding between fragments and charge and polarization densities in the fragments, and also from charge and polarization densities in the solvent and how they interact with the electrons and nuclei in the solute M described by its atomic groups or molecular fragments. The interactions among solvent molecules can be modeled with the functions already described for interacting nonbonding pairs. But the interactions in the solute M and between molecule M and the solvent require a full treatment of the electronic structure of M and how it is perturbed by the solvent. This is the subject of a following Chapter 7.

A similar treatment is applicable to a system where a molecule is chemically bonded to a surface, and the chemisorption complex can be considered to be a supermolecular system by analogy with the way energies are described in large molecules. The treatment is simplified when the atomic structure of the surface shows periodicity as in crystal surfaces.

4.6.2 Potential Energy Surfaces and Force Fields in Large Molecules

Using knowledge from electronic structure calculations, thermochemistry, and molecular spectroscopy of molecular fragments and their interactions, it is possible to construct potential energy functions of all the relevant atom positions in the system and to generate force fields from the gradients of the potential with respect to atomic displacements. These forces are basic to the calculation of the structure of new compounds, and to calculation of their molecular properties.

A large molecule containing many atoms of different types, as found especially in biomolecules, complex materials, and nanostructured surfaces, show complicated deformations when subject to forces resulting from interactions with other molecules or from electromagnetic radiation. Deformations may involve a bond stretching between two atoms, an angle bending between two adjacent bonds involving three atoms, an angle torsion between two adjacent atomic planes defined by four atoms, and buckling angles for an atom displaced relative to a plane defined by three atoms.

As long as the whole system is in a given electronic state and there are no electronic excitations involved, the potential energy can be given as a function of atomic positions, or as a function of internal variables dependent on several atomic positions, which when varied generate a potential energy surface (PES). The choices of variables depend on whether the molecular fragments composing the total system are chemically bonding or nonbonding. The potential energy for an isolated large molecule M can be written as

$$V_M = \left\langle \Psi_M^{(0)} \,|\, \hat{H}_M^{(0)} \,|\, \Psi_M^{(0)} \right\rangle = V^{(B)} + V^{(NB)}$$

with the first term a function of internal bond variables describing bonding fragments, and with the second term a function of atomic group or fragment positions for interacting nonbonding fragments. In detail, the bonding potential energy can be given in terms of the bond distances d_{ab} between atoms a and b, bending angle ϑ_{abc}, torsion angle $\gamma_{ab,\,cd}$ between two planes with a common edge, and buckling angle $\gamma_{a,\,bcd}$ between a plane and a line to a vertex. Provided displacements from equilibrium values are small, changes in lengths or angles can be used as variables, and deformation energies can be expanded in power series of displacements from equilibrium. We write

$$V^{(B)} = V_{str} + V_{bnd} + V_{tor} + V_{bck}$$

where V_{str} is the stretch component involving sums over two atoms, V_{bnd} is the bending components for sets of three atoms, V_{tor} the torsion component with sums over four atoms, and V_{bck} the buckling component also for four atoms in each term. These terms can be constructed from calculated or measured properties of isolated fragments. In detail, introducing force constants K for each type of displacement, equilibrium variable values, and a torsion integer index n,

$$V_{str} = \sum_{a<b} \frac{1}{2} K_{ab} \left(d_{ab} - d_{ab}^{(eq)} \right)^2$$

$$V_{bnd} = \sum_{a<b<c} \frac{1}{2} K_{abc} \left(\vartheta_{abc} - \vartheta_{abc}^{(eq)} \right)^2$$

$$V_{tor} = \sum_{a<b<c<d} \frac{1}{2} K_{ab,cd} \left[1 + \cos\left(n\gamma_{ab,cd} - \delta_{ab,cd}^{(eq)} \right) \right]$$

$$V_{bck} = \sum_{a<b<c<d} \frac{1}{2} K_{a,bcd} \left(\gamma_{a,bcd} - \gamma_{a,bcd}^{(eq)} \right)^2$$

where the summations involve only neighbor atoms. To this one must add energy terms for hydrogen bonding, if present, taken to involve atoms a, b (for H), and c, and expressed as a function of a variable $S_{ac} = \left| \vec{d}_{cb} - \vec{d}_{ab} \right| = d_{ab} d_{bc} \cos(\vartheta_{abc})$ of the form

$$v_{aHc} = \frac{1}{2} K_H \left(S_{ac} - S_{ac}^{(eq)} \right)^2 + K_H' \left(S_{ac} - S_{ac}^{(eq)} \right)$$

or alternatively as a potential function of the distance d_{aH} between the hydrogen and a nonbonding atom a, with a repulsive term varying as the inverse distance to the power of 12 and an attractive term with the inverse distance to the power of 10, with suitable parameters. When needed, improvements in accuracy of the PESs can be obtained by allowing coupling of two adjacent stretching bonds, in a potential energy $V_{str,str}(d_{ab}, d_{bc})$ containing three atom terms, and coupling of stretch and bending in a $V_{str,bnd}(d_{ab}, \vartheta_{abc})$ function.

The nonbonding potential energy involves short-range repulsive and long-range polarization terms that can be constructed for pairs of atomic groups or molecular fragments $\mathcal{I}$, $\mathcal{J}$ as $v_{IJ}^{(NB)}(R_{IJ}) = v_{IJ}^{(SR)}(R_{IJ}) + v_{IJ}^{(LR)}(R_{IJ})$ for a pair at relative distance R_{IJ}, with the repulsive SR term an exponential or inverse power function, and the LR interaction $v_{IJ}^{(LR)}(R_{IJ})$ containing induction, dispersion, and electrostatic (or Coulomb) terms deriving from charges and polarizabilities embedded in the group fragments. Using in particular the van der Waals LJ(12,6) function combining repulsion and dispersion energies with energy minimum ϵ_{IJ} at the distance σ_{IJ}, and collections of point charges c_j and polarizabilities α_j at each fragment $\mathcal{J}$ chosen to replicate the induction and Coulomb interactions, the nonbonding potential energy is

$$V^{(NB)} = \sum_{I,J} \epsilon_{IJ} \left[\left(\frac{\sigma_{IJ}}{R_{IJ}} \right)^{12} - 2 \left(\frac{\sigma_{IJ}}{R_{IJ}} \right)^{6} \right] + v_{IJ}^{(Coul)}\left(\{R_{ij}\}\right) + v_{IJ}^{(ind)}\left(\{R_{ij}\}\right)$$

$$v_{IJ}^{(Coul)}\left(\{R_{ij}\}\right) = \sum_{i \in I, j \in J} \frac{c_i c_j}{4\pi\varepsilon_0 \varepsilon_r R_{ij}} f_d\left(R_{ij}\right)$$

$$v_{IJ}^{(ind)}\left(\{R_{ij}\}\right) = -\sum_{i \in I, j \in J} \frac{c_i \alpha_j + \alpha_i c_j}{\left(4\pi\varepsilon_0 \varepsilon_r\right)^2 R_{ij}^{\ 4}} f_d\left(R_{ij}\right)$$

where ε_r is the dielectric constant of the medium. The Coulomb and induction terms have been given as sums of pair interactions between point charges and polarizabilities located at fragments $\mathcal{I}$ and $\mathcal{J}$ at the relative distance R_{ij}. They can also be cut-off at short distances using as shown suitable damping functions $f_d(R_{ij})$, previously introduced. The present force fields $V^{(B)} + V^{(NB)}$ and similar variations have been extensively parametrized on the basis of electronic structure calculations and empirical information [40–44]. The energy landscapes generated by these force fields can be shown as energy isocontours versus two bond variables, keeping the other ones constant. Figure 4.8 shows the definition of torsion angles φ and ψ for a polypeptide strand, and potential energy isocontours versus those angles.

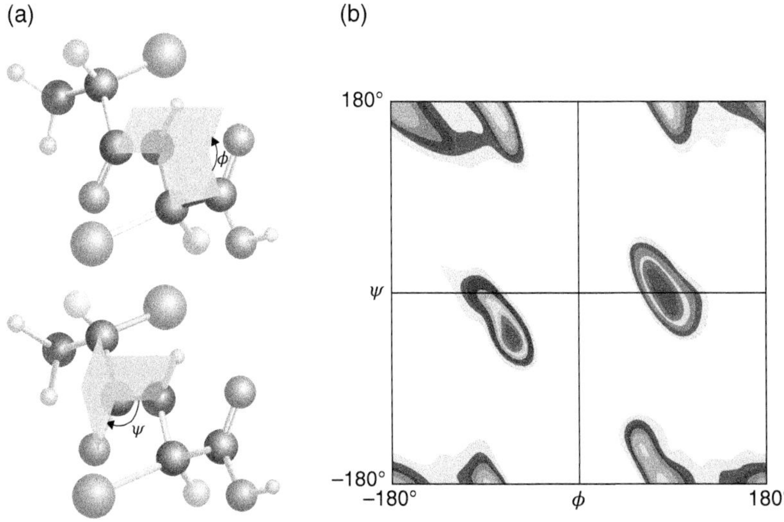

Figure 4.8 (a) Planes showing the definition of torsion angles ψ and φ between two peptide units in a polypeptide. (b) Potential energy isocontours for the polypeptide glycyl residue energy versus the torsion angles phi around the C—N bond and psi around the C—C bond (a Ramachandran diagram). Ovals lead to energy minima. *Source:* from Hovmoeller [45] and [46].

More accurately, the LR terms must be obtained from the perturbed wavefunction Ψ_M instead of $\Psi_M^{(0)}$, to account for interfragment interactions, and must also include electric multipole interactions between permanent and induced charges, which arise from intramolecular and intermolecular polarization of atoms and bonds. For the isolated molecule M, the resulting more accurate interactions are so-called polarizable force fields [47–51]. In addition, the total potential energy for the M–S liquid solution must account for the interaction of the molecule M with solvent molecules.

The M–S interaction requires knowledge of the wavefunction of M and calculation of the expectation value $\left\langle \Psi_M \mid \left(\hat{H}_{MS}^{(Coul)} + \hat{H}_{MS}^{(pol)} + \hat{H}_{MS}^{(rep)} \right) \mid \Psi_M \right\rangle$, where the Hamiltonian operators depend on the locations and charge distribution of each solvent species interacting with M. These interactions perturb M and involve repulsion, polarization, and Coulomb (electrostatic) terms as shown. To first order in the M–S interaction potentials and assuming that the solvent species have unchanged charge distributions, the expectation value can be obtained from the wavefunction $\Psi_M^{(0)}$ of the isolated M with its own permanent and induced charge distribution. Higher-order terms in a perturbation expansion $\Psi_M = \Psi_M^{(0)} + \Psi_M^{(1)} + \Psi_M^{(2)} + \ldots$ involve excited states of M and describe solute polarization, which include additional induction and dispersion terms, to first and second order in the M–S interaction. Further changes in potential energies arise from polarization of the solvent species.

Thermal equilibrium properties of the molecular system can be obtained from internal energies derived from the above potential energies plus the system kinetic energy at a given temperature. More detailed properties, such as isomerization rates, follow from molecular dynamics calculations involving force fields given by gradients of the potential functions with respect to atomic displacements [52, 53].

4.6.3 Potential Energy Functions of Global Variables Parametrized with Machine Learning Procedures

Given the structure of a many-atom system by a set of all its atomic position vectors $\boldsymbol{Q} = \left\{ \vec{R}_a \right\}$, a potential energy function $V(\boldsymbol{Q})$ for the ground electronic state can be written as a sum over embedded atoms, $V(\boldsymbol{Q}) = \sum_a V_a^{(emb)}(\boldsymbol{Q}), a = 1\, to\, N_{at}$, where each term depends not only on the position of atom a but also on the positions of atoms in the environment in which it is embedded. The PES may be known from calculations or measurements at a number of conformations. It can be given as a sum over embedded atom terms as shown, but each term must contain fitting parameters which describe their environment, and the total number of parameters can be very large.

It is possible to use artificial intelligence (AI) procedures making use of machine learning to train artificial neural networks (NN) which find all the needed parameters to fit the known PES $V(\boldsymbol{Q})$ at calculated conformations and to provide an interpolation for energies at other conformations. This was done for molecules chemically bonded to solid surfaces [54], for weakly bonded molecules [55], and for reaction PESs containing activation energies [56]. A number of improvements on the methodology and many more recent applications have been recently reviewed [57, 58], and will be briefly presented here with more details given in Chapter 7.

The early AI-NN procedure has been recently modified and extended to make it more efficient and general. Instead of using position coordinates as variables, the recent treatments have introduced global variables $G_\alpha(\boldsymbol{Q})$ forming a new set $\boldsymbol{G} = \{G_\alpha(\boldsymbol{Q})\}$ specifying the embedded atom positions and its environment with the correct symmetry, so that

$$V(\boldsymbol{Q}) = \sum\nolimits_a E_a^{(emb)}[\boldsymbol{G}(\boldsymbol{Q})]$$

The global variables can be chosen so that fittings to known values are physically invariant under collective translation and rotation of the rigid many-atom system, and can also include environmental symmetries such as they arise from the presence of identical atoms there. For example global (symmetry) variables $X_1, \ldots, X_8$ for the H_2/Pd(100) adsorbate have been chosen as functions of the six diatomic position variables (three for the center-of-mass of H_2 and three for its distance to the surface and the diatomic orientation angle) with X_4, X_5, X_7, X_8 taken as Fourier spatial components oscillating with the reciprocal lattice vector periodicity of the substrate surface [59]. Symmetry adapted global variables have also been constructed using permutational symmetry of identical atoms in treatments of reaction PES for H + H_2 and Cl + H_2 [60].

More generally, global symmetry functions for each embedded atom a have been defined in terms of the position coordinates of neighbor atoms within a finite sphere of chosen radius R_c. They are radial $G_a^{(rad)}$ and angle $G_a^{(ang)}$ variables dependent on two or three atom positions and given by [61]

$$G_a^{(rad)} = \sum\nolimits_{j \neq a}^{sph} \exp\left(-\eta R_{aj}^2\right) f_c(R_{aj})$$

$$G_a^{(ang)} = 2^{1-\zeta} \sum\nolimits_{j,k \neq a}^{sph} (1 + \cos\theta_{ajk}) \exp\left[-\eta\left(R_{aj}^2 + R_{ak}^2 + R_{jk}^2\right)\right] f_c(R_{aj}) f_c(R_{ak}) f_c(R_{jk})$$

with the sphere cut-off function $f_c(R) = [1 + \cos(\pi R/R_c)]/2$ for $R \leq R_c$ and zero past R_c, angle θ_{ajk} between the bonds aj and ak, and parameters η and ζ to be found.

The AI-NN procedure as presently used [57] involves at least an input NN layer with nodes μ containing each $G_a^{(\mu)}$ value, outputting their values into a hidden layer (number 1) of nodes m (as many as needed for training) coding partial energy components $x_{a,m}^{(1)}$ given by

$$x_{a,m}^{(1)} = b_m^{(1)} + \sum\nolimits_{\mu} G_a^{(\mu)} w_{\mu,m}^{(01)}$$

Here, the weights $w_{\mu,m}^{(01)}$ are parameters to be found, connecting the global variable in the input layer to node m in the hidden layer. This partial energy component is accepted with an activation (or acceptance) probability biased by the magnitude of $b_m^{(1)}$. The bias parameter is introduced to allow for an acceptance somewhere between zero and one for the output of node m, given by a sigmoid-shape (activation) function $f^{(1)}\left(x_{a,m}^{(1)}\right) = y_{a,m}^{(1)}$ varying from zero at minus infinity to one at plus infinity, with $b_m^{(1)}$ setting its value in the slope region and its acceptance probability. This function can be chosen, for example as $[1 + \tanh(x)]/2$.

When only one hidden layer is used in the NN, the final stage constructs the total energy of each embedded atom with

$$E_a^{(emb)} = \sum\nolimits_m y_{a,m}^{(1)}\left(\left\{G_a^{(\mu)}\right\}; b_m^{(1)}, \left\{w_{\mu,m}^{(01)}\right\}\right)$$

as a sum of the outputs from all the nodes of the hidden layer. The parameters $b_m^{(1)}$ and $\left\{w_{\mu,m}^{(01)}\right\}$ in the acceptance function follow from a NN training that optimizes the fitting of a set of reference PES values $E_{a,ref}^{(emb)}$ for given input atomic coordinates, using usually the mean absolute deviation (MAD)

$$MAD = \frac{1}{N_{at}} \sum\nolimits_a \left| E_{a,ref}^{(emb)} - E_{a,NN}^{(emb)} \right|$$

which must be minimized by the selection of parameters. An alternative is to minimize the root-mean-square error. Once the parameters are known, the interpolation of values to new atomic positions must be checked, and this is done using a validation set of previously generated PES values different from the ones used for training the NN. The total PES is given by the sum of embedded atom energies, $V = \sum_a E_a^{(emb)}$. The architecture of the NN is illustrated in Figure 4.9. Listing the number of nodes per layer and sizes of layers, this is a 2-3-1 NN.

Alternative AI-NN architectures have been explored, with different numbers of nodes in the input layer and more hidden layers in a deep-learning procedure. The training and validation sets of PES values have been generated with a variety

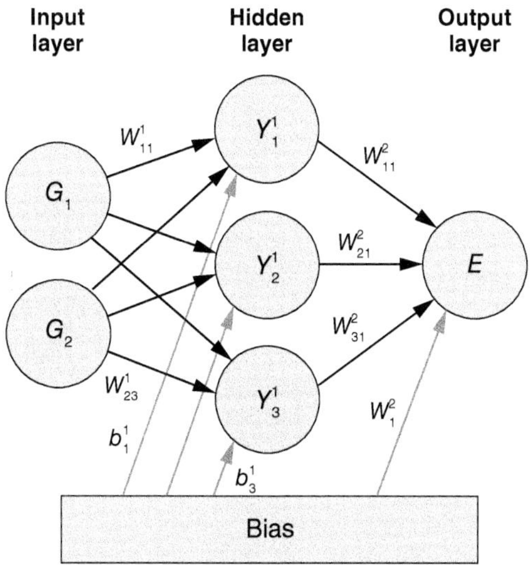

Figure 4.9 Architecture of a neural network with an input layer with $N = 2$ nodes containing initial assigned values of global variables coming from chosen atomic conformations. A hidden layer with $M = 3$ nodes and inputs from those variables, gives outputs with acceptances biased by an activation function. The outputs of the hidden layer nodes are added to generate the energy of an embedded atom in an output layer. *Source:* from [61].

of empirical and many-electron methods to test accuracies of the NN. As an example to test the AI-NN procedure in recent calculations, energies of the adsorbate in H_2/Pd(100) were generated with an empirical PES with 1560 training points and 5200 validation points and a 8-50-50-1 NN, which gave a validation error of 0.150 eV [59]. The energies of several solid structures of Si have been obtained using DFT [61], with 8200 PES training points and 800 validation points, for a NN containing 48 nodes in the input layer and two hidden layers with 40 nodes each (a 48-40-40-1 NN), giving a validation accuracy of 5 meV atom^{-1}. Another example is the calculation of the PES of vinyl bromide undergoing unimolecular decomposition [56], done generating energy sets with the MP4 many-electron treatment, with 1120 training points and 140 validation points, a 12-20-1 NN, and validation error of 0.001 eV.

The procedure just described can give very accurate fits to PESs with errors of the order of a few meV atom^{-1}. However, it needs to be reconsidered when charge transfer between atoms occurs as interatomic distances change, for example when an ionic bond changes into a neutral atom–atom interaction at large distances, such as in the ground electronic state of NaCl. Reconsideration may also be needed when long-range interactions are important, because dispersion energies may not be accounted by the training done with spheres of relatively small radius. A possible procedure in these cases is to increase the radius of the embedded atom spheres to verify that the training incorporates charge transfer and dispersion energies. This is however computationally demanding and other approaches need consideration. This is an active area of research.

Forces $F_{a,\xi}$ at atom a along the ξ-direction in the many-atom structure can be obtained as needed in molecular dynamics calculations from

$$F_{a,\xi} = -\frac{\partial V}{\partial R_{a,\xi}} = -\sum\nolimits_{a=1}^{N}\sum\nolimits_{\alpha=1}^{M}\frac{\partial E_a^{(emb)}}{\partial G_\alpha}\frac{\partial G_\alpha}{\partial R_{a,\xi}}$$

where N is the number of atoms and M the number of global symmetry functions. The fraction in the first factor is given by the NN architecture and the second factor follows from the choice of the global variables.

The procedure given here for interpolation of PESs allows for accurate fittings, improvement if additional calculated points are available for training of the NN, and is quite general for ground electronic states. It allows for description of bonding and nonbonding (weak or van der Waals) interactions and for systems undergoing reactions with broken and formed bonds. And it can provide results with the accuracy of the training set (such as DFT or CCSD) at the same computational cost as force-field calculations. The AI-NN procedure, however, requires preliminary calculations of PESs with standard treatments to be used for training and validation so that the AI-NN procedure is complementary to standard treatments of intermolecular forces using molecular properties or many-electron methods.

Given calculated PESs for a large class of compounds and interacting pairs of molecules, an AI-NN procedure can also be used to calculate new compounds and their interaction, provided they contain the same embedded atoms [62, 63]. This however must be done with caution and extensive validation, and some of these aspects are treated in Chapter 7.

References

1 Hirschfelder, J.O., Curtis, C.F., and Byron Bird, R. (1954). *Molecular Theory of Gases and Liquids*. New York: Wiley.

2 Margenau, H. and Kestner, N.R. (1971). *Intermolecular Forces*, 2e. Oxford, England: Pergamon Press.

3 Murrell, J.N., Carter, S., Farantos, S.C. et al. (1984). *Molecular Potential Energy Functions*. New York: Wiley.

4 Schatz, G.C. (1989). The analytical representation of electronic potential energy surfaces. *Rev. Mod. Phys.* 61: 669.

5 Stone, A.J. (2013). *The Theory of Intermolecular Forces*, 2e. Oxford, England: Oxford University Press.

6 Kaplan, I.G. (2006). *Intermolecular Interactions*, 81. New York: Wiley.

7 Herzberg, G. (1950). *Molecular Spectra and Molecular Structure I. Spectra of Diatomic Molecules*, 2e. Princeton NJ: Van Nostrand.

8 Lawley, K.P. (ed.) (1975). *Molecular Scattering* (Advan. Chem. Phys. vol. 30). New York: Wiley.

9 Maitland, G.C., Rigby, M., Brain Smith, E., and Wakeham, W.A. (1981). *Intermolecular Forces*. Oxford, England: Oxford University Press.
10 Jensen, F. (1999). *Introduction to Computational Chemistry*. New York: Wiley.
11 Karplus, M. and Porter, R.N. (1970). *Atoms and Molecules*. New York: W. A. Benjamin.
12 Kramer, H.L. and Herschbach, D.R. (1970). Combination rules for van der Waals force constants. *J. Chem. Phys.* 53: 2792.
13 Smith, F.T. (1972). Atomic distortion and the combining rule for repulsive potentials. *Phys. Rev. A* 5: 1708.
14 Koide, A., Meath, W.J., and Alltnatt, A.R. (1981). Second order charge overlap effects and damping functions for isotropic atomic and molecular inetractions. *Chem. Phys.* 58: 105.
15 Tang, K.T. and Toennies, J.P. (1984). An improved simple model of the van der Waals potential based on universal damping functions for dispersion coefficients. *J. Chem. Phys.* 80: 3726.
16 Siska, P.S., Parson, J.H., Schafer, T.P., and Lee, Y.T. (1971). Intermolecular potentials from crossed beam differential elastic scattering measurements III. He +He and Ne+Ne. *J. Chem. Phys.* 55: 5762.
17 Buck, U. (1975). Elastic scattering. *Adv. Chem. Phys.* 30: 313.
18 Smith, I.W.M. (1980). *Kinetics and Dynamics of Elementary Gas Reactions*. London, Great Britain: Butterworth.
19 Hirst, D.M. (1985). *Potential Energy Surfaces*. London, England: Taylor & Francis.
20 Gordon, M.D. and Secrest, D. (1970). Helium-atom-hydrogen-molecule potential energy surface employing the LCAO-MO-SCF and CI methods. *J. Chem. Phys.* 52: 120.
21 Edmonds, A.R. (1960). *Angular Momentum in Quantum Mechanics*. Princeton NJ, USA: Princeton University Press.
22 Levine, I.N. (2000). *Quantum Chemistry*, 5e. New Jersey: Prentice-Hall.
23 Mahoney, M.W. and Jorgensen, W.L. (2000). A five-site model for liquid water and the reproduction of the density anomaly by rigid, non-polarizable potential functions. *J. Chem. Phys.* 112: 8910.
24 Karplus, M. (1970). Potential energy surfaces. In: *Molecular Beams and Reaction Kinetics* (ed. C. Schlier), 320. New York: Academic Press.
25 Sato, S. (1955). Potential energy surface of the system of three atoms. *J. Chem. Phys.* 23: 2465.
26 Truhlar, D.G., Steckler, R., and Gordon, M.S. (1987). Potential energy surfaces for polyatomic reaction dynamics. *Chem. Rev.* 87: 217.
27 Polanyi, J.C. (1972). Some concepts in reaction dynamics. *Acc. Chem. Res.* 5: 161.
28 Bender, C.F., Pearson, P.K., O'Neil, S.V., and Schaeffer, H.F. III (1972). Potential energy surface including electron correlation for $F + H_2 \rightarrow FH + H$. I. Preliminary surface. *J. Chem. Phys.* 56: 4626.
29 Bender, C.F., O'Neil, S.V., Pearson, P.K., and Schaeffer, H.F. III (1972). Potential energy surface including electron correlation for $F + H_2 \rightarrow FH + H$: refined linear surface. *Science* 176: 1412.

30 Micha, D.A. (1985). General theory of reactive scattering. In: *Theory of Chemical Reaction Dynamics*, vol. II (ed. M. Baer), 181. Boca Raton, USA: CRC Press.
31 Kittel, C. (2005). *Introduction to Solid State Physics*, 8e. Hoboken, NJ: Wiley.
32 Tersoff, J. (1986). New empirical model for the structural properties of silicon. *Phys. Rev. Lett.* 56: 632.
33 McQuarrie, D.A. (2000). *Statistical Mechanics*. Sausalito CA: University Science Books.
34 Hansen, J.P. and McDonald, I.R. (1986). *Theory of Simple Liquids*, 2e. Orlando FL, USA: Academic Press.
35 Gordon, M.S., Fedorov, D.G., Pruitt, S.R., and Slipchenko, L.V. (2011). Fragmentation methods: a route to accurate calculations on large systems. *Chem. Rev.* 112: 632.
36 Israelachvili, J. (1992). *Intermolecular and Surface Forces*. San Diego CA: Academic Press.
37 Jeziorski, B., Moszynski, R., and Szalewicz, K. (1994). Perturbation theory approach to intermolecular potential energy surfaces of van der Waals complexes. *Chem. Rev.* 94: 1887.
38 Friesner, R.A. and Guallar, V. (2005). Ab initio quantum chemical and mixed quantum mechanics/molecular mechanics (QM/MM) methods for enzymatic catalysis. *Annu. Rev. Phys. Chem.* 56: 389.
39 Albaugh, A., Boateng, H.A., Bradshaw, R.T. et al. (2016). Advanced potential energy surfaces for molecular simulation. *J. Phys. Chem. B* (Feature Article) 120: 9811.
40 Case, D.A., Cheatham, T.E.I., Darden, T. et al. (2005). The AMBER biomolecular simulation programs. *J. Comput. Chem.* 26: 1668.
41 Brooks, B., Brooks, C.L., Mackerell, A.D. et al. (2009). CHARMM: the biomolecular simulation program. *J. Comput. Chem.* 30: 1545.
42 Arnautova, Y.A., Jagielska, A., and Scheraga, H.A. (2006). A new force field (ECEPP-05) for peptides, proteins, and organic molecules. *J. Phys. Chem. B* 110: 5025.
43 Van Der Spoel, D., Lindahl, E., Hess, B. et al. (2005). GROMACS: fast, flexible, and free. *J. Comput. Chem.* 26: 1701.
44 Allinger, N.L., Chen, K., and Lii, J.-H. (1996). An improved force field (MM4) for saturated hydrocarbonds. *J. Comput. Chem.* 17: 642.
45 Hovmöller, S., Zhou, T., and Ohlson, T. (2002). Conformations of amino acids in proteins. *Acta Cryst* D58: 768.
46 Atkins, P. and De Paula, J. (2010). *Physical Chemistry*, 9e. New York: W. H. Freeman and Co.
47 Kaminski, G.A., Stern, H.A., Berne, B.J. et al. (2002). Development of a polarizable force field for proteins via ab initio quantum chemistry: first generation model and gas phase test. *J. Comput. Chem.* 16: 1515.
48 Warshel, A., Kato, M., and Pisliakov, A.V. (2007). Polarizable force fields: history, test cases, and prospects. *J. Chem. Theory Comput.* 3: 2034.

49 Shi, Y., Xia, Z., Zhang, J. et al. (2013). Polarizable atomic multipole based AMOEBA force field for proteins. *J. Chem. Theory Comput.* 9: 4046.
50 Baker, C.M. (2015). Polarizable force fields for molecular dynamics simulations of biomolecules. *WIREs Comput. Mol. Sci* 5: 241.
51 Xie, W., Song, L., Truhlar, D.G., and Gao, J. (2008). The variational explicit polarization potential and analytical first derivative of energy: towards a next generation force field. *J. Chem. Phys.* 128: 234108.
52 Brooks, C.L.I., Karplus, M., and Montgomery-Pettitt, B. (1988). *Proteins: A Theoretical Perspective of Dynamics, Structure, and Thermodynamics*. New York: Wiley-Interscience.
53 Warshel, A. (1991). *Computer Modeling of Chemical Reactions in Enzymes and Solutions*. New York: Wiley.
54 Blank, T.B., Brown, S.D., Calhoun, A.W., and Doren, D.J. (1995). Neural network models of potential energy surfaces. *J. Chem. Phys.* 103: 4129.
55 Brown, D.F.R., Gibbs, M.N., and Clary, D.C. (1996). Combining ab initio computations, neural networks, and diffusion Monte carlo: an efficient method to treat weakly bound molecules. *J. Chem. Phys.* 105: 7597.
56 Raff, L.M., Malshe, M., Hagan, M. et al. (2005). Ab initio potential energy surfaces for complex, multichannel systems using modified novelty sampling and feedforward neural networks. *J. Chem. Phys.* 122: 084104–084101.
57 Behler, J. (2011). Neural network potential energy surfaces in chemistry: a tool for large scale simulations. *Phys. Chem. Chem. Phys.* 13: 17930.
58 Behler, J. (2016). Perspective: machine learning potentials for atomistic simulations. *J. Chem. Phys.* 145: 170901–170901.
59 Lorenz, S., Scheffler, M., and Gross, A. (2006). Description of surface chemical reactions using a neural network representation of the potential-energy surface. *Phys. Rev. B* 73: 115431.
60 Jiang, B. and Guo, H. (2013). Permutation invariant polynomial neural network approach to fitting potential energy surfaces. *J. Chem. Phys.* 139: 054112.
61 Behler, J. and Parrinello, M. (2007). Generalized neural network representation of high-dimensional potential-energy surfaces. *Phys. Rev. Lett.* 98: 146401–146401.
62 Smith, J.S., Isayev, O., and Roitberg, A.E. (2017). ANI-1: an extensible neural network potential with DFT accuracy at force-field computational cost. *Chem. Sci.* 8: 3192.
63 Ramakrishnan, R., Dral, P.O., Rupp, M., and von Lilienfeld, O.A. (2015). Big data meets quantum chemistry approximations: the delta-machine learning approach. *J. Chem. Theory Comput.* 11: 2087.

5

Intermolecular States

CONTENTS

Molecular Interactions: Concepts and Methods, First Edition. David A. Micha.

5.1 Molecular Energies for Fixed Nuclear Positions

5.1.1 Reference Frames

We must specify the positions of all the particles in our system in a suitable reference frame, with a chosen origin of coordinates and with axes oriented in a specified way. We begin with a laboratory reference frame or laboratory system $\mathcal{S}_L$ attached to a measuring device in the laboratory with axes x, y, and z oriented along directions given by unit vectors $\vec{n}_x, \vec{n}_y, \vec{n}_z$. It is however convenient in theoretical studies to work in a center-of-mass reference frame. To simplify, assume that only two nuclei a and b are present, of masses m_a and m_b, at positions $\vec{R}_a$ and $\vec{R}_b$, and that electron masses can be ignored to start with. The center of mass and relative positions are then

$$\vec{R}_{CMN} = \frac{\left(m_a\vec{R}_a + m_b\vec{R}_b\right)}{(m_a + m_b)}$$

$$\vec{R} = \vec{R}_b - \vec{R}_a$$

and we can translate the reference frame keeping axes parallel so that in a new reference frame $\mathcal{S}_{CMN}$ called the center-of-mass of the nuclei, space-fixed (CMN-SF) frame, we have $\vec{R}_{CMN} = 0$ and the (x, y, z) axes pointing along the previous unit vectors. A further convenient transformation, when we describe the properties of a molecule in an equilibrium conformation, is to reorient the axes to coincide with the principal axes of inertia of the nuclear framework, along new unit vectors $\vec{n}'_x, \vec{n}'_y, \vec{n}'_z$. This can be done specifying three Euler angles $(\alpha, \beta, \gamma) = \Gamma$ of the new frame called now the center-of-mass of the nuclei, body-fixed (CMN-BF) frame, designated as $\mathcal{S}'_{CMN}$. New coordinates (x', y', z') are related to the old ones by a 3×3 rotation matrix $\boldsymbol{A}(\alpha,\beta,\gamma) = \boldsymbol{R}_{z'}(\gamma)\boldsymbol{R}_y(\beta)\boldsymbol{R}_z(\alpha)$, a function of the Euler angles constructed from three axial rotations [1]. The three mentioned reference systems are shown in Figure 5.1.

In the case of intermolecular potentials, one usually finds two molecules A and B, each with its own CMN-BF reference frame $\mathcal{S}'_I$, $I = A, B$, with origins at their centers of mass and oriented by Euler angles Γ_I with respect to an overall CMN-SF frame located along the line joining the centers of mass of A and B. This is shown in Figure 5.2.

The remaining nuclear coordinates $\boldsymbol{q}_I$ are internal ones, such as bond distances, bond angles, and torsion angles. Therefore, the whole system of N_{nu} nuclei is described by the set of $3N_{nu}$ variables $\boldsymbol{Q} = \left(\vec{R}, \Gamma_A, \boldsymbol{q}_A, \Gamma_B, \boldsymbol{q}_B\right)$.

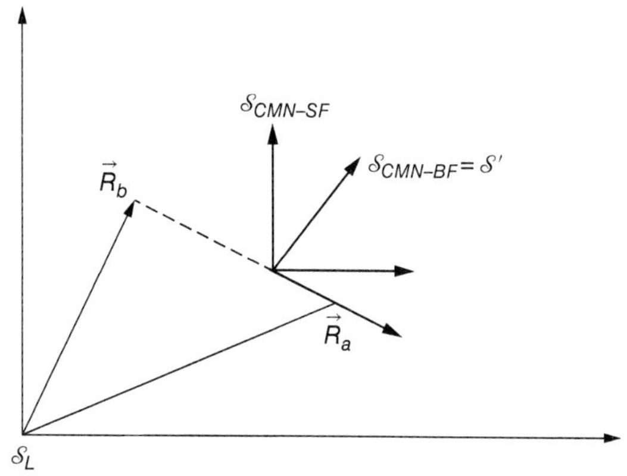

Figure 5.1 Reference frames for laboratory (L), center-of-mass of nuclei with space-fixed (CMN-SF) orientation of orthogonal axes, and center-of-mass of nuclei with body-fixed (CMN-BF) orientation of axes.

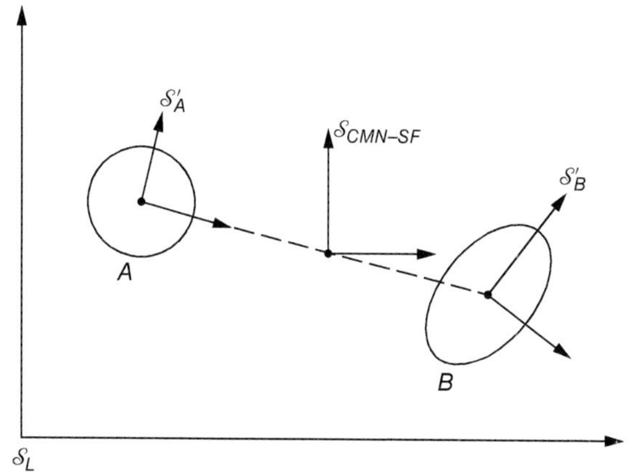

Figure 5.2 Two molecules *A* and *B* oriented in a common CMN-SF reference frame and with their own local reference frames.

The stationary state wavefunction Ψ of two interacting molecules is obtained from the Schrödinger equation

$$\left[\hat{\boldsymbol{P}}_Q^{\dagger}\frac{1}{2\boldsymbol{M}}\hat{\boldsymbol{P}}_Q + \hat{H}_Q\right]\Psi(\boldsymbol{Q},\boldsymbol{X}) = E\,\Psi(\boldsymbol{Q},\boldsymbol{X})$$

where we have introduced the electronic position and spin variables $\boldsymbol{x}_i = (\vec{r}_i, \zeta_i)$ of electron i and the sets $\boldsymbol{X} = (\boldsymbol{x}_1, \boldsymbol{x}_2, \ldots)$ for N_{el} electrons, the kinetic energy operator for the nuclei written in terms of $3N_{nu}$ mass weighted momentum operators $\hat{\boldsymbol{P}}_Q$ in a column $1 \times 3N_{nu}$ matrix and its adjoint $\hat{\boldsymbol{P}}_Q^{\dagger}$ row matrix, with $\boldsymbol{M}$ a $3N_{nu} \times 3N_{nu}$ matrix of masses, and the Hamiltonian operator $\hat{H}_Q$ for fixed nuclear positions. The wavefunction must satisfy normalization conditions for bound states or asymptotic conditions for scattering states. The solution to this equation can best be found with a preliminary treatment for the nuclei at fixed positions, leading to the equation

$$\hat{H}_Q\Phi(\boldsymbol{X};\boldsymbol{Q}) = E^{(el)}(\boldsymbol{Q})\,\Phi(\boldsymbol{X};\boldsymbol{Q})$$

where the electronic wavefunction $\Phi(X;\mathbf{Q})$ depends parametrically on nuclear position variables, and the electronic energy varies with those variables. This seems to indicate that all the details of a many-electron wavefunction must be known to obtain the system energy, but this is not the case since as shown next all what is needed is knowledge of the electronic density.

5.1.2 Energy Density Functionals for Fixed Nuclei

Starting with the electronic number density function $\rho(\vec{r};\mathbf{Q})$ per unit volume in a molecular system with fixed nuclear positions $\mathbf{Q}$, a knowledge of the electronic cusp positions and isocontour slopes of the density provide enough information to locate the nuclei and to find their nuclear charges. Therefore, the density uniquely defines the Coulomb potential energy due to N_{nu} nuclei with charges c_α, $\alpha = 1$ to N_{nu}, interacting with N_{el} electrons and among themselves, given by the Hamiltonian terms in

$$\hat{H}_{nu} = \hat{H}_{ne} + \hat{H}_{nn}$$

and appearing in a Hamiltonian operator containing also the electronic kinetic energy and electron–electron Coulomb interactions,

$$\hat{H}_Q = \hat{K}_{el} + \hat{H}_{ee} + \hat{H}_{nu}$$

Since the electronic kinetic and electron–electron repulsion terms are known, we expect on physical grounds that knowing the electronic density of the ground electronic state of the system will determine its Hamiltonian and therefore all its properties. This can be proved by the following theorem and its corollary [2]. The theorem states that the electronic energy $E(\mathbf{Q})$ of a nondegenerate ground electronic state is a unique functional of the electronic density ρ. To reproduce the proof, we introduce first the separation of electronic variables in the set $X = (\boldsymbol{x}_i, \boldsymbol{X}^{(i)})$ for N_{el} electrons, the normalized electronic wavefunction $\Phi(X;\mathbf{Q})$, and the electronic density as

$$\rho(\vec{r};\mathbf{Q}) = \int dX\, \Phi(X;\mathbf{Q})^* \left[\sum\nolimits_{i=1}^{N_{el}} \delta(\vec{r}-\vec{r}_i)\right] \Phi(X;\mathbf{Q}) = N_{el}P(\vec{r})$$

$$P(\vec{r}) = \sum\nolimits_{\zeta_i} \int dX^{(i)} |\Phi(X;\mathbf{Q})|^2_{\vec{r}_i=\vec{r}}$$

and note that the Coulomb interaction between nuclei and electrons

$$\hat{H}_{ne} = \sum_i v(\vec{r}_i),\ v(\vec{r}_i) = \frac{1}{4\pi\varepsilon_0}\sum\nolimits_{\alpha=1}^{N_{nu}} \frac{C_\alpha C_e}{|\vec{R}_\alpha - \vec{r}_i|}$$

has an expectation value

$$\langle \hat{H}_{ne} \rangle = \int d^3r\, \rho(\vec{r})\, v(\vec{r})$$

To prove the theorem by contradiction, we assume that for a given density $\rho(\vec{r})$, there may be two different nuclear potential energy operators $\hat{H}_{nu}$ and $\hat{H}'_{nu}$, which differ only in the magnitudes of the nuclear charges and lead to different states and energies. We use the variational theorem that gives an upper bound to the exact ground state energy, written for the primed quantities, to find

$$E'(\mathbf{Q}) = \left\langle \Phi'(\mathbf{Q}) \mid \hat{H}'_{Q} \mid \Phi'(\mathbf{Q}) \right\rangle < \left\langle \Phi(\mathbf{Q}) \mid \hat{H}'_{Q} \mid \Phi(\mathbf{Q}) \right\rangle$$
$$E'(\mathbf{Q}) < \langle \Phi(\mathbf{Q}) \mid (\hat{H}_{Q} - \hat{H}_{nu} + \hat{H}'_{nu}) \mid \Phi(\mathbf{Q}) \rangle$$
$$E'(\mathbf{Q}) < E(\mathbf{Q}) + \int d^3r\, \rho(\vec{r}) \left[v'(\vec{r}) - v(\vec{r}) \right] + H'_{nn} - H_{nn}$$

Reversing primed and unprimed quantities, the last three terms only change sign and when we add the two expressions they cancel and we find

$$E(\mathbf{Q}) + E'(\mathbf{Q}) < E'(\mathbf{Q}) + E(\mathbf{Q})$$

which is a contradiction resulting from the assumption that two different energies could arise from the same density. This establishes the one-to-one correspondence between the energy functional and the density. As a corollary, we find that if $E[\rho(\vec{r};\mathbf{Q})]$ is the ground state energy functional of the electronic density for a given nuclear conformation, then minimizing this functional leads to an upper bound to the exact ground state energy, which is yet a function of the nuclear position variables.

The potential energy function for species A and B at relative position $\vec{R}$ is obtained as

$$V_{AB}(\mathbf{Q}) = E[\rho(\vec{r};\mathbf{Q})] - E[\rho(\vec{r};\mathbf{Q})]_{R\to\infty}$$

provided the energy functional has been constructed to satisfy $E[\rho(\vec{r};\mathbf{Q})]_{R\to\infty} = E_A[\rho_A(\vec{r};\mathbf{Q})] + E_B[\rho_B(\vec{r};\mathbf{Q})]$.

This theorem justifies efforts to construct energy functionals of the electronic density and to obtain from them upper bounds to the exact ground state energy with a constrained variational procedure. A related alternative statement is that the energy can be written as a functional of a wavefunction that has been constrained to give the electronic density while minimizing the sum of electronic kinetic and Coulomb interaction energies [3]. This further justifies a procedure based on construction of a wavefunction constrained to give the density of a collection of independent electrons interacting in accordance with a Hamiltonian containing electron correlations [4]. Functionals can be constructed in different ways depending on whether atoms and molecules are chemically bonded or not. Two ways of generating them are the Kohn–Sham procedure [4], which makes use of a total density written as a sum of one-electron terms, and the Gordon–Kim procedure [5], giving the total density as a sum of the densities of molecular components of the whole system. Many details of the DFT treatments can be found in reference [6].

5.1.3 Physical Contributions to the Energy Density Functional

The Hohenberg–Kohn decomposition of the molecular energy of electrons and nuclei for fixed nuclear positions is [2]

$$E[\rho] = \int d^3r\, \rho(\vec{r})\nu(\vec{r}) + F[\rho] + H_{nn}$$

$$F[\rho] = \langle \Phi(\mathbf{Q}) \mid (\hat{K}_e + \hat{H}_{ee}) \mid \Phi(\mathbf{Q}) \rangle_\rho = V_{Coul} + G[\rho]$$

$$V_{Coul} = \frac{1}{2}\int d^3r d^3r' \frac{C_e{}^2}{4\pi\varepsilon_0 \,|\vec{r}-\vec{r}'|}\left[\rho(\vec{r})\rho(\vec{r}') - \delta(\vec{r}-\vec{r}')\rho(\vec{r})\right]$$

$$G[\rho] = G_{kin}[\rho] + G_{xc}[\rho]$$

where V_{Coul} is the classical Coulomb energy of interaction among all the electrons, a Hartree-like (purely electrostatic) energy, with the second term in V_{Coul} introduced to prevent electron–electron self-interactions. The two terms in $G[\rho]$ come from the electronic kinetic energy, and combined electronic exchange and correlation energies, with the second term containing the difference between the exact energy of the electron system and the Hartree energy. Once this functional is constructed, one can proceed to minimize it for variations of the electron density that conserve the number of electrons so that $\int d^3r\, \rho(\vec{r}) = N_{el}$. The variation of the constrained functional $E[\rho] - \mu \int d^3r\; \rho(\vec{r})$, where μ has the meaning of a chemical potential at absolute zero temperature, leads to the equation

$$\frac{\delta E}{\delta\rho(\vec{r})} - \mu = 0$$

The minimum for $E[\rho]$ provides an upper bound to the ground state electronic energy and the optimal density $\rho(\vec{r};\mu)$.

In the Kohn–Sham (K–S) procedure, the density is written as a sum of terms containing the spin-orbitals ψ_j from each electron, for spin up and down variables $\zeta = (+, -)$,

$$\rho(\vec{r}) = \sum\nolimits_{\zeta} \sum\nolimits_{j=1}^{N_{el}} \left|\psi_j(\vec{r}, \zeta)\right|^2 = \rho^{(+)}(\vec{r}) + \rho^{(-)}(\vec{r})$$

and these K–S functions are used to construct an antisymmetrized determinantal (independent-electron) wavefunction $\Phi^{(KS)}(\mathbf{X}; \mathbf{Q}) = \langle \mathbf{X} \mid \Phi^{(KS)}(\mathbf{Q})\rangle$ for fixed nuclear positions, from which single-electron kinetic energy and total functionals can be calculated as

$$\Phi^{(KS)}(\boldsymbol{X};\boldsymbol{Q}) = C\hat{A}\prod_j \psi_j(\vec{r}_j,\zeta_j)$$

$$K_s[\rho] = \left\langle \Phi^{(KS)}(\boldsymbol{Q}) \mid \hat{K}_e \mid \Phi^{(KS)}(\boldsymbol{Q}) \right\rangle_\rho$$

$$F[\rho] = V_{Coul}[\rho] + K_s[\rho] + E_{xc}[\rho]$$

The last term is an exchange-correlation functional (which also includes a contribution to correct the kinetic energy functional). This is parametrized and used in calculations. Improvements follow using separate functionals for up and down spins, needed to describe bonding between open shell species, and using functionals of both the density and its gradient, to account for inhomogeneous charge distributions.

The K–S spin-orbitals are obtained solving a one-electron eigenvalue equation

$$\left[-\frac{\hbar^2}{2m_e}\nabla^2 + v^{(KS)}(\vec{r},\zeta)\right]\psi_j(\vec{r},\zeta) = \epsilon_j\psi_j(\vec{r},\zeta)$$

where $v^{(KS)}(\vec{r},\zeta) = \delta E/\delta\rho(\vec{r},\zeta)$ is the K–S one-electron potential energy function. It is helpful to have in mind that the functional derivative of the energy can be defined in integral form by $E = \int d^3r\,\delta\rho(\vec{r})\,[\delta E/\delta\rho(\vec{r})]$ so that the physical dimensions of $\delta E/\delta\rho(\vec{r})$ are energy divided by density times volume element, again an energy quantity.

The Gordon–Kim (G–K) procedure, suitable for interacting systems that do not show chemical bonding, constructs the density of the pair AB from given separate densities $\rho_A(\vec{r})$ and $\rho_B(\vec{r})$, usually calculated at the Hartree–Fock level, as $\rho(\vec{r}) = \rho_A(\vec{r}) + \rho_B(\vec{r})$, and obtains potential energy functions for short and intermediate distances from functionals $G[\rho]$ of this density. Extensions have been introduced to also add long-range van der Waals interactions in terms of additional density changes $\Delta\rho(\vec{r})$ generated by displacing individual density components, and introducing restoring forces. More details of the K–S and G–K procedures can be found in Chapter 6 on many-electron treatments.

5.2 General Properties of Potentials

5.2.1 The Electrostatic Force Theorem

The electrostatic force theorem shows that once the electronic charge distribution is known from molecular (quantal) calculations, it is possible to calculate the forces on nuclei using classical electrostatics, applied to a collection of positive nuclear charges plus the negative charge distribution of the electrons [7].

Consider two nuclei a and b on the z-axis, with relative position $\vec{R}$, surrounded by electrons in a state Φ_k with energy E_k. This can be a ground or excited electronic state, and here we assume it is noninteracting with other states. We find the force on nucleus a by differentiation of the expectation value of the energy operator with respect to its position coordinate Z_a. We begin with the Schrödinger equation and

$$\frac{\partial}{\partial Z_a}\int dX\,\Phi_k\left(X;\vec{R}\right)^*\left[\hat{H}_R - E_k(R)\right]\Phi_k\left(X;\vec{R}\right) = 0$$

Expanding this expression, displaying the Coulomb terms in the Schrödinger equation and identifying the force on a, we have with $\vec{r}_{1a} = \vec{r}_1 - \vec{R}_a$,

$$F_{a,z} = -\frac{\partial E_k}{\partial Z_a} = -\frac{C_a C_b Z_{ab}}{4\pi\varepsilon_0 R^3} + \int d^3 r_1 \frac{C_e \rho\left(\vec{r}_1\right) C_a z_{1a}}{4\pi\varepsilon_0\, r_{1a}^3}$$

This is the electrostatic force theorem, formally identical to the corresponding classical mechanical expression, but now constructed from a quantal electronic density. The first term to the right is negative and shows the repulsive force on a from the nucleus b; the integrand in the second term is the Coulomb force between an electronic distribution at position $\vec{r}_1$ and nucleus a, with positive or negative integrand values depending on the location of the element of volume $d^3 r_1$ along the z-direction. If the element of volume of electrons is found between two parallel planes perpendicular to the axis through the nuclei, then it gives a positive force increment contributing to bonding. Far outside these planes, the volume will oppose bonding. The first region is called the bonding region, the second one is the antibonding region. They are separated by a surface where the electronic force densities in $F_{\mu,z} = \int d^3 r f_{\mu,z}\left(\vec{r}\right)$ at each nucleus $\mu = a, b$ are equal, or

$$f_{a,z}\left(\vec{r}\right) = f_{b,z}\left(\vec{r}\right)$$

This defines boundary surfaces, $B\left(\vec{r}\right) = f_{a,z}\left(\vec{r}\right) - f_{b,z}\left(\vec{r}\right) = 0$, showing the transition between an inner bonding region and two outer antibonding regions, one for each nucleus. Given an electronic charge distribution, this tells us whether the nuclei undergo a net attractive or repulsive force. The potential energy follows by integration of the relative force over the internuclear distance.

5.2.2 Electrostatic Forces from Approximate Wavefunctions

Electronic wavefunctions are usually known only within approximations obtained from a variational procedure or from perturbation theory. In these cases, the theorem must be applied considering the nature of the approximations.

For a trial wavefunction $\Phi\left(X;\vec{R},\lambda\right)$ dependent on a set of variational parameters $\lambda = \{\lambda_n\}$, the variational energy functional is

$$\mathcal{E}\left[\Phi\left(\vec{R},\lambda\right)\right] = \left\langle\Phi\left(\vec{R},\lambda\right)\middle|\hat{H}_R\middle|\Phi\left(\vec{R},\lambda\right)\right\rangle \Big/ \left\langle\Phi\left(\vec{R},\lambda\right)\middle|\Phi\left(\vec{R},\lambda\right)\right\rangle$$

optimized to obtain a minimal energy when the parameters are varied and calculated at the minimum for each distance R. This gives an energy $E_{min}(R) = \mathcal{E}\left[\Phi\left(\vec{R},\lambda_{min}\right)\right]$ from which the force at nucleus a follows from the gradient as

$$F_{a,z} = -\frac{\partial E_{min}}{\partial Z_a} = -\left(\frac{\partial E_{min}}{\partial Z_a}\right)_{\lambda_{min}} - \sum\nolimits_n \frac{\partial E_{min}}{\partial \lambda_n}\frac{\partial \lambda_n}{\partial Z_a}$$

where the first term to the right gives the force from a frozen trial function, while the second term gives an additional force contribution due to changes of parameters with the nuclear distance. The parameters may include coefficients of atomic orbitals in molecular orbitals, giving a MO distortion force; exponents in atomic orbitals, giving an AO distortion force; or coefficients in a combination of electronic configurations, giving a configuration interaction force.

When the electronic state Φ derives from a perturbation expansion, with the energy given by a perturbation series $E = E^{(0)} + E^{(1)} + E^{(2)} + \ldots$, the force appears also as a perturbation sum $F_{a,z} = F_{a,z}^{(0)} + F_{a,z}^{(1)} + F_{a,z}^{(2)} + \ldots$ obtained from the energy gradient, with the leading term as in the theorem expression but with the unperturbed electronic density $\rho^{(0)}\left(\vec{r}_1\right)$. Additional corrections involve density changes $\rho^{(1)}\left(\vec{r}_1\right)$, $\rho^{(2)}\left(\vec{r}_1\right)$, ... obtained from the perturbed wavefunction.

5.2.3 The Example of Hydrogenic Molecules

Consider a diatomic molecule with two electrons such as H_2, HD, HHe^+, ..., or a diatomic with only two active electrons outside closed shells of ion cores, such as LiH, with charge numbers ζ_μ, $\mu = a, b$, for the two cores. We can construct a two-electron wavefunction to describe the whole range of distances R and with the correct asymptotic form for $R \to \infty$. This can be done with molecular orbitals (MOs) written as linear combinations of atomic orbitals (AOs) with variable exponents. Two-electron determinants describe configurations and their combination provide the required total wavefunction with the correct asymptotic form. The electronic energy can be obtained from the expectation value of the electronic Hamiltonian in a variational minimization of the function of all the parameters, or fixing some parameters and varying the remaining ones. The electrostatic force is then obtained as shown above from the derivatives of all the varied parameters with respect to atomic distances.

The AO at center a in the two-electron systems contains an effective nuclear charge as the distance-dependent exponential $\zeta_a(R)$, varying from the united atom value for the diatomic to the value for the isolated atom. For a 1s orbital it is given by $\chi_a(\vec{r}) = (\zeta_a/a_0)^{3/2} 2\exp(-2\zeta_a r_a/a_0).(2\pi^{1/2})^{-1}$, where a_0 is the Bohr radius, and with r_a the distance from the electron to nucleus a. A molecular orbital formed as a linear combination of atomic orbitals (a MO-LCAO) is $\varphi_j(\vec{r}) = c_{aj}\chi_a(\vec{r}) + c_{bj}\chi_b(\vec{r})$, $j = 1, 2$, normalized to $\langle\varphi_j \mid \varphi_j\rangle = 1$ and with two variational parameters $c_{\mu j}(R)$ which can be obtained minimizing the diatomic electronic energy for fixed nuclei. A two-electron determinantal function is of form $\Phi_J(\boldsymbol{x}_1, \boldsymbol{x}_2) = |\varphi_j\vartheta_j(1)\varphi_k\vartheta_k(2)|$, where ϑ_j is a spin state (α *or* β), and a general two-electron state can be written as a superposition of configurations (a configuration interaction, CI)

$$\Phi^{(\mathrm{CI})}(\boldsymbol{x}_1, \boldsymbol{x}_2) = \sum\nolimits_J A_J \Phi_J(\boldsymbol{x}_1, \boldsymbol{x}_2)$$

with all terms of the same space and spin symmetry. The energy follows from the variational functional $\mathcal{E}\left[\Phi^{(CI)}\right] = E\left(\{\zeta_\mu, c_{\mu j}, A_J\}\right)$ and forces are given by

$$F_{a,z} = -\left(\frac{\partial E}{\partial Z_a}\right)_\lambda - \sum\nolimits_\mu \frac{\partial E}{\partial \zeta_\mu}\frac{\partial \zeta_\mu}{\partial Z_a} - \sum\nolimits_{\mu j} \frac{\partial E}{\partial c_{\mu j}}\frac{\partial c_{\mu j}}{\partial Z_a} - \sum\nolimits_J \frac{\partial E}{\partial A_J}\frac{\partial A_J}{\partial Z_a}$$

with $\lambda = \{\zeta_\mu, c_{\mu j}, A_J\}$ the set of variational parameters. This expression shows how the forces are affected by changing AO exponents, MO coefficients, and coefficients of CIs when these parameters have not been optimized by minimization of the energy. If instead they have been optimized, then the last three summations are zero and the force is given by the gradient of the energy calculated at the optimized parameter values.

5.2.4 The Virial Theorem

When interaction among particles are restricted to Coulomb forces, it is possible to find general relations between expectation values of kinetic and potential energies, known as virial theorems, applicable to situations where a physical system is in a stationary state. They have been applied to treatments of chemical bonding for cases where interatomic distances are fixed, or where the system is at equilibrium in an energy minimum.

The starting point is the Hellmann–Feynman theorem which follows immediately from the Schrödinger equation for a normalized quantum state $|\Psi(\lambda)\rangle$ dependent on a free parameter λ and satisfying the equation $\hat{H}(\lambda) \mid \Psi(\lambda)\rangle = E(\lambda) \mid \Psi(\lambda)\rangle$ for a molecular Hamiltonian $\hat{H}(\lambda)$ and with $E(\lambda)$ the eigenenergy. Differentiation of both sides with respect to λ, projecting on the left into $\langle\Psi(\lambda)|$, and using the equation and its adjoint, gives the Hellmann–Feynman relation

$$\frac{\partial E}{\partial \lambda} = \left\langle \Psi(\lambda) \middle| \frac{\partial \hat{H}}{\partial \lambda} \middle| \Psi(\lambda) \right\rangle$$

We consider a special case where the Hamiltonian is $\hat{H}_{el} = \hat{H}_{el}^{(kin)} + \hat{H}_{el}^{(Coul)}$, a sum of electronic kinetic and Coulomb electron–electron and electron–nucleus energies in $\hat{H}_{el}^{(Coul)} = \hat{H}_{ee} + \hat{H}_{en}$, for fixed nuclear positions and excluding the Coulomb interaction among nuclei. The corresponding electronic eigenstate is $|\Phi_{el}(R)\rangle$ with eigenenergy $E_{el}(R)$ for a relative position vector $\vec{R} = R\vec{n}_z$ along the z-axis. A parameter λ is introduced to scale the position coordinates of electrons, but not of the nuclei, writing $\lambda\vec{r}_j = \vec{s}_j$ for electron j and changing integration variables in expectation values to the set $\{\vec{s}_j\}$. The electronic energy becomes $E_{el}(\lambda) = \lambda^2 \langle \hat{H}_{el}^{(kin)} \rangle_R + \lambda \langle \hat{H}_{el}^{(Coul)}(\lambda R) \rangle_R$ and differentiation with respect to the parameter λ, using that

$$\frac{\partial}{\partial \lambda} \frac{1}{|\vec{s} - \lambda R \vec{n}_z|} = \left[\frac{\partial}{\partial(\lambda R)} \frac{1}{|\vec{s} - \lambda R \vec{n}_z|} \right] \frac{\partial(\lambda R)}{\partial \lambda}$$

in the expectation value for the Coulomb Hamiltonian, gives for $\lambda = 1$ a virial theorem valid for any fixed internuclear distance,

$$2\left\langle \hat{H}_{el}^{(kin)} \right\rangle_R + \left\langle \hat{H}_{el}^{(Coul)}(R) \right\rangle_R = -R\partial E_{el}/\partial R$$

where the Hellmann–Feynman theorem has been used for the right-hand side. This can be combined with $\langle \hat{H}_{el}^{(kin)} \rangle_R + \langle \hat{H}_{el}^{(Coul)}(R) \rangle_R = E_{el}(R)$ to obtain expectation values of separate electronic kinetic and Coulomb energies,

$$\left\langle \hat{H}_{el}^{(kin)} \right\rangle_R = -R\partial E_{el}/\partial R - E_{el}(R)$$

$$\left\langle \hat{H}_{el}^{(Coul)}(R) \right\rangle_R = 2E_{el}(R) + R\partial E_{el}/\partial R$$

showing how they depend on the internuclear distance. At equilibrium for the minimum of the electronic energy, where $\partial E_{el}/\partial R = 0$, one recovers the virial equality for Coulomb interactions,

$$2\left\langle \hat{H}_{el}^{(kin)} \right\rangle_{min} = -\left\langle \hat{H}_{el}^{(Coul)}(R_{min}) \right\rangle$$

relating the electronic kinetic and Coulomb energies for equilibrium nuclear positions. The expectation values for kinetic and Coulomb energies obtained from the virial relations, in the case of variable distances, are more accurate than their values when calculated from variationally optimized state averages. This follows because the errors in the wavefunction obtained in a variational calculation are of first order with respect to the exact wavefunction, but errors are only of second order for the optimized variational energy that appears in the virial relations.

The derivation for a diatomic system can easily be generalized to a polyatomic system where atoms are located at positions $\vec{R}_a$, $a = 1$ to N, in which case the virial relation is

$$2\left\langle \hat{H}_{el}^{(kin)} \right\rangle_{\boldsymbol{R}} + \left\langle \hat{H}_{el}^{(Coul)}(\boldsymbol{R}) \right\rangle_{\boldsymbol{R}} = -\sum\nolimits_a \vec{R}_a . \partial E_{el} / \partial \vec{R}_a$$

with $\boldsymbol{R} = \{\vec{R}_a\}$ the set of all atomic positions, and with electronic energy gradients calculated for each atomic displacement.

When the exact wavefunction is not known (as usual), then it is necessary to instead differentiate the previously introduced functional $\mathcal{E}\left[\Phi\left(\vec{R},\lambda\right)\right]$. Assuming that the approximate wavefunction has been normalized to $\left\langle \Phi\left(\vec{R},\lambda\right) \middle| \Phi\left(\vec{R},\lambda\right) \right\rangle = 1$, one finds that

$$\frac{\partial \mathcal{E}}{\partial \lambda} = \left\langle \Phi(\lambda) \middle| \frac{\partial \hat{H}}{\partial \lambda} \middle| \Phi(\lambda) \right\rangle + \left\langle \frac{\partial \Phi}{\partial \lambda} \middle| \hat{H}(\lambda) \middle| \Phi(\lambda) \right\rangle + \left\langle \Phi(\lambda) \middle| \hat{H}(\lambda) \middle| \frac{\partial \Phi}{\partial \lambda} \right\rangle$$

which can again be analyzed in terms of scaled kinetic and Coulomb components but with results depending now on how the states Φ depend on the parameter λ by construction.

5.2.5 Integral Form of the Virial Theorem

The Hellmann–Feynman theorem can be used in its integral version to obtain changes in a total energy due to changes in the Hamiltonian operator. Let the Hamiltonian between interacting species A and B be written as $\hat{H}(\lambda) = \hat{H}_A + \hat{H}_B + \lambda \hat{H}_{AB}$. Then provided the state $\Psi(\lambda)$ is an exact eigenstate with energy $E(\lambda) = \langle \Psi(\lambda) \mid \hat{H}(\lambda) \mid \Psi(\lambda) \rangle$, one finds that $\partial E / \partial \lambda = \langle \Psi(\lambda) \mid \hat{H}_{AB} \mid \Psi(\lambda) \rangle$ and integrating,

$$\Delta E = E - E_0 = \int_0^1 d\lambda \left\langle \Psi(\lambda) \middle| \hat{H}_{AB} \middle| \Psi(\lambda) \right\rangle$$

for the energy change due to the interaction of A and B, with $E(1) = E$ and $E(0) = E_0$.

However, when the state $\Psi(\lambda)$ is only approximately known as $\Phi(\lambda)$ from a variational or perturbative procedure, the relevant derivative $\partial \mathcal{E} / \partial \lambda$ contains additional terms and the integral form becomes

$$\Delta \mathcal{E} = \int_0^1 d\lambda \left[\left\langle \Phi(\lambda) \middle| \hat{H}_{AB} \middle| \Phi(\lambda) \right\rangle + \left\langle \left(\frac{\partial \Phi}{\partial \lambda} \right) \middle| \hat{H}(\lambda) \middle| \Phi(\lambda) \right\rangle + \left\langle \Phi(\lambda) \middle| \hat{H}(\lambda) \middle| \left(\frac{\partial \Phi}{\partial \lambda} \right) \right\rangle \right]$$

which cannot be further generally transformed insofar as $\partial\Phi/\partial\lambda$ depends on how the wavefunction was constructed. In the simplest special case, with the approximation $\Phi = \mathcal{A}_{AB}\Phi_A\Phi_B$, the antisymmetrized product of normalized noninteracting states of A and B, the result is $\Delta\mathcal{E} = \mathcal{E}_{AB} - \mathcal{E}_A - \mathcal{E}_B = \langle \mathcal{A}_{AB}\Phi_A\Phi_{BA} \,|\, \hat{H}_{AB} \,|\, \mathcal{A}_{AB}\Phi_A\Phi_B \rangle$, as expected.

The integral expressions can also be used when the wavefunction has been obtained from a density functional approximation, in which case the integrals become functionals of the electron density and of its derivative with respect to the parameter λ.

5.3 Molecular States for Moving Nuclei

5.3.1 Expansion in an Electronic Basis Set

A complete description of molecular states involves moving nuclei. These must be rigorously treated using quantum mechanics and wavefunctions dependent on the nuclear variables. The full Hamiltonian operator must contain also the kinetic energy operator for the nuclei, and one must allow for the interaction of electronic and nuclear degrees of freedom due to moving nuclei [8–14].

We work in a reference frame with space-fixed axes and its origin at the center of mass of the nuclei, assuming that electron masses are negligible (the center of mass of the nuclei frame CMN-SF). To simplify matters, we work here with only two nuclei a and b of masses m_a and m_b. Their reduced mass is $M = m_a m_b / (m_a + m_b)$ and their relative position is $\vec{R}$. The set of electron variables is given by $X = \{\vec{r}_i, \zeta_i\}$ in terms of the position vector and spin variables of electrons i.

Molecular states are described by solutions of the Schrödinger equation

$$\hat{H}\Psi = E\Psi$$

where $\hat{H}$, E, and Ψ are the Hamiltonian operator, energy, and state of the whole system of electrons and nuclei. The state Ψ must be normalized to describe a bound system, or must satisfy asymptotic boundary conditions for an unbound system at large R.

The Hamiltonian operator can be separated into the kinetic energy operator of the relative nuclear motion plus the Hamiltonian $\hat{H}_R$ for fixed nuclei, with the latter including the electronic kinetic energy, the Coulomb attraction of electrons to nuclei, the Coulomb repulsion among electrons and the repulsion among the nuclei, as

$$\hat{H} = -\frac{\hbar^2}{2M}\nabla_R^2 + \hat{H}_R$$

$$\hat{H}_R = \hat{K}_e + \hat{H}_{ee} + \hat{H}_{en} + \hat{H}_{nn}$$

If some of the electrons are found to form rigid ion cores, then the electron-nucleus interaction must be replaced by an electron-core interaction which is Coulombic at large distance and repulsive at short distance due to electronic exclusion; similarly the nucleus–nucleus interaction must be replaced by the core–core interaction.

Many-electron states of molecules can be constructed from many-electron atomic states, in an approach known as the valence-bond (VB) method, or from atomic orbitals combined to form molecular orbitals, in the so-called molecular orbital (MO) method, as we have seen with the example of H_2 in the previous Chapter 3.

The molecular wavefunction Ψ can be expanded in a basis of orthonormal electronic states Φ_k which are functions of the electronic variables and depend parametrically on the nuclear positions. Using the Dirac notation for electronic states we write $\Phi_k(\mathbf{X};\vec{R}) = \langle \mathbf{X} \mid \Phi_k(\vec{R})\rangle$ and work in a linear space of K states, introducing an orthonormal and complete basis set of such states for fixed nuclei, the expansion is

$$\Psi\left(\vec{R},\mathbf{X}\right) = \sum_{k=1}^{K} \Phi_k\left(\mathbf{X};\vec{R}\right)\psi_k\left(\vec{R}\right)$$

$$\left\langle \Phi_k\left(\vec{R}\right) \middle| \Phi_l\left(\vec{R}\right) \right\rangle = \delta_{kl}$$

$$\sum_{k=1}^{K} \left|\Phi_k\left(\vec{R}\right)\right\rangle\left\langle \Phi_k\left(\vec{R}\right)\right| = \hat{I}$$

The coefficients ψ_k are functions of the nuclear positions that describe the nuclear motions in the electronic states k. The bracket notation here and in what follows signifies integration and sum over all the electron variables, for fixed nuclei as shown. The last line gives an assumed completeness relation, with $\hat{I}$ the identity operator.

Several choices are possible for the basis set of electronic states, which lead to different electronic representations of potential energies and nuclear amplitudes. These representations are related by matrix transformations which can be conveniently introduced to simplify treatments of molecular spectroscopy and dynamics [15–19].

5.3.2 Matrix Equations for Nuclear Amplitudes in Electronic States

Introducing a matrix notation with rows and columns shown as boldface symbols, we have that

$$\left|\Psi\left(\vec{R}\right)\right\rangle = \left|\mathbf{\Phi}\left(\vec{R}\right)\right\rangle \boldsymbol{\psi}\left(\vec{R}\right)$$

where the first factor is a $1 \times K$ row of states Φ_k and the second factor is a $K \times 1$ column of amplitudes ψ_k. The Schrödinger equation can now be put in a matrix form. Operating on Ψ with the momentum operator $\hat{\vec{P}}_R = (\hbar/i)\nabla_R$, we find a sum of two terms, and using the completeness relation of the electronic basis in front of the second term, we have

$$\frac{\hbar}{i}\nabla_R\left|\Psi\left(\vec{R}\right)\right\rangle = \left|\mathbf{\Phi}\left(\vec{R}\right)\right\rangle\frac{\hbar}{i}\nabla_R\boldsymbol{\psi}\left(\vec{R}\right) + \left[\frac{\hbar}{i}\nabla_R\left|\mathbf{\Phi}\left(\vec{R}\right)\right\rangle\right]\boldsymbol{\psi}\left(\vec{R}\right)$$

$$= \left|\mathbf{\Phi}\left(\vec{R}\right)\right\rangle\left[\mathbf{I}\frac{\hbar}{i}\nabla_R + \mathbf{G}\left(\vec{R}\right)\right]\boldsymbol{\psi}\left(\vec{R}\right)$$

$$\mathbf{G}\left(\vec{R}\right) = \left(\frac{\hbar}{i}\right)\left\langle\mathbf{\Phi}\left(\vec{R}\right)\middle|\nabla_R\mathbf{\Phi}\left(\vec{R}\right)\right\rangle$$

where I is the identity matrix. The symbol $\left\langle\mathbf{\Phi}\left(\vec{R}\right)\right|$ represents a $K \times 1$ column matrix with the complex conjugate elements Φ_k^*. The elements $\vec{G}_{kl}$ of the momentum-coupling matrix **G** are vectors that have the meaning of momentum fluctuations resulting from the electronic transitions induced by nuclear displacements. This matrix is Hermitian, as can be seen by taking the gradient of the normalization condition for the electronic basis, giving

$$\vec{G}_{kl}\left(\vec{R}\right) = \left(\frac{\hbar}{i}\right)\left\langle\Phi_k\left(\vec{R}\right)\middle|\nabla_R\Phi_l\left(\vec{R}\right)\right\rangle = -\left(\frac{\hbar}{i}\right)\left\langle\Phi_l\left(\vec{R}\right)\middle|\nabla_R\Phi_k\left(\vec{R}\right)\right\rangle^* = \vec{G}_{lk}\left(\vec{R}\right)^*$$

Its diagonal elements are null when the electronic functions are real valued, as follows from $\nabla_R\left\langle\Phi_k\left(\vec{R}\right) \mid \Phi_k\left(\vec{R}\right)\right\rangle = 0$, but they are not necessarily zero when the functions are complex-valued.

Since the effect of operating with the gradient is to introduce a square matrix with vector-valued elements between the two factors making Ψ, operating twice with the gradient simply repeats the effect and one ends up with the vector product of two square matrices; the elements of this product matrix are scalar products of two vectors. The original Schrödinger equations takes the matrix form

$$\left\{\frac{1}{2M}\left[\mathbf{I}\frac{\hbar}{i}\nabla_R + \mathbf{G}\left(\vec{R}\right)\right]^2 + \mathbf{H}_R - \mathbf{I}E\right\}\boldsymbol{\psi}\left(\vec{R}\right) = 0$$

$$\mathbf{H}_R = \left\langle\mathbf{\Phi}\left(\vec{R}\right)\middle|\hat{H}_R\middle|\mathbf{\Phi}\left(\vec{R}\right)\right\rangle$$

where the second power of the square bracket signifies a scalar product of matrices with vector-valued elements. This is a set of coupled differential equations that must be solved with the appropriate boundary conditions for the $\boldsymbol{\psi}$ amplitudes, for bound or unbound (decaying or scattering) states.

It is of interest to analyze the vector product of the total momentum appearing in this equation. Expanding it,

$$\left[\mathbf{I}\frac{\hbar}{i}\nabla_R + \mathbf{G}\left(\vec{R}\right)\right]^2\boldsymbol{\psi}\left(\vec{R}\right) = \left\{-\hbar^2\nabla_R{}^2 + \frac{\hbar}{i}[2\mathbf{G}.\nabla_R + (\nabla_R.\mathbf{G})] + \mathbf{G}.\mathbf{G}\right\}\boldsymbol{\psi}\left(\vec{R}\right)$$

we find that the two terms in the square bracket to the right add up to a Hermitian operator but each of them is, separately, non-Hermitian. Hence, it is safer to keep them together as a sum.

5.3.3 The Flux Function and Conservation of Probability

The total nuclear density at position $\vec{R}$ is given by $\psi(\vec{R})^\dagger \psi(\vec{R})$, with the dagger meaning transpose and conjugate, which is time-independent. Consequently, one can define a matrix flux function $\mathbf{J}(\vec{R})$ whose divergence is zero, indicating conservation of the nuclear density at each position. To do this, first write the matrix differential equation for $\psi(\vec{R})$ in the form $\hat{D}_R\psi(\vec{R}) = 0$, where $\hat{D}_R$ is a matrix differential operator. Then multiplying to the left by $\psi(\vec{R})^\dagger$ and rearranging the equation

$$\psi(\vec{R})^\dagger \hat{D}_R\psi(\vec{R}) - \left[\hat{D}_R\psi(\vec{R})\right]^\dagger \psi(\vec{R}) = 0$$

with help of the relation $f\nabla_R^2 g = \nabla_R\cdot(f\nabla_R g) - (\nabla_R f)\cdot\nabla_R g$ and cancelling opposite terms, we find the equation for the divergence of the intermolecular flux matrix $\mathbf{J}(\vec{R})$, with vector valued elements $\vec{J}_{kl}$, as

$$\nabla_R\cdot\mathbf{J}(\vec{R}) = 0$$

$$\mathbf{J}(\vec{R}) = \frac{1}{M}\frac{\hbar}{i}[\psi^\dagger\nabla_R\psi - (\nabla_R\psi)^\dagger\psi] + \frac{1}{M}\psi^\dagger\mathbf{G}\psi$$

showing that probability is conserved, provided however that one must now include a contribution from the electronic momentum fluctuation in the definition of the flux.

5.4 Electronic Representations

5.4.1 The Adiabatic Representation

In this representation, we choose states $\Phi_k = \Phi_k^{(a)}$ so that they diagonalize the Hamiltonian for fixed nuclei,

$$\left\langle \Phi_k^{(a)} \middle| \hat{H}_R \middle| \Phi_l^{(a)} \right\rangle = E_k^{(a)}(\vec{R})\,\delta_{kl}$$

where we have specified the representation with a superscript a. The total molecular wavefunction is now given by

$$\Psi\left(\vec{R},\boldsymbol{X}\right)=\sum_{k=1}^{K}\Phi_k^{(a)}\left(\boldsymbol{X};\vec{R}\right)\psi_k^{(a)}\left(\vec{R}\right)$$

and the nuclear amplitudes satisfy corresponding matrix equations with the momentum and Hamiltonian matrices $\mathbf{G}^{(a)}$, $\mathbf{H}_R^{(a)}$ calculated in the adiabatic representation. The second one is diagonal by construction of the basis set.

5.4.2 Hamiltonian and Momentum Couplings from Approximate Adiabatic Wavefunctions

The adiabatic wavefunctions can be approximated by expanding them in a given basis set of N_D many-electron determinantal functions $D_J(\boldsymbol{X};\vec{R})=\langle \boldsymbol{X}\mid D_J(\vec{R})\rangle$, $J=1$ to N_D, generally dependent on the intermolecular position variables, with the expansion coefficients determined by a variational procedure. We have that

$$\Phi_k^{(a)}\left(\vec{R}\right)=\sum_J D_J\left(\vec{R}\right)C_{Jk}\left(\vec{R}\right)=\boldsymbol{D}\left(\vec{R}\right)\boldsymbol{C}_k\left(\vec{R}\right)$$

$$\boldsymbol{C}_k^{\dagger}\boldsymbol{\Delta}\,\boldsymbol{C}_l=\delta_{kl}$$

where $\boldsymbol{D}$ is a $1\times N_D$ row matrix and $\boldsymbol{C}_k\left(\vec{R}\right)$ is a $N_D\times 1$ column matrix representing the adiabatic functions, with elements dependent on the intermolecular position variables, while $\Delta\left(\vec{R}\right)=\langle \boldsymbol{D}\mid \boldsymbol{D}\rangle$ indicates the overlap matrix of the given basis. The coefficients follow from the diagonalization of the electronic Hamiltonian in the original basis,

$$\boldsymbol{H}_R\boldsymbol{C}_k\left(\vec{R}\right)=\boldsymbol{\Delta}\left(\vec{R}\right)\boldsymbol{C}_k\left(\vec{R}\right)E_k^{(a)}\left(\vec{R}\right)$$

and the matrix of the electronic Hamiltonian is diagonal because $H_{kl}^{(a)}=\boldsymbol{C}_k^{\dagger}\mathbf{H}_R\boldsymbol{C}_l=E_k^{(a)}\delta_{kl}$. The momentum coupling matrix has elements

$$\vec{G}_{kl}^{(a)}=\frac{\hbar}{i}\left[\boldsymbol{C}_k^{\dagger}\left\langle D_k\left(\vec{R}\right)\middle|\nabla_R D_l\left(\vec{R}\right)\right\rangle\boldsymbol{C}_l+\boldsymbol{C}_k^{\dagger}\boldsymbol{\Delta}\left(\nabla_R\boldsymbol{C}_l\right)\right]$$

where the first term describes the momentum coupling contributed by the displacement of the basis functions, and the second term gives the contribution by the changes of coefficients with positions. The gradients of coefficients in the second term can be obtained by differentiation of the secular equation for the coefficients, as

$$\nabla_R\boldsymbol{C}_l=\left(\boldsymbol{H}_R-E_l^{(a)}\boldsymbol{I}\right)^{-1}\left[\nabla_R\left(\boldsymbol{H}_R-E_l^{(a)}\boldsymbol{I}\right)\right]\boldsymbol{C}_l$$

which only involves gradients of adiabatic potential energies.

The equation of motion for the nuclear amplitudes in the adiabatic representation, with a diagonal matrix of adiabatic energy potentials $\boldsymbol{E}^{(a)}(\vec{R}) = \left[E_k^{(a)}\left(\vec{R}\right)\delta_{kl}\right]$, is

$$\left\{\frac{1}{2M}\left[\boldsymbol{I}\frac{\hbar}{i}\nabla_R + \boldsymbol{G}^{(a)}\left(\vec{R}\right)\right]^2 + \boldsymbol{E}^{(a)}\left(\vec{R}\right) - \boldsymbol{I}E\right\}\boldsymbol{\psi}^{(a)}\left(\vec{R}\right) = 0$$

which shows that the amplitudes are interacting only through the momentum couplings. Ways to generate nonadiabatic representations have been given in the literature [12, 13, 15, 16, 20].

5.4.3 Nonadiabatic Representations

Nonadiabatic representations can be defined in a variety of ways. A specific choice is the strictly diabatic or P-representation where new electronic states $\left|\Phi_k^{(d)}\left(\vec{R}\right)\right\rangle$ are constructed from given N_D basis functions such that

$$\left\langle\Phi_k^{(d)}\middle|\nabla_R\Phi_l^{(d)}\right\rangle = \delta_{kl}$$

so that there is no coupling of different electronic states due to displacements of nuclear positions. These new states can be related to the previous adiabatic ones by means of a $N_D \times N_D$ matrix transformation $\boldsymbol{A}(\vec{R})$. At a large intermolecular distance R_0, where neither electronic nor momentum interactions are present, both representations must coincide. We write in matrix notation the row equation

$$|\boldsymbol{\Phi}^{(d)}\left(\vec{R}\right)\rangle = |\boldsymbol{\Phi}^{(a)}\left(\vec{R}\right)\rangle\boldsymbol{A}\left(\vec{R};\vec{R}_0\right)$$

with $\boldsymbol{A}(\vec{R}_0;\vec{R}_0) = \boldsymbol{I}$ and the unitarity condition $\boldsymbol{A}^{\dagger} = \boldsymbol{A}^{-1}$. From the condition defining the diabatic basis set, we have the differential equation

$$\left(\frac{\hbar}{i}\right)\nabla_R\boldsymbol{A}\left(\vec{R};\vec{R}_0\right) + \boldsymbol{G}^{(a)}\left(\vec{R}\right)\boldsymbol{A}\left(\vec{R};\vec{R}_0\right) = 0$$

which can be solved to obtain the transformation matrix from given adiabatic momentum couplings. The total molecular wavefunction in the new basis set is

$$\Psi\left(\vec{R},\boldsymbol{X}\right) = \sum\nolimits_{k=1}^{M}\Phi_k^{(d)}\left(\boldsymbol{X};\vec{R}\right)\psi_k^{(d)}\left(\vec{R}\right) = \left\langle\boldsymbol{X}\,|\,\boldsymbol{\Phi}^{(d)}\left(\vec{R}\right)\right\rangle\boldsymbol{\psi}^{(d)}\left(\vec{R}\right)$$

with nuclear motion amplitudes satisfying

$$\left\{-\frac{\hbar^2}{2M}\nabla_R{}^2 + \boldsymbol{H}_R^{(d)} - \boldsymbol{I}E\right\}\boldsymbol{\psi}^{(d)}\left(\vec{R}\right) = 0$$

which shows them coupled through the nondiagonal electronic matrix $\boldsymbol{H}_R^{(d)} = \boldsymbol{A}^{\dagger}\boldsymbol{H}_R^{(d)}\boldsymbol{A}$. These couplings are typically smoother that the ones arising from momentum couplings so that the diabatic representation is frequently used to treat electronic rearrangements in molecular spectroscopy and collisions, usually introducing semiclassical approximations for the nuclear amplitudes [19, 21]. The transformation between adiabatic and diabatic representations in a space of N_D dimensions can be generally treated introducing an algebra of skew symmetric matrices [22].

5.4.4 The Two-state Case

A solution to the differential equation for $\boldsymbol{A}$ at all distances can be easily obtained for a simple case with two real-valued coupled electronic states, $\Phi_j^{(a)}$, $j = 1,2$, and only one radial variable. Introducing the gradient coupling in the adiabatic representation, $\tau^{(a)}(R) = \left\langle \Phi_1^{(a)} \mid \nabla_R \Phi_2^{(a)} \right\rangle$, the momentum coupling matrix has diagonal elements $G_{jj}^{(a)} = 0$ and off-diagonal ones $G_{12}^{(a)} = G_{21}^{(a)*} = (\hbar/i)\tau^{(a)}$ and by construction $H_{12}^{(a)} = H_{21}^{(a)} = 0$, $H_{jj}^{(a)}(R) = E_j^{(a)}(R)$.

We introduce a 2×2 transformation $\boldsymbol{A}(R; R_0)$ for a large R_0, where there are no couplings between states 1 and 2, such that $\boldsymbol{A}(R_0; R_0) = \boldsymbol{I}$, the identity matrix. When the two states and therefore $\tau^{(a)}$ are real valued, one can choose

$$\boldsymbol{A}(R; R_0) = \begin{bmatrix} \cos(\gamma) & \sin(\gamma) \\ -\sin(\gamma) & \cos(\gamma) \end{bmatrix}$$

It follows from the differential equation for $\boldsymbol{A}(R; R_0)$ that the angle $\gamma(R; R_0)$ is a function satisfying the differential equation $d\gamma/dR = \tau^{(a)}(R)$ with the boundary condition $\gamma(R_0; R_0) = 0$. Its solution

$$\gamma(R; R_0) = \int_{R_0}^{R} dR\, \tau^{(a)}(R)$$

can be used to construct the Hamiltonian matrix elements in the strictly diabatic representation, where $G_{jk}^{(d)} = 0$ for all elements, as

$$H_{11}^{(d)}(R) = E_1^{(a)}(R)\cos^2(\gamma) + E_2^{(a)}(R)\sin^2(\gamma),$$
$$H_{22}^{(d)}(R) = E_1^{(a)}(R)\sin^2(\gamma) + E_2^{(a)}(R)\cos^2(\gamma),$$
$$H_{12}^{(d)}(R) = \left[E_1^{(a)}(R) - E_2^{(a)}(R)\right]\sin(\gamma)\cos(\gamma).$$

These expressions allow us to consider what happens when the adiabatic potentials cross at a point $R_\times < R_0$, with R_0 large and $\tau^{(a)}(R)$ negligible there. Then for large R, one finds $H_{11}^{(d)}(R) = E_1^{(a)}(R)$ and $H_{22}^{(d)}(R) = E_2^{(a)}(R)$. Around the crossing point, $H_{12}^{(d)}(R)$ is small but the diagonal elements do not cross, and for even shorter distances, the diabatic elements differ significantly from the adiabatic energies.

We also find that the diagonal diabatic potential energies $H_{jj}^{(d)}(R)$ will cross whenever $\gamma = \pm(n/2 + 1/4)\pi$ and that this can happen more than once on the way to small R if the adiabatic momentum couplings are large. This is not physically meaningful and is a consequence of the definition of the strictly diabatic representation. An alternative more physically meaningful diabatic representation can be introduced by parametrizing the transformation angle to a form $\gamma(R;\alpha)$ with the parameter α chosen to minimize the momentum coupling in a new "m" representation [18]. This generates new smooth diabatic potentials $H_{jk}^{(m)}(R)$ and a new smaller momentum coupling $\tau^{(m)}(R;\alpha)$ which can be used for molecular spectroscopy and collision treatments in the new representation. Figure 5.3 shows an example constructed from Morse type potentials.

In any case, the new states are related to adiabatic ones through the transformation matrix as

$$|\Phi_1^{(d)}(\vec{R})\rangle = |\Phi_1^{(a)}(\vec{R})\rangle\cos(\gamma) - |\Phi_2^{(a)}(\vec{R})\rangle\sin(\gamma)$$

$$|\Phi_2^{(d)}(\vec{R})\rangle = |\Phi_1^{(a)}(\vec{R})\rangle\sin(\gamma) + |\Phi_2^{(a)}(\vec{R})\rangle\cos(\gamma)$$

This can be generalized when the coupling $\tau^{(a)}(R) = |\tau^{(a)}(R)|\exp[i\chi(R)]$ is a complex function, writing instead the diabatic states as combinations of the adiabatic basis functions $|\Phi_1^{(a)}(\vec{R})\rangle\exp[-i\chi(R)/2]$ and $|\Phi_2^{(a)}(\vec{R})\rangle\exp[i\chi(R)/2]$.

Extensions of the treatment to several variables and more than two electronic states have been presented in the literature [14]. For example the NaH molecule in its ground and excited $^1\Sigma^+$ states shows multiple pairs of avoided crossings, one for each excited potential energy that breaks into Na(nl) + H($1s$) after it pseudo-crosses the $Na^+ + H^-$ attractive potential energy as seen in Figure 5.4 [23].

5.4.5 The Fixed-nuclei, Adiabatic, and Condon Approximations

The previous description of alternative representations can give accurate results for properties such as molecular spectra and scattering cross sections by introducing a large number N_D of functions in the basis set of many-electron functions $\Phi_K(X;\vec{R})$ and by calculating nuclear amplitudes $\psi_K(\vec{R})$ as solutions of coupled differential equations. However, in many cases, it is sufficient to introduce approximations where some of the momentum couplings are neglected.

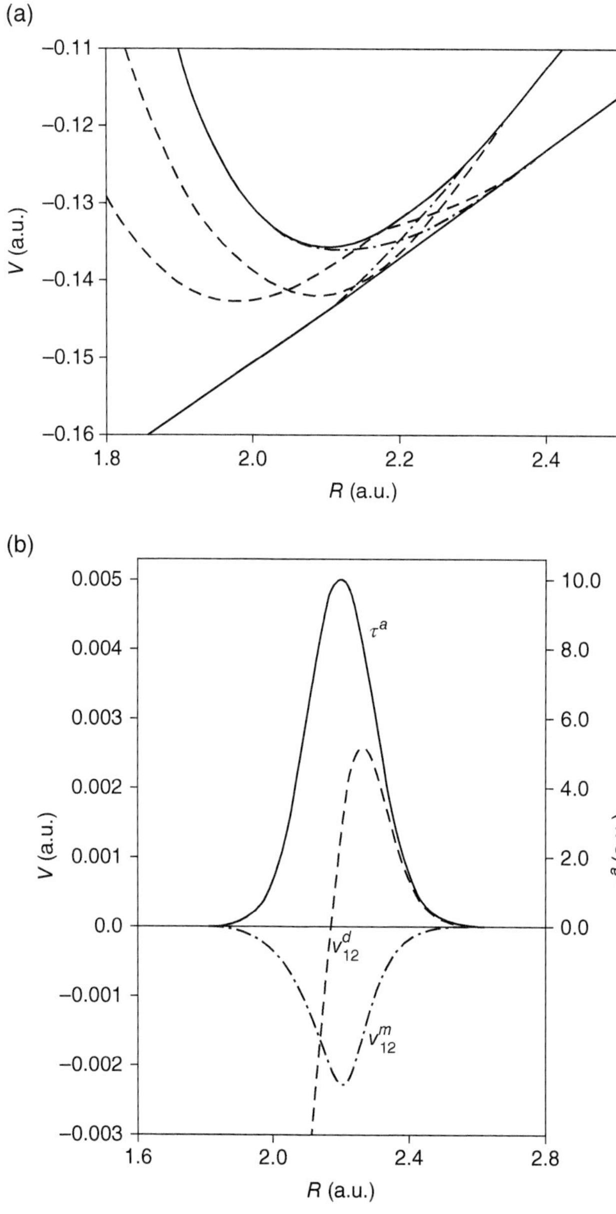

Figure 5.3 (a) Adiabatic (full line), diabatic (dash line) and minimum coupling (dash-dot line) potential energies for two states and (b) their couplings in the three representations. *Source:* from Ref. [18]. Reproduced with permission of John Wiley & Sons.

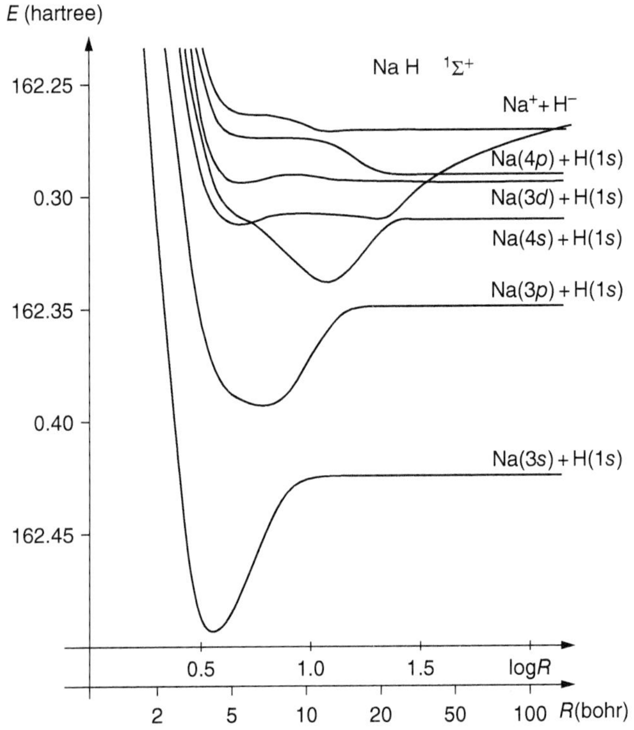

Figure 5.4 NaH potential energies showing multiple pairs of avoided crossings. *Source:* from Ref. [23]. Reproduced with permission of John Wiley & Sons.

The fixed-nuclei approximation (also called the Born–Oppenheimer approximation) starts with the adiabatic representation and sets all the momentum couplings $\vec{G}_{kl}^{(a)} = 0$ so that the equation for the nuclear motion in each electronic state k is

$$\left\{ -\frac{\hbar^2}{2M} \nabla_R^2 + E_k^{(a)}(R) - E \right\} \psi_k^{(a)}\left(\vec{R}\right) = 0$$

In the adiabatic approximation, one again starts with the adiabatic representation, but one keeps only one term in the expansion of the total molecular wavefunction so that

$$\Psi^{(ad)}\left(\vec{R}, X\right) = \Phi_k^{(a)}\left(X; \vec{R}\right) \psi_k^{(a)}\left(\vec{R}\right)$$

with $\Phi_k^{(a)}$ a real-valued normalized function for which $\left\langle \Phi_k^{(a)} \mid \nabla_R \Phi_k^{(a)} \right\rangle = 0$. When this is replaced in the Schrödinger equation one finds

$$\left\{-\frac{\hbar^2}{2M}\nabla_R{}^2 + E_k^{(a)}(R) + W_k^{(a)}(R) - E\right\}\psi_k^{(a)}\left(\vec{R}\right) = 0$$

with

$$W_k^{(a)}(R) = -\left(\frac{\hbar^2}{2M}\right)\left\langle\Phi_k^{(a)}\left(\vec{R}\right)\middle|\nabla_R^2\Phi_k^{(a)}\left(\vec{R}\right)\right\rangle = \left(\frac{\hbar^2}{2M}\right)\left\langle\nabla_R\Phi_k^{(a)}\left(\vec{R}\right)\middle|\nabla_R\Phi_k^{(a)}\left(\vec{R}\right)\right\rangle$$

a repulsive potential energy addition, with the last equality obtained by partial integration. Comparisons of the magnitudes of these approximations for the energy of the ground molecular state of hydrogen molecule isotopes can be found in [24]. There one finds that the ground state energy E_0 of the $X^1\Sigma_g^+$ molecular term for H_2, in cm^{-1} for $E_0/(hc)$ is 36 112.2, 36 118.0, and 36 114.7 for the fixed-nuclei, adiabatic, and nonadiabatic approximations, to be compared with the experimental value of 36 113.6 cm^{-1}. These differences are typical of corrections to equilibrium energy values. Considering that 1.0 cm^{-1} = 1.9864×10^{-23} J in the SI units and 1.0 cm^{-1} = 4.55634×10^{-6} au, it is found that the nonadiabatic corrections are meaningful in very high precision spectroscopy, but not essential for most work on intermolecular forces around equilibrium structures. The nonadiabatic couplings are however essential in many treatments of photoinduced or collision-induced electronic rearrangement.

Another nonadiabatic approximation useful in molecular spectroscopy, for a molecular system where internal variables remain near their equilibrium values $\boldsymbol{Q}_{eq}$ in the initial electronic state, is the so-called Condon approximation where a basis set of electronic functions $\Phi_k^{(C)}(\boldsymbol{X};\boldsymbol{Q}_{eq})$ is used in an expansion of the molecular state. Here all the expansion functions have distances kept constant at the equilibrium values $\boldsymbol{Q}_{eq}$ of internal coordinates, and the variation of energies with displacements is described expanding the electronic Hamiltonian as a function of displacements around the equilibrium conformation. This generates a vibronic Hamiltonian displaying couplings of electronic and nuclear degrees of freedom in molecular dynamics.

Given the set of coordinates $\boldsymbol{Q} = \{Q_\iota\}$, the total molecular wavefunction is expanded in the basis set of Condon states as

$$\Psi(\boldsymbol{Q},\boldsymbol{X}) = \sum\nolimits_{k=1}^{K}\Phi_k^{(C)}\left(\boldsymbol{X};\boldsymbol{Q}_{eq}\right)\psi_k^{(C)}(\boldsymbol{Q}),$$

while the electronic Hamiltonian is expanded around equilibrium values of internal variables in powers of displacements $\boldsymbol{q} = \boldsymbol{Q} - \boldsymbol{Q}_{eq}$, giving

$$\hat{H}_Q = \hat{H}_{eq} - \hat{\boldsymbol{F}}_{eq}.\boldsymbol{q} + \boldsymbol{q}^{\dagger}.\hat{\boldsymbol{G}}_{eq}.\boldsymbol{q}/2$$

in a matrix notation where $\boldsymbol{q}$ is a column matrix, $\hat{\boldsymbol{F}}_{eq}$ is a row matrix of force operators and $\hat{\boldsymbol{G}}_{eq}$ is a square matrix of potential energy curvatures, also operators. A vibronic description of dynamics can be based on a perturbation calculation of the coefficient functions $\psi_k^{(C)}$, starting with an expansion of the equation for $\Psi(\boldsymbol{Q}, \boldsymbol{X})$ in the basis set $\left\{\Phi_k^{(C)}(X; \boldsymbol{Q}_{eq})\right\}$. This leads to equations containing force matrix elements $F_{kl}^{(\mu)}$ multiplying the q_μ displacements and curvature matrix elements $G_{kl}^{(\mu\nu)}$ multiplying the products $q_\mu q_\nu$.

For a simple case of a diatomic, this amounts to the expansion

$$\Psi\left(\vec{R}, \boldsymbol{X}\right) = \sum_{k=1}^{K} \Phi_k^{(C)}\left(\boldsymbol{X}; \vec{R}_{eq}\right) \psi_k^{(C)}\left(\vec{R}\right)$$

and calculations of molecular energies as functions of the displacement $x = R - R_{eq}$ doing perturbation theory with the Hamiltonian perturbation operator $\left(\partial \hat{H}_R / \partial R\right)_{eq} x$. This leads to vibronic coupling among electronic-nuclear states, with electronic coupling matrix elements $\langle \Phi_k^{(C)} \mid \left(\partial \hat{H}_R / \partial R\right)_{eq} x \mid \Phi_l^{(C)} \rangle = -F_{kl}^{(x)} x$ and corresponding coupling of nuclear motion states through $\langle \psi_k^{(C)} \mid x \mid \psi_l^{(C)} \rangle$. The transition force $F_{kl}^{(x)}$ is different from zero only when the product of point group symmetry representations of states k and l, and of x, contains the totally symmetric representation.

The interaction of the molecular system with a light field coupled to the molecular electric dipole leads to spectral intensities calculated in terms of Frank–Condon overlaps $\langle \Phi_k^{(C)} \psi_k^{(C)} |\vec{D}| \Phi_l^{(C)} \psi_l^{(C)} \rangle \approx \vec{D}_{kl}(\vec{R}_{eq}) \langle \psi_k^{(C)} \mid \psi_l^{(C)} \rangle$, where $\vec{D}$ is the electric dipole vector operator and $\vec{D}_{kl}(\vec{R}_{eq}) = \langle \Phi_k^{(C)} |\vec{D} \mid \Phi_l^{(C)} \rangle_{eq}$. When the photoexcitation occurs around equilibrium, the vibronic displacement factor $\langle \psi_k^{(C)} \mid x \mid \psi_l^{(C)} \rangle$ is small for $k \neq l$ and one can use amplitudes $\psi_k^{(C)}(\vec{R})$ obtained from a Schrödinger equation with a potential energy function $H_{kk}^{(C)}(R)$ to obtain light absorption-emission rates.

5.5 Electronic Rearrangement for Changing Conformations

5.5.1 Construction of Molecular Electronic States from Atomic States: Multistate Cases

Given the state $\Psi(\boldsymbol{Q}, \boldsymbol{X})$ of a pair of molecules A and B as an expansion in a basis set of electronic wavefunctions $\Phi_k(\boldsymbol{X}; \boldsymbol{Q})$ (not necessarily adiabatic ones) which vary with a set $\boldsymbol{Q} = \left(\vec{R}, \boldsymbol{Q}'\right)$ of interatomic distances, its Schrödinger equation contains a potential energy matrix with elements $H_{kl}(\boldsymbol{Q})$ that are functions of distances, as well as a matrix of momentum coupling elements $G_{kl}(\boldsymbol{Q})$. It is usually possible to construct the wavefunction of the pair of molecules so that at large intermolecular distances, the molecular motions are independent and occur along well-defined separate electronic potential energy surfaces $E_{\kappa}^{(A)}(\mathbf{Q}_A)$ and $E_{\lambda}^{(B)}(\mathbf{Q}_B)$. However, as distances are decreased, the potential energy for each electronic state k originating in (κ, λ) becomes a function $E_k^{(AB)}(R, \mathbf{Q}_{int})$, which changes with the intermolecular distance and internal positions, excluding overall translations and rotations, and generates a potential energy surface (PES) showing as isocontours of energy in the multidimensional space of distances. Pairs of such PESs may show crossings or regions of minimal energy differences, which depend on what electronic representation is being used. They interact through nuclear momentum couplings, which also depend on the representation.

Special points can be identified looking for minimal values of differences in each pair of potential surfaces, as one varies the internal coordinates, so that $\left|E_k^{(AB)}(R, \mathbf{Q}_{int}) - E_l^{(AB)}(R, \mathbf{Q}_{int})\right| = Min$. This can be zero if there is a crossing along a surface in the hyperspace of internal coordinates, and its shape depends again on what electronic representation one is using. When dealing with a diatomic (and a single internal distance) one may find a crossing or single intersection at a distance R_0; for two or more internal variables one may find a conical intersection, crossing seams, or near crossings. All these are important in descriptions of electronic rearrangement.

5.5.2 The Noncrossing Rule

The procedure for calculating intermolecular states by first fixing nuclear positions leads to the construction of many-electron states Φ_k dependent parametrically on all the internal nuclear position variables. Starting with the total number $3N$ of atomic position variables for a system with N atoms, and indicating with $\mathbf{Q}_{RB}$ the set of rigid body variables (six for overall translations and rotations, or five for a linear conformation), the remaining internal variables form a set $\mathbf{Q}_{int} = \{Q_\nu, \nu = 1 \text{ to } N_{int}\}$ of $N_{int} = 3N - N_{RB}$ degrees of freedom. Matrix elements H_{kl} of the Hamiltonian operator for fixed nuclei, in an original electronic basis set $\{\Phi_k\}$ which may describe adiabatic or nonadiabatic electronic states and is taken to be orthonormalized, also depend on the nuclear variables. As they vary the diagonal elements H_{kk} and H_{ll} may cross or show avoided crossings at specific points or lines. These locations are important for the description of molecular spectra and interaction dynamics, showing the structure conformations where light absorption-emission may happen or where atomic rearrangement occurs for electronic transitions between k and l.

General statements can be made about the location and nature of interaction energies at these special points. To simplify, we consider first the case with only one internal degree of freedom, for two atoms at the internuclear distance R, and where only two states $k = 1, 2$ are involved. Examples are the lowest two $^1\Sigma$ states of LiF or of NaH, where ionic electronic distributions give states of lower energy around their ground equilibrium distance, but covalent distributions give lower energy states at larger distances.

A special point R_c of a crossing or avoided crossing is located by finding where the difference $|H_{11}(R) - H_{22}(R)|$ shows a minimum as R is varied. Writing $R = R_c + x$ and $H_{kl}(R_c) = H_{kl}^{(c)}$, matrix elements can be expanded around $x = 0$ as

$$H_{kl}(x) = H_{kl}^{(c)} + x\frac{\partial H_{kl}}{\partial R_c} + \frac{1}{2}x^2\frac{\partial^2 H_{kl}}{\partial R_c{}^2} + \cdots$$

with $H_{11}^{(c)} = H_{22}^{(c)}$ if there is a crossing when working with the original basis set. A nondiagonal Hamiltonian matrix element $H_{12}^{(c)}$ appears if the representation is nonadiabatic. Near the special point, an electronically adiabatic representation is generated by the linear combinations $\Phi_k^{(a)} = \Phi_1 C_{1k} + \Phi_2 C_{2k}$ with the coefficients obtained by diagonalizing the Hamiltonian matrix at each R so that the new elements are $H_{kl}^{(a)}(R) = \delta_{kl} E_k^{(a)}(x)$. These two energies are given by the well-known roots of the second-order equation $det[\boldsymbol{H} - E\boldsymbol{I}] = 0$ where 2×2 matrices are shown in boldface, as

$$E_{1,2}^{(a)}(x) = \frac{1}{2}[H_{11}(x) + H_{22}(x)] \pm \frac{1}{2}\left\{[H_{11}(x) - H_{22}(x)]^2 + 4|H_{12}(x)|^2\right\}^{1/2}$$

and become equal only when the radical $L(x) = [H_{11}(x) - H_{22}(x)]^2 + 4|H_{12}(x)|^2 = 0$, or when both $H_{11}(x) - H_{22}(x) = 0$ and $H_{12}(x) = 0$. These are two equations in x to be satisfied by a single solution root x_0, and there is no real valued answer, except if in a particular system one of the two equations is already obeyed for all x, such as when the states 1 and 2 are degenerate and $H_{11}(x) = H_{22}(x)$. For two different diagonal element functions of x and in the absence of accidental vanishing of their coupling H_{12}, the two adiabatic energies $E_{1,2}^{(a)}(x)$ are different for all $R = R_c + x$ around $x = 0$. This is the noncrossing rule for adiabatic potential energies.

In this case, the potentials show an avoided crossing and the difference between the new potential energies, $\Delta E(x) = E_1^{(a)}(x) - E_2^{(a)}(x)$, has a minimum at the gap located at x_m, where $\partial(\Delta E)/\partial x = 0$, equal to

$$\Delta E_m = \left[(\Delta H_m)^2 + 4|W_m|^2\right]^{1/2}$$

with $\Delta H_m = H_{11}(x_m) - H_{22}(x_m)$ and $W_m = H_{12}(x_m)$. Introducing derivatives $f_{kl}^{(c)} = \partial H_{kl}/\partial R_c$ and $g_{kl}^{(c)} = \partial^2 H_{kl}/\partial R_c^2$, the energies $E_{1,2}^{(a)}(x)$ can be expanded up to second order in x and are found to be opposing parabolas glancing at x_m.

The momentum coupling between the two adiabatic states is a special case of the previous relation described for approximate adiabatic states, here $|\Phi_k^{(a)}(R)\rangle = |\boldsymbol{\Phi}(R)\rangle \boldsymbol{C}_k(R)$ in matrix notation with $\boldsymbol{C}_k$ a 2×1 column matrix, giving

$$G_{12}^{(a)} = \frac{\hbar}{i}\left[\boldsymbol{C}_1^{\dagger}\langle\boldsymbol{\Phi}(R)\,|\,\nabla_R\boldsymbol{\Phi}(R)\rangle\,\boldsymbol{C}_2 + \boldsymbol{C}_1^{\dagger}\langle\boldsymbol{\Phi}(R)\,|\,\boldsymbol{\Phi}(R)\rangle\,(\nabla_R\boldsymbol{C}_2)\right]$$

$$\nabla_R\boldsymbol{C}_2 = \left(E_1^{(a)} - E_2^{(a)}\right)^{-1}\left[\nabla_R\left(\boldsymbol{H}_R - E_2^{(a)}\boldsymbol{I}\right)\right]\boldsymbol{C}_2$$

The denominator in the second line has a minimum value ΔE_a at the avoided crossing, showing that this is where the momentum coupling will be large.

Conditions under which $H_{12}(x) \equiv 0$ give situations with crossings and can be found from considerations of state symmetry. This happens if: (i) The two original states have different electronic spin quantum numbers; or (ii) the two original states belong to different irreducible representations of their point group symmetry. The groups are $\mathcal{D}_{\infty h}$ for homonuclear diatomics and $C_{\infty v}$ for heteronuclear diatomics. When the symmetry species (or labels) of the diatomic states are different, then $H_{12}(x) \equiv 0$, and if the $E_k(R) = H_{kk}(x)$ are equal at a real valued x_0 in the original representation, it follows that a crossing of adiabatic potential energies $E_{1,2}^{(a)}$ may occur at a distance $R = R_c + x_0$.

If instead the two states have the same spin quantum numbers and point group symmetry, then in general $H_{12}(x) \neq 0$. In this case, to find the root where adiabatic energies are equal one must require that the radical $L(x) = L_0 + xL_1 + x^2L_2 = 0$, giving in general a complex-valued root $x_0' + ix_0''$, and a crossing in the complex plane of position variables.

For a complex valued coupling $H_{12}(x) = |H_{12}(x)|\exp[i\alpha(x)]$, the two new adiabatic states around an avoided crossing are [25]

$$|\Phi_1^{(a)}(R)\rangle = |\Phi_1(R)\rangle\exp\left(-\frac{i\alpha}{2}\right)\cos(\beta) + |\Phi_2(R)\rangle\exp\left(\frac{i\alpha}{2}\right)\sin(\beta)$$

$$|\Phi_2^{(a)}(R)\rangle = -|\Phi_1(R)\rangle\exp\left(-\frac{i\alpha}{2}\right)\sin(\beta) + |\Phi_2(R)\rangle\exp\left(\frac{i\alpha}{2}\right)\cos(\beta)$$

with

$$\tan[2\beta(x)] = 2|H_{12}(x)|/[H_{11}(x) - H_{22}(x)]$$

taking a large value at the avoided crossing, leading there to a rapid change of $\beta(x)$ and rapid switch between states.

This analysis can be extended to cover interactions involving a spin-orbit couplings in a Hamiltonian, in which case the relevant point symmetry groups are extended groups with representations generated by the direct product of spin dependent and space-dependent basis functions. In particular double groups arise when orbital angular momentum states are extended to total angular momentum states to include half-integer spin states, and their representation labels can be used to derive conclusions from molecular symmetry arguments [26].

5.5.3 Crossings in Several Dimensions: Conical Intersections and Seams

The search for special intersections of potential energy surfaces (PESs) can be extended to systems with three or more atoms, involving more internal degrees of freedom. Even for three atoms, there are $N_{int} = 3$ or 4 degrees of freedom for the atomic positions and enough variables to satisfy the two equations that indicate an intersection of PESs. This fixes only two internal degrees of freedom. However, there are many rearrangements of atoms during electronic excitation where most degrees of freedom remain inactive and only two are changing. The two changing ones can be identified by searching for the configurations, where $|H_{11}(\{Q_\nu\}) - H_{22}(\{Q_\nu\})|$ is a minimum for a set $\{Q_\nu^{(m)}\}$ and where distortion forces for two of them, called here Q_a and Q_b, are small around $Q_a^{(m)}$ and $Q_b^{(m)}$. We first treat this case, with displacement variables $x = Q_a - Q_a^{(m)}$ and $y = Q_b - Q_b^{(m)}$, and all other internal variables set at their values for the minimum and omitted from arguments for now. Consider again just two original states $|\Phi_j\rangle$ in a diabatic representation with diagonal Hamiltonian matrix elements $H_{11}(x, y)$ and $H_{22}(x, y)$, and a nondiagonal $H_{12}(x, y)$ real valued for now but taken to be complex later on. A transformation to the adiabatic representation states $|\Phi_k^{(a)}\rangle$ gives the potential energy surfaces

$$E_{1,2}(x,y) = M(x,y) \pm N(x,y)$$

$$M(x,y) = \frac{1}{2}[H_{11}(x,y) + H_{22}(x,y)]$$

$$N(x,y) = \left\{[H_{11}(x,y) - H_{22}(x,y)]^2 + 4|H_{12}(x,y)|^2\right\}^{1/2}$$

The condition for intersection of PESs is $N(x, y) = 0$, or both $H_{11}(x, y) - H_{22}(x, y) = 0$ and $H_{12}(x, y) = 0$, which may be satisfied by roots (x_0, y_0), corresponding to variables $Q_\nu^{(0)}$ expected to be close to $Q_\nu^{(m)}$, $\nu = a, b$. When this happens for the original (diabatic set) matrix elements, the new energies $E_{1,2}(x, y)$ of the adiabatic states can be expanded around the roots up to second order in the displacements so that

$$E_{1,2}(x,y) = M^{(0,0)} + M^{(1,0)}x + M^{(0,1)}y + M^{(2,0)}x^2/2 + M^{(0,2)}y^2/2 + M^{(1,1)}xy \mp \left(N^{(1,0)}x + N^{(0,1)}y + N^{(2,0)}x^2/2 + N^{(0,2)}y^2/2 + N^{(1,1)}xy\right)$$

Each of these two energy functions gives an isocontour surface for each constant energy in the space of E versus x, y. Since they are quadratic forms, the

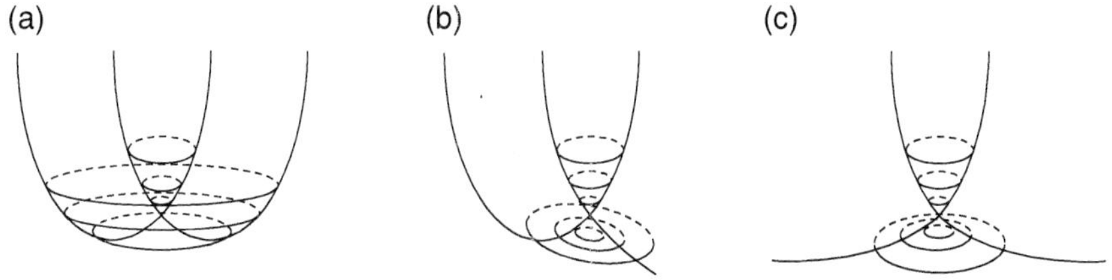

Figure 5.5 Illustration of three cases of conical intersections between two potential energy surfaces: (a) Bound-to-bound; (b) bound-to-unbound; (c) unbound-to-unbound.

isosurfaces are conical forms, and they touch at a degeneracy vertex, at the location of the roots (x_0, y_0) in a (x, y) plane. They can be opposing cones or collateral ones, depending on slopes along x and y. This in turn depends on what types of bonding surfaces are interacting. Figure 5.5 illustrates conical intersections for a triatomic system ABC, as can be expected coming from (a) isomerization, that is a bound-to-bound structure interaction; (b) from dissociation in bound-to-unbound structure changes; or also (c) from collisions as unbound-to-unbound structure events.

Insight on the nature of these conical intersections can be gained constructing simple models of the intersecting conical surfaces (or CISs) in a space E versus (x, y), where x and y are bond distances or angles, and the other $N_{int} - 2$ internal conformation variables are kept fixed. The intersection of two diabatic potential energies leads to CISs when coupled, and as the other variables are changed the conical shapes can be pictured as moving along seams in a hyperspace of $N_{int} - 2$ dimensions. Examples of models are given as follows.

a) Bound-to-bound CIS:

$$H_{11} = \frac{\omega_1^2}{2}\left(x + \frac{a}{2}\right)^2 + \frac{\omega_2^2}{2}y^2,\ H_{22} = \frac{\omega_1^2}{2}\left(x - \frac{a}{2}\right)^2 + \frac{\omega_2^2}{2}y^2 - \Delta,\ H_{12} = cy$$

This corresponds to two intersecting paraboloids and describes for example SH_2, of point group symmetry C_{2v}, in the excited states $1^1A''$ and $2^1A''$ [27], with x corresponding to the H-H distance r, and y for the distance R from S to the center of mass of HH. The third variable is the angle between $\vec{r}$ and $\vec{R}$ orientations, and is considered here to be chosen at a fixed value.

b) Bound-to-unbound CIS:

$$H_{11} = \frac{\omega_1^2}{2}\left(x + \frac{a}{2}\right)^2 + \frac{\omega_2^2}{2}y^2,\ H_{22} = A\exp[-\alpha(x - x_0)] + \frac{\omega_2^2}{2}y^2 - \Delta,$$

$$H_{12} = cy\exp\left[\frac{(x - x_0)^2}{\lambda^2} + \frac{(y - y_0)^2}{\mu^2}\right]$$

where the unbound motion occurs as the x variable increases. This applies to the dissociation of excited $HNO(^1\Delta)$ into $H(^2S) + NO(^2\Pi)$ [28], with the two variables corresponding to the H to NO distance (x) and the HNO bond angle (y).

c) Unbound-to-unbound CIS:

$$H_{11} = A\exp[-\alpha(x-x_0)],\ H_{22} = B\exp[-\beta(y-y_0)] - \Delta,$$

$$H_{12} = cy\exp\left[\frac{(x-x_0)^2}{\lambda^2} + \frac{(y-y_0)^2}{\mu^2}\right]$$

This applies to a triatomic reaction $A + BC \rightarrow AB + C$ with a fixed ABC angle and x and y representing the bond distances of initial and final diatomics, such as in collinear $H + H_2 \rightarrow H_2 + H$ reaction. Also, and more generally, a triatomic PES for three active electrons in an atom-diatomic reaction can be described by the LEPS surface as mentioned in Chapter 4. That is the lower PES of a pair generated by the combination of two valence-bond functions, and the two PESs form a conical intersection when the radical in the LEPS expression is null, as can be expected to also occur for other monovalent atom combinations at certain interatomic distances.

The diabatic to adiabatic transformation of electronic states is generated as before by coefficients dependent now on the two variables x and y. For a complex valued coupling $H_{12}(x, y) = |H_{12}(x, y)|\exp[i\alpha(x, y)]$, the two new adiabatic states are as before

$$|\Phi_1^{(a)}\rangle = |\Phi_1\rangle\exp\left(-\frac{i\alpha}{2}\right)\cos(\beta) + |\Phi_2\rangle\exp\left(\frac{i\alpha}{2}\right)\sin(\beta)$$

$$|\Phi_2^{(a)}\rangle = -|\Phi_1\rangle\exp\left(-\frac{i\alpha}{2}\right)\sin(\beta) + |\Phi_2\rangle\exp\left(\frac{i\alpha}{2}\right)\cos(\beta)$$

but now with

$$\tan[2\beta(x,y)] = 2|H_{12}(x,y)|/[H_{11}(x,y) - H_{22}(x,y)]$$

which is well defined away from the intersection point.

More accurately, one must consider all the internal degrees of freedom and work with the displacements $\boldsymbol{q} = \boldsymbol{Q} - \boldsymbol{Q}^{(m)}$, to allow for couplings of the displacements in the two variables Q_ν, $\nu = a, b$, with the remaining degrees of freedom, as all atoms are displaced from the minimum conformation. Indicating with $\bar{\boldsymbol{Q}}$ the set of all internal coordinates other than Q_a or Q_b, the two potential energy surfaces depend more generally also on the displacements $\bar{\boldsymbol{q}} = \bar{\boldsymbol{Q}} - \bar{\boldsymbol{Q}}^{(m)}$ and when these are not all zero the PESs have form $E_{1,2}(x,y,\bar{\boldsymbol{q}}) = M(x,y,\bar{\boldsymbol{q}}) \pm N(x,y,\bar{\boldsymbol{q}})$. One finds in general that $N(x_0,y_0,\bar{\boldsymbol{q}}) \neq 0$ at the previous roots, and the energy degeneracy at the vertex is lifted. This is likely to distort the conical energy surfaces touching at the vertex, changing them into glancing paraboloids. This avoided degeneracy is due to couplings to other variables within the adiabatic state description, and therefore different in origin from the avoided crossing in the one-dimensional case which arises from the two original electronic states having the same symmetry. The magnitude of the splitting at the vertex can be found doing perturbation theory with the vibronic perturbation energy operator generated by displacing internal variables from the reference values at the minimum conformation. Models can be presented in terms of adiabatic or nonadiabatic PESs [29]. Two adiabatic PESs are also subject to nonadiabatic momentum couplings $\vec{G}_{kl}^{(a)}$, with momentum components along the x and y directions, even if all other internal variables are assumed to be fixed.

Depending on the nature of the two variables, such as whether they describe distortions of a bond length or of a bond angle, the conical shapes may become flattened and in the limit of a breaking bond the degeneracy may become a line, showing as a seam for two intersecting PESs along the direction of the breaking bond. This is what happens, for example in the case of the PES for collinear $H^+ + H_2$ interacting with the PES for $H + H_2^+$ [20], and of the PES for $Ar^+ + H_2$ interacting with that of $Ar + H_2^+$ as shown in Figure 5.6a for the adiabatic potential energies and Figure 5.6b for their nonadiabatic momentum coupling $\tau = |\vec{G}_{kl}^{(a)}|/\hbar$ along the diatomic axis [30]. The crossing between the two lowest PESs can also be described as two intersecting potential energy lines in the plane of the energy E and the bond distance r in H_2 and in H_2^+, translated along a seam in a direction perpendicular to that plane, as the distance R between Ar^+ and H_2 is changing. The crossing in Figure 5.6a below 1.0 Å (or 0.1 nm) is transformed due to the momentum coupling along R, into an avoided crossing as shown in Figure 5.6b.

The two adiabatic PESs are also subject to nonadiabatic momentum couplings $\vec{G}_{kl}^{(a)}$, with momentum components along the x and y directions, even if all other internal variables are assumed to be fixed.

A special situation arises when the two original electronic states, 1 and 2, are degenerate with the same energy $H_{11}(x, y) = H_{22}(x, y)$ for all argument values. The degeneracy can be lifted due to a coupling $H_{12}(x, y) \neq 0$. Conditions under which this happens can be found assuming vibronic coupling and can be studied by expanding the electronic Hamiltonian operator of the system in powers of small displacements from the reference structure. Coupling matrix elements of the resulting force operator are different from zero if the direct product of the three representations of the point groups for the two original states and for any of the displacements contain the totally symmetric representation. In this case, the degeneracy can be lifted with new states and PESs found from the solution of the two-state eigenvalue equations. This analysis has been followed to arrive at the Jahn–Teller theorem [31], stating that in any symmetric nonlinear molecule with electronically degenerate states, displacement of atoms lead to symmetry distortions that decrease the original symmetry.

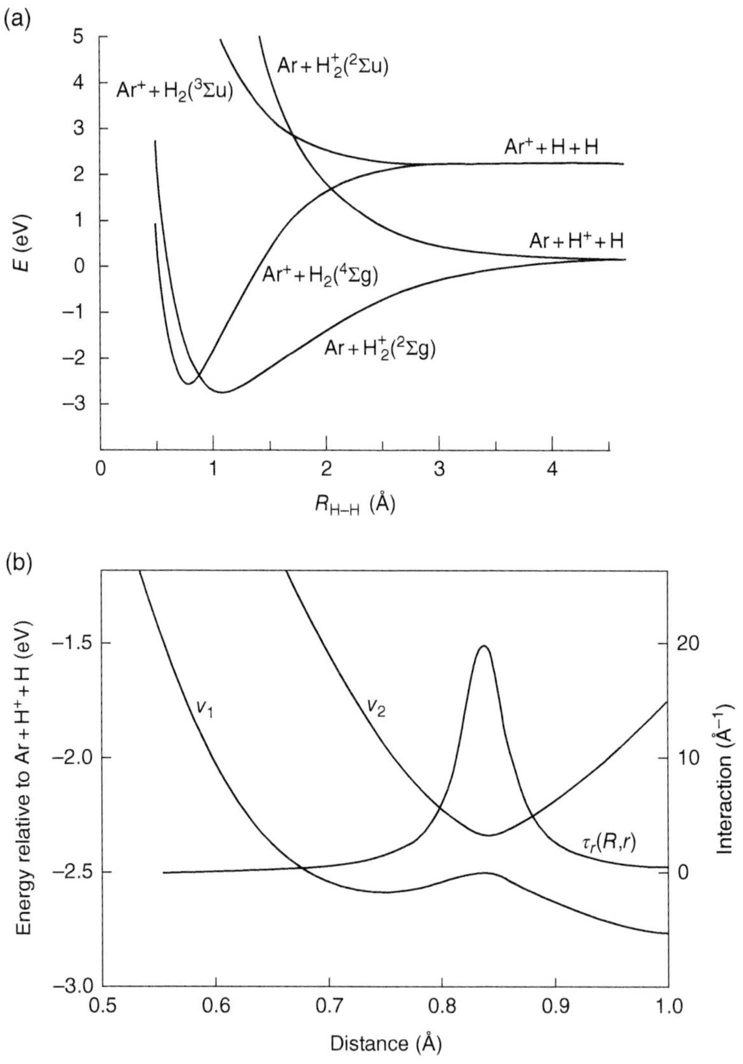

Figure 5.6 (a) Adiabatic potential energy surfaces for collinear $Ar^+ + H_2$ interacting with $Ar + H_2^+$ along the diatomic axis, as functions of the H to H distance and for large intermolecular Ar-H distances; (b) non-adiabatic momentum coupling between the two lowest PESs at a fixed intermolecular distance. *Source:* adapted from Ref. [30]. Reproduced with permission of Elsevier.

If the energy of the two degenerate states is, to begin with, a paraboloid showing a single minimum versus x and y, the theorem indicates that this form will be distorted leading to a more complex form with two minima in the (E, x) plane as shown in Figure 1.6. But now as a circle is traced in the (x, y) plane around the degeneracy point there are energy maxima and minima corresponding to lower symmetry structures [32]. An example is provided by the radical C_5H_5, of symmetry D_{5h} to begin with, and in doubly degenerate states of label $^2E_1''$. The coupling between the two states can be written in terms of the variables $x = \rho\cos(5\varphi)$ and $y = \rho\sin(5\varphi)$ and as the angle φ varies over a circle this shows energies of structures with the lower symmetry C_{2v}. The distortion may also arise from atomic momentum couplings between the two original states, involving derivatives in the x and y variables, again provided symmetry arguments indicate that the coupling is not zero.

Conical intersections of PESs can be expected to appear when dealing with photoexcited polyatomic molecules and may show up in large numbers when the molecular system is complex, as in polymers and biomolecules. Recent reviews have covered the role that CISs play in molecular spectroscopy and photoinduced dynamics [33], in isomerization [34], organic photochemistry [35], and inorganic compounds with transition metal atoms. The presence of CISs in the primary isomerization event in vision has been experimentally confirmed using ultrafast optical spectroscopy in the visible and near infrared [36]. It can be expected that many CISs will be present and linked through seams in photoinduced biomolecular dynamics.

5.5.4 The Geometrical Phase and Generalizations

The electronic wavefunctions introduced so far are real valued, but more generally, they can be constructed to also contain a phase factor making them complex valued. The phase factor can be obtained from the Schrödinger equation for the wavefunction, but it is nonunique. The phase appears to affect theoretical results when adiabatic electronically excited states are involved and may be needed to reproduce experimental results. This can be determined by doing calculations of dynamical properties and comparing results with experimental measurements of molecular spectra or collisional cross sections. The phase of an electronic wavefunction has been considered and used in dynamics calculations under the various names geometric phase, or Berry phase, or molecular Aharonov–Bohm phase [28, 37–41].

The role of the geometric phase is relevant at the level of the adiabatic approximation. The electronic wavefunction can be allowed to take complex values from a given phase factor $\eta_k^{(a)}(\mathbf{Q}) = \exp[i\gamma_k(\mathbf{Q})]$ with γ_k a real-valued geometrical phase, so that now

$$\Psi^{(ad)}(\mathbf{Q},\mathbf{X}) = \eta_k^{(a)}(\mathbf{Q})\,\Phi_k^{(a)}(\mathbf{X};\mathbf{Q})\psi_k^{(a)}(\mathbf{Q})$$

It appears as if a phase factor for the total molecular wavefunction should not have physical content. But here one must consider what happens when the nuclei undergo displacements through a hyperspace containing degenerate electronic wavefunction points, where other states with $l \neq k$ have the same energy $E_k(\boldsymbol{Q})$ and interfere quantum mechanically with the state above, and also what happens due to momentum couplings of electronic states.

Even when degeneracies and couplings are absent, the phase may alter numerical values of PESs and the related dynamics. From the definition of the momentum $\boldsymbol{G}_{kk}^{(a)}(\boldsymbol{Q})$, now for electronic state $\eta_k^{(a)}\Phi_k^{(a)}$, one finds that it is not any longer null but that instead $\boldsymbol{G}_{kk}^{(a)}(\boldsymbol{Q}) = \hbar\partial\gamma_k/\partial\boldsymbol{Q}$, a hyper-momentum-like quantity in the $\boldsymbol{Q}$ hyperspace, and that by integration along a line integral from point $\boldsymbol{Q}_0$ to $\boldsymbol{Q}$, the phase function factor is

$$\eta_k^{(a)}(\boldsymbol{Q}) = \exp\left[\frac{i}{\hbar}\int_{\boldsymbol{Q}_0}^{\boldsymbol{Q}} d\boldsymbol{Q}'.\boldsymbol{G}_{kk}^{(a)}(\boldsymbol{Q}')\right]$$

Further, replacing the above wavefunction in the Schrödinger equation and projecting it on $\eta_k^{(a)}(\boldsymbol{Q})^*\left\langle\Phi_k^{(a)}(\boldsymbol{Q})\right|$ with integration over electronic variables, one obtains

$$\frac{1}{2M}\left[\frac{\hbar}{i}\nabla_Q + \boldsymbol{G}_{kk}^{(a)}(\boldsymbol{Q})\right]^2 \psi_k^{(a)}(\boldsymbol{Q}) + \left[E_k^{(a)}(\boldsymbol{Q}) - E\right]\psi_k^{(a)}(\boldsymbol{Q}) = 0$$

The effect of $\boldsymbol{G}_{kk}^{(a)}(\boldsymbol{Q})$ on the nuclear wavefunction can be accounted for by writing it as $\psi_k^{(a)}(\boldsymbol{Q}) = \varphi_k^{(a)}(\boldsymbol{Q})\eta_k^{(a)}(\boldsymbol{Q})^*$ with $\varphi_k^{(a)}$ real valued. Applying the momentum operator within the square bracket above, in sequence, one finds that

$$\left[-\frac{\hbar^2}{2M}\nabla_Q^2 + E_k^{(a)}(\boldsymbol{Q}) - E\right]\varphi_k^{(a)}(\boldsymbol{Q}) = 0$$

Therefore, introduction of the phase factor in the electronic state must be compensated by a phase factor in the nuclear motion wavefunction describing the system dynamics, to be able to recover the form of the total wavefunction. If the electronic state is not degenerate, then the path in the line integral can be closed and deformed to disappear, corresponding to a choice of a null phase function. However, if the electronic state is degenerate with another one and they interact to create an intersection of PESs, such as the conical one, then there is a choice of paths around the intersection which will give nonzero phases, and from these, it follows that there are related changes in the energies of each of the interacting states. This has been carefully studied in examples of adiabatic transformations of a state for three identical nuclei [28, 39] and in many-state models analogous to a spin in a magnetic field [40].

An expression for the geometrical phase can be obtained introducing an integration path parameter s labeling positions along a chosen path $\boldsymbol{Q}(s)$. Writing the equation for the electronic state $\Upsilon_k = \eta_k\Phi_k$ as

$$\hat{H}_{Q(s)}|\Upsilon_k[\mathbf{Q}(s)]\rangle = E_k[\mathbf{Q}(s)]|\Upsilon_k[\mathbf{Q}(s)]\rangle$$

Differentiating it with respect to s, one can construct a state change at position $s_1 = s + \Delta s$ from

$$|\frac{\partial \Upsilon_k}{\partial s}\Big\rangle_1 = \sum_{l \neq k} |\Upsilon_l\rangle_1 \frac{\left\langle \Upsilon_l | \left(\frac{\partial \hat{H}_{Q(s)}}{\partial s}\right) | \Upsilon_k \right\rangle_1}{E_k[\mathbf{Q}(s_1)] - E_l[\mathbf{Q}(s_1)]}$$

followed by a second step to $s_2 = s + 2\Delta s$ giving

$$\left\langle \Upsilon_k | \frac{\partial \Upsilon_k}{\partial s} \right\rangle_2 = -\sum_{l \neq k} \frac{\left\langle \Upsilon_k | \left(\frac{\partial \hat{H}_{Q(s)}}{\partial s}\right) | \Upsilon_l \right\rangle_1 \left\langle \Upsilon_l | \left(\frac{\partial \hat{H}_{Q(s)}}{\partial s}\right) | \Upsilon_k \right\rangle_2}{\{E_k[\mathbf{Q}(s_1)] - E_l[\mathbf{Q}(s_1)]\}^2} = \left(\frac{d\gamma_k}{ds}\right)_2$$

This allows generation of the geometric phase in a step-wise procedure along a path, for a given electronic Hamiltonian and adiabatic basis set.

As a path moves around a CIS, the adiabatic state in the lowest surface will change with the parameter s and, after a closed loop in the x and y variables that brings them to their original values in $\tan(2\beta) = 2\,|\,H_{12}\,|\,/(H_{11} - H_{22})$, one finds that 2β will have changed by $\pm 2\pi$ and the adiabatic wavefunction coefficients $\cos(\beta)$ and $\sin(\beta)$ will have changed the wavefunction by a factor -1. This is an essential change indicating that one must incorporate the geometrical phase γ_k in electronic and nuclear wavefunctions in a treatment of the molecular dynamics so they compensate each other, to make sure that results are invariant when the physical system is distorted along a path until it returns to its original conformation.

Inspection of the matrix equation given above for $\psi_k^{(a)}$ shows that there is a similarity with the form of the Schrödinger equation for a charged particle in an electromagnetic field described by vector and scalar potentials $\vec{A}$ and φ, to be compared to $\mathbf{G}_{kk}^{(a)}(\mathbf{Q})$ and $E_k^{(a)}(\mathbf{Q})$. In the same way as a gauge transformation of the electromagnetic potentials preserves the form of the particle equation of motion, now the transformation generated by the geometric phase γ_k plays a similar role for the equation of a single adiabatic state. More generally, this can be extended for coupled adiabatic states by considering the previous coupled equations for moving nuclei and the column matrix $\boldsymbol{\psi}(\vec{R})$, which is also valid for a $\boldsymbol{\psi}(\mathbf{Q})$ in many variables, containing the square matrices $\mathbf{G}^{(a)}(\mathbf{Q})$ and $\mathbf{H}^{(a)}(\mathbf{Q})$. This suggests that the geometric phase in a molecular system can be generally treated using a unitary local gauge transformation $\mathbf{U}^{(a)}(\mathbf{Q})$ applied to the set of electronic states $\{|\Phi_k^{(a)}(\mathbf{Q})\rangle\}$. Corresponding transformations of $\mathbf{G}^{(a)}(\mathbf{Q})$ and $\mathbf{H}^{(a)}(\mathbf{Q})$ and of the nuclear motion states $\{\psi_k^{(a)}(\mathbf{Q})\}$ with the adjoint $\mathbf{U}^{(a)\dagger}$ and substitutions in the Schrödinger equation show that its form remains invariant under the new gauge transformation [42]. This generalizes the observation made above for a single adiabatic state and shows the way to construct geometric phase factors for many coupled states.

It can be expected that geometric phases in an adiabatic representation will play a quantitative role in the calculation of dynamical properties of electronically excited systems, such as optical spectra and collisional phenomena showing electronic rearrangement. This happens in the reaction $D + H_2 \rightarrow HD + H$, for which comparison of theory and experiment have shown the existence of a conical intersection and a need for a geometric phase in the calculation of the reaction cross section using adiabatic potential energy surfaces [43].

References

1 Arfken, G. (1970). *Mathematical Methods for Physicists*, 2e. New York: Academic Press.
2 Hohenberg, P. and Kohn, W. (1964). Inhomogeneous electron gas. *Phys. Rev. B* 136: 864.
3 Levy, M. (1979). Universal variational functionals of electron densities. *Proc. Nat. Acad. Sci. U. S. A* 76: 6062.
4 Kohn, W. and Sham, L.J. (1965). Self-consistent equations including exchange and correlation effects. *Phys. Rev. A* 140: 1133.
5 Gordon, R.G. and Kim, Y.S. (1972). Theory for the forces between closed shell atoms and molecules. *J. Chem. Phys.* 56: 3122.
6 Parr, R.G. and Yang, W. (1989). *Density Functional Theory of Atoms and Molecules*. Oxford, England: Oxford University Press.
7 Levine, I.N. (2000). *Quantum Chemistry*, 5e. Upper Saddle River: Prentice-Hall.
8 Born, M. and Huang, K. (1954). *Dynamical Theory of Crystal Lattices*, Appendix VIII. Oxford, England: Oxford university Press.
9 Hirschfelder, J.O., Curtis, C.F., and Bird, R.B. (1954). *Molecular Theory of Gases and Liquids*. New York: Wiley.
10 Hirschfelder, J.O. (ed.) (1967). *Intermolecular Forces* (Adv. Chem. Phys., vol. 12). New York: Wiley.
11 Nikitin, E.E. (1974). *Theory of Elementary Atomic and Molecular Processes in Gases*. London, England: Oxford University Press.
12 Micha, D.A. (1974). *Effective Hamiltonian Methods for Molecular Collisions* (Adv. Quantum Chem., vol. 71), 231. New York: Academic Press.
13 Garrett, B.C. and Truhlar, D.G. (1981). The coupling of electronically adiabatic states in atomic and molecular collisions. In: *Theoretical Chemistry, Part A*, vol. 6 (ed. D. Henderson), 216. New York: Academic Press.
14 Baer, M. (1985). The theory of electronic nonadiabatic transitions in chemical reactions. In: *Theory of Chemical Reaction Dynamics*, vol. II (ed. M. Baer), 219. Boca Raton, FL, USA: CRC Press.

15 Smith, F.T. (1969). Diabatic and adiabatic representations for atomic collision problems. *Phys. Rev.* 179: 111.
16 Baer, M. (1975). Adiabatic and diabatic representationsfor atom-molecule collisions: treatment of the collinear arrangement. *Chem. Phys. Lett.* 35: 112.
17 Mead, A.C. and Truhlar, D.G. (1982). Conditions for the definition of strictly diabatic electronic basis for molecular systems. *J. Chem. Phys.* 77: 6090.
18 Olson, J.A. and Micha, D.A. (1982). Electronic state representations at molecular potential pseudocrossings. *Int. J. Quantum Chem.* 22: 971.
19 Micha, D.A. (1983). A self-consistent eikonal treatment of electronic transitions in molecular collisions. *J. Chem. Phys.* 78: 7138.
20 Tully, J.C. (1976). Nonadiabatic processes in molecular collisions. In: *Dynamics od Molecular Collisions Part B* (ed. W.H. Miller), 217. New York: Plenum Press.
21 Cohen, J.M. and Micha, D.A. (1992). Electronically diabatic atom-atom collisions: a self-consistent eikonal approximation. *J. Chem. Phys.* 97: 1038.
22 Fernandez, F.M. and Micha, D.A. (1992). Time-evolution of molecular states in electronically diabatic phenomena. *J. Chem. Phys.* 97: 8173.
23 Bruna, P.J. and Peyerimhoff, S.D. (1987). *Excited-State Potentials* (Adv. Chem. Phys., vol. 67, Part I), 1. New York: Wiley.
24 Hirschfelder, J.O. and Meath, W.J. (1967). The nature of intermolecular forces. *Adv. Chem. Phys.* 12: 3.
25 Cohen-Tanoudji, C., Diu, B., and Laloe, F. (1977). *Quantum Mechanics*, vol. 1, Chap. IV. New York: Wiley.
26 Tinkham, M. (1964). *Group Theory and Quantum Mechanics*. New York: McGraw-Hill.
27 Matsunaga, N. and Yarkony, D.R. (1997). Energies and derivative couplings in the vicinity of a conical intersection II. CH2 and SH2. *J. Chem. Phys.* 107: 7825.
28 Herzberg, G. and Longuet-Higgins, H.C. (1963). Intersection of potential energy surfaces in polyatomic molecules. *Discuss. Faraday Soc.* 35: 77.
29 Worth, D.A. and Cederbaum, L.S. (2004). Beyond Born-Oppenheimer: molecular dynamics through a conical intersection. *Annu. Rev. Phys. Chem.* 55: 127.
30 Baer, M. and Beswick, J.A. (1979). Incorporation of electronically nonadiabatic effects into bimolecular reactive systems. III. The collinear Ar + H_2^+ system. *Phys. Rev. A* 19: 1559.
31 Jahn, H.A. and Teller, E. (1937). Stability of polyatomic molecules in degenerate electronic states. I. Orbital degeneracy. *Proc. R. Soc. A* 161: 220.
32 Herzberg, G. (1966). *Molecular Spectra and Molecular Structure III. Polyatomic Molecules*. New York: Van Nostrand Reinhold.
33 Domcke, W. and Yarkony, D.R. (2012). Role of conical intersections in molecular spectroscopy and photoinduced chemical dynamics. *Annu. Rev. Phys. Chem.* 63: 325.
34 Levine, B.G. and Martinez, T.J. (2007). Isomerization through conical intersections. *Annu. Rev. Phys. Chem.* 58: 613.

35 Robb, M.A. (2011). Conical intersections in organic photochemistry. In: *Conical Intersections: Theory, Computation, and Experiment* (eds. W. Domcke, D.R. Yarkony and H. Koeppel), 3. World Scientific Publishing Co.
36 Polli, D., Altoe, P., Weingart, O. et al. (2010). Conical intersection dynamics of the primary photoisomerization event in vision. *Nature* 467: 440.
37 Longuet-Higgins, H.C. (1975). The intersection of potential energy surfaces in polyatomic molecules. *Proc. R. Soc. Lond. A* 344: 147.
38 Stone, A.J. (1976). Spin-orbit coupling and the intersection of potential energy surfaces in polyatomic molecules. *Proc. R. Soc. Lond. A* 351: 141.
39 Mead, C.A. and Truhlar, D.G. (1979). On the determination of Born-Oppenheimer nuclear motion wavefunctions including complications due to conical intersections and identical nuclei. *J. Chem. Phys.* 70: 2284.
40 Berry, M.V. (1984). Quantal phase factors accompanying adiabatic changes. *Proc. R. Soc. Lond. A* 392: 45.
41 Aharonov, Y. and Bohm, D. (1959). Significance of the electromagnetic potential in the quantum theory. *Phys. Rev.* 115: 485.
42 Pacher, T., Mead, C.A., Cederbaum, L.S., and Koeppel, H. (1989). Gauge theory and quasidiabatic states in molecular physics. *J. Chem. Phys.* 91: 7057.
43 Kuppermann, A. and Wu, Y.-S.M. (1993). The geometric phase effect shows up in chemical reactions. *Chem. Phys. Lett.* 205: 577.

6

Many-Electron Treatments

CONTENTS

6.1 Many-Electron States

6.1.1 Electronic Exchange and Charge Transfer

A fundamental and general treatment of molecular interactions must be based on the quantal description of the many-electron system for given nuclear conformations, including Coulomb energies for charged particles and electronic

Molecular Interactions: Concepts and Methods, First Edition. David A. Micha.

spin-orbit coupling energies. Here the emphasis is on aspects of the many-electron theory for two interacting species A and B, as they are needed to obtain correct intermolecular energies at all relative distances between them. The presentation assumes familiarity with concepts of molecular electronic structure and properties as found in the literature [1–5].

Occupied and unoccupied electronic orbitals of two molecules A and B at relative distance R change at intermediate and short intermolecular distances to give energy levels of the supermolecule AB, as electronic orbitals of the two species combine and allow electrons to spread over the whole system. Electronic exchange involves combinations of the occupied molecular orbitals (MOs) of A with occupied MOs of B, while electron transfer involves combinations of occupied MOs of one species with unoccupied MOs of the other. In both cases, the species can be in their ground electronic states or in excited ones, as shown in Figure 6.1. The pair AB is mentioned here as a *compound* when A and B form bonds, or as a *complex* if they are not bonded.

Expectation values of extensive properties, such as the electronic energy $E(R)$ and the electron density $\rho(R)$, where R is the distance between the centers of mass of A and B, must go in the limit of large intermolecular distances into the sums of the fragments contributions. One must have for $R \to \infty$ that

$$E \approx E_A + E_B \text{ and } \rho \approx \rho_A + \rho_B$$

This will be called here *asymptotic consistency*, but is usually termed size consistency in the literature. Its requirement is essential in treatments of intermolecular forces. It guarantees that the potential energy of interaction at all R satisfies $V(R) = E(R) - (E_A + E_B) \approx 0$.

When the number of interacting identical species S is increased, as one proceeds from clusters to solids or fluids, the extensive properties (in the sense of statistical thermodynamics) must increase linearly. For a system S_N with a large number N of the stable species S, one has $E(S_N) \approx NE(S)$, as $N \to \infty$, where S_N may be a compound or a complex. This is *size extensivity*, a property needed in treatments of many-molecule systems.

Two general methods for intermolecular forces are the so-called supermolecule method, where the interacting species are considered to be a single many-electron system in a field of all the nuclei, and the many-atom method where the total many-electron system is instead treated as a collection of atomic many-electron systems. In both cases, one must properly treat the overall symmetry of the many-electron state with regard to electron exchange, its total electronic spin state, and its total point group or crystal group symmetry.

The methods described in what follows are applicable also to chemical reactions such as $A + B \to C + D$, where C and D are new species, involving the rearrangement of nuclei. Treatments of energetics require exploration of potential energy surfaces over wide ranges of atomic positions, to identify potential

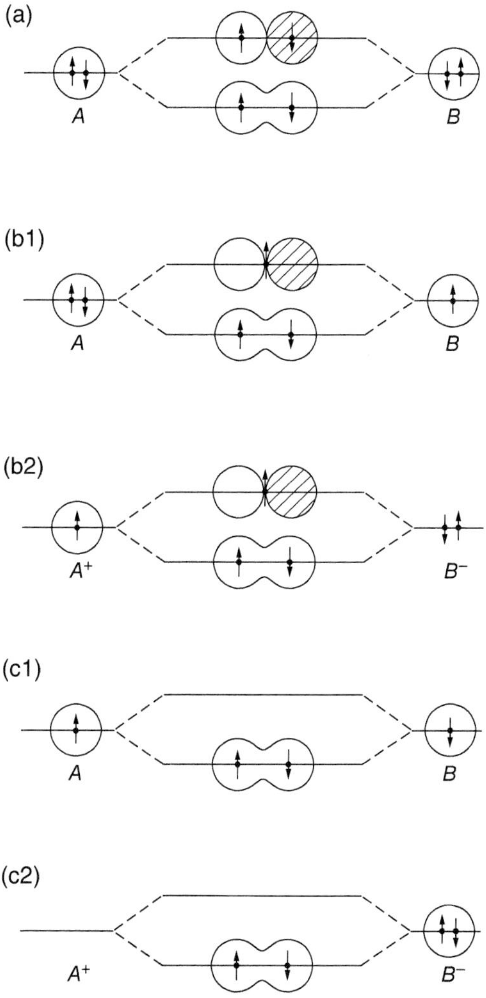

Figure 6.1 Pictorial description of electronic energy levels and their combination for interacting species *A* and *B*: (a) closed shell–closed shell; (b) closed–open shell; (c) open shell–open shell. Their molecular orbital (bonding and antibonding) occupations are shown at intermediate distances: (a, b1, c1) electronic exchange; (b2, c2) electron transfer.

energy barriers between regions of low energy. Much can be learned about allowed or disallowed reactions from qualitative treatments based on the shapes of molecular orbitals and the interactions between highest occupied and lowest unoccupied molecular orbitals (HOMOs and LUMOs) of the interacting species [6–8]. But accurate calculations of reaction energy barriers, stable conformations, and vibrational-rotational properties require many-electron treatments as described in what follows.

6.1.2 Many-Electron Descriptions and Limitations

In the supermolecule method, the basic idea is to introduce molecular spin-orbitals, which spread over all the atoms in the whole system, and then to construct many-electron states to calculate their properties. A self-consistent field (SCF) approximation such as the restricted Hartree–Fock (or RHF) treatment for the many-electron system in the supermolecule method, with doubly or singly occupied MOs constructed from linear combinations of atomic orbitals (LCAOs), is of limited applicability to intermolecular forces. It works for two interacting closed-shell systems such as He_2 for which SCF leads correctly to He + He asymptotically (although without their long-range van der Waals interaction), but gives incorrect results for two open-shell systems such as in H_2, for which the SCF wavefunction would asymptotically contain the electronic states of both H + H and $H^+ + H^-$ and would give incorrect potential energies. This can be improved using an unrestricted HF (or UHF) description, but this does not generate a molecular state of the correct spin symmetry. The SCF treatment for the interaction between an open-shell species and another with closed shells may also lead to asymptotically incorrect potential energies due to unphysical contamination of a neutral state with a charge transfer state, such as in ArF going to Ar + F and also $Ar- + F^+$. These are examples of appearances of "intruder states" at large distances, giving wrong asymptotic energies and requiring extensions of the SCF treatment. More general treatments must allow for asymptotic consistency and size extensivity and must be amenable to systematic improvement to a desired high accuracy.

Supermolecule treatments that do this are the configuration interaction (CI) treatment, many-body perturbation (MBPT) for electrons, and many-electron coupled clusters (CC) treatments. But even these must be carefully considered in applications to intermolecular forces to satisfy constrains and provide accuracy and would not necessarily account for van der Waals (vdW) interactions. They are treated in detail in Section 6.2.

In the many-atom treatment, electrons are located in spin orbitals, which extend only over each molecular component of the total system. Component many-electron functions are first constructed and then used to obtain the state of the whole system. This must be done so that electron exchange antisymmetry is properly incorporated in the whole many-electron function. Two such general procedures are the extended valence-bond method (EVB) and the symmetry-adapted perturbation theory (SAPT) of a many-atom system. They are treated in Section 6.3.

Both treatments introduce a many-electron wavefunction of all the electronic variables $\boldsymbol{X} = \{\boldsymbol{x}_j : j = 1 \ to \ N\}$ of N electrons, with $\boldsymbol{x}_j = (\vec{r}_j, \zeta_j)$ containing electron position and spin variables, and is a function that depends parametrically on the set of nuclear positions $\boldsymbol{Q}$, given as $\Phi(\boldsymbol{X}; \boldsymbol{Q})$. However, in what follows in this chapter the nuclear positions will be fixed and omitted, and the focus will be on electronic aspects. The total electronic wavefunction of a stationary state solution for a given electronic Hamiltonian $\hat{H}_Q = \hat{H}$ must be constructed to be antisymmetric with respect to electron exchange and will typically describe a system with given total spin quantum numbers S, $M_S(= -S, \ldots, +S)$, and with point group or crystal symmetry of type Γ, both obtained from the spin and space symmetries of the electronic Hamiltonian for fixed nuclei [3]. These conditions can be imposed by means of projection operators constructed from constant of motion of the Hamiltonian [9].

Antisymmetry is imposed applying the projection operator

$$\hat{\mathcal{O}}^{(AS)} = (N!)^{-1} \sum\nolimits_P (-1)^p \hat{P}$$

where $\hat{P}$ is a permutation among the N electrons, of parity p, and the summation extends over all different permutations, satisfying $\hat{P}\hat{H} = \hat{H}\hat{P}$. The self-adjoint property $\hat{P}^{\dagger} = \hat{P}$ implies that $\hat{\mathcal{O}}^{(AS)\dagger} = \hat{\mathcal{O}}^{(AS)}$ as well. Insofar as a product of two permutations is also a permutation in the set, it follows that $\hat{\mathcal{O}}_{AS}$ is idempotent, with $\hat{\mathcal{O}}^{(AS)}.\hat{\mathcal{O}}^{(AS)} = \hat{\mathcal{O}}^{(AS)}$ and that it also commutes with the Hamiltonian. Starting with an electronic function $F(X)$, a proper antisymmetrized state is usually constructed as $\Phi(\boldsymbol{X}) = \hat{\mathcal{A}}^{(N)} F(\boldsymbol{X})$, with the antisymmetrizer $\hat{\mathcal{A}}^{(N)} = (N!)^{\frac{1}{2}} \hat{\mathcal{O}}^{(AS)}$. A fundamental law of quantum mechanics for identical particles is that the wavefunction of identical fermions must be antisymmetric under their exchange even if the species A and B are far away.

Therefore, even at very large distances, the total wavefunction must be of the form $\Phi_{AB}(\boldsymbol{X}) \approx \hat{\mathcal{A}}^{(N)} F_A(\boldsymbol{X_A}) F_B(\boldsymbol{X_B})$ involving the total antisymmetrizer, while expectation values of an extensive property $\hat{M}_{AB}$ must satisfy the asymptotic condition $\langle \hat{M}_{AB} \rangle = \langle \Phi_{AB} | \hat{M}_{AB} | \Phi_{AB} \rangle \approx \langle \hat{M}_A \rangle + \langle \hat{M}_B \rangle$ as $R \to \infty$. This can be verified considering that the total antisymmetrizer can be expressed as a product of antisymmetrizers of the two species A and B times a complimentary operator $\hat{\mathcal{C}}^{(AS)} = \hat{I} + \sum\nolimits_T (-1)^t \hat{T}$ containing the identity $\hat{I}$ and transpositions $\hat{T}$ of electrons between A and B,

$$\hat{\mathcal{O}}^{(AS)} = N_A! N_B! (N!)^{-1} \hat{\mathcal{C}}^{(AS)} \hat{\mathcal{O}}_A^{(AS)} \hat{\mathcal{O}}_B^{(AS)}$$

or alternatively $\hat{\mathcal{A}}^{(N)} = \hat{\mathcal{A}}' \hat{\mathcal{A}}_A^{(N_A)} \hat{\mathcal{A}}_B^{(N_B)}$ with $\hat{\mathcal{A}}'$ a complimentary antisymmetrizer. When this operates at large distances R, the transpositions of electrons from, say, A moves its electrons to regions of B where expectation values with respect to states of A vanish because the wavefunctions of A fall exponentially with distance. Cross terms disappear, and the sum of asymptotic property values can be obtained. Therefore, as long as $\Phi_{AB}(\boldsymbol{X}) \approx \hat{\mathcal{A}}^{(N)} \Phi_A(\boldsymbol{X_A}) \Phi_B(\boldsymbol{X_B})$ in terms of wavefunctions for A and B with energies E_A and E_B, one finds for the total energy that $E = \left\langle \Phi_{AB} \mid \left(\hat{H}_A + \hat{H}_B + \hat{H}_{AB}^{(int)} \right) \mid \Phi_{AB} \right\rangle \approx E_A + E_B$ and the potential energy $V(R)$ vanishes at large distances.

Spin and space symmetries can be imposed starting from normal many-electron operators $\hat{N}$, with the property $\hat{N}.\hat{N}^{\dagger} = \hat{N}^{\dagger}.\hat{N}$, which generate symmetry transformation and are constants of motion insofar as $\hat{N}\hat{H} = \hat{H}\hat{N}$. Their eigenstates and eigenvalues in $\hat{N} \mid k\rangle = \nu_k \mid k\rangle$ can be used to define projection operators $\hat{O}_k = |k\rangle\langle k|$, which extract a component of correct symmetry from a wavefunction as $\Phi_k(\boldsymbol{X}) = \left\langle \boldsymbol{X} \mid \hat{O}_k \Phi \right\rangle = \langle \boldsymbol{X} \mid k\rangle \langle k \mid \Phi\rangle_X$, where it is shown that the second factor may also depend on some of the electronic variables. In applications where the constant of motion operator has only a discrete set of eigenvalues, a frequent situation, the projection operator can be conveniently constructed as

$$\hat{O}_k = \prod\nolimits_{l \neq k} \frac{\hat{N} - \nu_l}{\nu_k - \nu_l}$$

with the product extending over only different eigenvalues. It satisfies the orthogonality property $\hat{O}_k \hat{O}_l = \delta_{kl} \hat{O}_k$.

This is applicable to the angular momentum operator with components $\hat{M}_\xi$, $\xi = x, y, z$, whether this is a component of the spin $\vec{S}$ or of the orbital angular momentum $\vec{L}$. For the total spin operator with components $\hat{S}_\xi = \sum_j \hat{S}_{\xi j}$, of magnitude and projection $\hat{S}^2$ and $\hat{S}_z$ with eigenvalues $S(S+1)$ and $M_S = -S, -S+1, \ldots, S$, the projector $\hat{O}_{SM_S} = \hat{O}_S.\hat{O}_{M_S}$ can be written as [10]

$$\hat{O}_{SM_S} = (2S+1)\frac{(S+M_S)!}{(S-M_S)!}\sum\nolimits_{\mu=0}^{S_{max}-S} (-1)^{\mu} \frac{\hat{S}_{-}^{S-M_S+\mu}\, \hat{S}_{+}^{S-M_S+\mu}}{\mu!(2S+\mu+1)!}$$

in terms of spin-projection rising and lowering operators $\hat{S}_{\pm} = \hat{S}_x \pm i\hat{S}_y$, satisfying $\hat{S}_z \hat{S}_{\pm} = \hat{S}_{\pm}\left(\hat{S}_z \pm 1\right)$, in a form suitable for construction of spin-adapted many-electron states. They satisfy the orthogonality condition $\hat{O}_{SM_S}\hat{O}_{S'M_{S'}} = \delta_{SS'}\delta_{M_S M_{S'}}\hat{O}_{SM_S}$. At large distances, one can write $\hat{S}_{\pm} = \hat{S}_{A\pm} + \hat{S}_{B\pm}$ in terms of the operators for the species, which leads to products such as $\hat{S}_{A+}\hat{S}_{B-}$ in the projector. Provided the asymptotic states of A or B are nondegenerate, whether ground or excited states, the expectation value of these cross terms gives zero and one recovers a sum of uncoupled terms for A and B. If, however, the asymptotic states of A and B are degenerate, then it is necessary to analyze their expectation values to ascertain asymptotic behavior.

Relating to space symmetry, the constants of motion are operator movements $\hat{R}$ such as inversion, reflections, or rotations, which commute with the Hamiltonian and which form a symmetry group $\mathcal{G}$. Representation matrices, or their characters $\chi^{(k)}(\hat{R})$ for the k-th irreducible representation, can be used to construct projection operators $\hat{\mathcal{O}}(\Gamma^{(k)})$ for each irreducible representation $\Gamma^{(k)}$ as a sum over the group symmetry movements [3]. One such projector is

$$\hat{\mathcal{O}}\left(\Gamma^{(k)}\right) = \frac{d_k}{h}\sum_R \chi^{(k)}(\hat{R})^* \hat{R}$$

with h the order of the group and d_k the dimension of the representation. The orthogonality properties of characters ensures that $\hat{\mathcal{O}}(\Gamma^{(k)})\hat{\mathcal{O}}(\Gamma^{(l)}) = \delta_{kl}\hat{\mathcal{O}}(\Gamma^{(k)})$. The asymptotic form of this projector depends on how the overall symmetry is expressed in terms of the symmetries of the Hamiltonians of A and B. In some cases, such as for overall translation or reflection on a plane, the group is a direct product of those of A and B, $\mathcal{G} = \mathcal{G}_A \otimes \mathcal{G}_B$, with characters $\chi^{(k)}(\hat{R}) = \chi^{(k_A)}(\hat{R}_A)\chi^{(k_B)}(\hat{R}_B)$, which allow factorization of the projector, and a proper asymptotic decomposition will follow. However, if the total group involves also a complementary subgroup as in $\mathcal{G} = \mathcal{G}' \otimes \mathcal{G}_A \otimes \mathcal{G}_B$, then a further analysis of the asymptotic wavefunction and expectation values must be done to establish whether cross terms vanish to give the correct behavior [3].

Many-electron wavefunctions are sometimes variationally obtained by minimizing the total electronic energy but do not display the total symmetries of the system. In such cases, it is yet possible to extract the correct symmetry by using projection operators after the variation, in a postsymmetrization. Consider a variational but nonsymmetrized wavefunction Φ, which would contain symmetry-adapted components Φ_k as in $\Phi = \sum_k a_k \Phi_k$, assumed to be all normalized for simplicity. The variational energy is $E_{var} = \sum_k |a_k|^2 E_k$ with factors $|a_k|^2 \leq 1$. Therefore postsymmetrized components are found to provide energies E_k lower than the unsymmetrized original wavefunction. But while the original unsymmetrized wavefunction gives an upper bound to the calculated energy, the postsymmetrized one may not do so for the exact symmetrized energy and must be tested for accuracy of the postsymmetrized associated energy. This can be done, for example, by doing some calculations with prior symmetrization and otherwise the same conditions at selected nuclear conformations, to compare energies with results from postsymmetrization.

The interaction energy of open-shell or of electronically excited species presents additional complications because they frequently involve a number N_D of states Φ_J, $J = 1$ to N_D, of the whole system that are degenerate or nearly degenerate in energy, and are either eigenstates of the Hamiltonian for the energy $E^{(D)}$, with $\hat{H}\,|\,\Phi_J\rangle = E^{(D)}\,|\,\Phi_J\rangle$ or have very similar expectation values $\langle \Phi_J\,|\,\hat{H}\,|\,\Phi_J\rangle \cong E^{(D)}$. They define a multireference, or configuration interaction (CI), state space to be treated as a whole. In these cases, one must extend the usual treatments to include linear combinations of the Φ_J states in the CI multireference wavefunctions, or one must construct unperturbed multireference Hamiltonians in perturbation and coupled cluster treatments.

A wave operator treatment appears to be suitable for these extensions [11–13]. Introducing a multireference projection operator $\hat{\mathcal{P}} = \sum_J |\Phi_J\rangle\langle\Phi_J|$ onto the space of the degenerate states with energy $E^{(D)}$, and its complement $\hat{\mathcal{Q}} = \hat{I} - \hat{\mathcal{P}}$, the Schroedinger equation $\hat{H}|\Psi\rangle = E|\Psi\rangle$ can be formally solved writing $\Psi = \Phi^{(D)} + \Psi'$, with the first term a chosen reference state satisfying $\hat{\mathcal{P}}\Psi = \hat{\mathcal{P}}\Phi^{(D)}$, a generalization of intermediate normalization giving $\langle\Phi^{(D)}|\hat{\mathcal{P}}|\Psi\rangle = \sum_J |\langle\Phi^{(D)}|\Phi_J\rangle|^2 = N^{(D)}$. The energy is then $E = \langle\Phi^{(D)}|\hat{\mathcal{P}}\hat{H}|\Psi\rangle / N^{(D)}$. Defining a wave operator $\hat{\Omega}$ to give $\Psi = \hat{\Omega}\Phi^{(D)}$ as the solution of the Schrodinger equation, one finds that using the reduced resolvent operator

$$\hat{R}(E) = \left[\alpha\hat{\mathcal{P}} + \hat{\mathcal{Q}}\left(E - \hat{H}\right)\hat{\mathcal{Q}}\right]^{-1}$$

with α a convenient constant to be set to zero after calculations, the wavefunction is given by

$$\Psi = \hat{\Omega}\Phi^{(D)} = \left(\hat{\mathcal{P}} + \hat{\mathcal{Q}}\hat{R}(E)\hat{\mathcal{Q}}\hat{H}\hat{\mathcal{P}}\right)\Phi^{(D)}$$

as can be verified applying the inverse of $\hat{R}(E)$ to both sides of this equation. This multireference wave operator can be constructed using CI methods or perturbation expansions, described as follows.

Electronic antisymmetry and spin and space symmetries enforced by projection operators $\hat{\mathcal{O}}^{(AS)}$ and $\hat{\mathcal{O}}_k$ can all be maintained when the wave operator is applied, provided the projector $\hat{\mathcal{P}}$ has been symmetry-adapted so that $\hat{\mathcal{P}}\hat{\mathcal{O}}^{(AS)} = \hat{\mathcal{O}}^{(AS)}\hat{\mathcal{P}}$ and $\hat{\mathcal{P}}\hat{\mathcal{O}}_k = \hat{\mathcal{O}}_k\hat{\mathcal{P}}$. Then one finds that $\hat{\Omega}\hat{\mathcal{O}}^{(AS)} = \hat{\mathcal{O}}^{(AS)}\hat{\Omega}$ and $\hat{\Omega}\hat{\mathcal{O}}_k = \hat{\mathcal{O}}_k\hat{\Omega}$ insofar as the wave operator has been constructed from the full Hamiltonian.

The energy of the whole system is obtained as $E = \langle\Phi^{(D)}|\hat{\mathcal{P}}\hat{H}|\hat{\Omega}\Phi^{(D)}\rangle / N^{(D)}$, and the potential energy follows from $V = E - (E_A + E_B)$. For the purpose of generating potential energy surfaces going to the correct asymptotic limits, it is convenient, but not necessary, that the reference state is chosen as $\Phi^{(D)} = \hat{\mathcal{A}}'\Phi_A^{(D)}\Phi_B^{(D)}$ and that the wave operator for the pair of species A and B satisfies $\hat{\Omega}_{AB}\Phi^{(D)} \approx \hat{\mathcal{A}}'\hat{\Omega}_A\hat{\Omega}_B\Phi_A^{(D)}\Phi_B^{(D)}$ as $R \to \infty$, with $\hat{\mathcal{A}}'$ the complementary antisymmetrizer. In this case, the expectation value of an extensive property would go asymptotically to the sum of its values for the two separate species, and in particular $V = E - (E_A + E_B) \approx 0$, with asymptotic energies $E_I = \left[\langle\Phi^{(D)}|\hat{\mathcal{P}}\hat{H}|\hat{\Omega}\Phi^{(D)}\rangle\right]_I / N_I^{(D)}, I = A, B$.

6.1.3 Properties and Electronic Density Matrices

Many-electron operators can be classified as being one-electron, two-electron, or N-electron operators. Most physical properties are given by sums of one-, and two-electron operators and their expectation values can be obtained from one- and two-electron density operators and their matrices (DMs). They have been thoroughly considered in Refs. [2] and [14]. The operator of a property $\hat{A}$ in an N-electron system, such as the energy or an electric multipole, can be written as

$$\hat{A} = \hat{A}^{(1)} + \hat{A}^{(2)} + \cdots + \hat{A}^{(N)},$$
$$\hat{A}^{(1)} = \sum\nolimits_{1 \leq m \leq N} \hat{a}_m^{(1)},\ \hat{A}^{(2)} = \sum\nolimits_{1 \leq m < n \leq N} \hat{a}_{m,n}^{(2)} = \frac{1}{2}\sum\nolimits_{m \neq n} \hat{a}_{m,n}^{(2)}, \ldots$$

with $\hat{a}_m$ a one-electron operator containing in general the position and momentum variables of electron m, and matrix elements in the coordinate representation with states $|\boldsymbol{x}_m\rangle = |\vec{r}_m, \zeta_m\rangle$ written as $\langle \boldsymbol{x}_m | \hat{a}_m | \boldsymbol{x}'_m \rangle = a_m(\boldsymbol{x},\boldsymbol{x}')$, and so on.

The expectation value of each term in a state with wavefunction $\Phi_J(1, 2, \ldots, N)$ is given by

$$\left\langle \Phi_J | \hat{A}^{(1)} | \Phi_J \right\rangle = \int d(1)d(1')a^{(1)}(1,1')\rho_J^{(1)}(1';1),$$

$$\rho_J^{(1)}(1';1) = N\int d(2)\cdots d(N)\Phi^*{}_J(1',2,\ldots,N)\Phi_J(1,2,\ldots,N)$$

$$\left\langle \Phi_J | \hat{A}^{(2)} | \Phi_J \right\rangle = \int d(1)d(1')d(2)d(2')a^{(2)}(1,2,1',2')\rho_J^{(2)}(1',2';1,2),$$

$$\rho_J^{(2)}(1',2';1,2) = \frac{N(N-1)}{2}\int d(3)\cdots d(N)\Phi^*{}_J(1',2',3,\ldots,N)\Phi_J(1,2,3,\ldots,N)$$

and so on. Here $\rho_J^{(1)}$ and $\rho_J^{(2)}$ are one- and two-electron reduced density functions for this state in the coordinate representations of reduced density operators (RDOps) $\hat{\rho}_J^{(1)}$ and $\hat{\rho}_J^{(2)}$ with the argument (1) a short form for $(\boldsymbol{x}_1)$ and with $\int d(1)(\ldots) = \int d^3r_1 \sum_{\zeta_1 = \pm}(\ldots)$. They have been defined here with normalization $tr_1\left(\hat{\rho}^{(1)}\right) = \int d(1)\rho^{(1)}(1;1) = N,\ tr_2\left(\hat{\rho}^{(2)}\right) = \int d(1)d(2)\rho^{(2)}(1,2;1,2) = N(N-1)$ $/2$, the number of different electron pairs, and so on. This is a common normalization [14] but not universal. Transition averages $\left\langle \Phi_J | \hat{A}^{(n)} | \Phi_K \right\rangle$ can similarly be expressed in terms of transition density matrices $\rho_{JK}^{(n)}$ [14]. The electronic reduced density operators are usually constructed in a basis set of one-electron functions for computational purposes.

In particular, the N-electron Hamiltonian, including here only kinetic energies of electrons, and Coulomb electron–electron $v_{m,n}$ and electron–nuclei interaction V_m energies, is given by

$$\hat{H} = \hat{H}^{(1)} + \hat{H}^{(2)},\ \hat{H}^{(1)} = \sum\nolimits_{1 \le m \le N} \hat{h}_m,\ \hat{H}^{(2)} = \sum\nolimits_{1 \le m < n \le N} v_{m,n}$$

where $\hat{h}_m = \hat{K}_m + V_m$ is the sum of kinetic energy plus electron attraction to all nuclei (or ion cores), and $v_{m,n} = c_e^2 / |\vec{r}_m - \vec{r}_n|$. Total electronic energies as functions of atomic positions for a state J are found from

$$E_J = \int d(1)d(1')h(1,1')\rho_J^{(1)}(1';1) + \int d(1)d(2)v(1,2)\rho_J^{(2)}(1,2;1,2)$$

insofar as the Coulomb potential energy is local so that $\langle 1, 2 \mid v_{1,2} \mid 1', 2' \rangle = \delta(1-1')\delta(2-2')v(1,2)$. The density function $\rho_J^{(1)}(1';1)$ can be obtained from $\rho_J^{(2)}(1',2;1,2)$ integrating over (2) so that the total energy is a linear functional of $\rho_J^{(2)}$.

When the interacting pair of species A and B move apart as $R \to \infty$, each containing N_A and N_B electrons, the reduced densities for one electron must physically go to

$$h(1,1')\rho_J^{(1)}(1';1) \approx \left[h(1,1')\rho_J^{(1)}(1';1)\right]_A + \left[h(1,1')\rho_J^{(1)}(1';1)\right]_B$$

with $tr_1\left(\rho_{J,A}^{(1)}\right) = N_A$ and $tr_1\left(\rho_{J,B}^{(1)}\right) = N_B$, and with the corresponding limit for the two-electron function $\rho_J^{(2)}(1,2;1,2)$. One then finds that $E_J \approx E_{J,A} + E_{J,B}$. This, however, is not always the case in approximations, depending on how the state J has been constructed, and it must be verified in applications to intermolecular forces.

In the particular case when the many-electron state is a single determinantal function, as in the Hartree–Fock method, with $\Phi_J(1,2,\ldots,N) = D(1,2,\ldots,N) = \hat{\mathcal{A}}\psi_1(1)\psi_2(2)\ldots\psi_N(N) = (N!)^{-1/2}\det\left[\psi_j(n)\right]$, constructed from N orthonormal spin orbitals (SOs) ψ_j, one finds, recalling that the determinant of a product of matrices equals the product of their determinants, the special results

$$\rho_{HF}^{(1)}(1';1) = \sum\nolimits_{j=1}^{N} \psi_j(1')^* \psi_j(1)$$

$$\rho_{HF}^{(2)}(1',2';1,2) = \frac{1}{2}\left[\rho_{HF}^{(1)}(1';1)\rho_{HF}^{(1)}(2';2) - \rho_{HF}^{(1)}(1';2)\rho_{HF}^{(1)}(2';1)\right]$$

Here the first line shows that each SO in the determinant appears with an occupation number of 1, and the second line indicates that the total energy can be obtained from just $\rho_{HF}^{(1)}$, insofar as the total HF energy follows from

$$E_{HF} = \int d(1)d(1')h(1,1')\rho_{HF}^{(1)}(1';1) + \int d(1)d(2)v(1,2)\rho_{HF}^{(2)}(1,2;1,2)$$

The density operators $\hat{\rho}^{(1)}$ and $\hat{\rho}^{(2)}$ for a general state J can be expanded in an orthonormal basis set $\{u_k(\boldsymbol{x}_m)\} = \{u_k(m)\}$ of one-electron functions $u_k(m) = \langle m \mid k\rangle$. This gives in the Dirac notation

$$\hat{\rho}^{(1)} = \sum\nolimits_{k,k'} |k\rangle \rho^{(1)}(k,k')\langle k'|$$

$$\hat{\rho}^{(2)} = \sum\nolimits_{k_1,k_1'} \sum\nolimits_{k_2,k_2'} |k_1 k_2\rangle \rho^{(2)}(k_1,k_2,k_1' k_2')\langle k_1' k_2'|$$

providing reduced density matrix (RDM) elements in the basis set.

These operators have *natural spin orbitals* (or NSOs) eigenstates and eigenvalues (or occupation numbers), which can be obtained by diagonalizing the matrices with a unitary transformation to a new basis set $\left\{u_l^{(1)}(m)\right\}$ [14]. For the one-electron RDOp, this gives

$$\hat{\rho}^{(1)} = \sum\nolimits_l n_l^{(1)} \left| u_l^{(1)} \right\rangle \left\langle u_l^{(1)} \right|$$

where the NSO occupation numbers $n_l^{(1)}$ satisfy $\sum_l n_l^{(1)} = N$ so that $0 \le n_l^{(1)} \le 1$. These occupation numbers provide a useful criterion for choosing a basis set of optimal expansion SOs in many-electron systems. The molecular NSOs with large occupation numbers can be expected to be the most relevant for description of electronic correlation. Given a subset of NSOs as a basis set, it can be systematically improved by adding to the set the NSOs with subsequent large occupation numbers, to achieve convergence of properties. The NSOs can be obtained at different levels of accuracy from wavefunctions containing electronic correlation, and they can be tested to determine whether they stay stable as levels of correlation are increased, to generate a computationally useful basis set. This selection procedure can be done separately for the component species A and B, and the NSOs from each one can be combined into a larger basis set to be used for the whole AB system.

6.1.4 Orbital Basis Sets

The calculation of electronic orbitals as functions of electron positions requires solving partial differential equations for the orbitals. This is usually done introducing basis sets to expand the orbitals, with expansion coefficients extracted by solving coupled matrix equations. The choices of basis sets are guided by requirements of accuracy and feasibility in the calculations of properties. Delocalized electron orbitals (MOs) ψ_j in a molecular system can be expanded as linear combinations of atomic orbitals (LCAOs) $\chi_\mu(\vec{r})$, with the index μ containing the atom location and quantum numbers of core (inner shell) or valence (outer shell) electrons, and may include AOs hybridized or polarized by interatomic interactions. These have overlaps $S_{\mu\nu} = \langle \chi_\mu | \chi_\nu \rangle$ components of the matrix $\boldsymbol{S}$ and, if they form in practice a complete set, then they also satisfy the identity decomposition $\sum_{\mu,\nu} |\chi_\mu\rangle (S^{-1})_{\mu\nu} \langle\chi_\nu| = \hat{I}$. Orbitals of large systems like solids or polymers can also be alternatively expanded in a basis set of plane wavefunctions $\phi_{\vec{k}}(\vec{r})$ for wavevectors $\vec{k}$ forming a grid in reciprocal space. Important computational considerations in the choice of a basis set are how fast can electronic integrals be done with the basis set and how large are the matrices that must be stored and multiplied in calculations of properties. The size of the basis set is important for achieving high accuracy of potential energies, and consistent basis sets must be used when subtracting the energies of components A and B from the total energy of a complex $A + B$, to obtain changes of energies with structural parameters.

Numerous basis sets of AOs have been introduced in the literature and implemented for computational work [3, 4, 15]. The original basis sets derived from hydrogenic orbitals, products of radial functions times angular functions, and have used Cartesian coordinates (x, y, z) or spherical coordinates (r, ϑ, φ) with origin at the atomic nucleus positions. In the so-called (nlm) Slater-type orbitals (or STOs), a radial factor $R_n(r) = r^{n-1}\exp(-\zeta r)$ containing an exponential parameter ζ multiplies spherical harmonics $Y_{lm}(\vartheta, \varphi)$ for the angular dependence. The nodeless radial functions are combined into contractions with fixed coefficients to describe the physical radial functions. The more popular and recent basis sets are made up of Gaussian radial functions (or GTOs), which simplify the calculation of one- and two-electron integrals. The primitive radial factors are of form $\exp(-\zeta r^2)$ times a power of the radius, multiplying spherical harmonics, or instead the exponential multiplies powers $x^a y^b z^c$ of Cartesian coordinates with $a + b + c = l$. The quantum number l corresponds to an occupied l-subshell (s-, p-, d- ...) or to a subshell with higher l if there is a need to describe polarization or hybridization of shells.

Products of primitive GTOs $g_{a,b,c}(x, y, z; \zeta) = \exp(-\zeta r^2)x^a y^b z^c$ located at different centers can be rewritten as combinations of GTOs centered at intermediate points, a fact that speeds the calculation of electron integrals. But GTOs do not accurately describe the electronic wavefunction cusp at an atomic position, and they decrease with r much faster than the physical exponential functions. These shortcomings are compensated using: (i) primitives with more than one value of ζ combined to give double-zeta (or DZ), triple- (TZ), quadruple- (QZ), or n-zeta (n-Z) functions; (ii) contracted sums of gaussians (CGTO) where N_l primitives of a given l are combined with fixed coefficients to form $n_l < N_l$ CGTOs. A set such as 6-31G, suitable for compounds with first and second row elements, contracts six primitive GTOs for inner shells and treats outer shells with a contraction of three primitives plus another primitive with a smaller exponent parameter ζ, for a more spread out orbital. Parameters are obtained from atomic HF calculations. Hydrogen and He orbitals of 31G type have exponentials further adjusted for molecular calculations.

The introduction of primitives with several ζ values allows for electronic rearrangement during molecular interactions where electronic densities move closer to or further from atomic positions as interatomic distances change. In addition, it is usually necessary to describe polarization of the electron clouds as it occurs in induction and dispersion forces. This requires adding primitives with values of l larger than that for the highest occupied shells, such as adding p-type outer shell AOs to a Na atom in a molecular interaction. The basis sets are then called DZP, TZP, and so on, with additional explicit nomenclature to indicate whether the polarization orbitals are added only for atoms in the second and higher rows of the periodic table or also for hydrogen atoms [4].

With so many considerations relevant to the choice of a basis set, it is convenient to devise a systematic method for improvement of a basis set. One such method is to construct atomic natural orbitals (or ANOs) from accurate atomic wavefunctions [4]. The procedure starts with diagonalization of the first-order density matrix for each isolated atom as obtained from an accurate treatment of its electron correlation, usually by means of an atomic configuration-interaction calculation. This leads to ANOs with decreasing occupation numbers as measures of their importance. The ordering shows which orbitals are more important to describe electronic correlation energy and polarization of the atomic cloud. Convergence of atomic properties may require a large number of ANOs, even though it can be expected that features of molecular interactions can be accurately described with a relatively small set of orbitals similar to the ANOs.

More compact basis sets of orbitals have been derived as alternatives to ANOs, constructed to provide accurate atomic correlation energies, and are called *correlation-consistent* (or cc) basis sets, or correlation- and polarization-consistent (cc-p) basis sets [4, 16, 17]. When done for valence shells, they are called cc-pVDZ, cc-pVTZ, and so on. Basis sets of this sort are suitable for calculations of intermolecular forces, while the ones that do not contain polarization functions fail to account for induction and dispersion forces. In addition, accurate calculation of intermolecular forces at large distances may require a treatment including diffuse (widespread) electron distributions of A and B. This can be done adding to the basis set other functions with smaller exponent parameters ζ for each angular momentum quantum number l. The resulting basis sets are labeled with *aug* as in aug-cc-pVDZ.

Accurate calculations of intermolecular forces involve differences of large energies for the pair AB minus energies of A and B, giving small potential energy $V(R)$ values changing with the relative distance R between A and B and also with their internal conformations given by coordinates $\boldsymbol{Q}_A$ and $\boldsymbol{Q}_B$. This requires that the energy terms must have comparable accuracy. Two procedures must be implemented to achieve this. One is that the energies of systems AB, A and

B must be calculated with the same type of basis set. The other relates to the size of the basis sets for the complex and for each component. Indicating with $E_S[R, \mathbf{Q}_S; \{\chi_S\}]$ an electronic energy obtained at relative distance R with the basis set $\{\chi_S\}$ suitable for species $S = A, B$, and AB, it would appear as if the potential energies should be $E_{AB}[R, \mathbf{Q}_A, \mathbf{Q}_B; \{\chi_A\} \cup \{\chi_B\}]$, obtained from the union of the component basis sets, minus $E_A[\{\chi_A\}]$ and $E_B[\{\chi_B\}]$ from each separate basis set. However, this means that at each distance R, the total energy is being calculated with a more extensive (or more complete) basis set for the complex and less so for the components, giving an inaccurate potential energy $V'(R, \mathbf{Q}_A, \mathbf{Q}_B)$. The resulting error is called the *basis set superposition error*. It can be corrected by calculating the components energy with the larger (union) basis set at the conformations of the components for each distance R, as a *counterpoise* (CP) correction energy

$$\Delta E_{CP}(\mathbf{Q}_A, \mathbf{Q}_B) = E_A[\mathbf{Q}_A; \{\chi_A\} \cup \{\chi_B\}] + E_B[\mathbf{Q}_B; \{\chi_A\} \cup \{\chi_B\}] - E_A[\{\chi_A\}] - E_B[\{\chi_B\}]$$

giving a more accurate potential energy $V(R, \mathbf{Q}_A, \mathbf{Q}_B) = V'(R, \mathbf{Q}_A, \mathbf{Q}_B) + \Delta E_{CP}(\mathbf{Q}_A, \mathbf{Q}_B)$ [4, 18].

Basis sets suitable for extended systems frequently use instead plane wave expansions for valence shells and delocalized orbitals, introduced together with a description of inner shells in terms of atomic pseudopotentials [19]. Plane waves $\phi_{\vec{k}}(\vec{r}) = (\Omega)^{-1/2}\exp(i\vec{k}.\vec{r})$ for a system of volume Ω with a set of wavevectors $\vec{k}$ forming a grid $\{\vec{k}_{\mathbf{n}}\}, \mathbf{n} = (n_1, n_2, n_3)$, in reciprocal space, must be dense enough to describe spatial variations of the electronic density. The upper limit $N_j^{(B)}$ of the grid numbers n_j, in $1 \le n_j \le N_j^{(B)}$, is increased to account for highly oscillating orbitals in real space. This provides a basis set suitable for calculations of interaction forces in large systems, and properties such as the total energy of a crystalline solid as a function of the size of its unit cell volume. Inner- and outer-shell atomic orbitals are differently treated, with inner-shell orbitals $\phi_l^{(PS)}(\vec{r})$ introduced to construct atomic pseudopotential operators $\hat{V}_l^{(PS)}$ in which valence electrons move [20]. Expansions of one-electron wavefunctions in plane wave functions, and fast Fourier transformations between grids in real and reciprocal spaces, provide one-electron valence states and energies for many-electron treatments of intermolecular forces in solids and at solid surfaces. The same grid in reciprocal space can be used for a total system AB and for its isolated components A and B to obtain accurate energy differences.

Expanding MOs $\psi_j(\vec{r})$ as linear combinations of AOs $\chi_\mu(\vec{r})$, with $N_{bf}^{(a)}$ basis functions for atom $1 \le a \le N_{at}$ and a total number of basis functions N_{bf}, means that

$$\psi_j(\vec{r}) = \sum_{\mu=1}^{N_{bf}} c_{j\mu}\chi_\mu(\vec{r})$$

and that MO integrals $\langle \psi_j | \hat{a}^{(1)} | \psi_k \rangle$ and $\langle \psi_j \psi_k | \hat{a}^{(2)} | \psi_l \psi_m \rangle$ for one- and two-electron operators are sums with AO matrix elements

$$\int d(1)d(1')\chi_\mu(1)^* a^{(1)}(1,1')\chi_\nu(1') = \langle \mu | \hat{a}^{(1)} | \nu \rangle$$

$$\int d(1)d(1') \int d(2)d(2')\chi_\kappa(1)^* \chi_\mu(2)^* a^{(2)}(1,2;1',2')\chi_\lambda(1')\chi_\nu(2') = \langle \kappa\mu | \hat{a}^{(2)} | \lambda\nu \rangle$$

which depend on two or four AO indices. This allows an estimation of how computing times and data storage scale with the number N_{bf} of basis functions, also considering invariance under permutation of integration variables, which gives $N_{bf}^2/2$ and $N_{bf}^4/8$ for the number of integrals of one- and two-electron operators. Transformation of two-electron integrals from AOs to MOs requires adding over four loops with the transformation coefficients multiplying AO integrals. This can be done in a sequence of four summations, each scaling as N_{bf}^5. It is therefore important to work with efficient basis sets to decrease computing times and data storage for two-electron integrals. The scaling power of N_{bf} can be decreased for local two-electron operators, such as the Coulomb e-e energy, introducing criteria to discard integrals when the overlap products $\chi_\kappa(1)^*\chi_\mu(1)$ are very small for distant atomic nuclei. This brings down the effective scaling for two-electron integrals to below N_{bf}^3.

6.2 Supermolecule Methods

6.2.1 The Configuration Interaction Procedure for Molecular Potential Energies

Here we present only an outline of the CI method as it relates to the calculation of intermolecular forces. Extensive presentations can be found in references [5, 15, 21, 22]. Starting with supermolecular spin orbitals (SOs) as products of space and spin functions, $\psi_j(\vec{r}_n, \zeta_n) = \varphi_j^{(\eta)}(\vec{r}_n)\eta(\zeta_n) = \langle n | j \rangle$, with the notation $| n \rangle = | \vec{r}_n, \zeta_n \rangle$ for the space and spin variables of electron $n = 1$ *to* N, and $\eta = \alpha, \beta$ for spin up and down states, a single reference state is given by an electronic determinantal wavefunction $D_0(1,2,\ldots,N) = \hat{\mathcal{A}}\psi_1(1)\psi_2(2)\ldots\psi_N(N) = (N!)^{-1/2}\det[\langle n | j \rangle]$ for N electrons and occupied SOs j = 1 to N, also shown in a short notation as $D_0 = |\psi_1\psi_2\ldots\psi_N|$. As written, the SO ψ_j displays a different orbital factor $\varphi_j^{(\eta)}$ for different spins. In applications to closed-shell systems, with an even number of electrons, electrons with up and down spins are located at the same space orbitals, and this leads to a wavefunction with good total spin quantum numbers $S = 0$, $M_S = 0$. But for different orbitals for different spins, the unrestricted determinant does not generate an eigenfunction of total spin.

However, application of the spin projection operator $\hat{O}_{SM_S}$ gives a spin-adapted configuration state function (or CSF) $|\Phi_{SM_S0}\rangle = \hat{O}_{SM_S}|D_0\rangle$, which is in general a combination of determinantal functions. The reference CSF for given spin quantum numbers will be called Φ_0.

Other CSFs are generated by removal of an occupied SO i and addition of an unoccupied SO a from each determinant in the CSF to form a new singly excited state $|\Phi_i^a\rangle = \hat{c}_a^\dagger \hat{c}_i|\Phi_0\rangle$, written in terms of fermion creation and annihilation operators $\left(\hat{c}_i^\dagger, \hat{c}_i\right)$ with the anticommutation relations

$$\hat{c}_j^\dagger \hat{c}_i + \hat{c}_i \hat{c}_j^\dagger = \delta_{ij},\ \hat{c}_j^\dagger \hat{c}_i^\dagger + \hat{c}_i^\dagger \hat{c}_j^\dagger = 0,\ \hat{c}_j \hat{c}_i + \hat{c}_i \hat{c}_j = 0$$

Doubly excited CSFs $\left|\Phi_{ij}^{ab}\right\rangle$, and generally n-excited ones, can be similarly constructed. A CI wavefunction is given by the combination of the reference CSF plus singly excited (S) ones, plus doubly (D) excited ones, and so on. This can be described within the wave operator formalism of Section 6.1.2, starting with the projector $\hat{\mathcal{P}} = |\Phi_0\rangle\langle\Phi_0|$ valid for a single reference state. Single excitation determinants add up to a wave operator term $\hat{\Omega}^{(1)}$, doubly excited ones add to $\hat{\Omega}^{(2)}$ and so on, leading to

$$\Psi = \left(\hat{I} + \hat{\Omega}^{(1)} + \hat{\Omega}^{(2)} + \cdots\right)\Phi_0$$

$$\hat{\Omega}^{(1)} = \sum\nolimits_{i,a} A_i^a \hat{c}_a^\dagger \hat{c}_i,\ \hat{\Omega}^{(2)} = \sum\nolimits_{i,j,a,b} A_{ij}^{ab}\, \hat{c}_b^\dagger\, \hat{c}_a^\dagger\, \hat{c}_j\, \hat{c}_i, \ldots$$

where the coefficients $A_{ij\ldots}^{ab\ldots}$ are functions of the energy E and must be found for each intermolecular distance.

Keeping only $\hat{\Omega}^{(1)}$ and $\hat{\Omega}^{(2)}$ in the expansion leads to the CISD treatment. This is sufficient for many applications at finite distances, but would not satisfy the asymptotic consistency test. Indeed, if one treats the species A and B to the same accuracy, with $\Psi^{(S)} = \left(\hat{I} + \hat{\Omega}^{(1)} + \hat{\Omega}^{(2)} + \cdots\right)_S \Phi_0^{(S)}$, $S = A, B$, for each species, one finds that the asymptotic product $\Psi_A\Psi_B$ contains also wave operator terms of the form $\hat{\Omega}^{(3)} = \hat{\Omega}_A^{(1)}\hat{\Omega}_B^{(2)} + \hat{\Omega}_A^{(2)}\hat{\Omega}_B^{(1)}$ and $\hat{\Omega}^{(4)} = \hat{\Omega}_A^{(2)}\hat{\Omega}_B^{(2)}$, which are absent in the total wavefunction Ψ as constructed. This inconsistency can be avoided by including all excitations in the expansion of Ψ up to $\hat{\Omega}^{(N)}$, the last nonzero operator with the largest number of excitations. This generates the so-called full-CI wavefunction, which, however, is usually too demanding of computing time for applications.

To make calculations more affordable and yet asymptotically consistent, a valid procedure is to use a complete active space self-consistent field (or CASSCF) treatment [23]. Here we again consider, to begin with, the case of a single reference CSF constructed from SOs $\psi_j(\vec{r}_n, \zeta_n) = \psi_j(n)$ obtained now as self-consistent field (SCF) solutions of the Hartree–Fock (or HF) equation

$$\hat{f}_{HF}(n)\psi_j(n) = \epsilon_j^{(HF)}\psi_j(n), \hat{f}_{HF} = \hat{h} + \hat{J} - \hat{K}$$

where $\hat{f}_{HF}$ is the one-electron Fock operator and $\epsilon_j^{(HF)}$ is the energy of the SO [3]. The term $\hat{h}$ contains the electron n kinetic energy operator and the Coulomb attraction energy of the electron to all the nuclei (or atomic ion cores). The Coulomb and exchange HF operators are given by

$$\hat{J}(n)\psi_j(n) = \sum_k^{occ}\left[\int d(1)v(1,n)|\psi_k(1)|^2\right]\psi_j(n)$$

$$\hat{K}(n)\psi_j(n) = \sum_k^{occ}\left[\int d(1)\psi_k(1)^* v(1,n)\hat{t}_{1n}\psi_k(1)\right]\psi_j(n)$$

with $v(1, n)$ the Coulomb energy between electrons 1 and n, and $\hat{t}_{1n}$ the transposition operator exchanging the variables of electrons 1 and n in the electron integrals: $\hat{t}_{1n}\psi_k(1)\psi_j(n) = \psi_k(n)\psi_j(1)$. The integrals here are over space variables and also indicate sums over spin variables $\zeta = \pm 1$. The SO energies are given by

$$\epsilon_j^{(HF)} = \langle j|\hat{h}|j\rangle + \sum_k^{occ}\langle jk|v(1-\hat{t})|jk\rangle$$

where alternatively $\langle jk|v(1-\hat{t})|jk\rangle = \langle jk||jk\rangle$. A special case (RHF) follows when the orbitals are doubly occupied with up and down spins and integrals are over only space variables, in which case there are twice as many Coulomb integrals as exchange integrals.

For the ground electronic state at each distance R, a reference determinantal function will usually contain doubly occupied orbitals with up and down spins for the lower energies, and singly occupied orbitals of higher energy, chosen to generate the CSF with correct spin quantum numbers. The HF equation gives not only physically meaningful occupied SOs for the supermolecule, but also excited (although usually unphysical) SOs, which can be used to form a basis set for expansion of the kinetic energy and Coulomb energy terms in the Hamiltonian.

The SCF SOs can be obtained expanding them in a basis set of atomic SOs $\chi_\mu(\vec{r}, \zeta)$, both factorized into space and spin parts. Introducing the expansions $\psi_j = \sum_\mu a_{\mu j}\chi_\mu$, the Fock operator matrix elements $f_{\mu\nu} = \langle\chi_\mu|\hat{f}|\chi_\nu\rangle$ and overlaps $S_{\mu\nu} = \langle\chi_\mu|\chi_\nu\rangle$, the one-electron equations are solved finding the expansion coefficients that satisfy

$$\mathbf{f}\mathbf{a}_j = \epsilon_j\mathbf{S}\mathbf{a}_j$$

where $\mathbf{a}_j$ is a column matrix with elements $a_{\mu j}$ for each supermolecular SO j. This is self-consistently done for the occupied SOs with $j = 1$ to N, which then provide the Fock matrix elements in an equation that can be solved for excited (unoccupied) SOs, also called virtual orbitals.

The HF treatment can be applied to two interacting closed-shell species, such as He + He or Ar + H_2, by choosing as the reference state a determinantal function with doubly occupied molecular orbitals giving a restricted HF (or RHF) wavefunction. Such a determinant can properly describe their physical ground state also as $R \to \infty$. However, that choice would not be acceptable for a system with open shells, such as H + H or H + Li. Then a doubly occupied determinant like $D_{RHF} = |\sigma\alpha, \sigma\beta|$ for two valence electrons, with $|\sigma\rangle = c_A|\chi_A\rangle + c_B|\chi_B\rangle$ a linear combination of atomic orbitals with the symmetry of the diatomic, and with the same MO for up and down spins, contains a spurious determinantal term at large distances corresponding to $H^+ + H^-$ or $Li^+ + H^-$, and gives the wrong asymptotic energy. This problem can be partly solved introducing an unrestricted determinant in the unrestricted HF (or UHF) treatment. In this example, it amounts to using orbitals $|\sigma^\eta\rangle = c_A^\eta|\chi_A\rangle + c_B^\eta|\chi_B\rangle$, with $\eta = \alpha, \beta$ to form $D_{UHF} = |\sigma^\alpha\alpha, \sigma^\beta\beta|$. The four coefficients must satisfy normalization and orthogonality of the two MOs. This leaves one free parameter to optimize the UHF energy, and its value changes with R to represent the electronic rearrangement taking place as R grows. The resulting orbital densities are found to be different for up and down spins, allowing each electron to move toward one of the two nuclei, and the pair then breaks up into the correct ground state of the neutral species.

The UHF determinant is not an eigenfunction of the total spin operators, but this can be extracted with the corresponding projection operator in a postsymmetry procedure. Alternatively, the projector can be first applied to a determinantal function to construct an unrestricted CSF, to be optimized. Results of calculations for H_2 are shown in Figure 6.2. The UHF behavior is qualitatively correct but must be improved to agree with accurate results, using CI or PT treatments.

An alternative choice to UHF for open-shell systems is the restricted open-shell HF (or ROHF) choice, where most electrons are located in doubly occupied orbitals, which are the same for up and down spin. Remaining electrons are located in orbitals that may depend on the spin orientation, in which case the ROHF wavefunction is not an eigenfunction of the total spin operators, and its spin components must be extracted by means of a projection operator, or constructed from a superposition of determinants.

The RHF and UHF procedures can be applied to a supermolecule as needed to assure that the breakup product $A + B$ at large distances is in the proper physical states. Returning to the CI treatment, this can make use of SOs optimized with RHF or UHF to construct excited configurations. The orbitals in the supermolecule basis set are classified as inactive if they remain doubly occupied as the

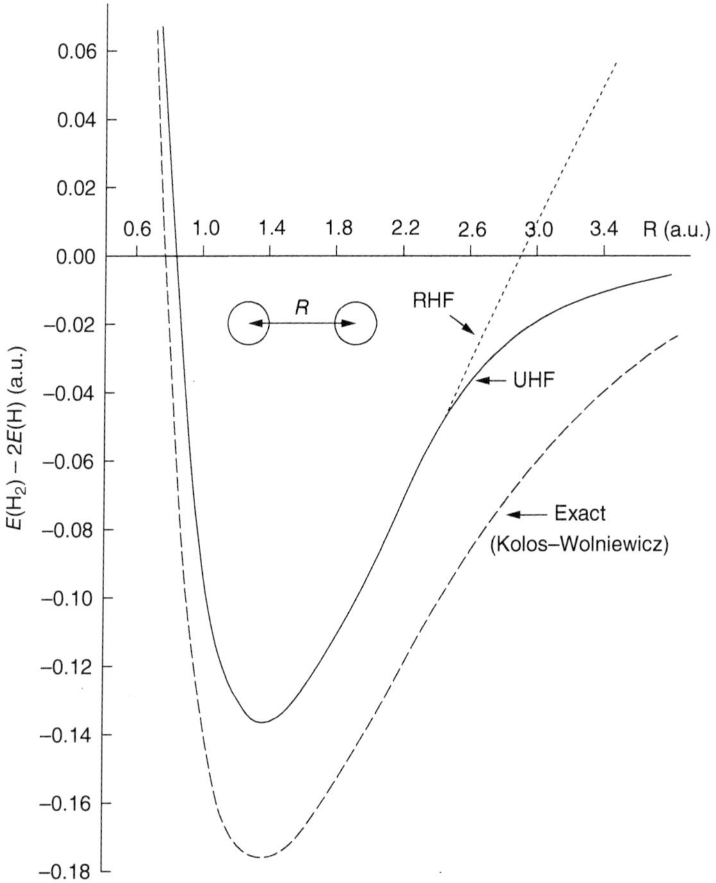

Figure 6.2 Results from RHF, UHF, and CI calculations of potential energy versus interatomic distance for H_2, using the basis set 6-31G**. *Source:* from Ref. [5]. Reproduced with permission of Dove Publishing.

molecular system changes conformation, typically core orbitals, and as active SOs if they have single occupancy or are unoccupied to begin with at each conformation.

A full-CI expansion is generated by promotion of electrons from all active occupied SOs to all unoccupied and partly occupied ones. Doing this for all conformations involves all possible n-excitation wave operator terms and guarantees that a supermolecule at intermediate distances and its asymptotic molecular fragments are described with the same accuracy. The many-electron equation $\hat{H}\Psi = E\Psi$ containing single excited CSFs $\Phi_i^a = \hat{c}_a^\dagger \hat{c}_i \Phi_0$, and similarly for higher excitations, must be solved for the lowest energy E and the associated state

$$\Psi = \Phi_0 + \sum\nolimits_{i,a} A_i^a \Phi_i^a + \sum\nolimits_{i,j,a,b} A_{ij}^{ab} \Phi_{ij}^{ab} + \cdots = \mathbf{\Phi A}$$

in a matrix notation with the row matrix $\mathbf{\Phi}$ of CSFs and the column matrix $\mathbf{A}$ of expansion coefficients. This can be done searching for the variational minimum of $\langle \Psi | \hat{H} | \Psi \rangle = \mathbf{A}^{\dagger} \langle \mathbf{\Phi} | \hat{H} | \mathbf{\Phi} \rangle \mathbf{A} = \mathbf{A}^{\dagger}\mathbf{HA}$ subject to the normalization $\langle \Psi | \Psi \rangle = \mathbf{A}^{\dagger} \langle \mathbf{\Phi} | \mathbf{\Phi} \rangle \mathbf{A} = 1$ in a well-known procedure leading to the matrix eigenvalue equation $(\mathbf{H} - E\mathbf{I})\mathbf{A} = \mathbf{0}$, which needs to be solved only for the lowest eigenvalue as a function of conformations and in particular as a function of R, if one wants only the ground state PES.

A more flexible and accurate procedure is the multiconfiguration SCF (or MCSCF) treatment where the CI coefficients $\mathbf{A}$ and also the HF coefficients $\mathbf{a}$ are varied to achieve self-consistency at the lowest energy. This gives orbitals adapted to optimize not only the reference configuration, but also useful to optimize its mixture with excited ones, and leads to more realistic excited SOs for the supermolecule.

The CASSCF expansion is a MCSCF treatment involving all the electron excitations among active SOs, and guarantees that the potential energy $V(R)$ from the difference of total energies goes properly to zero at large distances, provided the reference CSF Φ_0 correctly describes the asymptotic states of A and B at large R. The limitation here is that the number of configurations to be generated increases exponentially with the number of active SOs. In fact, if N_{act} is the number of active orbitals, N the number of active electrons, and S the total spin quantum number, the number of complete active space functions to be generated is [23]

$$N_{CAS} = \frac{2S+1}{N_{act}+1} \frac{(N_{act}+1)!}{(N/2-S)!(N_{act}+1-N/2+S)!} \times \frac{(N_{act}+1)!}{(N/2+S+1)!(N_{act}+1-N/2-S-1)!}$$

a very large number when N_{act} is larger than 10, which prevents calculations of this type for large molecular systems or large electronic excitation.

The calculation of interaction potential energies gets more demanding when there are state degeneracies or electronically excited states, or where a single CSF Φ_0 does not properly describe the asymptotic A and B states. In these cases, one must use a treatment based on multireference states. Instead of a single reference CSF Φ_0, one must consider a set $\{\Phi_J, J = 1\ to\ N_D\}$ of CSFs, introduce the projection operator $\hat{\mathcal{P}} = \sum_J |\Phi_J\rangle\langle\Phi_J|$, and proceed to construct multireference CI (or MR-CI) states, to calculate energies.

An example is provided by the treatment of $H_2 \rightarrow H + H$ in its ground electronic state ${}^1\Sigma_g^+$ with a minimal MR-CI containing only two configurations made up of diatomic orbitals $|\sigma_{g,u}\rangle = C_{g,u}(|1s_A\rangle \pm |1s_B\rangle)$ constructed from hydrogenic 1 second orbitals of A and B atoms, where g, u stand for gerade and ungerade symmetry. The configurations $\Phi_1 = |\sigma_g\alpha, \sigma_g\beta|$ and $\Phi_2 = |\sigma_u\alpha, \sigma_u\beta|$ are two determinants that can be taken as a multireference pair interacting to give the correct neutral species asymptotically, with a ground (G) state

$$\Psi_G = A_{1G}\left|\sigma_g\alpha,\sigma_g\beta\right| + A_{2G}\left|\sigma_u\alpha,\sigma_u\beta\right| \approx C_G(\left|1s_A\alpha, 1s_B\beta\right| + \left|1s_A\beta, 1s_B\alpha\right|)$$

insofar as $A_{1G} \approx -A_{2G}$ for $R \to \infty$, while a treatment based only on Φ_1 leads to an incorrect mixture containing a state of $H^+ + H^-(1s^2)$ in addition to the correct H(1s) + H(1s) state. Alternatively, the $H_2 \to H + H$ process can be described with a single unrestricted determinantal function, which employs different orbitals for different spins, with parameters to be varied.

The formalism presented above, based on the multistate projection operator, shows how one can proceed to generate needed excited CSFs from the multireference ones by transferring electrons from active occupied SOs into unoccupied ones. The treatment has been made feasible by introducing energy selection criteria so that only excited CSFs that are energetically coupled to within a selected energy range are included in the expansion. This so-called MRD-CI procedure has produced many potential energy surfaces of high accuracy [24]. The limitations of the MR-CI treatment, on account of the formal absence of asymptotic consistency, have been covered in reference [25].

6.2.2 Perturbation Expansions

Many-body perturbation theory (MBPT) and its applications to an N-electron system are based on the standard perturbation theory for a nondegenerate or a degenerate reference state, and a choice of a zeroth-order Hamiltonian $\hat{H}^{(0)}$ that allows construction of a basis set $\left\{\Phi_J^{(0)}\right\}$ of unperturbed states. Details of the general theory can be found in references [3–5, 25]. The treatment can be derived for a single reference determinantal wavefunction constructed from SOs, $D_0(1,2,\ldots,N) = \hat{\mathcal{A}}\psi_1(1)\psi_2(2)\ldots\psi_N(N)$ obtained from $\hat{f}\psi_j = \epsilon_j\psi_j$, where $\hat{f}$ is a convenient one-electron effective operator giving physically meaningful energies ϵ_j, SOs ψ_j orthonormalized with $\langle\psi_j|\psi_k\rangle = \delta_{jk}$, and a suitable density of occupied and unoccupied states for an expansion basis set. A standard choice for $\hat{f}$ is the mentioned Fock operator $\hat{f}_{HF}$, which provides physically acceptable occupied states but usually unsuitable unoccupied states. An alternative is the $\hat{f}_{KS} = \hat{h} + \hat{J} - \hat{v}_{xc}$ Kohn–Sham (KS) operator, which can be chosen to generate the SOs of density functional theory [1]. Other basis sets may involve the occupied HF SOs and a set of complimentary (orthogonalized) basis functions suitable for the treatment of excited SOs [14].

The unperturbed (zeroth-order) Hamiltonian is $\hat{H}_0 = \hat{F} = \sum_{1 \le m \le N} \hat{f}_m$ with the eigenstate equation $\hat{F} D_0 = E_0^{(0)} D_0$ and the zeroth-order energy

$$E_0^{(0)} = \sum_j^{occ} \epsilon_j$$

obtained from one-electron energies of occupied SOs. This is a sum of approximate ionization energies and does not correspond physically to the total energy, but must be corrected by higher order energy terms. The perturbation operator $\hat{H}' = \hat{H}^{(1)} + \hat{H}^{(2)} - \hat{F}$ is the fluctuation (or correlation) potential energy operator containing one- and two-electron terms and giving perturbation energies $E_0^{(r)}$ of order r. The PT generates the terms in $\Psi = \left(\hat{I} + \hat{\Omega}^{(1)} + \hat{\Omega}^{(2)} + \cdots\right) D_0$ and $E = E_0^{(0)} + E_0^{(1)} + E_0^{(2)} + \cdots$ by recurrence at each order, using for convenience the intermediate normalization $\langle D_0 | \Psi \rangle = 1$. These can be obtained from D_0 and a set of determinants $D_J = D_i^a, D_{ij}^{ab}, \ldots,$ with single, double, ..., excitations. The energies to first and second orders of a many-body PT (or MBPT) can be obtained for a pair of species A and B from a wave operator $\hat{\Omega}_{AB}^{(MB)}$ and are given by

$$E_0^{(1)} = \left\langle D_0 | \hat{H}' | D_0 \right\rangle = \sum_j^{occ} \left(\left\langle j | \hat{h} | j \right\rangle - \epsilon_j \right) + \sum_{j<k}^{occ} \langle jk | v(1-\hat{t}) | jk \rangle$$

$$E_0^{(2)} = \sum_{J \neq 0} \frac{\left| \left\langle D_0 | \hat{H}' | D_J \right\rangle \right|^2}{E_0 - E_J}$$

$$= \sum_i^{occ} \sum_a^{unocc} \frac{\left| \left\langle D_0 | \left(\hat{H}^{(1)} + \hat{H}^{(2)} \right) | D_i^a \right\rangle \right|^2}{\epsilon_i - \epsilon_a}$$

$$+ \sum_{i<j}^{occ} \sum_{a<b}^{unocc} \frac{\left| \left\langle D_0 | \hat{H}^{(2)} | D_{ij}^{ab} \right\rangle \right|^2}{\epsilon_i + \epsilon_j - \epsilon_a - \epsilon_b}$$

where $\hat{t}$ is the transposition operator exchanging the variables of electrons 1 and 2 in the electron integrals so that $\hat{t}_{12} \psi_j(1) \psi_k(2) = \psi_j(2) \psi_k(1)$, occupied SO indices run between 1 and N, and unoccupied ones between $N + 1$ and the largest value in the basis set. The numerators can be evaluated in terms of SO using the Condon–Slater rules for matrix elements of operators between determinantal wavefunctions [1].

These and higher order energies can be analyzed in terms of Feynman diagrams and organized to contain only linked diagrams in each order. The wavefunction is then written as

$$\Psi_{MBPT} = D_0 + \sum_{k=1}^{\infty} \left[\left(\hat{R}_0 \hat{H}' \right)^k D_0 \right]_{link} = \exp\left(\hat{T}_{cnc} \right) D_0$$

where $\hat{R}_0$ is the unperturbed resolvent, the summation is restricted to linked Feynman diagrams, and the operator $\hat{T}_{cnc}$ is a sum of connected terms. The corresponding MBPT energy then satisfies the size extensivity condition to all orders [25].

As a special case, it is possible to consider the Moller-Plesset PT (or MPPT) where the unperturbed Hamiltonian is the Fock operator, a sum of one-electron effective Hamiltonians $\hat{f} = \hat{f}_{HF}$, with

$$\hat{f}_{HF}(m) = \hat{h}(m) + \hat{J}(m) - \hat{K}(m)$$

containing the HF Coulomb $(\hat{J})$ and exchange $(\hat{K})$ operators. The zeroth-order energy

$$E_0^{(0)} = \sum\nolimits_j^{occ} \epsilon_j^{(HF)}$$

and the first-order correction

$$E_0^{(1)} = -\frac{1}{2}\sum\nolimits_{j<k}^{occ} \langle jk \mid v(1-\hat{t}) \mid jk \rangle$$

reconstruct the HF energy, $E_0^{(0)} + E_0^{(1)} = E_0^{(HF)}$, a meaningful total energy. The first summation in the second-order energy involving single excited states is zero in the HF treatment, insofar as $\left\langle D_0 \mid \left(\hat{H}^{(1)} + \hat{H}^{(2)}\right) \mid D_i^a \right\rangle = \left\langle a \mid \hat{f}_{HF} \mid i \right\rangle = 0$, per the so-called Brillouin theorem [4].

The remaining summation in the second-order energy gives the leading correlation energy involving excited SOs. However, the excited (unoccupied) SOs generated in HF correspond formally to an electron moving in a field of N electrons, instead of N-1 electrons as happens for the occupied SOs. As a result, they usually describe electrons that are less bound and are more delocalized than accurate excited SOs. Calculation of interaction energies may require a larger basis set and inclusion of higher orders in the MPPT expansion, as compared to expansions with more realistic excited SOs, or otherwise they may lead to large errors in intermolecular forces at large distances. Furthermore perturbation terms must be carefully constructed to avoid unlinked diagrams and non-extensive results.

The asymptotic consistency requirement $E \approx E_A + E_B$ for a pair of species A and B, essential for intermolecular forces, is not necessarily satisfied by MBPT. This is because the reference state must satisfy $|D_0\rangle \approx \hat{\mathcal{A}}'|D_{0,A}\rangle|D_{0,B}\rangle$, a correct asymptotic reference state made up of products of states of A and B, and in addition, the perturbation treatment must construct the wave operator $\hat{\Omega}_{AB}^{(MB)}$ so that the energy as obtained from $\hat{H}_{AB}\hat{\Omega}_{AB}^{(MB)}|D_0\rangle \approx (E_A + E_B)\hat{\Omega}_A^{(MB)}\hat{\Omega}_B^{(MB)}\hat{\mathcal{A}}'|D_{0,A}\rangle|D_{0,B}\rangle$ gives accurate energies for A and B. This must be verified in a treatment of the intermolecular forces.

Another aspect affecting the accuracy of intermolecular forces relates to the order of perturbation for the pair AB and each component. For example, if a treatment is done up to second order in the MPPT2 treatment, one finds that each operator $\hat{\Omega}_S^{(MB)}, S = A, B$ will introduce two-electron excitations so that the product $\hat{\Omega}_A^{(MB)}\hat{\Omega}_B^{(MB)}$ will generate up to four excitations. However, the operator $\hat{\Omega}_{AB}^{(MB)}$ to the same order would only involve double excitations and would therefore miss some of the asymptotic states. This is avoided in the MBPT treatment based on linked diagrams and the connected $\hat{T}_{cnc}$ operator, such as the coupled-cluster treatment, which consistently generates each perturbation order for the pair AB.

6.2.3 Coupled-Cluster Expansions

Consistency in the treatment of excitations can be achieved within the coupled cluster (CC) treatment [25, 26]. Here the wave operator takes an exponential form and one writes

$$\Psi = \hat{\Omega}^{(CC)} D_0 = \exp\left(\hat{T}\right) D_0$$

with the excitation operator $\hat{T} = \hat{T}^{(1)} + \hat{T}^{(2)} + \cdots + \hat{T}^{(N)}$ where $\hat{T}^{(n)}$ contains only n electronic excitations, and with the intermediate normalization $\langle D_0 \mid \Psi \rangle = 1$. Expansion of the exponential leads to the appearance of terms like $\left(\hat{T}^{(n)}\right)^m \left(\hat{T}^{(n')}\right)^{m'}$ for integers $0 \le m \le \infty$, involving $1 \le (n.\ m) + (n'.\ m') \le N$ excitations since no more than N SOs may be excited. The excitation operator terms $\hat{T}^{(n)}$ can be expanded in a basis set of SOs classified as occupied or unoccupied, introducing excitation amplitudes for each term as in

$$\hat{T}^{(1)} = \sum_i^{occ} \sum_a^{unocc} t_i^a \hat{c}_a^\dagger \hat{c}_i,\ \hat{T}^{(2)} = \sum_{i<j}^{occ} \sum_{a<b}^{unocc} t_{ij}^{ab}\, \hat{c}_b^\dagger \hat{c}_a^\dagger \hat{c}_j \hat{c}_i, \ldots$$

One finds, with the expansion of the exponential operator, a sum of configurations as in the CI method, but here their superposition is generated by the CC excitation terms in a convenient way, which involves fewer transition amplitudes than the number of coefficients in the CI. The unknowns in the CC treatment are $E, \left\{t_i^a\right\}, \left\{t_{ij}^{ab}\right\}, \ldots,$ and these can be extracted by projecting (or taking moments of) the equation $\exp\left(-\hat{T}\right)\hat{H}\exp\left(\hat{T}\right) \mid D_0\rangle = E \mid D_0\rangle$ on the Dirac bras $\left\langle D_0 \right|, \left\langle D_i^a \right|, \left\langle D_{ij}^{ab} \right|, \ldots$. This gives the same number of nonlinear algebraic equations as there are unknowns, which can be solved by iteration. An analysis of many-electron terms in $E = \langle D_0 \mid \exp\left(-\hat{T}\right)\hat{H}\exp\left(\hat{T}\right) \mid D_0\rangle$ after an expansion in commutators of $\hat{H}$ and $\hat{T}$ shows that only linked diagrams appear, assuring size extensivity [25].

Relating to asymptotic behavior, there are two convenient consequences of the exponential form in the CC treatment. In the limit for $R \to \infty$, one finds that $\hat{T}_{AB} \approx \hat{T}_A + \hat{T}_B$ and $\hat{\Omega}_{AB}^{(CC)} \approx \hat{\Omega}_A^{(CC)} \hat{\Omega}_B^{(CC)}$. This together with a choice of reference state satisfying $|D_0\rangle \approx \hat{\mathcal{A}}' |D_{0,A}\rangle |D_{0,B}\rangle$ warrants that the potential energy of interaction is $V(R) = E(R) - (E_A + E_B) \approx 0$ as desired. It may require using an unrestricted determinantal function with different orbitals for different spins, although for a special system such as H_2 a RHF reference determinant is sufficient. Secondly, choosing a level of expansion such as CCSD, with single and double excitations included in the excitation operator, it follows that all excitations, up to the maximum N, N_A and N_B, will be respectively present in the wavefunctions of the pair AB and the fragments A and B, so that the energy differences should be better balanced than in the MPPT treatment. These reasons make the CC treatment a more reliable procedure for the calculation of intermolecular forces using supermolecule SOs.

The expansion containing only $\hat{T}^{(1)}$ and $\hat{T}^{(2)}$ produces the CCSD treatment, which gives accurate energies for many molecules near their equilibrium conformation. A better treatment also adds $\hat{T}^{(3)}$ to generate triplet excitations and CCSDT wavefunctions and energies. However, this level of calculations is very demanding of computer time and data storage, and an alternative is to incorporate only part of the triple excitations by means of perturbation theory, in a CCSD(T) treatment. This improves potential energies over wider ranges of R around equilibrium, but can lead to erroneous results at large R when excitation energies appearing in denominators of the third-order PT become very small at large R. The asymptotic behavior can be corrected using an UHF reference state, as shown in Figure 6.3 for N_2.

The CC approach has been extended to excited states by means of the equation-of-motion CC (or EOM-CC) treatment. It preserves the role of the non-Hermitian Hamiltonian operator $\exp(-\hat{T})\hat{H}\exp(\hat{T}) = \bar{H}$, introduced to obtain the ground state Ψ_0 of energy E_0, in the calculation of excited states $\Psi_K = \hat{R}_K \Psi_0$ with energies E_K. These states can be generated by excitation operators

$$\hat{R}_K = r_0 \hat{I} + \sum\nolimits_{i,a} r_i^a \hat{c}_a^\dagger \hat{c}_i + \sum\nolimits_{i,j,a,b} r_{ij}^{ab} \hat{c}_b^\dagger \hat{c}_a^\dagger \hat{c}_j \hat{c}_i + \cdots$$

where the terms are assumed to apply in the normal order (of creation operator followed by annihilation operator to the right) as done with the terms in the excitation operator $\hat{T}$. Since operators in $\hat{R}_K$ and $\hat{T}$ are all of excitation type, this gives the commutator $[\hat{R}_K, \hat{T}] = 0$. It follows from $\hat{H}\hat{R}_K | \Psi_0\rangle = E_K \hat{R}_K | \Psi_0\rangle$ and from $\Psi_0 = \exp(\hat{T}) D_0$ that [25]

$$[\bar{H}, \hat{R}_K] \, |D_0\rangle = (\Delta E_K - \Delta E_0) \hat{R}_K \, |D_0\rangle$$

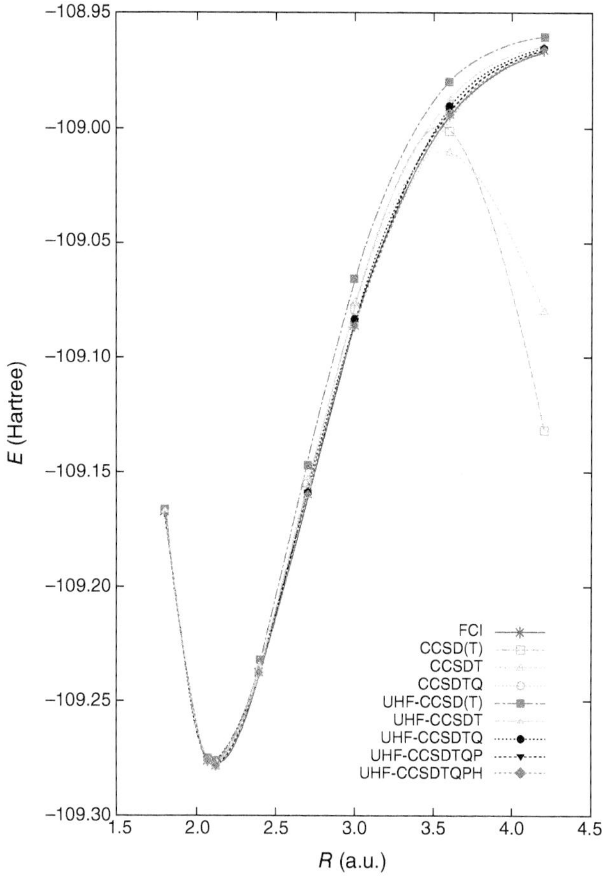

Figure 6.3 Potential energy curves versus interatomic distance for the ground state of the N_2 molecule. Results from CC are compared to full CI using a cc-pVDZ basis set. Incorrect behavior of CC at large distances can be corrected using a UHF reference state. *Source:* from Ref. [27]. Reproduced with permission of American Physical Society.

where relative energies are defined by $\Delta E_K = E_K - E_{ref}$ with $E_{ref} = \langle D_0 \mid \hat{H} \mid D_0 \rangle$. Solving this equation for ΔE_K and for the coefficients in $\hat{R}_K$ as functions of intermolecular distances provides PESs for excited states. This is an active area under development, involving excited states of open-shell systems and bond dissociation and related computational algorithms [27–31].

The CC derivations above have assumed that only a single reference state is needed. When a reference state is degenerate or close to others in energy, one can return to the treatment using the more general wave operator in

$\Psi = \hat{\Omega}\Phi^{(D)} = \left(\hat{\mathcal{P}} + \hat{Q}\hat{R}(E)\hat{Q}\hat{H}\hat{\mathcal{P}}\right)\Phi^{(D)}$ including a multireference projection operator $\hat{\mathcal{P}} = \sum_J |\Phi_J\rangle\langle\Phi_J| = \sum_J \hat{\mathcal{P}}_J$ for many-electron states $J = 1$ *to* N_D. This has been the subject of extensions of the CC treatment [32] and has been computationally implemented [25]. The multireference set of states is made up of determinants with double or single occupancy. The MR-CC extension consists of introducing partial excitation operators called here $\hat{T}_J$ to be applied to each state $|\Phi_J\rangle$ in the multireference set. Each of them contains a sum over many-electron excitations. The wave operator is then of the form

$$\hat{\Omega} = \sum_J \exp\left(\hat{T}_J\right)\hat{\mathcal{P}}_J$$

and its application into the $\hat{\mathcal{P}}$ subspace can be done by first defining an effective Hamiltonian $\hat{H}^{(eff)}$ operating only within the $\hat{\mathcal{P}}$ subspace with matrix elements $H_{KJ}^{(eff)} = \langle\Phi_K|\hat{H}\hat{\Omega}|\Phi_J\rangle = \langle\Phi_K|\hat{H}\exp\left(\hat{T}_J\right)|\Phi_J\rangle$. The perturbation corrected energies E_J in $\hat{H}\Psi_J = E_J\Psi_J, J = 1$ *to* N_D, and excitation amplitudes like $t_i^a(J)$ in each partial excitation operator follow from projections of the kets

$$\exp\left(-\hat{T}_J\right)\hat{H}\exp\left(\hat{T}_J\right)|\Phi_J\rangle = \sum_K \exp\left(-\hat{T}_J\right)\exp\left(\hat{T}_K\right)|\Phi_K\rangle H_{KJ}^{(eff)}$$

onto the bras $\langle\Phi_J|, \langle\Phi_i^a|, \left\langle\Phi_{ij}^{ab}\right|, \ldots,$ and diagonalization of $H_{KJ}^{(eff)}$ with a linear transformation in the $\hat{\mathcal{P}}$ subspace.

This procedure allows consideration of dissociation and intermolecular forces in the *AB* system when the $\hat{\mathcal{P}}$ subspace contains all the determinantal functions needed to construct correct asymptotic states of *A* and *B*. For example, in the case of $H_2 \rightarrow H + H$, it is convenient to choose the $\hat{\mathcal{P}}$ subspace as made of states with the configurations $\Phi_1 = |\sigma_g\alpha, \sigma_g\beta|$ and $\Phi_2 = |\sigma_u\alpha, \sigma_u\beta|$, which are close in energy for large internuclear distances. The MR-CC procedure can then give the correct potential energy surface at all distances. Dissociation in other diatomic systems including multiple bonds has been reviewed in reference [27]. Accurate results over all interatomic distances have been obtained for F_2, N_2, and HF. Results so far indicate that using a MR-CC approach, or alternatively using an UHF reference state, provides ways to correctly treat bond dissociation. This is an area of theoretical research yet under development.

The scaling of MBPT and CC calculations with the size N_{bf} of the basis set can be described separating the MO basis set into two sets of occupied and unoccupied MO functions numbering $N_{bf}^{(o)}$ $(\leq N)$ and $N_{bf}^{(u)}$, respectively. Excitation operators fix the indices of occupied MOs so that transformations from the AO basis to the MO basis involve fewer summation loops. Transforming a four index (two-electron) integral like $\left\langle D_0|\hat{H}^{(2)}|D_{ij}^{ab}\right\rangle$ in MBPT or t_{ij}^{ab} in a CC treatment, for fixed *i* and *j*, involves summations scaling only as $\left(N_{bf}^{(u)}\right)^2$, which must

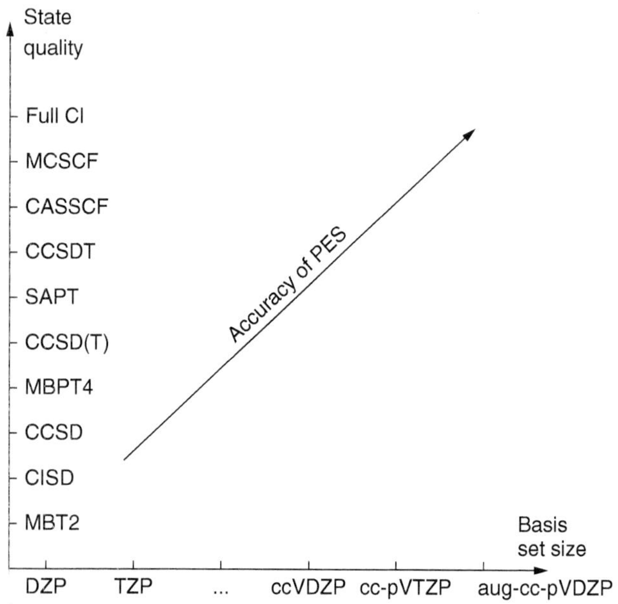

Figure 6.4 Diagram with increasing basis set quality, containing polarization terms, going from left to right on the horizontal axis, and increasing many-electron quality going up as needed for the calculation of accurate potential energy surfaces (PESs).

be done $\left(N_{bf}^{(o)}\right)^2$ times, an effort much smaller than $(N_{bf})^4$ for a general four-index quantity. Analysis of the scaling in CCSD(T) shows it growing as $\left(N_{bf}^{(o)}\right)^3\left(N_{bf}^{(u)}\right)^4$. Further analysis of scaling is found in the literature [4, 15, 25].

The quality of electronic structure calculations for intermolecular forces depends on the type of treatment being used and the size of the basis set in expansions. Figure 6.4 shows what treatments of electronic structure account for asymptotic and size consistency and what basis sets are large enough to account for electron transfer, electronic localization, and polarization.

6.3 Many-Atom Methods

6.3.1 The Generalized Valence-Bond Method

Alternatives to the supermolecule approach are provided by treatments that start from the states and SOs of species A and B before interaction, and construct the potential energies of the $A + B$ pair from them. Electrons are assumed to be localized around species A and B separately, and their charge densities are then allowed to distort, polarize, or transfer between species. One of these treatments derives from the valence-bond (or VB) method for electronic structure, generalized to deal with electron transfer and polarization. The GVB method has been mathematically developed, and more recently has been reintroduced as a viable computational method [2, 33, 34]. It is appealing for studies of intermolecular forces because it provides a direct route to construction of reference states with the correct asymptotic behavior.

Total wavefunctions for N electrons are written as products of a spin function $\Theta^{(J)}_{SM_S}(\zeta_1,\ldots,\zeta_N)$ of spin variables for quantum numbers (S, M_S) times a space function of position variables $F_J(\vec{r}_1,\ldots,\vec{r}_N)$ for many-electron states J, describing single or multiple bonds with paired electrons in the case of bonding between A and B, or describing their nonbonding interactions. The space function is constructed from atomic orbitals of each species and the pairing of spin and space functions is done so that as $R \to \infty$, one finds the correct asymptotic limit factorized, with the total wavefunction

$$\begin{aligned}\Phi(1,\ldots,N) &= \hat{A}\sum_J F_J(1,\ldots,N)\Theta^{(J)}_{SM_S}(1,\ldots,N)\\ &\approx \sum_{KL} C_{KL}\hat{A}F_K(1,\ldots,N_A)\Theta^{(K)}_{S_A,M_{SA}}(1,\ldots,N_A)F_L(1,\ldots,N_B)\Theta^{(L)}_{S_B,M_{SB}}(1,\ldots,N_B)\end{aligned}$$

In a simple example with one valence electron per species A and B in atomic orbitals (AOs) $\chi_p(\vec{r}_1)$ and $\chi_q(\vec{r}_2)$ forming a single bond, the standard covalent VB wavefunction, written here for a homonuclear diatomic, is the single bond function with $F^{(cov)}(1, 2) = \chi_p(1)\chi_q(2)$ and

$$\Phi^{(cov)}(1,2) = \frac{\chi_p(1)\chi_q(2) + \chi_p(2)\chi_q(1)}{\left(2 + 2\,S^2_{ab}\right)^{1/2}}\Theta_{0,0}(1,2)$$

$$\Theta_{0,0}(\zeta_1,\zeta_2) = [\alpha(\zeta_1)\beta(\zeta_2) - \alpha(\zeta_2)\beta(\zeta_1)]/2^{1/2}$$

with $S_{pq} = \langle\chi_p\,|\,\chi_q\rangle \approx 0$ the overlap integral of the two nonorthogonal AOs, going to zero at large distances. This wavefunction clearly gives the correct asymptotic behavior.

When the interacting species dissociate instead into ions with the two electrons at the same site, the two-electron state contain $F^{(ion)}(1, 2) = \chi_u(1)\chi_u(2)$, $u = p, q$, and is

$$\Phi^{(ion)}(1,2) = \frac{\chi_p(1)\chi_p(2) + \chi_q(1)\chi_q(2)}{\left(2 + 2\,S^2_{pq}\right)^{1/2}}\Theta_{0,0}(1,2)$$

again giving the correct limit. An accurate treatment of the potential energy for all distances R combines the *cov* and *ion* states.

A pair of species with multiple bonds can be similarly described with a product of bond functions constructed with the correct total electronic antisymmetry.

Generalizations can be derived using more flexible orbitals ϕ_μ, which are yet centered at each atom or fragment μ, but have each AO mixed with the orbitals of other fragments, present as small components. Another generalization involves using all the spin functions, which can be generated for N electron spins, instead of just the spin-pairing per bond.

The generalized VB (or GVB) treatment [33] replaces the AO χ_p with a more flexible localized orbital ϕ_p written as a linear combination of functions with coefficients varied to optimize the system energy. It can be chosen to be a combination of AOs all centered at the same fragment A, to account for hybridization and polarization, or it can be a combination of orbitals at center A with some mixing of orbitals at B to account for electron transfer between them. Furthermore, orbitals localized at A can be the original ones nonorthogonal to those of B, or they can be transformed by symmetric orthonormalization into orbitals predominantly centered at A or B [14]. Orthonormalization allows for simpler expressions in matrix elements of operators, but dilutes the physical significance of spin pairing of orbitals in the description of bonds.

The simple spin pairing shown above in $\Theta_{0,0}(1, 2)$ is only one of several pairings of electron spin that are acceptable in polyatomic systems with multiple bonds. The treatment can be made rigorous introducing a set of N-electron spin functions $\Theta^{(J)}_{SM_S}(1,\ldots,N)$ for given quantum numbers (S, M_S) and making use of properties of the symmetric group $\mathcal{S}_N$ of all permutation operators $\hat{\mathcal{P}}_N$ of electron variables in N-variable functions [2]. Insofar as these permutations and the projection operator $\hat{\mathcal{O}}_{SM_S}$ for given S, M_S commute, it follows that $\hat{\mathcal{P}}_N\Theta^{(J)}_{SM_S} = \sum_K \Theta^{(K)}_{SM_S} P_{JK}$, with P_{JK} a matrix element of an irreducible matrix representation of the symmetric group [2]. These irreducible representations and their characters are well known, and they can be used to construct all the needed spin functions for each M_S. The space functions $F_J(1, \ldots, N)$ multiplying the spin functions in the expansion of total states Φ can be written as products of N localized orbitals ϕ_u, $u = a, b$, of a basis set chosen to account for electronic rearrangements.

Among the possible total states, the ones relevant to intermolecular forces go asymptotically to products $\hat{\mathcal{A}}'\Phi^{(A)}_K\Phi^{(B)}_L$ with $\Phi^{(A)}_K$ and $\Phi^{(B)}_L$ states of the species A and B constructed from the same orbitals. The number of spin functions created in the unitary group procedure above is

$$f_S^N = \frac{(2S+1)N!}{(N/2+S+1)!(N/2-S)!}$$

and it can get quite large for a large number N of active electrons, so that this general procedure is in practice limited to systems with few active electrons for all the intermolecular conformations of interest in $A + B$.

6.3.2 Symmetry-Adapted Perturbation Theory

It is appealing to derive a treatment of molecular interactions at all relative distances using component states like $\Phi_K^{(A)}$ and $\Phi_L^{(B)}$, eigenstates of the Hamiltonians $\hat{H}_A$ and $\hat{H}_B$, and functions of electron variables 1, ..., N_A and $N_A + 1$, ..., $N_A + N_B$, respectively. They are the ones that appear at large distances and are likely to be good approximations to component states also at intermediate distances when one or both are closed-shell systems and their bonding is nonexistent or weak. The main obstacle to such treatments, however, is that the total Hamiltonian $\hat{H}$ for the pair AB has symmetries different from those of the component Hamiltonians. The most essential is electronic antisymmetrization of the total states $\Phi_J^{(AB)}$, which for the pair AB must include not only component antisymmetrizers $\hat{\mathcal{A}}_S^{(N_S)}, S = A, B$, satisfying $\hat{\mathcal{A}}_S^{(N_S)}\Phi_K^{(S)} = \Phi_K^{(S)}$, but also a complimentary antisymmetrizer $\hat{\mathcal{A}}' = N_A!N_B!(N!)^{-1}\hat{\mathcal{C}}^{(AS)} = N_A!N_B!(N!)^{-1}\left(\hat{\mathcal{I}} - \hat{\mathcal{T}}\right)$ containing electron transposition operators in $\hat{\mathcal{T}}$ between A and B. The total antisymmetrizer $\hat{\mathcal{A}}^{(N)} = \hat{\mathcal{A}}'\hat{\mathcal{A}}_A^{(N_A)}\hat{\mathcal{A}}_B^{(N_B)}$ provides the symmetry-adapted $\Phi_J^{(AB)} = \hat{\mathcal{A}}'\Phi_K^{(A)}\Phi_L^{(B)}$. In addition, the total Hamiltonian has spatial and spin symmetries generally different from those of the component Hamiltonians. Therefore, a PT treatment with the Hamiltonian partitioning $\hat{H} = \hat{H}_0 + \hat{H}', \hat{H}_0 = \hat{H}_A + \hat{H}_B, \hat{H}' = \hat{H}_{AB}^{(int)}$, where $\hat{H}_0$ is the unperturbed Hamiltonian, must be done explicitly imposing the total symmetry on the wavefunction perturbation terms. This can be achieved by total postsymmetrization after a PT has been done with only component symmetries, in a so-called weak symmetry-adapted PT (SAPT), or more rigorously by total presymmetrization before the perturbation expansion is done, in a strong SAPT [35, 36].

The total wavefunction can be obtained from the previously introduced wave operator in

$$\Psi = \hat{\Omega}\Phi^{(D)} = \left(\hat{\mathcal{P}} + \hat{\mathcal{Q}}\hat{R}(E)\hat{\mathcal{Q}}\hat{H}\hat{\mathcal{P}}\right)\Phi^{(D)}$$

with $\hat{R}(E) = \left[\alpha\hat{\mathcal{P}} + \hat{\mathcal{Q}}\left(E - \hat{H}\right)\hat{\mathcal{Q}}\right]^{-1}$, which is as written valid for any partition of the total Hamiltonian and implies an intermediate normalization. Provided the multireference projection operator $\hat{\mathcal{P}}$ commutes with total symmetry projection operators $\hat{\mathcal{O}}$, which can be done by choosing a basis set of symmetry-adapted total functions, this wave operator can be used to generate all perturbation orders for the wavefunction and the energy, by recurrence. Expanding the resolvent $\hat{R}(E)$ in terms of the unperturbed resolvent $\hat{R}_0(E_0)$, for the noninteracting $A + B$ pair with total energy E_0, and using the perturbation $\hat{H}_{AB}^{(int)} = \lambda\hat{H}_{AB}^{(1)}$ [11], the resulting Rayleigh–Schrodinger expansion of energy and wavefunction in powers of λ can then be implemented computationally.

At large distances between the components, it is possible to ignore the complementary electron transpositions and to obtain electrostatic, induction, and dispersion energies from PT as done in previous chapters. But the transpositions must be accounted for if one wants to have smooth potential energies as the distance R is decreased, maintaining the same PT for each conformation of the AB pair. Indicating results from unsymmetrized PT as $E_{dir}(R)$, the total energy with a generalized intermediate normalization is $E(R) = \langle \Phi^{(D)} \mid \hat{\mathcal{P}}\hat{H} \mid \Psi \rangle / N^{(D)} = E_{dir}(R) + E_{exch}(R)$ with the first term, also called E_{pol}, obtained from the perturbation treatment excluding the transpositions in $\hat{\mathcal{T}}$, and the second term arising from the effects of transpositions. The potential energy for a many-atom conformation (R, $\mathbf{Q}_A$, $\mathbf{Q}_B$) is given by

$$
\begin{aligned}
V(R,\mathbf{Q}_A,\mathbf{Q}_B) &= E(R,\mathbf{Q}_A,\mathbf{Q}_B) - [E_A(\mathbf{Q}_A) + E_B(\mathbf{Q}_B)] \\
&= V_{dir}(R,\mathbf{Q}_A,\mathbf{Q}_B) + V_{exch}(R,\mathbf{Q}_A,\mathbf{Q}_B).
\end{aligned}
$$

The direct and exchange energies have been analyzed in detail and developed for computational work. To simplify here we consider a single reference state $\Phi_0^{(A)}\Phi_0^{(B)}$ of energy $E_0 = E_0^{(A)} + E_0^{(B)}$. The direct, or "pol," term contains to first order the classical electrostatic (or Coulomb) interaction energy and to second order induction and dispersion energies,

$$
V_{dir}^{(1)}(AB) = V_{elst}(AB), V_{dir}^{(2)}(AB) = V_{ind}(A) + V_{ind}(B) + V_{disp}(AB)
$$

The corresponding terms in the exchange V_{exch} are gathered from perturbation terms containing transpositions for each order in the perturbation as

$$
V_{exch}^{(1)}(AB) = V_{exch-elst}(AB), V_{exch}^{(2)}(AB) = V_{exch-ind}(AB) + V_{exch-disp}(AB)
$$

where $V_{exch-elst}$ is a correction to the classical Coulomb interaction containing the effect of electron exchange. The second-order energies involving sums over excited states of A and B can be rewritten in terms of their dynamical polarizabilities as done in the Casimir–Polder treatment of dispersion forces [37]. Higher order terms for direct and exchange energies have been similarly analyzed [35, 36].

The implementation of this SAPT to calculate intermolecular forces requires in addition a procedure for correcting the component wavefunctions $\Phi_0^{(S)}$ and energies $E_0^{(S)}$, $S = A, B$, when as usual these have been obtained in approximate treatments for many-electron systems. They are intramonomer electronic correlation corrections that can be incorporated by means of perturbation treatments of each species. Starting with known eigenfunctions $\Phi_K^{(S)}$ of an approximate Hamiltonian $\hat{F}_S$ with eigenenergies $E_K^{(S)}$, an additional perturbation treatment can be done with $\hat{F} = \hat{F}_A + \hat{F}_B$ as the zeroth-order Hamiltonian and the difference $\hat{W} = \hat{H}_0 - \hat{F} = \hat{W}_A + \hat{W}_B$ as the perturbation energy operator. Usual choices for $\hat{F}$ are the Hartree–Fock Hamiltonian or a sum of one-electron Hamiltonians derived from density functional theory. In these cases, the operator $\hat{W}$ is a kind of correlation energy giving corrections beyond HF or DFT mean values. This intramonomer PT can be done separately for each species, but it is more accurate to keep the same energy term order for both of them, insofar they are equally present in energy differences. The whole computational procedure therefore involves a double PT with perturbations $\hat{H}_{AB}^{(int)} = \kappa \hat{H}_{AB}^{(1)}$ and $\hat{W} = \lambda \hat{W}^{(1)}$ giving expansions in the products of order parameters $\kappa^k \lambda^l$, with k and l integers. The double PT gives energies

$$V(R) = \sum_{k=1}^{K} \sum_{l=0}^{L} \kappa^k \lambda^l \, V^{(k,l)}(R)$$

where K is the highest affordable perturbation order for the molecular interaction energy, usually $K = 3$, and L is the largest order in the intramolecular perturbation, which within a CC treatment can include some multiple excitations of the species A and B to all orders.

The SAPT combined with HF molecular functions, in SAPT(HF) treatments, involves some simplifications resulting from the absence of singly excited matrix elements, insofar as $\langle a | \hat{f}_{HF} | i \rangle = 0$, so that $V_{dir}^{(k,1)}(R) = 0$. Electronic exchange between A and B and correlation in A and B, however, distort their HF charge distributions and add to the intramolecular corrections. The SAPT(HF) treatment has been successfully applied to closed-shell systems such as $(H_2O)_N$, and Ar + H_2O. They have been obtained going to second order ($K = 2$) in the intermolecular interaction, which scales as N_{bf}^7 with the number of basis functions [36, 38]. The formalism is also suitable for interactions involving open-shell species provided unrestricted HF states are chosen as reference ones in a SAPT(UHF) treatment.

When $\hat{F}$ is chosen from a DFT treatment, the present PT is named SAPT(DFT). This has been found to give accurate results provided that: (i) the second-order interaction energies are obtained from dynamical polarizabilities, which correctly account for the coupling of one-electron KS excitations, as in time-dependent DFT, called SAPT(CKS); (ii) high-quality basis sets are used for atomic orbitals; and (iii) density functionals are preferably of the hybrid type containing some exact long range electron exchange energy. Results have been published for dimers $(He)_2$, $(Ne)_2$, $(H_2O)_2$ and $(CO_2)_2$ [39], and also for the benzene dimers in a variety of conformations.

Nonadditivity in three-body systems has also been considered within SAPT, which extends to intermediate and shorter distances the arguments presented in Chapter 3 for long-range nonadditivity of pair interactions, by incorporation of electron exchange [40]. The nonadditive terms originate already in first-order

PT due to exchange, with $E_{non-add}(ABC) = E^{(1)}_{exch}(ABC) + E^{(2)}_{ind}(ABC) + E^{(2)}_{ind-exch}(ABC) + E^{(2)}_{dsp-exch}(ABC) + E^{(3)}_{ind}(ABC) + E^{(3)}_{dsp}(ABC) + \cdots$ where each term is an intrinsic three-body function. Combining SAPT with DFT, it has been possible to calculate nonadditive energy terms for trimers of He, Ar, water, and benzene [41].

In addition to giving accurate results for a variety of *AB* pairs and *ABC* triplets, the SAPT treatment provides a systematic procedure for analyzing electrostatic, induction, and dispersion contributions to intermolecular forces as functions of atomic conformations, including electron exchange, even for many-body systems. This is useful in the development of approximations where accurate electronic energies are calculated first at short and intermediate intermolecular distances, to which long-range induction and dispersion energies are added *a posteriori* to cover all distances. In these treatments, it is necessary to avoid overcounting electronic correlation energies, which are sometimes duplicated in initial and added terms, something that can be avoided with a consistent SAPT treatment over all intermolecular distances.

6.4 The Density Functional Approach to Intermolecular Forces

6.4.1 Functionals for Interacting Closed- and Open-Shell Molecules

The original density functional theory, stating that the ground state energy of an electronic system is a unique functional $E[\rho(\vec{r})]$ of the total electronic density $\rho(\vec{r})$ at each position $\vec{r}$ [42], has been applied to the calculation of molecular properties and to intermolecular forces involving one or two closed-shell species, where chemical bond formation does not occur, within the KS and Gordon–Kim treatments as described in previous chapters [43–45]. Those treatments, involving local density (LD) functionals, have been subsequently extended with semilocal functionals by including the density gradient and local kinetic energy, and with hybrid functionals combining the exact Hartree–Fock exchange energy with an approximate local or semilocal form. They give reasonably accurate descriptions of short- and intermediate-range distances but do not account for the long-range vdW forces, which physically arise from electronic density fluctuations within the interacting species. The long-range forces can be added as corrections, but this must be done introducing short-range damping factors multiplying the added dispersion energies in a physically meaningful way. A more satisfactory procedure is to extend local or semilocal DFT treatments to include nonlocal functional terms in a consistent way. This must be done, however, while avoiding double counting the vdW forces, which are partly present in semilocal functionals at intermediate distances and might be erroneously added again when introducing long-range nonlocal terms in a functional.

Improvements to the semilocal treatments are needed to deal with dissociation of bonds between two open-shell species and with the long-range interactions. These come from: (i) separating up and down electronic spin densities, to be able to describe the asymptotic limit when a bond is dissociated and the two fragments are open-shell systems with different numbers of up and down spins; (ii) correcting the uniform gas expressions to account for the inhomogeneous nature of the densities; (iii) requiring correct behavior of the exchange one-electron potential to avoid electron self-interaction errors and incorrect one-electron potential energies at large interelectronic distances; and (iv) allowing for local electronic density fluctuations at large distances. These improvements have been covered in many reviews, and here the focus is on aspects directly related to intermolecular forces as they change with distances all the way to separation of nonbonding species or to breaking of bonds.

Recent developments relevant to molecular interactions have *climbed the accuracy ladder* of functionals from the original LD approximation involving only the electronic density $\rho(\vec{r})$, up to the generalized gradient approximation (GGA) including the gradient $\nabla\rho(\vec{r})$, and more recently the meta-GGA, which also includes a local kinetic energy density $\tau(\vec{r})$ constructed from local electronic momenta. The resulting functionals can generally be called *semilocal* (or *sl*). Also relevant are the hybrid functionals, which mix the exchange energy $E_x^{(sl)}$ as a semilocal functional with the accurate Hartree–Fock exchange energy $E_x^{(HF)}$, in ways that can provide better thermochemical energies [46], correct electronic excitation energies, crystal lattice constants and band gaps in solids [47, 48], or asymptotically correct electronic long-range potential energies [49–51].

The original treatment by Kohn and Sham (K–S) in terms of density amplitudes (or KS spin-orbitals) $\psi_j(\vec{r},\zeta)$ has been generalized to include interactions of open-shell species where bonds are formed or broken, by instead introducing a ground state functional of a local 2×2 spin density matrix $\rho_{\eta\eta'}(\vec{r})$ with spin indices $\eta=\alpha,\beta$, that is a unique functional provided the external potential energy matrix in which the electrons move is local, of the type $V_{\eta\eta'}^{(\mathrm{ext})}(\vec{r})$ [52]. A simplified version with a diagonal external potential energy matrix was implemented at the level of a local spin density functional (LSDF) of spin densities $\rho_{\eta\eta}(\vec{r})=\rho^{(\eta)}(\vec{r})$ and was shown to correctly describe the dissociation of a H_2 molecule into two ground state H atoms. This is done optimizing the LSDs so that $\rho^{(\alpha)}(\vec{r})$ and $\rho^{(\beta)}(\vec{r})$ go asymptotically into separate electronic densities located at the H atoms as needed, something that is not possible using a single LD. The formalism with an LSDF has been applied also to other molecules to calculate their spectral properties and their dissociation into open-shell species [53].

The LSDF approach describes near-range interaction energies and also long-range electrostatic and induction forces, but not the dispersion (or vdW) forces that arise from electronic density fluctuations and require introduction of dynamical susceptibilities. Efforts to extend DFT to include dispersion forces have involved recasting the long range electron–electron interactions in terms of charge density operators and dynamical susceptibilities for each species A and B, and approximating the susceptibilities in a variety of ways. This provides a treatment suitable for extended systems, which contains as special cases vdW interactions derived from multipolar polarizabilities present in the susceptibilities. We consider in what follows a case where electronic charge distributions of A and B do not overlap, avoiding electronic exchange or transfer, and only interact through Coulomb forces.

Dispersion energy functions of intermolecular and intramolecular position variables can be combined with near-range functionals obtained from DFT and containing all energies (electrostatic, exchange, near-range correlation, long-range induction) except for the vdW dispersion energy. This is a very active area of research where several DFT-plus-vdW treatments have been proposed and tested by comparison with accurate results from CCSD(T) or SAPT calculations for interaction potentials $V(R, \mathbf{Q})$. Here the focus is on methods applicable to large classes of chemical systems, described as many-atom structures A and B containing sets of atoms $\{a\}$ and $\{b\}$ with their centers of mass at a relative distance R in a body-fixed reference frame, and with internal variables $\mathbf{Q} = \left\{\vec{R}_a, \vec{R}_b\right\}$ referred to their C-of-M.

We have seen in a previous chapter that the dispersion energy can be separately treated and calculated from the dynamical susceptibilities of the two species, or from their multipolar polarizabilities. The aim here is to add them to the semilocal energies while avoiding double counting of long-range interactions and to ascertain how important these are for development of potential energy surfaces valid at all distances. A variety of ways to do this can be termed DFT-D treatments, with the D label signifying that dispersion energies are to be added to DFT results. This has led recently to treatments where the semilocal DFT energies $E^{(sl)}(R, \mathbf{Q})$ at near-range distances (rapidly decreasing at large distances) and the nonlocal dispersion (vdW) energies $E_{dsp}(R, \mathbf{Q})$ at large distances are combined to give total energies as

$$E(R,\mathbf{Q}) = E^{(sl)}(R,\mathbf{Q}) + E_{dsp}^{(nl)}(R,\mathbf{Q}) f_d(R,\mathbf{Q})$$

for all intermolecular distances. It restricts the dispersion contribution to the long range by multiplying it times a suitable damping function $f_d(R, \mathbf{Q})$, whereas before $f_d(R) \approx 1$, for $R \to \infty$ and $f_d(R) \approx 0$, for $R \to 0$ [54–57]. An alternative that provides a more clear-cut separation of a dispersionless functional energy and an energy obtained from the long-range dispersion energy uses results from accurate SAPT to identify dispersion-like contributions to a density functional.

Subtraction of those contributions provides a dispersionless density functional (or dlDF) to which a long-range, separately calculated dispersion energy with an appropriate cut-off, can be added. It is followed by fitting energy parameters to reproduce interaction energies in groups of compounds [58]. This, however, requires additional calculations done with the SAPT many-electron approach as a preliminary task. Details on the DFT-D treatments relying on calculations at the *sl*–functional level for two interacting many-atom systems are presented in a following chapter.

An alternative to the explicit addition of long-range vdW energies to semilocal energies is to develop heavily parametrized energy functionals containing one-electron terms that can be adjusted to reproduce vdW energies at large distances. Functionals have been recently introduced with one-electron semilocal potential energy terms $v_{xc}^{(sl)}(\vec{r})$ constructed to include long-range interaction energies, also combined with meta-hybrid GGA functionals [59]. Hybrid functionals contain a chosen amount of orbital-dependent exchange energy as derived in the Hartree–Fock treatment, combined with the standard DFT exchange as a functional of density, and are described in Section 6.4.3 [46]. Another approach has introduced, within functionals, additional one-electron dispersion-corrected atom-centered potentials (DCACPs) $v^{(ECP)}(\vec{r},\vec{r}\,') = v^{(sl)}(\vec{r})\delta(\vec{r}-\vec{r}\,') + v^{(nl)}(\vec{r},\vec{r}\,')$ containing both semilocal and nonlocal terms [60]. In both treatments, the one-electron potential energies have been parametrized and tested by reproducing accurately known dispersion energies for atom pairs and have been used in calculations for large many-atom systems.

Semilocal potentials for electrons of a given spin $\sigma = \alpha, \beta$ depend not only on the electronic density and its derivatives (as in the generalized gradient or GGA approximations) that can describe long-range electrostatic and induction energies, but depend also on local electron momentum densities $-i\hbar\nabla\varphi_{j\sigma}^{(KS)}(\vec{r}) = \vec{\pi}_{j\sigma}^{(KS)}(\vec{r})$, derivatives of KS orbitals $\varphi_{j\sigma}^{(KS)}$ of a given spin component. This describes to some extent density fluctuations as they appear in interaction dispersion energies. Exchange-correlation functionals are of the form $\mathcal{E}_{xc,\sigma}[\rho,s,\tau]$ with $s(\vec{r}) = |\nabla\rho|/\left[2k_F\rho(\vec{r})\right]$ a dimensionless function of the density gradient $\nabla\rho(\vec{r})$, where $k_F(\vec{r}) = \left[3\pi^2\rho(\vec{r})\right]^{1/3}$ is a local Fermi wavenumber, and with $\tau_\sigma = \sum_j \pi_{j\sigma}^{(KS)}(\vec{r})^2/(2m_e)$ a kinetic energy density. Fitting parameters in the total energy functional $\mathcal{E}_{tot,\sigma}[\rho,s,\tau]$ to reproduce values in large sets of compound properties can then provide some of the polarization (induction and dispersion) interactions around equilibrium atomic conformations, including nonbonding structures and molecular crystals. This program has been implemented, for example, within the M06 [59] and MN12 [61] software packages, which have also been extended in hybrid versions with screened electronic exchange suitable for applications to large molecules and solids [62].

6.4.2 Electronic Exchange and Correlation from the Adiabatic-Connection Relation

Yet another approach introduces nonlocal functionals based on the adiabatic connection fluctuation-dissipation (ACFD) relation giving a formally exact expression for the electron–electron correlation energy in terms of the dynamical susceptibility. It leads to the introduction of a nonlocal expression for the correlation energy. This can be written as $E_c(R,\mathbf{Q}) = E_c^{(sl)}(R,\mathbf{Q}) + E_c^{(nl)}(R,\mathbf{Q})$ with the first term coming from a semilocal functional and the second term given by a nonlocal functional modified by subtraction of the homogeneous-electron correlation component to avoid double counting it. This second term can be constructed from the susceptibilities as a known functional of the electronic density so that no parametrization from molecular properties is needed. It goes at large distances to the known form of the dispersion energy in terms of susceptibilities, and in addition, it can be obtained as a smooth function of distance for the near-range without introduction of a separate damping function. The approach has generated a treatment called vdW-DF in various versions, suitable for applications to large as well as small molecules and to physisorption and chemisorption at solid surfaces [63–65]. It has also been implemented with a more detailed and accurate parametrization of dynamical polarizabilities in the so-called VV10 treatment [65].

It is of interest to summarize the ACFD formalism as it provides a compact form for the exact exchange-correlation energy and may open a route to new nonlocal density functionals. That can be done by means of the adiabatic connection theorem of many-electron theory, closely related to the Hellmann–Feynman (or H–F) theorem for molecular forces. The total Hamiltonian for interacting electrons can be redefined to apply to an electron–electron Coulomb energy where the squared charge c_e^2 is replaced by λc_e^2, with a parameter $0 \leq \lambda \leq 1$. This can be interpreted also as the interaction of electrons at far positions $\vec{r}/\lambda$ for small $\lambda \ll 1$, coming in as the parameter increases to its value $\lambda = 1$ for the physical system. The total Hamiltonian is $\hat{H}_\lambda = \hat{H}^{(1)} + \lambda\hat{H}^{(2)}$, a sum of one-electron and two-electron terms, with

$$\hat{H}^{(1)} = \sum\nolimits_{1\leq m\leq N} \hat{h}_m,\ \hat{H}^{(2)} = (1/2)\sum\nolimits_{m\neq n} c_e^2/\,|\vec{r}_m - \vec{r}_n|$$

where $\hat{h}_m$ contains the kinetic energy of the m-th electron, its attraction energy to positive ion cores, and the energy coming from the repulsion among the ions, with the latter independent of electron variables. When the one-electron term also contains some mean-field contribution from electron–electron interactions, such as a functional of electron density, this must be subtracted from $\lambda\hat{H}^{(2)}$.

Differentiating the eigenstate equation $\hat{H}_\lambda\Psi_\lambda = E_\lambda\Psi_\lambda$ with respect to λ and projecting on the normalized state Ψ_λ, we find as in Chapter 5 that

$\partial E_\lambda / \partial \lambda = \left\langle \Psi_\lambda \,|\, \hat{H}^{(2)} \,|\, \Psi_\lambda \right\rangle$, the H–F relation. Its integral form for $0 \le \lambda \le 1$ can be written considering that $E_{\lambda=0} = E_s^{(1)}$, the sum of single-electron kinetic and nuclear attraction energies, and $E = E_{\lambda=1}$ is the exact energy, so that the integral form of the H–F relation is

$$E = E_s^{(1)} + \int_0^1 d\lambda \left\langle \Psi_\lambda \,|\, \hat{H}^{(2)} \,|\, \Psi_\lambda \right\rangle$$

The exact energy $E = E_{\lambda=1}$ can be expressed in terms of a two-electron density $\rho_\lambda^{(2)}(1,2;1,2)$ obtained as a function of the λ-parameter values. To derive its form, one can start with the expression for $\hat{H}^{(2)}$ written in terms of the density operator

$$\hat{\rho}^{(1)}(\vec{r}) = \sum\nolimits_{1 \le m \le N} \delta\left(\vec{r} - \hat{\vec{r}}_m\right)$$

where the electron position is being treated as an operator and one performs integrals over the electron variables in $\vec{r}$. The double summation in $\left\langle \Psi_\lambda \,|\, \hat{H}^{(2)} \,|\, \Psi_\lambda \right\rangle$ must be constructed while avoiding the interaction of each electron with itself. To this effect, first write

$$\hat{H}^{(2)} = (1/2) \int d^3 r_1 \int d^3 r_2 \, \frac{c_e^2}{|\vec{r}_1 - \vec{r}_2|} \sum\nolimits_{m \ne n} \delta\left(\vec{r}_1 - \hat{\vec{r}}_m\right) \delta\left(\vec{r}_2 - \hat{\vec{r}}_n\right)$$

$$\sum\nolimits_{m \ne n} \delta\left(\vec{r}_1 - \hat{\vec{r}}_m\right) \delta\left(\vec{r}_2 - \hat{\vec{r}}_n\right) = \sum\nolimits_{m,n} \delta\left(\vec{r}_1 - \hat{\vec{r}}_m\right) \delta\left(\vec{r}_2 - \hat{\vec{r}}_n\right) - \delta\left(\vec{r}_1 - \vec{r}_2\right) \sum\nolimits_m \delta\left(\vec{r}_1 - \hat{\vec{r}}_m\right)$$

where the double sum to the right includes $m = n$. This gives

$$\sum\nolimits_{m \ne n} \delta\left(\vec{r}_1 - \hat{\vec{r}}_m\right) \delta\left(\vec{r}_2 - \hat{\vec{r}}_n\right) = \hat{\rho}^{(1)}(\vec{r}_1) \hat{\rho}^{(1)}(\vec{r}_2) - \delta\left(\vec{r}_1 - \vec{r}_2\right) \hat{\rho}^{(1)}(\vec{r}_1)$$

to be replaced in the double integral. Its expectation value in the state Ψ_λ can also be obtained from the two-electron density function $\rho_\lambda^{(2)}(1,2;1,2)$ as previously done for $\lambda = 1$, with the short-hand notation $(1) = (\vec{r}_1, \zeta_1)$ including space and spin variables. For the spin-independent electron–electron interaction energy, a summation over spin variables can be carried out and the expectation value is found to contain the two-electron pair density $P_\lambda^{(2)}(\vec{r}_1, \vec{r}_2) = \sum_{\zeta_1, \zeta_2} \rho_\lambda^{(2)}(1,2;1,2)$ in

$$\left\langle \Psi_\lambda \,|\, \hat{H}^{(2)} \,|\, \Psi_\lambda \right\rangle = (1/2) \int d^3 r_1 \int d^3 r_2 \, \frac{c_e^2}{|\vec{r}_1 - \vec{r}_2|} P_\lambda^{(2)}(\vec{r}_1, \vec{r}_2)$$

$$P_\lambda^{(2)}(\vec{r}_1, \vec{r}_2) = \left\langle \Psi_\lambda \,|\, \hat{\rho}^{(1)}(\vec{r}_1) \hat{\rho}^{(1)}(\vec{r}_2) \,|\, \Psi_\lambda \right\rangle$$

$$-\delta(\vec{r}_1-\vec{r}_2)\left\langle\Psi_\lambda\,|\,\hat{\rho}^{(1)}(\vec{r}_1)\,|\,\Psi_\lambda\right\rangle$$

From this and the integral form of the H–F relation, it follows that

$$E=E_s^{(1)}+\frac{1}{2}\int_0^1 d\lambda\int d^3r_1\int d^3r_2\,\frac{c_e^2}{|\vec{r}_1-\vec{r}_2|}P_\lambda^{(2)}(\vec{r}_1,\vec{r}_2)$$

This can be compared to the Hartree energy that contains the classical electron–electron interaction as

$$E_H=E_s^{(1)}+\frac{1}{2}\int_0^1 d\lambda\int d^3r_1\int d^3r_2 P_\lambda^{(1)}(\vec{r}_1)\,\frac{c_e^2}{|\vec{r}_1-\vec{r}_2|}P_\lambda^{(1)}(\vec{r}_2)$$

where $P_\lambda^{(1)}(\vec{r})=\left\langle\Psi_\lambda\,|\,\hat{\rho}^{(1)}(\vec{r})\,|\,\Psi_\lambda\right\rangle$. The difference $E-E_H=E_{xc}$ is the exchange-correlation energy of DFT, which is therefore given by

$$E_{xc}=\frac{1}{2}\int_0^1 d\lambda\int d^3r_1\int d^3r_2\,\frac{c_e^2}{|\vec{r}_1-\vec{r}_2|}\left[P_\lambda^{(2)}(\vec{r}_1,\vec{r}_2)-P_\lambda^{(1)}(\vec{r}_1)P_\lambda^{(1)}(\vec{r}_2)\right]$$

Further introducing the density fluctuation operators

$$\Delta_\lambda\hat{\rho}^{(1)}(\vec{r})=\hat{\rho}^{(1)}(\vec{r})-P_\lambda^{(1)}(\vec{r})$$

and the spatial correlation $\Delta P_\lambda^{(2)}(\vec{r}_1,\vec{r}_2)$, one finds the equivalent expression

$$E_{xc}=\frac{1}{2}\int_0^1 d\lambda\int d^3r_1\int d^3r_2\,\frac{c_e^2}{|\vec{r}_1-\vec{r}_2|}\left[\Delta P_\lambda^{(2)}(\vec{r}_1,\vec{r}_2)-\delta(\vec{r}_1-\vec{r}_2)P_\lambda^{(1)}(\vec{r}_1)\right]$$

$$\Delta P_\lambda^{(2)}(\vec{r}_1,\vec{r}_2)=\left\langle\Psi_\lambda\,|\,\Delta_\lambda\hat{\rho}^{(1)}(\vec{r}_1)\,\Delta_\lambda\hat{\rho}^{(1)}(\vec{r}_2)\,|\,\Psi_\lambda\right\rangle$$

so that $\Delta P_\lambda^{(2)}(\vec{r}_1,\vec{r}_2)$ is the spatial correlation function of two density fluctuations.

The next step in the proof of the ACDF theorem relates the correlation of fluctuations of a pair of electrons to the dynamical susceptibility of the pair [66]. The function $P_{\lambda,xc}^{(2)}(\vec{r}_1,\vec{r}_2)=\Delta P_\lambda^{(2)}(\vec{r}_1,\vec{r}_2)-\delta(\vec{r}_1-\vec{r}_2)P_\lambda^{(1)}(\vec{r}_1)$ can be re-expressed in terms of a dynamical susceptibility $\chi_{\lambda,xc}(\vec{r}_1,\vec{r}_2;\omega)$ of the many-electron system, dependent on the frequency ω, as an integral over an imaginary-valued argument $i\omega$, to obtain an expression used within DFT [52, 53, 63, 67, 68].

The relation follows from response theory applied to the calculation of changes in $\hat{\rho}^{(1)}(\vec{r},t)$ induced by an external electric potential $\phi_{ext}(\vec{r}\,',t')$ starting at time $t'=0$ and null before then, as it couples to the density with energy $\hat{H}_{ext}^{(1)}(t')=\int d^3r'\phi_{ext}(\vec{r}\,',t')\hat{\rho}^{(1)}(\vec{r}\,',t')$ and with the system initially in its ground electronic eigenstate $|0_\lambda\rangle$ of the Hamiltonian operator $\hat{H}_\lambda$. This has been done for the response of a molecular charge density in Chapter 2, solving for an excited state $\Psi_\lambda(t)$, which can be used here to write the density change $\Delta_\lambda\rho^{(1)}(\vec{r},t) = \langle\Psi_\lambda(t)\,|\,\hat{\rho}^{(1)}(\vec{r})\,|\,\Psi_\lambda(t)\rangle - P_\lambda^{(1)}(\vec{r})$ as

$$\Delta_\lambda\rho^{(1)}(\vec{r},t) = \int d^3r' \int_0^t dt'\chi_\lambda(\vec{r},t;\vec{r}',t')\phi_{ext}(\vec{r}',t')$$

$$\chi_\lambda(\vec{r},t;\vec{r}',t') = -(i/\hbar)\langle 0_\lambda\,|\left[\hat{\rho}^{(1)}(\vec{r},t)\hat{\rho}^{(1)}(\vec{r}',t') - \hat{\rho}^{(1)}(\vec{r}',t')\hat{\rho}^{(1)}(\vec{r},t)\right]|\,0_\lambda\rangle$$

where χ_λ is a scaled response function containing $\hat{\rho}^{(1)}(\vec{r},t) = \exp\left[(i/\hbar)\hat{H}_\lambda t\right]$ $\hat{\rho}^{(1)}(\vec{r},0)\exp\left[-(i/\hbar)\hat{H}_\lambda t\right]$, a Hermitian operator. It has the same form if rewritten in terms of the density fluctuations $\Delta_\lambda\hat{\rho}^{(1)}$ instead of the densities $\hat{\rho}^{(1)}$. The lower time integration limit can be extended to $-\infty$, and insofar as the ground state is stationary, the susceptibility depends only on $\tau = t - t'$. It can also be written as an imaginary-part form

$$\chi_\lambda(\vec{r},\vec{r}';\tau) = \frac{2}{\hbar}\mathcal{I}m\left[\langle 0_\lambda\,|\,\Delta_\lambda\hat{\rho}^{(1)}(\vec{r},\tau)\,\Delta_\lambda\hat{\rho}^{(1)}(\vec{r}',0)\,|\,0_\lambda\rangle\right]$$

in a compact expression. Its Fourier transform is

$$\tilde{\chi}_\lambda(\vec{r},\vec{r}';\omega) = \int_{-\infty}^{\infty} d\tau\chi_\lambda(\vec{r},\vec{r}';\tau)\exp(i\omega\tau)$$

It contains the Fourier transformed $\widetilde{\Delta_\lambda\hat{\rho}^{(1)}}(\vec{r},\omega)$ operator in

$$\tilde{\chi}_\lambda(\vec{r},\vec{r}';\omega) = (2/\hbar)\mathcal{I}m\left[\langle 0_\lambda\,|\,\widetilde{\Delta_\lambda\hat{\rho}^{(1)}}(\vec{r},\omega)\,\Delta_\lambda\hat{\rho}^{(1)}(\vec{r}',0)\,|\,0_\lambda\rangle\right]$$

The calculation of density changes is conveniently done introducing a retarded (time-forward) susceptibility $\chi_\lambda^{(+)}(\tau) = \theta(\tau)\chi_\lambda(\tau)$ where $\theta(\tau)$ is the step function null for negative arguments. This gives

$$\Delta_\lambda\rho^{(1)}(\vec{r},t) = \int d^3r' \int_{-\infty}^{\infty} d\tau\chi_\lambda^{(+)}(\vec{r},\vec{r}';\tau)\phi_{ext}(\vec{r}',t-\tau)$$

A Fourier transform from τ to ω of this time convolution leads to

$$\Delta_\lambda\tilde{\rho}^{(1)}(\vec{r},\omega) = \int d^3r'\tilde{\chi}_\lambda^{(+)}(\vec{r},\vec{r}';\omega)\tilde{\phi}_{ext}(\vec{r}',\omega)$$

showing that $\tilde{\chi}_\lambda^{(+)}(\vec{r},\vec{r}';\omega) = \int_{-\infty}^{\infty} d\tau\chi_\lambda^{(+)}(\vec{r},\vec{r}';\tau)\exp(i\omega\tau)$ is the scaled dynamical susceptibility, related to the transform of the response through $\tilde{\chi}_\lambda^{(+)}(\vec{r},\vec{r}';\omega) = \int_0^{\infty} d\omega'\tilde{\chi}_\lambda(\vec{r},\vec{r}';\omega-\omega')\tilde{\theta}(\omega')$. Therefore the dynamical response $\tilde{\chi}_\lambda(\vec{r},\vec{r}';\omega)$ provides both total energies and also the time evolution of the electronic density.

The function $\tilde{\chi}_\lambda(\vec{r},\vec{r}\,';\omega)$ can be related to the density-fluctuation correlation $\Delta P_\lambda^{(2)}(\vec{r}_1,\vec{r}_2)$ appearing in the energy, in a series of steps starting with an expansion of $\Delta_\lambda\hat{\rho}^{(1)}(\vec{r},\tau)$ in the basis set $\{|J_\lambda\rangle\}$ of eigenstates of $\hat{H}_\lambda$ with energies $\{|E_{\lambda,J}\rangle\}$ [68], designed to extract $\langle 0_\lambda | \Delta_\lambda\hat{\rho}^{(1)}(\vec{r})\Delta_\lambda\hat{\rho}^{(1)}(\vec{r}\,') | 0_\lambda\rangle$ from an integral of $\tilde{\chi}_\lambda(\vec{r},\vec{r}\,';\omega)$ over frequencies. Introducing the identity decomposition $\hat{I} = \sum_J |J_\lambda\rangle\langle J_\lambda|$ between the operators in the response function χ_λ, the matrix element

$$\langle 0_\lambda | \Delta_\lambda\hat{\rho}^{(1)}(\vec{r},\tau) | J_\lambda\rangle = \exp[i(E_{\lambda,0}-E_{\lambda,J})\tau/\hbar]\langle 0_\lambda | \Delta_\lambda\hat{\rho}^{(1)}(\vec{r},0) | J_\lambda\rangle$$

can be Fourier transformed and introduced in $\tilde{\chi}_\lambda(\vec{r},\vec{r}\,';\omega)$ to obtain its imaginary part. An integration over frequencies of $\tilde{\chi}_\lambda(\vec{r},\vec{r}\,';i\omega)$ leads to

$$-\frac{\hbar}{\pi}\int_0^\infty d\omega\,\tilde{\chi}_\lambda(\vec{r},\vec{r}\,';i\omega) = \langle 0_\lambda | \Delta_\lambda\hat{\rho}^{(1)}(\vec{r},0)\,\Delta_\lambda\hat{\rho}^{(1)}(\vec{r}\,',0) | 0_\lambda\rangle$$

as needed. This relation can alternatively be derived from the analytical properties of $\tilde{\chi}_\lambda(\vec{r},\vec{r}\,';i\omega)$ in the complex frequency plane and its integration in a closed path [66].

The final expression for exchange correlation is then

$$E_{xc} = \frac{1}{2}\int_0^1 d\lambda \int d^3r_1 \int d^3r_2\,\frac{c_e^2}{|\vec{r}_1-\vec{r}_2|}\left[-\frac{\hbar}{\pi}\int_0^\infty d\omega\,\tilde{\chi}_\lambda(\vec{r}_1,\vec{r}_2;i\omega)\right.$$
$$\left. -\,\delta(\vec{r}_1-\vec{r}_2)P_\lambda^{(1)}(\vec{r}_1)\right]$$

This allows calculation of E_{xc} from the dynamical susceptibility, and if the latter is constructed as a functional of the electron density $\rho_\lambda(\vec{r}) = P_\lambda^{(1)}(\vec{r})$, then a formally exact expression has been obtained for the functional $E_{xc}[\rho_\lambda(\vec{r})]$.

Applications to the calculation of intermolecular forces between species A and B at relative distance R require a knowledge of $\tilde{\chi}_\lambda$ and E_{xc} over each distance R and in particular for large R. In this case, it is known that the dispersion energy $E_{dsp}(R)$ is given in terms of susceptibilities $\tilde{\chi}^{(S)}(\vec{r},\vec{r}\,';i\omega)$ of $S = A, B$ by a Casimir–Polder expression, and $E_{xc}[R;\rho_\lambda(\vec{r})]$ must correctly give $E_{dsp}(R)$ at large distances. An important distinction in the two asymptotic forms (the one above and Casimir–Polder's) is that each spatial integral in the above E_{xc} extends over all space including both species, while the spatial integrals in the Casimir–Polder expression extend only into nonoverlapping regions containing A or B. The connection can be done introducing localized basis functions around A and B, to partition the spatial integrals in E_{xc}. This is presented in some detail in the next chapter on extended systems.

The connection of the ACDF relation with DFT was pointed out and used in early molecular calculations [52] and reviewed [53] some time ago and has been the basis for the vdW–DF formulation of density functional, which has provided many results in applications to complexes involving molecules and surfaces, adsorbates, and their potential energy changes with distances [55, 63, 69]. One such result for two interacting benzene molecules in a complex with parallel planar structures is shown in Figure 6.5. Here the semilocal (near-range) energies were obtained from generalized gradient density functional as shown, and the vdW interaction was added as described in a vdW–DF treatment.

The factorization $\Delta_\lambda \hat{\rho}^{(1)}(\vec{r}) = \hat{\rho}^{(1)}(\vec{r}) - P_\lambda^{(1)}(\vec{r}) = P_\lambda^{(1)}(\vec{r})\Delta_\lambda \hat{\varphi}(\vec{r})$, which introduces a relative density fluctuation operator $\Delta_\lambda \hat{\varphi}(\vec{r})$ at all distances R between A and B, leads to the expression $\Delta P_\lambda^{(2)}(\vec{r}_1, \vec{r}_2; R) = P_\lambda^{(1)}(\vec{r}_1)\Phi_{\lambda,xc}^{(2)}(\vec{r}_1, \vec{r}_2; R)P_\lambda^{(1)}(\vec{r}_2)$ where the two-electron function $\Phi_{\lambda,xc}^{(2)}(\vec{r}_1, \vec{r}_2; R)$ accounts for exchange correlation arising from density fluctuations at each relative distance R. This factorization can be used to bring E_{xc} to a form similar to the one developed to account for long-range vdW interactions in terms of local polarizabilities,

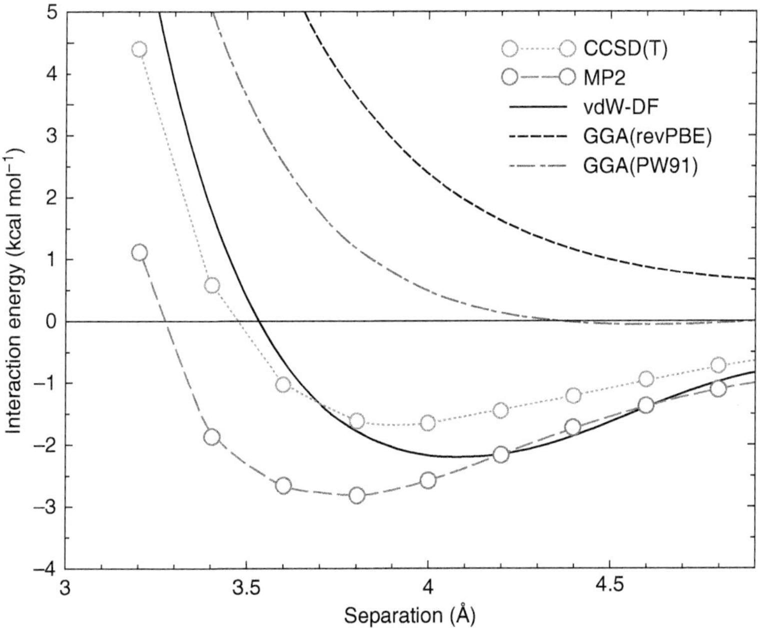

Figure 6.5 Interaction energy versus distance between the planes of two benzene molecules in a "sandwich" conformation. *Source:* from Ref. [63]. Reproduced with permission of Institute of Physics.

but here we have an expression that is applicable to all distances and contains both semilocal and nonlocal contributions to the energy. Several treatments have been based on this expression, which can in principle provide a seamless connection between the long-range dispersion energy given by a Casimir–Polder integral, and semilocal DFT energies for intermediate and short ranges [55, 65, 67]. This is presently an active area of research, and the following chapters present some related aspects relevant to interaction energies between extended systems and at solid surface.

6.4.3 Issues with DFT, and the Alternative Optimized Effective Potential Approach

The absence of a constructive proof for the density functional has led to the development of many versions of DFT. The many density functionals in the literature have been subjected to continuous tests and improvements within two procedures: imposing a large number of constrains on the functionals; and comparing DFT calculated properties of a large set of molecules with accurate results generated by many-electron methods including correlation. An example of imposition of many constrains is found in the construction of a recent strongly constrained functional called SCAN [70]. Several comparisons of DFT molecular structures and properties, and of interaction energies of molecular pairs, have been done with known values from experiment or from accurate non-DFT calculations, collected for several groups of compounds [59, 71–73]. Comparisons of properties such as vibrational frequencies of molecules, and polarizabilities of molecular dimers, that are relevant to calculation of intermolecular forces, give some idea of which functionals may work for potential energy surfaces and which ones can be discarded. Generally, the more accurate functionals are meta-GGA functionals containing a hybrid of semilocal exchange and Hartree–Fock exchange terms. Much, however, remains to be done to improve the accuracy of DFT methods for molecular interactions.

The original Hohenberg–Kohn treatment of functionals is rigorous for the ground electronic state of a nondegenerate energy level and for the lowest nondegenerate energy of a given symmetry. It is therefore limited to states with zero or maximal total electronic spin when the Hamiltonian is spin-free. Its generalization including spin densities in addition to charge densities allows treatments of open-shell molecular pairs in states with a good projection quantum number M_S for the electronic spin, but the electronic states do not correspond to good total spin quantum numbers S [52]. A way around this limitation is to construct a determinantal wavefunction with the K–S orbitals of lowest energy and to project out of it the state with the correct spin quantum numbers using the projector $\hat{\mathcal{O}}_{SM_S}$ introduced in Section 6.1, in a postsymmetrization procedure.

The issue of degeneracies or near-degeneracies can be treated within DFT introducing a functional of a statistically averaged density function $\bar{\rho}_W(\vec{r})$ constructed as a weighted sum over a set of interacting states of similar energy. The functional can be chosen to be the grand-canonical ensemble of a system of electrons under thermodynamical constrains given by a medium at temperature $T = (k_B\beta)^{-1}$ and chemical potential μ [74], or can be built as an average over energies $E_1, E_2, \ldots E_D$ with weights $W = \{w_d, d = 1 \; to \; D\}$ [43, 75]. In each case, the many-electron states are given, within a DFT treatment, as determinantal wavefunctions with K–S orbitals $\psi_{d,j}^{(KS)}$ and the resulting average density function is a K–S sum

$$\bar{\rho}_W(\vec{r}) = \sum_{d=1}^{D} \sum_{j=1}^{\infty} w_{d,j} \left| \psi_{d,j}^{(KS)}(\vec{r}) \right|^2$$

with fractional occupation numbers $w_{d,j}$ of K–S orbitals, instead of the previous occupation numbers of 1 or 0. These procedures provide average energies as functions of interatomic distances, and they can be extended to include perturbation energies arising from external fields. Properties relevant to intermolecular forces, such as electric dipoles and polarizabilities, energy gradients, and vibrational frequencies can then be obtained by taking derivatives of perturbed energies with respect to applied fields [4].

An important concern in DFT treatments of intermolecular forces relates to the appearance of electron self-interaction errors resulting from approximations in density functionals. In a wavefunction treatment of a many-electron system at the level of the Hartree–Fock approximation, it becomes clear that terms in the HF one-electron potential energy operator, in the notation of Section 6.1,

$$\hat{v}^{(HF)}(1) = \hat{J}(1) - \hat{K}(1) = \sum_{k}^{occ} \left[\int d(1')\psi_k(1')^* v(1',1)\left(1 - \hat{t}_{1'1}\right)\psi_k(1') \right]$$

containing the Coulomb and exchange energy operators $\hat{v}_{Coul}^{(HF)} = \hat{J}$ and $\hat{v}_x^{(HF)} = -\hat{K}$ partly cancel so that an electron in an occupied orbital, generated by HF or KS methods, does not interact with itself, and also that asymptotically for large r_1, $\hat{v}_{Coul}^{(HF)}(1)\psi_l(1) \approx \psi_l(1)N_e c_e^2/r_1$ while $\hat{v}_x^{(HF)}(1)\psi_l(1) \approx -\psi_l(1)c_e^2/r_1$, for occupied ψ_l, with the sum correctly containing $(N_e - 1)c_e^2 r^{-1}$. But when the exchange-correlation functional $E_{xc}[\rho]$ is approximated, that cancellation does not occur because the one-electron DFT exchange potential $v_x(1) = \delta E_x/\delta\rho(1)$ does not cancel the Coulomb self-interaction in the DFT one-electron potential energy $v_{Coul}(1)$. As a consequence, the KS one-electron potential energy $v^{(KS)}(\vec{r}) = v_{Coul}(\vec{r}) + v_x(\vec{r}) + v_c(\vec{r})$ deviates from the correct $(N_e - 1)c_e^2 r^{-1}$ behavior at large distances and is unphysical near each ion core. This affects the values of molecular potential energies for dissociation and charge transfer and properties like energy-band gaps in semiconductors and polarization interactions in molecular crystals. Calculations of potential energy surfaces must be done using functionals that at least decrease the self-interaction error and preferably lead to correct one-electron potential functions over all distances.

Another complication in the application of DFT to intermolecular forces arises from the continuous dependence of the energy on the electron density $\rho(\vec{r})$, which integrates to the total number of electrons N_e for two interaction species A and B. When these interact and exchange electrons, as in the dissociation of a compound AB into $A^+ + B^-$, the density must split at large distances as $\rho(\vec{r}) \approx \rho^{(A)}(\vec{r}) + \rho^{(B)}(\vec{r})$ with each term integrating to the correct number of electrons, $N_e^{(A)}$ and $N_e^{(B)}$. However, this is not assured in DFT, where a functional of the AB pair density may lead to asymptotic densities integrating to noninteger values $N_e^{(S)} \pm \delta N_e$ of electrons and to unphysical charges in the fragments $S = A$ and B. This problem is related to the absence of discontinuities in functionals as electron densities are changed, since the energy functionals change continuously with electronic density variations, while physically one must instead have sharp energy changes for varying integer number of electrons in each fragment. This may lead to inaccurate calculations for electron ionization energies and for proton affinities.

At a more formal level, the appearance of fractional numbers of electrons when bonds are broken can be reinterpreted by means of an ensemble DFT (or EDFT), where a fractional number $\bar{N}$ is considered to result from a weighted average of energy functionals for two species with integer number of electrons N and $N + 1$. The ensemble density is chosen as $\bar{\rho} = (1-\nu)\rho_N + \nu\rho_{N+1}$, with an interpolating variable ν obtained from a variational procedure [76]. This approach provides a foundation for chemical reactivity theory with a consistent use of chemical reactivity indices [43] for species with physically correct integer number of electrons.

The heavily parametrized functionals already mentioned in connection with DFT calculations, including dispersion energies, have also been constructed to minimize errors from electronic self-energy interactions and from continuous variations of the number of electrons, and provide an alternative treatment giving more accurate dissociation energies, proton affinities, reaction barriers, and vibrational constants [59, 77].

Two approaches have been developed to correct for self-interaction and to impose the correct asymptotic limit for the electron potential energy function. They are the introduction of hybrid functionals containing the exact exchange potential energy, and treatments based on functionals of orbitals obtained from an *optimized effective potential* (or OEP) with the correct asymptotic behavior. Here they are briefly reviewed as they relate to molecular interactions.

Hybrid exchange-correlation functionals containing a mixture of DFT and HF exchange functionals take the form

$$E_{xc} = E_{xc}^{(LSD)} + a_0\left(E_x^{(HF)} - E_x^{(LSD)}\right) + a_x \Delta E_x^{(GGA)} + a_c \Delta E_c^{(GGA)}$$

with three coefficients a chosen to fit sets of known results, and LSD and GGA signifying the local spin-density and generalized gradient approximations. They were introduced and tested for thermodynamically relevant properties like atomization energies, ionization and proton affinity energies [46]. This approach was further developed to avoid fitting parameters [78] and to include a dependence of the functional on the electronic kinetic energy density [48] again free of adjustable parameters. This is potentially more useful, for the calculation of intermolecular forces, than using parametrized functionals with parameters fit to properties of molecules around equilibrium interatomic positions, insofar as in principle parameters are physically expected to change with intermolecular distances.

The calculation of $E_x^{(HF)}$ can be very demanding for extended systems such as solids and biomolecules. An efficient procedure has been developed to decrease the computational demands for $E_x^{(HF)}$. The electron–electron Coulomb interaction is separated for this purpose into short-range and long-range e–e distances as

$$\frac{1}{r} = \frac{1-\mathrm{erf}(\omega r)}{r} + \frac{\mathrm{erf}(\omega r)}{r}$$

in terms of the error function erf, which goes from zero at $r = 0$ to one at large distances with a slope controlled by the parameter ω, so that the first term to the right is short-range (SR) and the the second term is long-range (LR). The semilocal (*sl*) exchange energy can be separated into ωSR and ωLR terms and can be combined with $E_x^{(HF,SR)}$ and a *sl* correlation as $E_{xc} = aE_x^{(HF,SR)} + (1-a)E_x^{(sl,\omega SR)} + E_x^{(sl,\omega LR)} + E_c^{(sl)}$ [47]. Introducing a screened electron–electron Coulomb interaction at short distances, with a functional $E_x^{(HF,SR)}$ containing the HF exchange only at short electronic ranges (SR) where the density is larger, while keeping the DFT exchange at long ranges (LR) leads to computational times shorter than for other hybrid functional treatments, and provided good energy gaps for semiconductors and molecular crystal properties [79].

This mixture of HF and *sl* exchange energies can give good results for bond energies and lengths around equilibrium interatomic distances in solids, but it does not assure a correct $-c_e^2 r^{-1}$ asymptotic form of the KS one-electron $v_x(\vec{r})$ potential. The correct asymptotic form, or electronic long-range corrected (LC) behavior, can be recovered modifying hybrid functionals so that the long-range exchange energy functional is equal to the Hartree–Fock expression, with [49–51]

$$E_{xc}^{(LC-sl)} = E_x^{(HF,\omega LR)} + a\,E_x^{(HF,\omega SR)} + E_x^{(sl,\omega SR)} + E_c^{(sl)}$$

from which the DFT one-electron potential energy $v_{xc}^{(LC-sl)}(\vec{r}) = \delta E_{xc}^{(LC-sl)} / \delta\rho(\vec{r})$ is found to behave correctly at large electronic distances, insofar as it involves asymptotically only the exact HF potential. This is important for treating intermolecular forces when electron transfer is present. The choices of exchange and correlation functionals to be used in the above expression can be made based on needs for accuracy and efficiency, and need not be of semilocal form, with the above parameter a chosen to reproduce known sets of accurate values.

An alternative general treatment is to recast the theory for a system with N electrons introducing a functional $E\left[\left\{\psi_j(\vec{r},\zeta)\right\}\right]$ of a set of j = 1 *to* N spin-orbitals ψ_j describing electrons moving in an effective potential $v(\vec{r})$, with the orbitals given as functionals of $v(\vec{r})$. This can be derived from the relation $\delta E/\delta v(\vec{r}) = 0$, and optimized by energy minimization, to obtain an optimized effective potential $v^{(OEP)}(\vec{r})$. This was done in early work to calculate the OEP for the Hartree–Fock functional, and from it energies for atoms [80, 81]. It has been the subject of recent developments and applications to molecules [82], some of which are covered here as they can be applied to intermolecular forces.

Given an accurate many-electron wavefunction, it is possible to calculate from it $\rho(\vec{r})$ and $v(\vec{r})$ to compare their form with results from DFT functionals. The deviation of $v^{(KS)}(\vec{r})$ from a physically correct one-electron potential $v(\vec{r})$ can be drastic. The OEP procedure instead gives a new $v^{(OEP)}(\vec{r})$, which is more physical and accurate, allowing calculation of more reliable PESs [83].

A general derivation of the equations for a possibly spin-dependent $v^{(OEP)}(\vec{r},\zeta)$ starts by choosing an explicit functional $E^{(OEP)}\left[\left\{\psi_j(\vec{r},\zeta)\right\}\right]$ of spin orbitals, constructed from a wavefunction treatment. A suitable choice is the HF functional $E^{(HF)}\left[\left\{\psi_j(\vec{r},\zeta)\right\}\right] = \sum_j^{occ}\langle j|\hat{h}|j\rangle + \sum_{j<k}^{occ}\langle jk|v^{(Coul)}(1-\hat{t})|jk\rangle$, in the notation of Section 6.2 with the index j = (p, σ) containing both orbital and spin quantum numbers. More accurately, an extension containing electron correlation can be constructed from many-electron perturbation theory, so that variation of the total energy gives a single-electron reference potential $v_s(\vec{r},\zeta) \approx -c_e{}^2 r^{-1}$, with the proper asymptotic form. Using the chain rule for functional differentiation, with the notation $\boldsymbol{x} = (\vec{r},\zeta)$ [84],

$$\frac{\delta E}{\delta v_s(\boldsymbol{x})} = \sum_j \int d\boldsymbol{x}' \left\{ \frac{\delta E^{(OEP)}}{\delta \psi_j(\boldsymbol{x}')} \frac{\delta \psi_j(\boldsymbol{x}')}{\delta v_s(\boldsymbol{x})} + c.c. \right\} = 0$$

and the differential in the second factor to the right can be obtained from the equation

$$\left[-\frac{\hbar^2}{2m}\nabla^2 + v_s(\boldsymbol{x}) \right] \psi_j(\boldsymbol{x}) = \varepsilon_j \psi_j(\boldsymbol{x})$$

when adding a perturbation $\delta v_s(\boldsymbol{x})$, in terms of the equation resolvent (or electronic Green function) $G_j(\boldsymbol{x}, \boldsymbol{x}')$ giving $\delta\psi_j(\boldsymbol{x}') = \int d\boldsymbol{x}\, G_j(\boldsymbol{x}', \boldsymbol{x})\delta v_s(\boldsymbol{x})$. The first differential factor can be expressed as $\delta E^{(OEP)}/\delta\psi_j(\boldsymbol{x}') = \psi_j(\boldsymbol{x}')^* u_j(\boldsymbol{x}')$ with the orbital potential energy $u_j(\boldsymbol{x}')$ known by initial construction. This leads to an integral equation for the potential $v_s^{(OEP)}(\boldsymbol{x})$ which can be solved when this is expanded in a basis set, or is taken to be a functional of the electron density $\rho(\boldsymbol{x})$ as in the HK formalism. In this case, it can be decomposed into its Coulomb electronic, plus external, plus exchange-correlation terms as

$$v_s^{(OEP)}(\boldsymbol{x}) = v_{Coul}(\boldsymbol{x}) + v_{ext}(\boldsymbol{x}) + v_{xc}^{(OEP)}[\rho(\boldsymbol{x})]$$

with the last term appearing in

$$v_{xc}^{(OEP)}(\boldsymbol{x}) = \frac{\delta E_{xc}^{(OEP)}}{\delta\rho(\boldsymbol{x})} = \sum\nolimits_j \int d\boldsymbol{x}' \int d\boldsymbol{x}'' \left\{ \frac{\delta E_{xc}^{(OEP)}}{\delta\psi_j(\boldsymbol{x}')} \frac{\delta\psi_j(\boldsymbol{x}')}{\delta v_s(\boldsymbol{x}'')} \frac{\delta v_s(\boldsymbol{x}'')}{\delta\rho(\boldsymbol{x})} + c.c. \right\}$$

to be calculated from the functional derivative $\chi_s(\boldsymbol{x}, \boldsymbol{x}') = \delta\rho(\boldsymbol{x})/\delta v_s(\boldsymbol{x}')$, which has the physical meaning of an electronic response function or susceptibility [82, 84, 85]. This procedure has been followed to obtain OEP $v_s^{(OEP)}(\boldsymbol{x}; \boldsymbol{Q})$ for many-atom systems as functions of interatomic distances $\boldsymbol{Q}$, and from them accurate potential energy functions for molecules and dimers such as F_2, He_2, $HeBe^{2+}$, Ne_2, and Be_2, involving electronic rearrangement and long-range interaction energies [86, 87].

Another alternative development can be based on functionals that, instead of using the local properties of the electron gas, use the electron pair distribution $g^{(2)}(1, 2)$. This readily incorporates the Fermi hole and the correlation hole and can be made to change with atomic locations to represent an inhomogeneous system and its changes with interatomic distances. The functionals must, however, be constructed so that they are consistent with constrains derived from many-electron antisymmetrized wavefunctions, a nontrivial task [2, 88].

6.5 Spin-Orbit Couplings and Relativistic Effects in Molecular Interactions

6.5.1 Spin-Orbit Couplings

Relativistic quantum mechanics requires going beyond the Coulomb interaction energies among electrons and nuclei, as it takes into account the change of mass of electrons moving with speeds close to c, the speed of light, and the need for an electron equation of motion invariant under the Lorentz transformation of relativistic mechanics. For a single electron, this is contained in the Dirac equation [12] and leads to the appearance of electron spin terms in the

Hamiltonian of a molecular system. A single electron moving in a field of nuclei is described by electron states containing four components in a spinor. Of these, two are large components and two are small ones, and the coupled equations for the four amplitudes can be rearranged to display an effective equation for the large components containing a spin-orbit coupling. In the limit of small electron velocities compared with the speed of light c, this gives a differential equation containing the Pauli matrices for two spatial components with up and down spins, and the form of a one-electron Hamiltonian containing the coupling of the electron spin with its orbital motion in the field of atomic centers. Relativistic mass effects are present in the inner shells of heavy atoms where electron speeds are substantial fractions of c [4]. As a result, the size of inner shells in heavy atoms can shrink, which in turn affects the shape of potential energy surfaces and vibrational frequencies of polyatomic systems. This can be treated with the introduction of relativistic effective core potentials for heavy atoms, which are also needed in studies of electronic properties of solids [20, 89].

Interaction energies involving pairs of electrons in terms of two components spinors are covered by the Breit–Pauli Hamiltonian [90–92], which bring additional terms into the Hamiltonian for many-electron treatments, and in particular the spin-other-orbit couplings between an electron spin at an atomic center and electron orbitals at other centers, as well as coupling of the electronic spins with the rotational angular momentum from moving nuclei. Effects of spin-orbit couplings on intermolecular and interatomic forces were covered some time ago [93–96], and many-electron calculations were done using Dirac–Fock SCFs [97–99], and correlated CI, and MCSCF treatments with effective relativistic atomic pseudopotentials [100] giving potential energy functions of interatomic distance for diatomics containing heavy atoms, such as Au_2 and Pb_2. Modern MBPT, CC, and SAPT methods have also been extended and computationally implemented to include spin-orbit couplings. Spin-orbit effects in computed molecular properties have been reviewed in recent work [4, 101], but much remains to be done on intermolecular forces especially when metal atoms or heavy atoms are involved.

The modern methods described in this chapter are also applicable to molecular systems including spin-orbit couplings, working with two-component spinors. The Hamiltonians must, however, be extended to include spin-dependent terms. In the Breit–Pauli treatment, spin-orbit one- and two-electron Hamiltonian terms are, with $\alpha = e^2/(4\pi\varepsilon_0\hbar c)$ the fine structure constant,

$$\hat{H}_{SO} = \frac{\alpha^2}{2}\left\{\sum\nolimits_{i=1}^{N_{el}}\sum\nolimits_{u=1}^{N_{at}}\frac{Z_u\, c_e^2}{r_{iu}{}^3}\left(\vec{r}_{iu}\times\vec{p}_i\right)\vec{s}_i - \sum\nolimits_{i\neq j}\frac{c_e^2}{r_{ij}^3}\left(\vec{r}_{ij}\times\vec{p}_i\right).\left(\vec{s}_i+2\vec{s}_j\right)\right\}$$

for electrons i and j moving in a framework of nuclei u with position, momentum, and spin $\vec{r}_i$, $\vec{p}_i$, and $\vec{s}_i$, respectively, and where $\vec{r}_{iu} = \vec{r}_i - \vec{r}_u$ and $\vec{r}_{ij} = \vec{r}_i - \vec{r}_j$. In electronic structure calculations, this Hamiltonian is frequently approximated by a sum of one-electron terms with an effective nuclear charges $Z_u^{(eff)}$ obtained from a mean-field approximation and fit to atomic fine structure levels that include atomic spin-orbit energies,

$$\hat{H}_{SO} \cong \frac{\alpha^2}{2} \sum_{i=1}^{N_{el}} \sum_{u=1}^{N_{at}} \frac{Z_u^{(eff)} c_e^2}{r_{iu}^3} \vec{l}_{iu} . \vec{s}_i = \sum_{u=1}^{N_{at}} \hat{H}_{SO}^{(u)}$$

with $\vec{l}_{iu} = \vec{r}_{iu} \times \vec{p}_i$ the orbital angular momentum of an electron in the field of a nucleus, giving a sum of atomic terms. Each electronic term $\vec{l}_{iu} . \vec{s}_i$ does not commute with the orbital and spin components $\hat{l}_{i,\xi}$ or $\hat{s}_{i,\xi}, \xi = x, y, z$, but it does commute with their sum, the total electronic angular momentum components $\hat{j}_{i,\xi} = \hat{l}_{i,\xi} + \hat{s}_{i,\xi}$ for nucleus u so that its quantum numbers (j, m_j) can be used to label the q-th atomic spin-orbital located at nucleus u as $|q\, j\, m_j\rangle_u$, a function of electron position and spin variables $(\vec{r}, \zeta)$. For relativistic effective atomic pseudopotentials $\Delta U_{u,l}^{(REP)}(r)$, with l an atomic orbital quantum number, chosen to depend only on the radial distance to each nucleus u, one finds that $\hat{H}_{SO}^{(u)} = \sum_l \Delta U_{u,l}^{(REP)}(r) \left(|l j m_j\rangle\langle l j m_j| \right)_u$ provides a useful parametrization for potential energy calculations [100]. Reference [102] analyzes how these pseudopotentials are related to the detailed Breit–Pauli Hamiltonian in molecular interactions. Insofar as the pseudopotentials are obtained from expectation values with many-electron wavefunctions, they are found to depend on the atomic conformation of the AB pair, with atomic positions $\mathbf{Q} = (R, \mathbf{Q}')$ appearing as parameters in $\Delta U_{u,l}^{(REP)}(r; R, \mathbf{Q}')$. In the limit of large R, one finds that $\hat{H}_{SO}(R, \mathbf{Q}') \approx \hat{H}_{SO}^{(A)}(\mathbf{Q}_A) + \hat{H}_{SO}^{(B)}(\mathbf{Q}_B)$ as needed.

6.5.2 Spin-Orbit Effects on Interaction Energies

Given the total Hamiltonian including spin terms and dependent on nuclear positions, it is possible to calculate potential energy surfaces using the many-electron treatments previously described. This usually requires using multireference versions applicable to degenerate or near-degenerate reference states insofar as total spin multiplets have the same energy before allowing for spin-orbit couplings. The introduction of spin-orbit couplings splits energy levels of degenerate states into groups with similar energies, which must be treated with comparable accuracy. Atomic states $\langle \vec{r}, \zeta \,|\, q j m_j \rangle = \psi_{qjm_j}(\vec{r}, \zeta)$ can be re-expressed in terms of factorized spin orbitals $\chi_\mu^{(\eta)}(\vec{r})\eta(\zeta)$ to be used in the calculation of one- and two-electron integrals. For pairs AB containing only light atoms in the second and third rows of the periodic table, spin-orbit coupling energies are small compared to electron–electron Coulomb energies, and energy changes due to spin-orbit couplings can be obtained from MR-MBPT perturbation theories as functions of interatomic distances. For molecular pairs with heavier atoms, spin-orbit energies can be comparable or larger than electron–electron energies, and it is necessary to obtain energies from MRCI, MRSCF, or MRCC treatments, working with determinantal wavefunctions constructed from orbitals dependent on electron spin projection quantum numbers.

Treatments of large many-atom systems using density functionals have also been extensively developed and computationally implemented to incorporate spin-orbit coupling energies, particularly for polyatomic systems and solids containing heavy atoms [20, 96, 103, 104].

The dynamics of many-electron states are frequently affected in essential ways by coupling of electron spins with nuclear motions, mediated by spin-orbit couplings. Two examples are intersystem crossings where electrons in singlet states of the pair *AB* rearrange into triplet states due to spin recouplings compensated by orbital and nuclear motion changes, and singlet fission where the pair *AB* in a singlet state is perturbed and undergoes a transition into a state where separate *A* and *B* species are found in triplet states. States of the *AB* pair changing with interatomic distances can be labeled by total electronic angular momentum quantum numbers (J, M_J),which result from addition of all the electronic spins and orbital angular momenta. When spin-orbit coupling is small, it is possible to label the pair's unperturbed states by total electronic spin and orbital quantum numbers (S, M_S) and (L, M_L) in an L–S coupling scheme. When the spin-orbit coupling is large, it is instead necessary to first couple one-electron spin and orbital quantum numbers in a *j-j* scheme [3].

Spin-orbit coupling has an indirect effect on intermolecular potential energies in the *AB* pair. The orbital motion of electrons is affected by the electric fields of the nuclear charges, which vary as the distance between *A* and *B* changes. For a diatomic, this means that the electronic charge distribution has rotational symmetry around the internuclear axis and that electrons have axial angular momenta. Spin-orbit coupling transfers this axial symmetry, as it changes in strength with intermolecular distance, to the spin of the many-electron system, and this leads to spin recoupling as the intermolecular distance changes.

In the L–S scheme, addition of electron spins in Cartesian components $\hat{S}_\xi = \sum_i s_{i,\xi}$, $\xi = x, y, z$, provides the total electronic spin operators $\hat{S}^2$ and $\hat{S}_z$, which commute with the spin-free Hamiltonian and provide good quantum numbers in the absence of spin-orbit couplings. When spin terms are included in the Hamiltonian, the total spin components must be added to total orbital momenta $\hat{L}_\xi$ to obtain the electronic angular momenta $\hat{J}_\xi = \hat{S}_\xi + \hat{L}_\xi$. One must in addition account for the coupling of the electronic angular momenta to the rotational momenta of the nuclei, $\vec{L}_{nu}$, also called $\vec{R}$ in the literature of molecular spectroscopy, giving a total (electronic plus nuclear) angular momentum $\vec{J}_{tot} = \vec{L} + \vec{S} + \vec{L}_{nu}$. In an interaction pair $A + B$, the angular momenta $\vec{J}_{tot}$, $\vec{L}$ and $\vec{S}$ have values along the intermolecular axis labeled, in units of $\hbar$, as Ω, Λ, and Σ, respectively, with $\Omega = \Lambda + \Sigma$. These provide quantum numbers to identify PESs as atomic distances vary. Whether they are good quantum numbers at a given intermolecular distance R depends on the relative magnitude of the spin-orbit coupling energy compared to the electronic rotational energy around the intermolecular axis. The relative values depend on the intermolecular distance, with the electronic axial rotational energy being larger at small distances and disappearing at large distances.

Recouplings in polyatomic systems, changing these quantum numbers and spin multiplicity and the shape of intermolecular PESs, can be analyzed similarly to how it is done for diatomics, where different cases have been classified as Hund's cases: (i) where the electronic angular momentum rotates around the intermolecular axis and spin-orbit coupling is large enough to force the spin to also rotate axially, so that both Λ and Σ are good quantum numbers; (ii) where spin-orbit coupling is weak and spin is not coupled to the axial field; (iii) where spin-orbit coupling is very strong and $\vec{J} = \vec{L} + \vec{S}$ rotates axially to give a good Ω; and (iv) when it is very weak [3, 105, 106]. Recouplings and interacting PESs can therefore be expected as the intermolecular distance R changes.

References

1 Levine, I.N. (2000). *Quantum Chemistry*, 5e. New York: Prentice-Hall.

2 McWeeney, R. (1989). *Methods of Molecular Quantum Mechanics*, 2e. London, England: Academic Press.

3 Atkins, P.W. and Friedman, R.S. (1997). *Molecular Quantum Mechanics*. Oxford, England: Oxford University Press.

4 Jensen, F. (2001). *Introduction to Computational Chemistry*. New York: Wiley.

5 Szabo, A. and Ostlund, N.S. (1982). *Modern Quantum Chemistry*. New York: Macmillan.

6 Woodward, R.B. and Hoffmann, R. (1970). *The Conservation of Orbital Symmetry*. New York: Academic Press.

7 Pearson, R.G. (1976). *Symmetry Rules for Chemical Reactions*. New York: Wiley.

8 Turro, N.J. (1978). *Modern Molecular Photochemistry*. Menlo Park, CA: Benjamin-Cummings.

9 Lowdin, P.O. (1962). Normal constants of motion in quantum mechanics treated by projection techniques. *Rev. Mod. Phys.* 34: 520.

10 Lowdin, P.O. (1964). Angular momentum wavefunctions constructed by projection operators. *Rev. Mod. Phys.* 36: 966.

11 Lowdin, P.O. (1962). Studies in perturbation theory IV. Solution of eigenvalue problems by a projection operator formalism. *J. Math. Phys.* 3: 969.

12 Messiah, A. (1962). *Quantum Mechanics*, vol. 2. Amsterdam: North-Holland.

13 Micha, D.A. (2017). Quantum partitioning methods for few-atom and many-atom dynamics. In: *Advances in Quantum Chemistry*, vol. 74, 107. London (England): Elsevier.
14 Lowdin, P.O. (1959). Correlation problem in many-electron quantum mechanics. I. Different approaches and some current ideas. In: *Advances in Chemical Physics*, vol. II, 207. New York: Interscience.
15 Hehre, W.J., Radom, L., Schleyer, P.V., and Pople, J.A. (1986). *AB INITIO Molecular Orbital Theory*. New York: Wiley.
16 Claudino, D., Gargado, R., and Bartlett, R.J. (2016). Coupled-cluster based basis sets for valence correlation calculations (Erratum in vol. 145, p. 019901-1). *J. Chem. Phys.* 144: 104106–104101.
17 Claudino, D. and Bartlett, R.J. (2018). Coupled-cluster based basis sets for valence correlation calculations. New primitives and frozen atomic natural orbitals. *J. Chem. Phys.* 149: 064105–064101.
18 Chalasinski, G. and Szszesniak, M.M. (2000). State of the art and challenges of the ab initio theory of intermolecular interactions. *Chem. Rev.* 100: 4227.
19 Ashcroft, N.W. and Mermin, N.D. (1976). *Solid State Physics*. London, England: Thomson.
20 Martin, R.M. (2004). *Electronic Structure: Basic Theory and Practical Methods*. Cambridge, England: Cambridge University Press.
21 Schaefer, H.F. (1972). *The Electronic Structure of Atoms and Molecules*. Reading, MA, USA: Addison-Wesley.
22 Schaefer, H.F. (1977). *Methods of Electronic Structure Theory. Volume 3: Modern Theoretical Chemistry* (eds. W.H. Miller and H.F. Schaefer). New York: Plenum.
23 Roos, B.O. (1987). The complete active space self-consistent field method and its applications in electronic structure calculations. In: *Advances in Chemical Physics*, vol. 69 (ed. K.P. Lawley), 399. New York: Wiley.
24 Bruna, P.J. and Peyerimhoff, S.D. (1987). Excited-state potentials. In: *Advances in Chemical Physics, Part I*, vol. 67 (ed. K.P. Lawley), 1. New York: Wiley.
25 Shavitt, I. and Bartlett, R.J. (2009). *Many-Body Methods in Chemistry and Physics*. Cambridge, England: Cambridge University Press.
26 Bartlett, R.J. (1981). Many-body perturbation theory and coupled cluster theory for electronic correlation in molecules. *Annu. Rev. Phys. Chem.* 32: 359.
27 Bartlett, R.J. and Musial, M. (2007). Coupled-cluster theory in quantum chemistry. *Rev. Mod. Phys.* 79: 291.
28 Stanton, J.F. and Gauss, J. (2003). A discussion of some problems associated with the quantum mechanical treatment of open-shell molecules. In: *Advances in Chemical Physics*, vol. 125, 101. Hoboken, NJ: Wiley-Interscience.
29 Hirata, S., Fan, P.D., Auer, A.A. et al. (2004). Combined coupled-cluster and many-body perturbation theory. *J. Chem. Phys.* 121: 12197.
30 Krylov, A.I. (2008). *Equation-of -Motion Coupled-Cluster Methods for Open-Shell and Electronically Excited Species* (*Annu. Rev. Phys. Chem*, vol. 59), 433. Palo Alto, CA: Annual Reviews Inc.

31 Bartlett, R.J. (2012). Coupled-cluster theory and its equation-of-motion extensions. *WIREs Comput. Mol. Sci.* 2: 126.
32 Jeziorski, B. and Monkhorst, H.J. (1981). Coupled cluster method for multideterminantal reference states. *Phys. Rev. A* 24: 1668.
33 Goddard, W.A.I. and Harding, L.B. (1978). generalized valence bond description of bonding in low-lyingstates of molecules. *Annu. Rev. Phys. Chem.* 29: 363.
34 Dunning, T.H., Xu, L.T., Takeshita, T.Y., and Lindquist, B.A. (2016). Insight into the electronic structure of molecules from generalized valence bond theory. *J. Phys. Chem. A* 120: 1763.
35 Jeziorski, B., Moszynski, R., Ratkiewitcz, A. et al. (1993). *SAPT: A Program for Many-Body Symmetry-Adapted Perturbation Theory Calculations of Intermolecular Interaction Energies* (METTEC-94, vol. B) (ed. E. Clementi), 79. Cagliari, Italy: STEF.
36 Jeziorski, B., Moszynski, R., and Szalewicz, K. (1994). Perturbation theory approach to intermolecular potential energy surfaces of van der Waals complexes. *Chem. Rev.* 94: 1887.
37 Casimir, H.B.G. and Polder, D. (1948). The influence of retardation on the London-van der Waals forces. *Phys. Rev.* 73: 360.
38 Szalewicz, K. (2012). Symmetry-adapted perturbation theoryof intermolecular forces. *WIREs Comput. Mol. Sci.* 2: 254.
39 Misquitta, A.J., Podeszwa, R., Jeziorki, B., and Szalewicz, K. (2005). Intermolecular potentials based on SAPT with dispersion energies from time-dependent density functional calculations. *J. Chem. Phys.* 123: 214103–214101.
40 Lotrich, V.F. and Szalewicz, K. (1997). Symmetry-adapted perturbation theory of three-body nonadditivity of intermolecular interaction energy. *J. Chem. Phys.* 106: 9668.
41 Podeszwa, R. and Szalewicz, K. (2007). Three-body symmetry-adapted perturbation theory based on the Kohn-Sham description of the monomers. *J. Chem. Phys.* 19: 194101.
42 Hohenberg, P. and Kohn, W. (1964). Inhomogeneous electron gas. *Phys. Rev. B* 136: 864.
43 Parr, R.G. and Yang, W. (1989). *Density Functional Theory of Atoms and Molecules*. Oxford, England: Oxford University Press.
44 Trickey, S.B. (ed.) (1990). *Advances in Quantum Chemistry. Volume 21: Density Functional Theory of Many-Fermion Systems*. San Diego, CA, USA: Academic Press.
45 Lundqvist, S. and March, N.H. (eds.) (1983). *Theory of the Inhomogeneous Electron Gas*. New York: Plenum Press.
46 Becke, A.D. (1993). Density-functional theormochemistry III. The role of exact exchange. *J. Chem. Phys.* 98: 5648.
47 Heyd, J., Scuseria, G.E., and Ernzerhof, M. (2003). Hybrid functionals based on a screened Coulomb potential. *J. Chem. Phys.* 118: 8207.

48 Perdew, J.P., Tao, J., Staroverov, V.N., and Scuseria, G.E. (2004). Meta-generalized gradient approximation: explanation of a realistic nonempirical density functional. *J. Chem. Phys.* 120: 6898.
49 Iikura, H., Tsuneda, T., and Yanai, T.H.K. (2001). A long-range corrected scheme for generalized-gradient approximation exchange functionals. *J. Chem. Phys.* 115: 3540.
50 Vydrov, O.A. and Scuseria, G.E. (2006). Assessment of a long-range corrected hybrid functional. *J. Chem. Phys.* 125: 234109.
51 Chai, J.-D. and Head-Gordon, M. (2008). Systematic optimization of long-range corrected hybrid density functionals. *J. Chem. Phys.* 128: 084106.
52 Gunnarsson, O. and Lundqvist, B.I. (1976). Exchange and correlation in atoms, molecules, and solids by the spin density functional formalism. *Phys. Rev. B* 13: 4274.
53 Jones, R.O. and Gunnarsson, O. (1989). The density functional formalism, its applications and prospects. *Rev. Mod. Phys.* 61: 689.
54 Grimme, S., Hansen, A., and Brandenburg, J.G.C. (2016). Dispersion corrected mean-field electronic structure methods. *Chem. Rev.* 116: 5105.
55 Hermann, J., DiStasio, R.A.J., and Tkatchenko, A. (2017). First-principles models for van der Waals interactionsin molecules and materials: concepts, theory, and applications. *Chem. Rev.* 117: 4714.
56 Sato, T. and Nakai, H. (2009). Density functional method including weak interactions: dispersion coefficients based on the local response approximation. *J. Chem. Phys.* 131: 224104.
57 Sato, T. and Nakai, H. (2010). Local response dispersion method. II. Generalized multicenter interactions. *J. Chem. Phys.* 133: 194101.
58 Podeszwa, R. and Szalewicz, K. (2012). Communication: density functional theory overcomes the failure of predicting intermolecular interaction energies. *J. Chem. Phys.* 136: 161102–161101.
59 Zhao, Y. and Truhlar, D.G. (2008). The M06 suite of energy functionals systematic testing of four M06 functionals and 12 other functionals. *Theor. Chem. Accounts* 120: 215.
60 von Lilienfeld, O.A., Tavernelli, I., Rothlisberger, U., and Sebastiani, D. (2005). Performance of atom-centered potentials for weakly bonded systems using density functional theory. *Phys. Rev. B* 71: 195119.
61 Peverati, R. and Truhlar, D. (2012). Exchange-correlation functionals with good accuracy for both structural and energetic properties while depending only on the density and its gradient. *J. Chem. Theory Comput.* 8: 2310.
62 Peverati, R. and Truhlar, D.G. (2012). Screened exchange density functionals with broad accuracy for chemistry and solid state physics. *Phys. Chem. Chem. Phys.* 14: 16187.
63 Langreth, D.C., Lundqvist, B.I., Chakarova-Kaeck, S.D. et al. (2009). A density functional for sparse matter. *J. Phys. Condens. Matter* 21: 084203.

64 Bortolani, V., March, N.H., and Tosi, M.P. (eds.) (1990). *Interaction of Atoms and Molecules with Solid Surfaces*. New York: Plenum Press.
65 Vydrov, O.A. and Van Voohis, T. (2010). Nonlocal van der Waals density functional: the simpler the better. *J. Chem. Phys.* 133: 244103–244101.
66 Nozieres, P. and Pines, D. (1999). *The Theory of Quantum Liquids*. Cambridge, MA: Perseus.
67 Dobson, J. F. and T. Gould, "Calculation of dispersion energies," *J. Phys. Condens. Matter*, vol. 24, p. 073201, 2012.
68 Dobson, J.F. (2012). Dispersion (van der Waals) forces and TDDFT. In: *Fundamentals of Time-Dependent Density Functional Theory* (eds. M.A.L. Marques, N.T. Maitra, F.M.S. Nogueira, et al.), 417. Berlin: Springer-Verlag.
69 Berland, K., Cooper, V.R., Lee, K. et al. (2015). van der Waals forces in density functional theory: a review of the vdW-DF method. *Rep. Prog. Phys.* 78: 066501.
70 Sun, J., Ruzsinszky, A., and Perdew, J. (2015). Strongly constrained and appropriately normed semilocal density functional. *Phys. Rev. Lett.* 115: 036402.
71 Staroverov, V.N., Scuseria, G.E., Tao, J., and Perdew, J.P. (2003). Comparative assesment of a new nonempirical density functional: molecules and hydrogen bonded complexes. *J. Chem. Phys.* 119: 12129.
72 Xu, X. and Goddard, W.A. (2004). Bonding properties of the water dimer: a comparative study of density functional theories. *J. Phys. Chem. A* 108: 2305.
73 Taylor, D.E., Angyan, J.G., Galli, G. et al. (2016). Blind test of density functional based methods on intermolecular interaction energies. *J. Chem. Phys.* 145: 124105–124101.
74 Mermin, N.D. (1965). Thermal properties of the inhomogeneous electron gas. *Phys. Rev.* 137: A1441.
75 Gross, E.K.U., Oliveira, L.N., and Kohn, W. (1988). Density functional theory for ensembles of fractionally occupied states. I. Basic formalism. *Phys. Rev. A* 37: 2809.
76 Cohen, M.H. and Wassermann, A. (2007). On the foundations of chemical reactivity theory. *J. Phys. Chem. A* 111: 2229.
77 Zhao, Y. and Truhlar, D.G. (2008). Density functionals with broad applicability in chemistry. *Acc. Chem. Res.* 41: 157.
78 Perdew, J.P., Burke, K., and Ernzenhof, M. (1996). Generalized gradient approximation made simple. *Phys. Rev. Lett.* 77: 3865.
79 Heyd, J. and Scuseria, G.E. (2004). Efficient hybrid density functional calculations in solids: assessment of the Heyd – Scuseria – Ernzerhof screened Coulomb hybrid functional. *J. Chem. Phys.* 121: 1187.
80 Talman, J.D. and Shadwick, W.F. (1976). Optimized effective atomic central potentials. *Phys. Rev. A* 14: 36.
81 Krieger, J.B., Li, Y., and Iafrate, G.J. (1995). Recent developments in Kohn-Sham theory for orbital dependent exchange-correlation energy functionals. In: *Density Functional Theory* (eds. E.K.U. Gross and R.M. Dreizler), 191. New York: Plenum Press.

82 Kuemmel, S. and Kronik, L. (2008). Orbital-dependent density functionals: theory and applications. *Rev. Mod. Phys.* 80: 3.
83 Bartlett, R.J. (2010). Ab initio DFT and its role in electronic structure theory. *Mol. Phys.* 108: 3299.
84 Grabowski, I., Hirata, S., Ivanov, S., and Bartlett, R.J. (2002). Ab initio density functional theory: OEP-MBPT(2). A new orbital dependent correlation functional. *J. Chem. Phys.* 116: 4415.
85 Heaton-Burgess, T. and Yang, W. (2008). Optimized effective potentials from arbitrary basis sets. *J. Chem. Phys.* 129: 194102–194101.
86 Bartlett, R.J., Lotrich, V.F., and Schweigert, I.V. (2005). Ab initio density functional theory: the best of both worlds? *J. Chem. Phys.* 123: 062205–062201.
87 Lotrich, V.F., Bartlett, R.J., and Grabowski, I. (2005). Intermolecular potential energy surfaces of weakly bound dimers computed from ab initio density functional theory: the right answer for the right reason. *Chem. Phys. Lett.* 405: 43.
88 Ayers, P.W. and Levy, M. (2005). Using the Kohn–Sham formalism in pair density-functional theories. *Chem. Phys. Lett.* 415: 211.
89 Hay, P.J. and Martin, R.L. (1998). Theoretical studies of the structures and vibrational frequencies of actinide compounds using relativistic effective core potentials. *J. Chem. Phys.* 109: 3875.
90 Bethe, H.A. and Salpeter, E.E. (1957). *Quantum Mechanics of One- and Two-electron Atoms*. Berlin: Springer-verlag.
91 Hirschfelder, J.O., Curtis, C.F., and Byron Bird, R. (1954). *Molecular Theory of Gases and Liquids*. New York: Wiley.
92 Langhoff, S.R. and Kern, W.C. (1977). Molecular fine structure and spectroscopy by ab initio methods. In: *Applications of Modern Theoretical Chemistry* (ed. H.F. Schaefer), 381. New York: Plenum Press.
93 Hirschfelder, J.O. and Meath, W.J. (1967). The nature of intermolecular forces. *Adv. Chem. Phys.* 12: 3.
94 Chang, T.Y. (1967). Moderately long range intermolecular forces. *Rev. Mod. Phys.* 39: 911.
95 Richards, W.G., Trivedi, H.P., and Cooper, D.L. (1981). *Spin-Orbit Coupling in Molecules*. Oxford, England: Oxford University Press.
96 Marian, C.M. (2001). Spin-orbit coupling in molecules. In: *Reviews in Computational Chemistry*, vol. 17 (eds. K.B. Lipkowitz and D.B. Boyd). Wiley-VHC.
97 Pyykko, P. (1978). Relativistic quantum chemistry. In: *Advanced Quantum Chemistry*, vol. 11, 353. New York: Academic Press.
98 Pyykko, P. (1988). Relativistic effects in structural chemistry. *Chem. Rev.* 88: 563.
99 Pisani, L. and Clementi, E. (1994). Relativistic Dirac-Fock calculations for closed shell molecules. *J. Comput. Chem.* 15: 466.

100 Balasubramanian, K. and Pitzer, K.S. (1987). Relativistic quantum chemistry. In: *Advances in Chemical Physics. Volume 67: Ab Initio Methods in Quantum Chemistry Part I* (ed. K.P. Lawley), 287. New York: Wiley.

101 Fedorov, D.G., Koseki, S., Schmidt, M.W., and Gordon, M.S. (2003). Spin-orbit coupling in molecules: chemistry beyond the adiabatic approximation. *Int. Rev. Phys. Chem.* 22: 551.

102 Fedorov, D.G. and Gordon, M.S. (2000). A study of the relative importance of one and two-electron contributions to spin–orbit. *J. Chem. Phys.* 112: 5611.

103 Hafner, J. (2008). Ab initio simulation of materials using VASP: density functional theory and beyond. *J. Comput. Chem.* 29: 2044.

104 Kleinschmidt, M., Tatchen, J., and Marian, C.M. (2002). Spin-orbit coupling of DFT/MRCI wavefunctions: method, test calculations and applications to thiophene. *J. Comput. Chem.* 23: 824.

105 Herzberg, G. (1950). *Molecular Spectra and Molecular Structure I. Spectra of Diatomic Molecules,* 2e. Princeton NJ: Van Nostrand.

106 Herzberg, G. (1966). *Molecular Spectra and Molecular Structure III. Polyatomic Molecules.* New York: Van Nostrand Reinhold.

7

Interactions Between Two Many-Atom Systems

CONTENTS

7.1 Long-range Interactions of Large Molecules

7.1.1 Interactions from Charge Density Operators

The interactions of atoms and molecules with an extended many-atom system such as a solid surface, atomic cluster, molecular crystal, polymer, or a biomolecule can be described by means of electronic charge densities and polarization

Molecular Interactions: Concepts and Methods, First Edition. David A. Micha.

densities per unit volume and Hamiltonian operators involving space integrals over charge density operators [1–7]. This bypasses the need for expansions of interaction energy operators in terms of electrical multipoles of high order. Long-range interaction energies follow from perturbation theory or from many-body treatments in terms of response functions.

The form of the interaction energy operator between two species A and B, now considered to be extended systems containing many atoms, has been given in Sections 2.1 and 2.4 and in Section 3.2 in terms of charge density operators $\hat{c}_A(\vec{r})$ and $\hat{c}_B(\vec{r})$, defined as functions of positions over the whole region of interactions. The two species are given by the locations of the nuclei in each one, labeled by the index a for A and by b for B. They are then defined by the collection $\boldsymbol{Q}_A = \{\vec{r}_a\}$ of atomic position vectors for A and similarly $\boldsymbol{Q}_B$ for B, which are parameters in the density operators. Those density expressions are used here in applications to extended systems such as clusters, polymers, and solid surfaces interacting with molecules or among themselves. The expectation values of the density operators, for given electronic states of A and B, occupy regions of space with a variety of shapes. Some may show cavities in which a species may be located, so that a distance R between centers of mass may not be meaningful as a separation distance. Two examples are shown in Figure 7.1.

We consider two species A and B with density boundaries sufficiently far away so that electron exchange is improbable and does not affect the properties of the system and can be omitted. In this case, electrons $i = 1$ to N_A can be assigned to species A, and electrons $j = N_A + 1$ to $N_A + N_B$ are assigned to species B. Their relative distance r_{ij} remains larger than some distance d and their Coulomb interaction remains finite, insofar electrons i and j are well separated during the interaction of the two many-electron systems. This treatment, however, must be reconsidered when electron exchange is relevant.

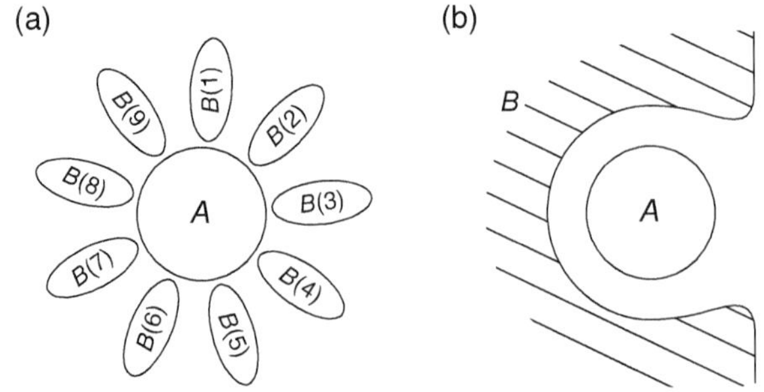

Figure 7.1 (a) Molecule A in a solvent B containing molecules $B(n)$; (b) A in a cavity of a large molecule B.

It is possible to avoid expansions of interaction energies in terms of multipoles, by instead using the charge density operator of all the charges C_I, $I = a, i$, in species A, given as functions of space positions $\vec{r}$ by

$$\hat{c}_A(\vec{r}) = \sum_I C_I \delta\left(\vec{r} - \hat{\vec{r}}_I\right) = \hat{c}_A^{(nu)}(\vec{r}) + \hat{c}_A^{(el)}(\vec{r})$$

$$\hat{c}_A^{(nu)}(\vec{r}) = \sum_a C_a \delta\left(\vec{r} - \hat{\vec{r}}_a\right), \hat{c}_A^{(el)}(\vec{r}) = C_e \sum_i \delta\left(\vec{r} - \hat{\vec{r}}_i\right)$$

a sum of nuclear and electronic charge densities, and similarly for species B. The nuclear charge density $\hat{c}_A^{(nu)}$ is, for fixed nuclei, simply a function of nuclear positions, but the electronic charge density $\hat{c}_A^{(el)}$ is a one-electron operator in the space of many-electron wavefunctions.

The Hamiltonian of species A and B can be written in terms of nuclear and electronic charge densities as

$$\hat{H}_A = \hat{K}_A^{(e)} + \hat{H}_A^{(en)} + \hat{H}_A^{(ee)}$$

$$\hat{K}_A^{(e)} = -\frac{\hbar^2}{2m_e}\sum_i \nabla_i^2, \hat{H}_A^{(en)} = \frac{1}{4\pi\varepsilon_0}\int d^3r \int d^3r' \frac{\hat{c}_A^{(el)}(\vec{r})\, \hat{c}_A^{(nu)}(\vec{r}')}{|\vec{r} - \vec{r}'|}$$

$$\hat{H}_A^{(ee)} = \frac{1}{4\pi\varepsilon_0}\int d^3r \int d^3r' \frac{\hat{c}_A^{(el)}(\vec{r})\, \hat{c}_A^{(el)}(\vec{r}') - \delta(\vec{r} - \vec{r}')\, \hat{c}_A^{(el)}(\vec{r})}{|\vec{r} - \vec{r}'|}$$

where the form of the electron–electron Coulomb energy accounts for the absence of interaction of each electron with itself. A similar expression applies to species B.

The Coulomb interaction energy operator for all charges in A interacting with all those in B, but without electron exchange between them, is

$$\hat{H}_{AB}^{(Coul)}(\mathbf{Q}_A, \mathbf{Q}_B) = (4\pi\varepsilon_0)^{-1}\int d^3r \int d^3s\, \hat{c}_A(\vec{r}; \mathbf{Q}_A)\, |\vec{r} - \vec{s}|^{-1}\, \hat{c}_B(\vec{s}; \mathbf{Q}_B)$$

which can be expanded as $\hat{H}_{AB}^{(Coul)} = \hat{H}_{AB}^{(nn)} + \hat{H}_{AB}^{(n,e)} + \hat{H}_{AB}^{(e,n)} + \hat{H}_{AB}^{(ee)}$ to display types of interactions of nuclear and electronic charges. Insofar the nuclear charge distributions are dependent on nuclear positions for fixed nuclei, and the electronic distributions are operators dependent on the electronic position operators, the second and third terms are one-electron operators dependent on all nuclear positions while the last term is a two-electron operator.

In more detail, $\hat{H}_{AB}^{(nn)}$ is just the Coulomb energy of interacting nuclear charge distributions for A and B,

$$H_{AB}^{(nn)}(\mathbf{Q}_A, \mathbf{Q}_B) = (4\pi\varepsilon_0)^{-1}\int d^3r \int d^3s\, c_A^{(nu)}(\vec{r}; \mathbf{Q}_A)\, |\vec{r} - \vec{s}|^{-1}\, c_B^{(nu)}(\vec{s}; \mathbf{Q}_B)$$

which can alternatively be written as the Coulomb energy of nuclear point charges at their positions.

The electronic-nuclear interaction Hamiltonian energy is

$$\hat{H}_{AB}^{(ne)} = \int d^3r\, c_A^{(nu)}\left(\vec{r};\mathbf{Q}_A\right)\hat{\varphi}_B\left(\vec{r}\right) + \int d^3s\, \hat{\varphi}_A\left(\vec{s}\right) c_B^{(nu)}\left(\vec{s};\mathbf{Q}_B\right)$$

$$\hat{\varphi}_B\left(\vec{r}\right) = \frac{1}{4\pi\varepsilon_0}\int d^3s\, \frac{\hat{c}_B^{(el)}\left(\vec{s}\right)}{\left|\vec{r}-\vec{s}\right|},\ \hat{\varphi}_A\left(\vec{s}\right) = \frac{1}{4\pi\varepsilon_0}\int d^3r\, \frac{\hat{c}_A^{(el)}\left(\vec{r}\right)}{\left|\vec{r}-\vec{s}\right|}$$

where $\hat{\varphi}_B\left(\vec{r}\right)$ is the electric field created by electronic charges of B at the location of the nuclear charges of A, and similarly for $\hat{\varphi}_A\left(\vec{s}\right)$. This Hamiltonian is a one-electron operator and its matrix elements can be obtained from one-electron density operators $\hat{\rho}_A^{(1)}$ and $\hat{\rho}_B^{(1)}$.

The electron–electron interaction is

$$\hat{H}_{AB}^{(ee)} = \left(4\pi\varepsilon_0\right)^{-1}\int d^3r \int d^3s\, \hat{c}_A^{(el)}\left(\vec{r}\right)\left|\vec{r}-\vec{s}\right|^{-1}\hat{c}_B^{(el)}\left(\vec{s}\right)$$

and does not depend explicitly on the nuclear positions. This Hamiltonian is a two-electron operator, and calculation of its matrix elements between many-electron states of A and B involve the two-electron density operator $\hat{\rho}_{AB}^{(2)} = \hat{\rho}_A^{(1)}\hat{\rho}_B^{(1)}$, derived from the separate many-electron states of A and B, insofar they do not exchange electrons. The matrix elements do change with nuclear distances.

These one- and two-electron expressions appear in the total Hamiltonian in a form useful for calculations involving both ground and excited electronic states for fixed atomic positions.

7.1.2 Electrostatic, Induction, and Dispersion Interactions

Long-range electrostatic, induction, and dispersion energies are obtained as shown in Chapter 3 from first- and second-order perturbation theory, using basis sets $|J_A\rangle$ and $|K_B\rangle$ for species A and B. The electrostatic interaction energy of A and B in their ground electronic states is

$$E_{els}^{(AB)}\left(\mathbf{Q}_A,\mathbf{Q}_B\right) = \left\langle 0_A, 0_B \mid \hat{H}_{AB}^{(Coul)} \mid 0_A, 0_B \right\rangle$$

Writing the interaction Hamiltonian operator in terms of charge density operators shows that the electrostatic energy contains their expectation values in the ground state

$$c_0^{(A)}\left(\vec{r}\right) = \left\langle 0_A \mid \hat{c}_A\left(\vec{r}\right) \mid 0_A \right\rangle = \sum\nolimits_a C_a\delta\left(\vec{r}-\hat{\vec{r}}_a\right) + C_e\rho_0^{(A)}\left(\vec{r}\right)$$

where $\rho_0^{(A)}(\vec{r})$ is the ground state electronic density of A, and similarly with $c_0^{(B)}(\vec{s})$ for B. Its form is

$$E_{els}^{(AB)}(\mathbf{Q}_A,\mathbf{Q}_B) = (4\pi\varepsilon_0)^{-1}\int d^3r\int d^3s\, c_0^{(A)}(\vec{r};\mathbf{Q}_A)\left|\vec{r}-\vec{s}\right|^{-1} c_0^{(B)}(\vec{s};\mathbf{Q}_B)$$

and can be further decomposed into Coulomb interaction energies of nuclei and electrons in A and B, $E_{els}^{(AB)} = E_{0,0}^{(nn)} + E_{0,0}^{(ne)} + E_{0,0}^{(ee)}$.

Induction and dispersion energies given by

$$E_{ind}^{(AB)}(\mathbf{Q}_A,\mathbf{Q}_B) = -\sum\nolimits_{J\neq 0}\frac{\left|\left\langle 0_A,0_B \mid \hat{H}_{AB}^{(Coul)} \mid J_A,0_B\right\rangle\right|^2}{E_J^{(A)} - E_0^{(A)}} - \sum\nolimits_{K\neq 0}\frac{\left|\left\langle 0_A,0_B \mid \hat{H}_{AB}^{(Coul)} \mid 0_A,K_B\right\rangle\right|^2}{E_K^{(B)} - E_0^{(B)}}$$

$$E_{dsp}^{(AB)}(\mathbf{Q}_A,\mathbf{Q}_B) = -\sum\nolimits_{J\neq 0}\sum\nolimits_{K\neq 0}\frac{\left|\left\langle 0_A,0_B \mid \hat{H}_{AB}^{(Coul)} \mid J_A,K_B\right\rangle\right|^2}{E_J^{(A)} - E_0^{(A)} + E_K^{(B)} - E_0^{(B)}}$$

contain transition values $c_{0J}^{(A)}(\vec{r}) = \langle 0_A \mid \hat{c}_A(\vec{r}) \mid J_A\rangle$ of the charge density operators for A, and similarly for B, with

$$c_{0J}^{(A)}(\vec{r}) = c_{nu}^{(A)}(\vec{r})\delta_{0J} + C_e\rho_{0J}^{(A)}(\vec{r}),\; c_{0K}^{(B)}(\vec{s}) = c_{nu}^{(B)}(\vec{s})\delta_{0K} + C_e\rho_{0K}^{(B)}(\vec{s})$$

where contributions from nuclei and electrons have been separated. Here $C_e\rho_{0J}^{(A)}(\vec{r}) = \left\langle 0_A \mid \hat{c}_A^{(el)}(\vec{r}) \mid J_A\right\rangle$ is an electronic charge excitation, and similarly for B, obtained from their many-electron states. The second-order energies $E_{ind}^{(AB)} = E_{0,0}^{(ind)}(AB^*) + E_{0,0}^{(ind)}(A^*B)$ and $E_{dsp}^{(AB)} = E_{0,0}^{(dsp)}(A^*B^*)$ involve integrals like

$$\left\langle 0_A,0_B \mid \hat{H}_{AB}^{(Coul)} \mid J_A,0_B\right\rangle = (4\pi\varepsilon_0)^{-1}\int_A d^3r\int_B d^3s C_e\rho_{0J}^{(A)}(\vec{r})\left|\vec{r}-\vec{s}\right|^{-1} c_0^{(B)}(\vec{s})$$

for the interaction of charges of B with electronic density fluctuations in A, appearing in the induction energy, and integrals

$$\left\langle 0_A,0_B \mid \hat{H}_{AB}^{(Coul)} \mid J_A,K_B\right\rangle = (4\pi\varepsilon_0)^{-1}\int_A d^3r\int_B d^3s C_e\rho_{0J}^{(A)}(\vec{r})\left|\vec{r}-\vec{s}\right|^{-1} C_e\rho_{0K}^{(B)}(\vec{s})$$

for excitation density-excitation density couplings in the dispersion energy. These integrals appear in summations over usually unknown electronic excited states of A^* and B^*, and it is of practical interest to find ways to calculate the perturbation corrections while avoiding sums over excited states. Also, large many-atom systems may involve degenerate ground states all of the same energy, requiring modified treatments for perturbation of degenerate states. In addition, large systems may have vanishing or very small excitation energies $E_J^{(A)} - E_0^{(A)}$ and $E_K^{(B)} - E_0^{(B)}$, which lead to large perturbation energy corrections and require alternative treatment. These problems can sometimes be avoided introducing static and dynamical susceptibilities functions of excitation frequencies. Instead of using the molecular excited states, the summations in second-order perturbation terms can best be calculated as integrals over their frequencies with susceptibilities obtained from a variety of electronic basis sets.

7.1.3 Population Analyses of Charge and Polarization Densities

Given the above energy expressions, calculation of the energies must be done introducing many-electron states $\Phi_{J_A}(1,2,\ldots,N_A) = \langle 1,2,\ldots,N_A \mid J_A \rangle$ for A and $\Phi_{K_B}(N_A+1, N_A+2,\ldots,N_A+N_B)$ for B. Insofar no electron exchange or transfer needs consideration, the combined states for the AB pair are given by $|J_A, K_B\rangle = |J_A\rangle \otimes |K_B\rangle$.

Transition density values $c_{JJ'}^{(A)}(\vec{r}) = \langle J_A \mid \hat{c}_A(\vec{r}) \mid J_A' \rangle$ of the charge density operators for A, and similarly for B, are

$$c_{JJ'}^{(A)}(\vec{r}) = c_{nu}^{(A)}(\vec{r})\delta_{JJ'} + C_e\rho_{el,JJ'}^{(A)}(\vec{r}),\ c_{KK'}^{(B)}(\vec{s}) = c_{nu}^{(B)}(\vec{s})\delta_{KK'} + C_e\rho_{el,KK'}^{(B)}(\vec{s})$$

where $C_e\rho_{el,JJ'}^{(A)}(\vec{r}) = \left\langle J_A \mid \hat{c}_A^{(el)}(\vec{r}) \mid J_A' \right\rangle$ is an electronic charge density fluctuation, and similarly for B, obtained from their many-electron states. Transition densities $\rho_{el,JJ'}^{(A)}$ can be obtained from one-electron transition density functions $\rho_{JJ'}^{(1)}(1,1')$, for states J and J' of A with N_A electrons. They are defined as an extension of the single-state one-electron density functions in Chapter 6, by means of

$$\rho_{JJ'}^{(1)}(1;1') = N_A \int d(2)\cdots d(N_A)\,\Phi_J(1,2,\cdots,N_A)^*\,\Phi_{J'}(1',2,\cdots,N_A)$$

with (1) standing for space and spin electron variables $(\vec{r}_1,\zeta_1) = \boldsymbol{x}_1$, and so that

$$\rho_{el,JJ'}^{(A)}(\vec{r}) = \sum\nolimits_\zeta \rho_{JJ'}^{(1)}(\vec{r},\zeta;\vec{r},\zeta)$$

after adding over the spin variables.

Densities take simple forms when the J and J' states are N_A-electron determinantal wavefunctions containing orthonormal molecular spin-orbitals (or MSOs) $\psi_j(\boldsymbol{x}_n) = \langle \boldsymbol{x}_n \mid j \rangle$, where the indices n and j can be taken to label rows and columns of a $N_A \times N_A$ matrix. Given determinantal functions $\Phi_J(1, 2,\ldots, N_A)$

$= (N_A!)^{-1/2}\det[\langle \boldsymbol{x}_n | j\rangle]$ and $\Phi_{J'}(1',2',\ldots,N'_A) = (N'_A)^{-1/2}\det[\langle \boldsymbol{x}'_n | j'\rangle]$ constructed from a basis set of N orthonormal SOs, one finds

$$\rho^{(1)}_{JJ'}(\boldsymbol{x}_1;\boldsymbol{x}'_1) = \frac{1}{(N_A-1)!}\int d(2)\cdots d(N_A)\det\left[\sum\nolimits_{n=1}^{n=N_A}\langle j|\boldsymbol{x}_n\rangle\langle \boldsymbol{x}'_n|j'\rangle\right]_{2=2',\ldots,N_A=N'_A}$$

obtained recalling that the determinant of a product of matrices equals the product of their determinants, and doing next the integrals over electron variables (2), …, (N_A). This one-electron density is different from zero only if the many-electron states J and J' are equal or if they differ by only one MSO. General treatments of transition density functions, also for nonorthogonal basis sets, can be found in references [8, 9].

Furthermore, when the MSOs $\psi_j(\boldsymbol{x})$ are written as linear combinations of a basis set of functions $\xi_\mu(\vec{r},\zeta)$, generally nonorthogonal, the one-electron transition density functions can be expanded as

$$\rho^{(1)}_{JJ'}(\vec{r},\zeta;\vec{r}',\zeta') = \sum\nolimits_{\mu}\sum\nolimits_{\mu'}\xi_\mu(\vec{r},\zeta)^* P_{JJ'}(\mu,\mu')\xi_{\mu'}(\vec{r}',\zeta')$$

with expansion coefficients $P_{JJ'}(\mu,\mu')$ determined by the form of the many-electron states J. The basis functions $\xi_\mu(\vec{r},\zeta)$ can be chosen as atomic spin orbitals (ASOs) to form MSO-LCAO orbitals, or as delocalized functions.

Choosing a basis set of ASO μ for A, with overlaps $\langle \xi_\mu | \xi_{\mu'}\rangle = S_{\mu\mu'}$, and similarly for $\rho^{(1)}_{KK'}$ with AOs $\xi_\nu(\vec{r},\zeta)$ for B, it is possible to perform a population analysis to decompose the densities into terms giving atomic populations and atom–atom bond orders. Labeling ASOs centered at the nuclei a of A, with orbital and spin quantum numbers (n, l, m, m_s) by $\mu = (a, n, l, m, m_s)$, atomic components $P_{JJ'}(a)$ of transition densities at each center a are obtained adding all terms where μ and μ' contain only a. This can be quite generally done introducing projection operators

$$\hat{P}_a = \sum\nolimits_{\mu\subset a}\sum\nolimits_{\mu'\subset a}|\xi_\mu\rangle\left(\mathbf{S}^{-1}\right)_{\mu\mu'}\langle \xi_{\mu'}|$$

which satisfy $\hat{P}_a\,\hat{P}_{a'} = \delta_{aa'}\hat{P}_a$ and add up to the identity. Applied to the density operator, they select its components localized at centers a in A and give

$$\hat{\rho}^{(1)}_{JJ'} = \sum\nolimits_a P_{JJ'}(a)\hat{P}_a + \sum\nolimits_{a\neq a'}\hat{P}_a P_{JJ'}(a,a')\hat{P}_{a'}$$

where $P_{JJ'}(a)$ is a transition component at location a, while bond (aa') components $P_{JJ'}(a,a')$ of the transition density are given by the sums that contain $a \neq a'$. There is a similar decomposition for B with $\rho^{(1)}_{KK'}$ containing the ASO ξ_ν. This in turn allows for decomposition of interaction energies for AB into atom–atom, atom–bond, and bond–bond intermolecular components, useful when treating the whole system as made up of molecular fragments.

A second choice for the basis functions $\xi_\mu(\vec{r},\zeta)$ is provided by plane waves times spin functions $\Omega^{-1/2}\exp(-i\vec{q}_n.\vec{r})\vartheta(\zeta)$ with quantized wavevectors satisfying periodic boundary conditions in a large volume Ω. This is a convenient choice when the many-atom systems are very large, when they display structural periodicity, or when electrons are delocalized, and is further considered in what follows.

7.1.4 Long-range Interactions from Dynamical Susceptibilities

Induction energies arise from the static susceptibilities of A and B, coupled respectively to charges in B and A. The static electrical susceptibility of B in its ground electronic states is given by

$$\chi_0^{(B)}(\vec{s},\vec{s}\,') = -2\sum_{K\neq 0} \frac{c_{0K}^{(B)}(\vec{s})\, c_{0K}^{(B)}(\vec{s}\,')^*}{E_0^{(B)} - E_K^{(B)}}$$

a positive-valued tensor, in terms of which the induction energy at A due to the polarization of B, shown as AB^*, is given by

$$E_{ind}^{(AB^*)} = -\frac{1}{2}\int d^3r \int d^3r' \int d^3s \int d^3s' \frac{c_0^{(A)}(\vec{r})\, \chi_0^{(B)}(\vec{s},\vec{s}\,')\, c_0^{(A)}(\vec{r}\,')}{(4\pi\varepsilon_0)^2\, |\vec{r}-\vec{s}|\,|\vec{r}\,'-\vec{s}\,'|}$$

with the same integration limits in A and B. This corresponds to a total (nuclear plus electronic) static charge in a volume d^3r of A at location $\vec{r}$ interacting with an electronic charge fluctuation $c_{0K}^{(B)}(\vec{s})$ within d^3s of B, followed by the reverse interaction of a charge fluctuation in B with an element of total static charge in A. An expression for the other induced energy $E_{ind}^{(A^*B)}$ follows by interchanging A and B functions and arguments and contains $\chi_0^{(A)}(\vec{r},\vec{r}\,')$.

The dispersion energy resulting from charge density fluctuations in A and B, shown here as $E_{dsp}^{(A^*B^*)}$, involves the susceptibility of the pair AB and all its pair excitation energies $E_J^{(A)} - E_0^{(A)} + E_K^{(B)} - E_0^{(B)}$ in the denominator of the perturbation expansion of the dispersion energy. This appears to intertwine the properties of A and B. However, as done in Section 3.4 for the dispersion interaction of two molecules due to their electric dipole fluctuations, which could be expressed in terms of their separate dynamical dipole polarizabilities, it is possible here to transform the AB pair static susceptibility into an integral over separate dynamical susceptibilities of the two extended systems A and B, now without a restriction to particular multipole polarizabilities. We again use the Casimir–Polder integral (for positive a, b)

$$\frac{2}{\pi}\int_0^\infty \frac{du\, a\, b}{(a^2+u^2)(b^2+u^2)} = \frac{1}{a+b}$$

with $a = E_J^{(A)} - E_0^{(A)}$ and $b = E_K^{(B)} - E_0^{(B)}$ and introduce the dynamical susceptibility

$$\chi_0^{(A)}\left(\vec{r}, \vec{r}\,'; \omega\right) = -2\sum\nolimits_{J\neq 0} \frac{c_{0J}^{(A)}\left(\vec{r}\right)\left(E_0^{(A)} - E_J^{(A)}\right) c_{0J}^{(A)}\left(\vec{r}\,'\right)^*}{\left(E_0^{(A)} - E_J^{(A)}\right)^2 - (\hbar\omega)^2}$$

which leads, with the choice $\hbar\omega = u$, to the convenient form

$$E_{dsp}^{(A^*B^*)} = -\frac{\hbar}{2\pi}\int d^3r \int d^3r' \int d^3s \int d^3s' \int_0^{\infty} d\omega \frac{\chi_0^{(A)}\left(\vec{r}, \vec{r}\,'; i\omega\right)\chi_0^{(B)}\left(\vec{s}, \vec{s}\,'; i\omega\right)}{(4\pi\varepsilon_0)^2 |\vec{r} - \vec{s}||\vec{r}\,' - \vec{s}\,'|}$$

written in terms of the susceptibilities for imaginary-valued frequencies.

The susceptibilities have so far been written as sums over excited many-electron states of A and B. However, these states are seldom known for extended many-atom systems. This difficulty can be avoided writing the susceptibility in operator forms using resolvents, which can be calculated in a convenient basis set. The summation in $\chi_0^{(A)}\left(\vec{r}, \vec{r}\,'; i\omega\right)$ over states $|J_A\rangle$ with energies $E_J^{(A)}$ can be rewritten in a compact form using that $\hat{H}_A\,|J_A\rangle = E_J^{(A)}\,|J_A\rangle$ and introducing the projection operator $\hat{P}_A = \hat{I} - |\,0_A\rangle\langle 0_A\,| = \sum_{J\neq 0}|J_A\rangle\langle J_A\,|$. This allows us to write

$$\chi_0^{(A)}\left(\vec{r}, \vec{r}\,'; i\omega\right) = -2\langle 0_A\,|\,\hat{c}_A\left(\vec{r}\right)\hat{P}_A \frac{E_0^{(A)} - \hat{H}_A}{\left(E_0^{(A)} - \hat{H}_A\right)^2 + (\hbar\omega)^2}\hat{P}_A\hat{c}_A\left(\vec{r}\,'\right)|\,0_A\rangle$$

which can now be calculated with an alternative many-electron basis set. Properties of this susceptibility and ways to calculate it and to provide upper and lower bounds to it as a function of the variable u have been the subject of detailed research [10–12]. The procedure is similar to the one followed in Section 3.4 for molecular interactions. The static susceptibilities in the induction energy are found for $\omega = 0$.

Two particularly simple forms are obtained from a closure approximation, and from an interpolation of $\chi_0^{(A)}(i\omega)$ between its values at $\omega = 0$ and $\omega \to \infty$. In a closure approximation, the operator $\left(E_0^{(A)} - \hat{H}_A\right)\hat{P}_A$ in the dynamical susceptibility is replaced by $\left(E_0^{(A)} - E_{exc}^{(A)}\right)\hat{P}_{A,}$ which contains an excitation energy $E_{exc}^{(A)}$ to be treated as an adjustable parameter chosen to reproduce the value of the dispersion energy $E_{0,0}^{(dsp)}(A^*A^*)$, extracted from measurements or calculations. With this approximation, the susceptibility becomes

$$\chi_{0,exc}^{(A)}(\vec{r},\vec{r}\,';i\omega) = 2F_{exc}^{(A)}(\omega)G_{cc}^{(A)}(\vec{r},\vec{r}\,')$$

$$G_{cc}^{(A)}(\vec{r},\vec{r}\,') = \langle 0_A|\hat{c}_A(\vec{r})\hat{P}_A\hat{c}_A(\vec{r}\,')|0_A\rangle = \left[\langle 0_A|\hat{c}_A(\vec{r})\hat{c}_A(\vec{r}\,')|0_A\rangle - c_0^{(A)}(\vec{r})c_0^{(A)}(\vec{r}\,')\right]$$

with the frequency in the factor $F_{exc}^{(A)}(\omega) = |E_0^{(A)} - E_{exc}^{(A)}| / \left[\left(E_0^{(A)} - E_{exc}^{(A)}\right)^2 + (\hbar\omega)^2\right]$, which multiplies the charge–charge correlation function $G_{cc}^{(A)}(\vec{r},\vec{r}\,')$ for the ground state expectation value of charge densities at two positions. This contains one- and two-electron operators averaged only over a known many-electron ground state. The choice $E_{exc}^{(A)} = E_1^{(A)}$, the first excited state of A, gives an upper bound to this approximate susceptibility. The correlation function provides the static susceptibility through $\chi_0^{(A)}(\vec{r},\vec{r}\,';0) = 2G_{cc}^{(A)}(\vec{r},\vec{r}\,')/|E_0^{(A)} - E_{exc}^{(A)}| = \chi_0^{(A)}(\vec{r},\vec{r}\,')$, with the parameter $\hbar\omega^{(A)} = |E_0^{(A)} - E_{exc}^{(A)}|$ to be extracted from calculations or measurements.

With a similar approximation for $\chi_0^{(B)}(i\omega)$, it is possible to estimate the induction and dispersion energies of the AB pair. Induction energies are given by interactions between permanent charges and expectation values of charge–charge fluctuations, however involving just the ground states of A and B, and information on their energies for AA and BB. By analogy with the treatment of interaction energies for two molecules in the dipole approximation, it follows that here we can also introduce a combination rule giving the dispersion energy of AB in terms of the values for AA and BB. By analogy with the treatment for two interacting molecules leading to combination rules for the dispersion energy [13], here for two large many-atom systems, it is convenient to introduce the limit for $\omega \to \infty$, $\chi_{0,exc}^{(A)}(\vec{r},\vec{r}\,';i\omega) \approx 2|E_0^{(A)} - E_{exc}^{(A)}|\,G_{cc}^{(A)}(\vec{r},\vec{r}\,')/(\hbar\omega)^2 = (\omega^{(A)}/\omega)^2\chi_0^{(A)}(\vec{r},\vec{r}\,';0)$. The limits at $\omega = 0$ and $\omega \to \infty$ can be combined in an interpolated approximate form

$$\beta_0^{(A)}(\vec{r},\vec{r}\,';\omega) = \chi_0^{(A)}(\vec{r},\vec{r}\,')\left[1 + \left(\frac{\omega}{\omega^{(A)}}\right)^2\right]^{-1}$$

where $\beta_0^{(A)}(\vec{r},\vec{r}\,';\omega) = \chi_0^{(A)}(\vec{r},\vec{r}\,';i\omega)$. The parameter $\omega^{(A)}$ can be obtained from $E_{0,0}^{(dsp)}(A^*A^*)$, containing a calculated $G_{cc}^{(A)}(\vec{r},\vec{r}\,')$ and written in terms of $\beta_0^{(A)}(\omega)$. Doing the same for $\omega^{(B)}$ obtained from $E_{0,0}^{(dsp)}(B^*B^*)$, these two parameters and the correlation functions $G_{cc}^{(A)}(\vec{r},\vec{r}\,')$ and $G_{cc}^{(B)}(\vec{s},\vec{s}\,')$ can be used to obtain $E_{0,0}^{(dsp)}(A^*B^*)$ from a combination rule, here for extended systems.

7.2 Energetics of a Large Molecule in a Medium

7.2.1 Solute–Solvent Interactions

The total energy $E_{0,slt}^{(AB)}$ in a solution (slt) of a solute molecule A in a medium B (which can be a liquid or solid solvent [slv], a cavity, or a solid surface) must include the energy $E_0^{(AB)}(\mathbf{Q}_A, \mathbf{Q}_B)$ of A interacting with B, plus the interaction energy $E_{0,slv}^{(B)}$ of all solvent molecules or medium fragments in B,

$$E_{0,slt}^{(AB)}(\mathbf{Q}_A, \mathbf{Q}_B) = E_0^{(AB)}(\mathbf{Q}_A, \mathbf{Q}_B) + E_{0,slv}^{(B)}(\mathbf{Q}_B)$$

The terms to the right can be separately considered and are simpler to construct when electrons are not exchanged between A and B, and they are also sufficiently separated so that short-range repulsion does not distort them. This is first done as follows.

The electrostatic, induction, and dispersion energies for A in a given electronic state $|0_A\rangle$ interacting with a species B in a general electronic state $|K_B\rangle$ can be obtained from potential energy operators $\hat{V}_{AB}^{(els)}, \hat{V}_{AB}^{(ind)}$, and $\hat{V}_{AB}^{(dsp)}$ arising from the charges in A interacting with electric potentials originating in B. The electrostatic potential energy operator $\hat{V}_{AB}^{(els)}(\vec{r}; \mathbf{Q}_A, \mathbf{Q}_B)$ involves the charge density operator $\hat{c}^{(A)}(\vec{r}; \mathbf{Q}_A)$ of A multiplying the electric scalar potential $\varphi_B^{(els)}(\vec{r}; \mathbf{Q}_B)$ resulting from the stationary charge distribution of B in its ground state $|0_B\rangle$. Additional potential energy operators $\hat{V}_{AB}^{(ind)}(\vec{r}, \vec{r}\,'; \mathbf{Q}_A, \mathbf{Q}_B)$, and $\hat{V}_{AB}^{(dsp)}(\vec{r}, \vec{r}\,'; \mathbf{Q}_A, \mathbf{Q}_B)$ can be constructed from the charge distributions $\hat{c}_A(\vec{r}; \mathbf{Q}_A)$ and $\hat{c}_A(\vec{r}\,'; \mathbf{Q}_A)$ at two locations linked by reaction potential operators $\hat{\varphi}_{AB}^{(ind)}(\vec{r}, \vec{r}\,'; \mathbf{Q}_A, \mathbf{Q}_B)$ and $\hat{\varphi}_{AB}^{(dsp)}(\vec{r}, \vec{r}\,'; \mathbf{Q}_A, \mathbf{Q}_B)$ generated by density fluctuations when B is excited into states $|K_B\rangle$. These reaction potentials can be obtained as shown below, or from time-dependent response theory [6].

This treatment is convenient when describing a large solute molecule A in a medium B composed of many solvent molecules treated as molecular fragments [14], or when B is a medium involving many atoms in a complex structure described by distributed multipoles and polarizabilities [15]. The external electric potentials generated by B can then be considered as one- and two-electron operators to be added to the Hamiltonian $\hat{H}_A$ of A in a treatment requiring a many-electron description of A but treating B only in terms of its electric multipoles and polarizabilities. The energy of the solute A in its ground electronic state in a medium provided by B is then obtained to first order in the potentials as

$$E_0^{(AB)}(\mathbf{Q}_A, \mathbf{Q}_B) = \left\langle 0_A \,|\, \hat{H}_A + \hat{V}_{AB}^{(els)} + \hat{V}_{AB}^{(ind)} + \hat{V}_{AB}^{(dsp)} \,|\, 0_A \right\rangle$$

which is valid here only provided the electronic charge distributions of species A and B in their ground and excited states do not overlap. The electric potentials can be derived comparing the above terms with the known forms of terms in $E_0^{(AB)}(\mathbf{Q}_A,\mathbf{Q}_B) = E_0^{(A)}(\mathbf{Q}_A) + E_{els}^{(AB)}(\mathbf{Q}_A,\mathbf{Q}_B) + E_{ind}^{(AB)}(\mathbf{Q}_A,\mathbf{Q}_B) + E_{dsp}^{(AB)}(\mathbf{Q}_A,\mathbf{Q}_B)$. The external potentials $\hat{V}_{AB}^{(ext)}$, with *ext* = *els, ind, dsp,* can also be used in a perturbation treatment of states of A to obtain more accurate interaction energies when they are large compared to internal energies in A.

From the previous ground state first-order perturbation energy, one finds

$$E_{els}^{(AB)}(\mathbf{Q}_A,\mathbf{Q}_B) = \int d^3r \langle 0_A \mid \hat{c}_A(\vec{r};\mathbf{Q}_A) \mid 0_A \rangle \varphi_{B,0}^{(els)}(\vec{r};\mathbf{Q}_B)$$

$$\varphi_{B,0}^{(els)}(\vec{r};\mathbf{Q}_B) = \frac{1}{4\pi\varepsilon_0}\int d^3s \frac{c_0^{(B)}(\vec{s};\mathbf{Q}_B)}{|\vec{r}-\vec{s}|}$$

$$\hat{V}_{AB}^{(els)}(\vec{r};\mathbf{Q}_A,\mathbf{Q}_B) = \hat{c}_A(\vec{r};\mathbf{Q}_A)\varphi_{B,0}^{(els)}(\vec{r};\mathbf{Q}_B)$$

which is just the classical electrostatic interaction between the ground state charge distributions $c_0^{(A)}(\vec{r}) = \langle 0_A \mid \hat{c}_A(\vec{r};\mathbf{Q}_A) \mid 0_A \rangle$ and the electrical potential $\varphi_{B,0}^{(els)}$ created by $c_0^{(B)}(\vec{s}) = \langle 0_B \mid \hat{c}_B(\vec{s};\mathbf{Q}_B) \mid 0_B \rangle$.

The second-order energies lead to forms of the desired induction and dispersion electric potentials generated by B, after rewriting

$$\left|\left\langle 0_A,0_B \mid \hat{H}_{AB}^{(Coul)} \mid J_A,K_B \right\rangle\right|^2 = \left\langle 0_A,0_B \mid \hat{H}_{AB}^{(Coul)} \mid J_A,K_B \right\rangle \left\langle J_A,K_B \mid \hat{H}_{AB}^{(Coul)} \mid 0_A,0_B \right\rangle$$

and introducing the resolvents

$$\hat{R}_A^{(0)} = -\sum_{J\neq 0} \frac{|J_A\rangle\langle J_A|}{E_J^{(A)} - E_0^{(A)}},\quad \hat{R}_B^{(0)} = -\sum_{K\neq 0} \frac{|K_B\rangle\langle K_B|}{E_K^{(B)} - E_0^{(B)}},$$

to obtain a compact expression for the induction electrical potential (a reaction potential) $\hat{\varphi}_{AB,0}^{(ind)}(\vec{r},\vec{r}\,';\mathbf{Q}_A,\mathbf{Q}_B)$, as

$$\hat{\varphi}_{AB,0}^{(ind)}(\vec{r},\vec{r}\,') = \hat{\varphi}_{A^*B,0}^{(ind)}(\vec{r},\vec{r}\,') + \hat{\varphi}_{AB^*,0}^{(ind)}(\vec{r},\vec{r}\,')$$

$$\hat{\varphi}_{A^*B,0}^{(ind)}(\vec{r},\vec{r}\,') = \frac{1}{(4\pi\varepsilon_0)^2}\hat{R}_A^{(0)}\int d^3s \frac{c_0^{(B)}(\vec{s})}{|\vec{r}-\vec{s}|}\int d^3s' \frac{c_0^{(B)}(\vec{s})}{|\vec{r}\,'-\vec{s}\,'|}$$

$$\hat{\varphi}_{AB^*,0}^{(ind)}(\vec{r},\vec{r}\,') = \frac{1}{(4\pi\varepsilon_0)^2}|0_A\rangle\langle 0_A|\int d^3s\int d^3s' \left\langle 0_B \left| \frac{\hat{c}_B(\vec{s}\,')}{|\vec{r}\,'-\vec{s}\,'|}\hat{R}_B^{(0)}\frac{\hat{c}_B(\vec{s})}{|\vec{r}-\vec{s}|}\right| 0_B \right\rangle$$

which as expected depends on the conformations of both A and B. The term $\hat{\varphi}^{(ind)}_{A^*B,0}$ corresponds to permanent charges of B interacting with charge fluctuations of A, which can be constructed from the charge distribution or multipoles of B, and also involves excited states of A. The term $\hat{\varphi}^{(ind)}_{AB^*,0}$ involves excitations of B that interact with the permanent charges of A and can be constructed from polarizabilities of B. This electric reaction potential is a two-electron term giving an induction potential energy

$$\hat{V}^{(ind)}_{AB}\left(\vec{r},\vec{r}\,';\mathbf{Q}_A,\mathbf{Q}_B\right) = \hat{c}_A\left(\vec{r};\mathbf{Q}_A\right)\hat{\varphi}^{(ind)}_{AB,0}\left(\vec{r},\vec{r}\,';\mathbf{Q}_A,\mathbf{Q}_B\right)\hat{c}_A\left(\vec{r}\,';\mathbf{Q}_A\right)$$

to be added to the Hamiltonian of A. It contributes an induction energy

$$E^{(AB)}_{ind}\left(\mathbf{Q}_A,\mathbf{Q}_B\right) = \int d^3r\int d^3r'\left\langle 0_A \mid \hat{V}^{(ind)}_{AB}\left(\vec{r},\vec{r}\,';\mathbf{Q}_A,\mathbf{Q}_B\right) \mid 0_A\right\rangle$$

The dispersion electric potential, containing the resolvent

$$\hat{R}^{(0)}_{AB} = -\sum\nolimits_{J\neq 0}\sum\nolimits_{K\neq 0}\frac{|J_A,K_B\rangle\langle J_A,K_B|}{E^{(A)}_J - E^{(A)}_0 + E^{(B)}_K - E^{(B)}_0}$$

is similarly given by the electric potential operator

$$\hat{\varphi}^{(dsp)}_{AB,0}\left(\vec{r},\vec{r}\,'\right) = \frac{1}{4\pi\varepsilon_0}\int d^3s\int d^3s'\left\langle 0_B\left|\frac{\hat{c}_B\left(\vec{s}\,'\right)}{\left|\vec{r}\,'-\vec{s}\,'\right|}\hat{R}^{(0)}_{AB}\frac{\hat{c}_B\left(\vec{s}\right)}{\left|\vec{r}-\vec{s}\right|}\right|0_B\right\rangle$$

which is again a compact expression, suitable for calculations in an extended system B from its polarizability function or from expansions in multipole components. This leads to the potential energy operator

$$\hat{V}^{(dsp)}_{AB}\left(\vec{r},\vec{r}\,';\mathbf{Q}_A,\mathbf{Q}_B\right) = \hat{c}_A\left(\vec{r};\mathbf{Q}_A\right)\hat{\varphi}^{(dsp)}_{AB,0}\left(\vec{r},\vec{r}\,';\mathbf{Q}_A,\mathbf{Q}_B\right)\hat{c}_A\left(\vec{r}\,';\mathbf{Q}_A\right)$$

and the dispersion energy

$$E^{(AB)}_{dsp}\left(\mathbf{Q}_A,\mathbf{Q}_B\right) = \int d^3r\int d^3r'\left\langle 0_A \mid \hat{V}^{(dsp)}_{AB}\left(\vec{r},\vec{r}\,';\mathbf{Q}_A,\mathbf{Q}_B\right) \mid 0_A\right\rangle$$

arising from the coupling of many-electron density fluctuations in A with the dynamical polarizabilities of B.

These expressions are especially useful in applications to solvation energetics, when one wants to calculate the solvation energy of species A interacting with many solvent molecules in a liquid or solid B, or when B is an extended many-atom structure enveloping A. Introducing many-electron states of A and B, the electrostatic energy can be written in terms of the densities $\rho^{(A)}_{el,0}\left(\vec{r}\right)$ and $\rho^{(B)}_{el,0}\left(\vec{s}\right)$, and induction and dispersion energies follow from transition densities $\rho^{(A)}_{el,0J}\left(\vec{r}\right)$

and $\rho_{el,0K}^{(B)}(\vec{s})$. However, it is then necessary to make sure there are no contributions from overlapping electronic states of A and B, which would lead to small values of $|\vec{r}-\vec{s}|$ in integrands, and inaccurate energies.

For decreasing distances between solute and solvent, one can proceed somewhat as done when damping long-range interactions for small distances to avoid unphysical superposition of forces. Here there is no single intermolecular distance R to construct a damping function, but the problem can be avoided replacing $|\vec{r}-\vec{s}|^{-1}$ with a damped charge–charge interaction function such as $f_d(|\vec{r}-\vec{s}|)|\vec{r}-\vec{s}|^{-1}$ with

$$f_d(|\vec{r}-\vec{s}|;d,\alpha) = \frac{\exp(\alpha|\vec{r}-\vec{s}|)-1}{\exp(\alpha|\vec{r}-\vec{s}|)-\exp(\alpha d)} \approx 0 \; for \; |\vec{r}-\vec{s}| \to 0$$

where d is the range over which the function is damped as $|\vec{r}-\vec{s}|$ decreases, and with α giving the slope of the transition region. This function goes to 1.0 for large $|\vec{r}-\vec{s}| \gg d$ and preserves the values of the electric potentials at these large distances. The calculation of energies with this damped function requires numerical evaluation of the integrals over $\vec{r}$ and $\vec{s}$.

7.2.2 Solvation Energetics for Short Solute–Solvent Distances

When the distances between the boundaries of A and B are small and their electronic charges overlap, it is necessary to treat the pair AB as a single entity, to account for two effects: (i) the combined Pauli repulsion of electronic distributions and interelectronic exchange energy, resulting from the requirement of wavefunction antisymmetry, and (ii) electron transfer between A and B. This can be done with additional effective one-electron potential energy operators $\hat{V}_{AB}^{(exc-rep)}(\vec{r};\mathbf{Q}_A,\mathbf{Q}_B)$ and $\hat{V}_{AB}^{(ctr)}(\vec{r};\mathbf{Q}_A,\mathbf{Q}_B)$. They can be obtained from Hartree–Fock or SAPT treatments [16, 17] and from approximate treatments of the electronic states in the AB, A^+B^-, and A^-B^+ pairs [14, 18]. The total energy $E_{0,slt}^{(AB)}$ for A in a solution must also include the energy $E_{0,slv}^{(B)}$ of the solvent including all its interaction energies, as in

$$E_{0,slt}^{(AB)}(\mathbf{Q}_A,\mathbf{Q}_B) = E_0^{(AB)}(\mathbf{Q}_A,\mathbf{Q}_B) + E_{0,slv}^{(B)}(\mathbf{Q}_B)$$

$$E_0^{(AB)}(\mathbf{Q}_A,\mathbf{Q}_B) = \left\langle 0_A \mid \hat{H}_A + \hat{V}_{AB}^{(exc-rep)} + \hat{V}_{AB}^{(ctr)} + \hat{V}_{AB}^{(els)} + \hat{V}_{AB}^{(ind)} + \hat{V}_{AB}^{(dsp)} \mid 0_A \right\rangle$$

$$E_{0,slv}^{(B)}(\mathbf{Q}_B) = E_{0,exc-rep}^{(B)}(\mathbf{Q}_B) + E_{0,ctr}^{(B)}(\mathbf{Q}_B) + E_{0,els}^{(B)}(\mathbf{Q}_B) + E_{0,ind}^{(B)}(\mathbf{Q}_B) + E_{0,dsp}^{(B)}(\mathbf{Q}_B)$$

The one-electron potential energy operators can be constructed to reproduce interaction energies in $E_{0,exc-rep}^{(AB)}(\mathbf{Q}_A,\mathbf{Q}_B) = \left\langle 0_A \mid \hat{V}_{AB}^{(exc-rep)} \mid 0_A \right\rangle$ and

$E_{0,ctr}^{(AB)}(\mathbf{Q}_A,\mathbf{Q}_B) = \left\langle 0_A \mid \hat{V}_{AB}^{(ctr)} \mid 0_A \right\rangle$, with these energies obtained in suitable approximations.

A reliable treatment of exchange-repulsion and charge transfer at short distances can be obtained from a self-consistent field (or Hartree–Fock) treatment of the electronic structure of the pair *AB*, insofar the short-range region of electronic interactions is dominated by Coulomb and exchange electron energies. In the notation of Chapter 6, the H–F energy of species *A* is

$$E_0^{(A)}(HF) = \sum_{j_A}^{occ} \left(\left\langle j_A \mid \hat{h} \mid j_A \right\rangle - \epsilon_{j_A} \right) + \sum_{j_A < k_A}^{occ} \langle j_A k_A \mid v(1-\hat{t}) \mid j_A k_A \rangle$$

in terms of MSOs $|j_A\rangle$ and $|k_A\rangle$ of *A*, and similarly for *B*. The H–F energy of the pair *AB* is given as

$$E_0^{(AB)}(HF) = \sum_{j}^{occ} \left(\left\langle j \mid \hat{h} \mid j \right\rangle - \epsilon_j \right) + \sum_{j<k}^{occ} \langle jk \mid v(1-\hat{t}) \mid jk \rangle$$

where MSOs $|j\rangle = |j_A\rangle + |j_B\rangle$ and $|k\rangle = |k_A\rangle + |k_B\rangle$ extend over both *A* and *B*. This contains the H–F energies of each component *A* and *B*, and also their electrostatic, induction, and dispersion interaction energies to the extent they appear in the H–F treatment. The needed exchange-repulsion energy is the remaining energy after all these have been subtracted, so that

$$\begin{aligned} E_{0,exc-rep}^{(AB)} &= \left\langle 0_A \mid \hat{V}_{AB}^{(exc-rep)} \mid 0_A \right\rangle \\ &= E_0^{(AB)}(HF) - E_0^{(A)}(HF) - E_0^{(B)}(HF) - E_{0,els}^{(AB)}(HF) - E_{0,ind}^{(AB)}(HF) \\ &\quad - E_{0,dsp}^{(AB)}(HF) \end{aligned}$$

which provides the form of $\hat{V}_{AB}^{(exc-rep)}$ as an operator between states containing MSOs $|j_A\rangle$ and $|k_A\rangle$ of *A*. The charge transfer energy and one-electron potential energy operator in $E_{0,ctr}^{(AB)}(\mathbf{Q}_A,\mathbf{Q}_B) = \left\langle 0_A \mid \hat{V}_{AB}^{(ctr)} \mid 0_A \right\rangle$ can be constructed in the same way from H–F energies for charge states A^+B^- or A^-B^+. But here it is necessary to use spin-polarized MSOs, consisting of different orbitals for different spins, to account for the open-shell structures of the ions.

Calculation of all these energy terms is clearly a major undertaking, limited by computational speed and digital storage. Progress is being made to obtain results by partitioning the large systems into smaller, more accessible, components. This has been computationally implemented within an effective fragment potential (or EFP) treatment of the medium *B* and its interactions with the solute *A* [14], and has been generalized to include exchange-repulsion and charge transfer in calculations of intermolecular forces [19]. It has been applied to amino acids and to aqueous solutions of organic molecules.

Another implementation has developed force fields including polarization effects, named X-Pol to begin with, and has been applied to water clusters and polypeptide chains [20]. It considers the effect of charge distributions within B as they polarize A and change its force field. In the context of the previous equations, it amounts to incorporation of the electrical potential $\varphi_{B,0}^{(els)}(\vec{r};\mathbf{Q}_B)$ in calculations of the electronic structure of A, by means of a variational procedure for the interaction energy that includes the potential energies from charges in B, and is a functional of the orbitals in A [21].

In the present treatment, polarization effects on A have been included to second order in $\varphi_{B,0}^{(els)}$, insofar the electrostatic energy $E_{0,0}^{(els)}(\mathbf{Q}_A,\mathbf{Q}_B)$ contains its first-order effect and the reaction potential $\hat{\varphi}_{A*B,0}^{(ind)}(\vec{r},\vec{r}\,') = \varphi_{B,0}^{(els)}(\vec{r})\hat{R}_A^{(0)}\varphi_{B,0}^{(els)}(\vec{r}\,')$ accounts for the second-order effect within $E_{0,0}^{(ind)}$. Terms in the total energy $E_{0,slt}^{(AB)}$ of the molecular solution include in addition the dispersion energy $E_{0,0}^{(dsp)}$ of interaction, which arises from mutual polarization of A and B.

7.2.3 Embedding of a Molecular Fragment and the *QM/MM* Treatment

Large systems can be treated selecting groups of atoms, or molecular fragment, in a large many-atom system to be described in detail by means of many-electron wavefunctions, while simplifying the treatment of the environment to which the selected atoms are either bonded, or the treatment of the environment that perturbs the selected atoms through long-range forces. The selected atoms may be described as a molecular fragment with peripheral bonds saturated with chosen atoms (or links) such as hydrogen, or more generally can be treated with embedding methods where the fragment is self-consistently coupled to its environment. The embedding procedure has computational advantages when the environment is treated by means of a force field (within molecular mechanics, or *MM*) while the selected fragment is treated with wavefunctions in a quantum mechanical (or *QM*) way. In this *QM/MM* treatment, the computational effort for an environment (also called the subsidiary system SS or outer system O) containing $N_{At}^{(env)}$ atoms scales linearly with this number, while a fragment (called a primary system PS or inner system I) is described with $N_{fcn}^{(frg)} = \sum_{At}^{(frg)} n_{fcn}^{(At)}$ electronic basis functions, which is proportional to the number of atoms in the group and a sum over the number of basis functions per atom as shown, and scales in principle as $N_{fcn}^{(frg)}$ to a power of 3 or larger depending on the quality of the many-electron treatment [22]. The large powers in scalings with the number of basis functions can, however, be decreased using a variety of computational and software procedures for calculation of electron integrals, an active area of research [23]. This has recently allowed *QM* treatments of very large fragments, with sizes that can be increased to obtain convergence in total energy values.

Similar aspects have been treated in the previous section on the interactions of a molecule in a solution, as they relate to long-range forces but without bonding. Here more generally one is dealing with strong coupling due to bonding between a cluster of atoms and its environment. The *QM/MM* treatments originated on work dealing with dynamics and reactions in biomolecular systems [24–26]. and has been extended in several ways, for example, introducing models with a transition region between the *QM* region of the fragment and the *MM* region of the environment [27]. The challenge here is to identify the boundaries of the regions and to allow for changes of the electronic distribution in the *QM* region and of the force field in the *MM* region, due to electronic rearrangements or chemical reactions within the selected *QM* region. Recent developments have been reviewed as they apply to molecular complexes and to biomolecules [28–31] and are briefly summarized here.

To be consistent with our notation, we let A indicate the selected molecular fragment (or PS), including a set of link atoms capping bonds, and let B indicate the surrounding medium (or SS). We show with labels *QM* or *MM* that a many-electron treatment or force field description is being done. The total energy can be obtained starting with the *MM* treatment of the whole system and correcting it with the detailed energy change for the subsystem A,

$$E^{(AB)}(QM/MM) = E^{(AB)}(MM) + E^{(A)}(QM) - E^{(A)}(MM)$$

The calculation of the MM energies are done as described in Chapter 4, with a sum of terms for bond distance, bond angles, dihedral angles, electrostatic, induction, and dispersion energies, parametrized from independent calculations. The remaining term $E^{(A)}(QM)$ is obtained from an electronic wavefunction $\Phi_0^{(A)}$ calculated for the state of interest from the Hamiltonian operator $\hat{H}_A$ of A. This is called a *subtraction scheme* or *mechanical embedding* and is straightforward to implement numerically and to improve if needed by increasing the size of the PS region. It assumes that the force field in the SS region is unchanged as the two regions interact, and even if electronic rearrangement occurs in A.

However, physically one knows that all three energy terms depend on all the atomic positions and are functionals of the total electronic density of the pair AB. Therefore, charge rearrangement or chemical reactions in A (the PS), involving bond breaking or formation, affect B (the SS) by polarizing the force field and require some self-consistent coupling between the two regions. This is done with an *additive scheme* or *electrostatic embedding* writing instead that

$$E^{(AB)} = E^{(A)}(QM) + E^{(B)}(MM') + E_{int}^{(AB)}(QM/MM')$$

with $E^{(B)}(MM')$ obtained from a polarizable force field in B (show as MM') and the interaction $E_{int}^{(AB)}(QM/MM')$ obtained from the quantal expectation value of the energy coupling operator between A and B for many-electron states $\Phi_J^{(AB)}$. This must be calculated from the full Hamiltonian $\hat{H}_{AB} = \hat{H}_A + \hat{H}_B + \hat{H}_{AB}^{(Coul)}$ in our previous notation. Here the B (or SS) region is treated as a static but polarized force field so that $\hat{H}_B = E^{(B)}(MM')$ is an energy value, and the AB coupling $\hat{H}_{AB}^{(Coul)} = \hat{V}_{AB}^{(Coul)}$ is a potential energy operator that depends on the electronic and nuclear position variables of A, but is only a parametric function of atomic positions and polarizable charge densities in B. Therefore, states $\Phi_J^{(AB)}$ satisfying

$$\left[\hat{H}_A + \hat{V}_{AB}^{(Coul)} + E^{(B)}(MM')\right]\Phi_J^{(AB)} = E_J^{(AB)}\Phi_J^{(AB)}$$

can be generated as functions of the electronic variables in A, with parameters given by the atomic positions and charge densities in B. In turn, the charge densities in B must be allowed to polarize as electronic rearrangement or chemical reactions occur in A, and to alter the force field in B. This is considered in more detail in what follows relating to density functional methods for a large system.

7.3 Energies from Partitioned Charge Densities

7.3.1 Partitioning of Electronic Densities

The energetics of two large interacting molecular structures, as they are present in organic chemistry and biochemistry, can be described in terms of smaller molecular fragments or functional groups of atoms. The energetics of materials and interfaces can instead be treated taking advantage of periodicity of structures or of the presence of smooth charge densities due to electron delocalization. These treatments have in common the introduction of convenient decompositions of electronic densities for the structures, which can be done in general by expansions of densities and polarizabilities in suitable basis sets. The starting point in such partitioning involves the structural and electronic properties of the free fragments. But this must be modified to account for changes in the properties due to interactions among the fragments in each molecule. The interaction is strong when fragments are chemically bonded, and the property changes must be obtained from knowledge of the many-electron wavefunction of the molecule.

An extensively developed and used treatment introduces expansions of densities and polarizabilities in a basis set of functions $\chi_\mu^{(s)}(\vec{r}) = R_{lm}(\vec{r}-\vec{r}_s)$ $\exp\left[-\zeta(\vec{r}-\vec{r}_s)^2\right]$ with the angular dependence of the $(lm) = \mu$ solid spherical harmonic R_{lm} localized around selected points $\vec{r}_s$ situated at atoms and bonds. This leads to treatments of energies involving distributed density multipoles and distributed multipolar polarizabilities [15]. The one-electron density

$\rho(\vec{r}) = \sum_{\zeta} \rho^{(1)}(\vec{r},\zeta;\vec{r},\zeta)$ from a many-electron state of molecule A can be rewritten as $\rho(\vec{r}) = \sum_{s\mu,t\nu} P^{(s,t)}_{\mu,\nu} \chi^{(s)}_{\mu}(\vec{r}) \chi^{(t)}_{\nu}(\vec{r})$ where $P^{(s,t)}_{\mu,\nu}$ is a density matrix in the localized basis set, which must be obtained from the known many-electron wavefunction of A. The product of orbitals at locations s and t can be expressed as an expansion around a new location $\vec{r}_a$ as

$$\chi^{(s)}_{\mu}(\vec{r}) \chi^{(t)}_{\nu}(\vec{r}) = \sum_{a,\lambda} C^{(st,a)}_{\mu\nu,\lambda} \chi^{(a)}_{\lambda}(\vec{r})$$

with known coefficients $C^{(st,a)}_{\mu\nu,\lambda}$, which contain by construction information about the molecular charge distribution. One can then derive multipole amplitudes $Q^{(a)}_{lm}$ from expansions

$$\rho(\vec{r}) = \sum_{a,lm} Q^{(a)}_{lm} \chi^{(a)}_{lm}(\vec{r})$$

where $\chi^{(a)}_{lm}$ is localized at a position $\vec{r}_a$ in A, and rotates as a spherical tensor. Doing the same for another molecule B, the interaction Hamiltonian energy for the AB molecular pair can be written as a bilinear expression in $\hat{Q}^{(a)}_{lm}$ and $\hat{Q}^{(b)}_{pq}$ operators physically arising from the energy of interaction of multipoles at $\vec{r}_a$ in A with multipoles at $\vec{r}_b$ in B,

$$\hat{H}_{AB} = \sum_{a,lm} \sum_{b,pq} \hat{Q}^{(a)}_{lm} T^{(a,b)}_{lm,pq} \hat{Q}^{(b)}_{pq}$$

with coupling coefficients $T^{(a,b)}_{lm,pq}$ fit to calculated energies for each separate fragment pair ab and containing a damping factor so that $|T^{(a,b)}_{lm,pq}| \approx 0$ as $|\vec{r}_a - \vec{r}_b|$ becomes small. This has been done and tested for many systems and requires using rotational and translational properties of the functions $\chi^{(a)}_{lm}$ as relative positions and orientations of A (and B) are changed. The calculations can be accurate provided they account for changes in values of properties going from free fragments to fragments interacting with their neighbors, but can be laborious [15].

A simpler but less general procedure decomposes the one-electron density by means of weights generated from the electronic densities of atoms in A and B [32]. The required two items of information are the spherically averaged electronic densities $\bar{\rho}^{(a)}(\vec{r})$ of each free atom a appearing in the molecule A, and the electronic density $\rho^{(A)}(\vec{r})$ of the whole molecule. A set of atomic weight functions is defined by

$$w_a(\vec{r}) = \frac{\bar{\rho}^{(a)}_{free}(\vec{r})}{\sum_a \bar{\rho}^{(a)}_{free}(\vec{r})}$$

adding as $\sum_a w_a(\vec{r}) = 1$. They provide a first version of functions to be used in integrals over space to coarse-grain the densities as sums over atomic components, by means of

$$\int d^3r F(\vec{r}) = \sum_a \int d^3r F(\vec{r}) w_a(\vec{r}) = \sum_a F_a$$

for a generic function $F(\vec{r})$. The free atom densities can be corrected to account for bondings in A using knowledge of the molecular density, by defining the bonded atom densities $\rho_{bnd}^{(a)}(\vec{r}) = w_a(\vec{r})\rho^{(A)}(\vec{r})$. The atomic deformation density is given by $\Delta\rho_{bnd}^{(a)}(\vec{r}) = \rho_{bnd}^{(a)}(\vec{r}) - \bar{\rho}_{free}^{(a)}(\vec{r})$ and the molecular deformation density is $\Delta\rho^{(A)}(\vec{r}) = \rho^{(A)}(\vec{r}) - \sum_a \bar{\rho}_{free}^{(a)}(\vec{r})$, with both functions providing insight on the changes due to bonding.

The weights can be used to decompose molecular properties into atomic fragments and to define effective values for each fragment. For example, the physical proportionality between a fragment electronic volume and its polarizability has led to the introduction of an effective static dipolar polarizability for atomic fragment a as

$$\alpha_{eff}^{(a)}(0) = \alpha_{free}^{(a)}(0) \frac{\int_A d^3r\, w_a(\vec{r})\rho^{(A)}(\vec{r}) \left|\vec{r} - \vec{r}_a\right|^3}{\int_A d^3r\, \rho^{(A)}(\vec{r}) \left|\vec{r} - \vec{r}_a\right|^3}$$

This together with interpolation over frequencies as described in Chapter 3 can provide effective dynamical polarizabilities $\alpha_{eff}^{(a)}(\omega)$. They have been extensively used to calculate van der Waals interactions in AB from sums over pairs (ab), with encouraging results involving small and large molecules interacting with solid surfaces [33].

7.3.2 Expansions of Electronic Density Operators

Doing expansions for density operators, instead of expanding their state-averaged density functions of positions, allows for a general treatment valid for both ground and excited electronic states and for couplings between states. The electronic density operator $\hat{c}_A^{(el)}(\vec{r}) = C_e \sum_i \delta\left(\vec{r} - \hat{\vec{r}}_i\right)$ can be re-expressed in terms of density components obtained by expanding it in a complete and orthonormal basis set $\{f_\alpha^{(A)}(\vec{r})\}$ of functions of electronic positions, with α a triplet of indices, satisfying $\int d^3r f_\alpha^{(A)}(\vec{r}) f_{\alpha'}^{(A)}(\vec{r})^* = \delta_{\alpha\alpha'}$ and $\sum_\alpha f_\alpha^{(A)}(\vec{r}) f_\alpha^{(A)}(\vec{r}\,')^* = \delta(\vec{r} - \vec{r}\,')$. These functions can be chosen to partition real space into manageable regions, for example, using a coarse-graining treatment for spatial integration, or can be chosen to partition large molecules into fragments. Expansion of the charge density operator for A gives

$$\hat{c}_A^{(el)}(\vec{r}) = \sum\nolimits_\alpha \hat{c}_{A\alpha}^{(el)} f_\alpha^{(A)}(\vec{r}),\ \hat{c}_{A\alpha}^{(el)} = C_e \sum\nolimits_i f_\alpha^{(A)}\left(\hat{\vec{r}}_i\right)^*$$

with operator-valued coefficients, which reconstruct the electronic charge density operator. The completeness of a set is in practice only approximately satisfied and must be verified by increasing the number of functions in the finite set to reach convergence. Each coefficient $\hat{c}_{A\alpha}^{(el)}$ is a one-electron partial density operator extending over a region of space selected by the shape of the density related $f_\alpha^{(A)}$ function, which allows a breakup of the density into components chosen to describe certain molecular fragments, or certain properties of an extended system. The density expansions for A, and for B with a basis set $\left\{f_\beta^{(B)}(\vec{s})\right\}$, transform the energy integrals over space into sums, more convenient for numerical work.

The matrix elements

$$\left\langle J_A \,|\, \hat{c}_{A\alpha}^{(el)} \,|\, J_A' \right\rangle = C_e \int d^3r f_\alpha^{(A)}(\vec{r}) \rho_{JJ'}^{(A)}(\vec{r}) = C_e \rho_{\alpha,JJ'}^{(A)}$$

and similarly for B in its many-electron basis set $\{|K_B\rangle\}$, that appear in the interaction energies, can be obtained from the many-electron functions. The matrix elements of one-electron Hamiltonian terms contain only $\rho_{\alpha,JJ'}^{(A)}$ and $\rho_{\beta,KK'}^{(B)}$, while matrix elements of two-electron terms contain products $\rho_{\alpha,JJ'}^{(A)}\rho_{\beta,KK'}^{(B)}$. When using a density basis set of localized functions, a population analysis of $\rho_{\alpha,JJ'}^{(A)}$ in terms of atomic orbitals will show that many of its terms will be small and negligible when the location of AOs are far removed from the location of $f_\alpha^{(A)}(\vec{r})$.

The interaction energy operator can again be written as $\hat{H}_{AB}^{(Coul)} = \hat{H}_{AB}^{(nn)} + \hat{H}_{AB}^{(ne)} + \hat{H}_{AB}^{(ee)}$, with the first term given by the Coulomb repulsion of nuclear charges (or atomic ion cores) and the other two terms decomposed into contributions from electronic charge components. In particular, the electronic–nuclear interaction Hamiltonian energy is

$$\hat{H}_{AB}^{(ne)} = (4\pi\varepsilon_0)^{-1}\left[\sum\nolimits_{\alpha,b} \hat{c}_{A\alpha}^{(el)} I_{\alpha b}^{(AB)}(\vec{s}_b) C_b + \sum\nolimits_{\beta,a} C_a I_{a\beta}^{(AB)}(\vec{r}_a) \hat{c}_{B\beta}^{(el)}\right]$$

$$I_{\alpha b}^{(AB)}(\vec{s}_b) = \int d^3r \frac{f_\alpha^{(A)}(\vec{r})}{|\vec{r}-\vec{s}_b|},\ I_{a\beta}^{(AB)}(\vec{r}_a) = \int d^3s \frac{f_\beta^{(B)}(\vec{s})}{|\vec{r}_a-\vec{s}|}$$

where the $I_{\alpha b}^{(AB)} C_b$ and $C_a I_{a\beta}^{(AB)}$ integrals are electric potentials generated by nuclei of charges C_a and C_b interacting with electronic density distributions.

This Hamiltonian is a one-electron operator and its matrix elements can be obtained from one-electron density components $\rho_\alpha^{(A)}$ and $\rho_\beta^{(B)}$.

The electron–electron interaction is

$$\hat{H}_{AB}^{(ee)} = (4\pi\varepsilon_0)^{-1}\sum\nolimits_{\alpha,\beta}\hat{c}_{A\alpha}^{(el)}J_{\alpha\beta}^{(AB)}\hat{c}_{B\beta}^{(el)}$$

$$J_{\alpha\beta}^{(AB)} = \int d^3r\int d^3s f_\alpha^{(A)}(\vec{r})\frac{1}{|\vec{r}-\vec{s}|}f_\beta^{(B)}(\vec{s})$$

with the integral J in the second line giving the interaction of two densities. This Hamiltonian is a two-electron operator and its calculation involves two-electron density components $\rho_{\alpha\beta}^{(AB)} = \rho_\alpha^{(A)}\rho_\beta^{(B)}$ derived from the many-electron states of A and B separately, insofar they do not exchange electrons. These one- and two-electron expressions appear in the total Hamiltonian in a form useful for calculations involving both ground and excited electronic states for fixed atomic positions. The choices of the density basis sets are made to limit the number of (α, β) terms needed for convergence.

Interaction energies between A and B in their ground electronic states involve the total charge densities

$$c_0^{(A)}(\vec{r}) = c_{nu}^{(A)}(\vec{r}) + \sum\nolimits_\alpha c_{\alpha,0}^{(A)}f_\alpha^{(A)}(\vec{r})$$

containing the expectation values $c_{\alpha,0}^{(A)} = \left\langle 0_A\,|\,\hat{c}_{A\alpha}^{(el)}\,|\,0_A\right\rangle$, and also involve dynamical susceptibilities

$$\chi_0^{(A)}(\vec{r},\vec{r}\,';i\omega) = \sum\nolimits_\alpha\sum\nolimits_{\alpha'} f_\alpha^{(A)}(\vec{r})\chi_0^{(A)}(\alpha,\alpha';i\omega)f_{\alpha'}^{(A)}(\vec{r}\,')$$

with a similar expression for B. Electrostatic, induction, and dispersion energies follow from densities $c_{\alpha,0}^{(A)}$ and susceptibilities $\chi_0^{(A)}(\alpha,\alpha';i\omega)$, and similar functions for B.

The electrostatic interaction energy for A and B in their ground electronic states is $E_{els}^{(AB)} = E_{nn}^{(AB)} + E_{ne}^{(AB)} + E_{ee}^{(AB)}$, obtained from the corresponding Hamiltonians as averages with the operators $\hat{c}_{A\alpha}^{(el)}$ replaced by ground state values $c_{\alpha,0}^{(A)}$ and similarly for B. Induction and dispersion energies follow from related replacements, so that, for example,

$$E_{ind}^{(AB^*)} = -\frac{1}{2(4\pi\varepsilon_0)^2}\sum\nolimits_{\alpha\alpha'}\sum\nolimits_{\beta\beta'} c_{\alpha,0}^{(A)}J_{\alpha\beta}^{(AB)}\chi_0^{(B)}(\beta,\beta')J_{\beta'\alpha'}^{(AB)}c_{\alpha',0}^{(A)}$$

and corresponding expressions for $E_{ind}^{(A^*B)}$ and $E_{dsp}^{(A^*B^*)}$.

Depending on the structure of the extended system, it is convenient to introduce basis sets that allow for decomposition of the total system into physical fragments, or allow for expansion of the density function into spatial components. Integrals over the susceptibility position variables can be converted into sums over basis set indices. As mentioned in Chapter 2, two convenient basis sets involve expansions in localized oscillator functions of position variables, or expansions in Fourier (plane wave) components of the whole system.

7.3.3 Expansion in a Basis Set of Localized Functions

Given positions $\vec{r}_a$ and $\vec{r}_b$ within the two species A and B, one choice for $f_a^{(A)}(\vec{r})$ is to use at position $\vec{r}_a$ in A a selected set of N_a harmonic oscillator functions

$$\left\{ u_m^{(a)}(\xi;\lambda) = \lambda^{\frac{1}{2}} C_m \exp\frac{\lambda^2(\xi-\xi_a)^2}{2} H_m[\lambda(\xi-\xi_a)], m = 1 \, to \, N_a \right\}$$

with $H_m(x)$ the Hermite polynomial [34], containing displacements along directions $\xi = x, y, z$, at each location $\vec{r}_a$, and with optimized exponent parameters λ, which may vary with direction to account for local anisotropy. The spatial basis functions $u_{m_x}^{(a)}(x)u_{m_y}^{(a)}(y)u_{m_z}^{(a)}(z) = u_{\boldsymbol{m}}^{(a)}(\vec{r})$ with ($\boldsymbol{m}$ = m_x, m_y, m_z) can be used to expand the charge distribution as

$$\hat{c}_A^{(el)}(\vec{r}) = \sum\nolimits_{a \in A} \sum\nolimits_{\boldsymbol{m}} \hat{c}_{\boldsymbol{a}}^{(el)}(\boldsymbol{m}) u_{\boldsymbol{m}}^{(a)}(\vec{r})$$

where the coefficients $\hat{c}_{\boldsymbol{a}}^{(el)}(\boldsymbol{m})$ provide a coarse-graining representation of the charge densities, convenient for further treatment of interaction energies in a many-atom system.

Oscillator functions in $\hat{c}_A^{(el)}$ at different locations a and a' are not orthonormal and their overlap integrals $S_{\boldsymbol{m},\boldsymbol{m}'}^{(a,a')} = \langle u_{\boldsymbol{m}}^{(a)} \mid u_{\boldsymbol{m}'}^{(a')} \rangle$, involving different nuclear positions, must be incorporated into calculations. These overlaps are, however, short-ranged in (a,a') due to the localized nature of the basis functions. The expansion coefficients are obtained multiplying the above equation to the left times $u_{\boldsymbol{m}'}^{(a')}$ and integrating over space to obtain in matrix notation with $\boldsymbol{S} = \left[S_{\boldsymbol{m},\boldsymbol{m}'}^{(a,a')} \right]$,

$$\hat{c}_{\boldsymbol{a}}^{(el)}(\boldsymbol{m}) = \sum\nolimits_{a' \in A} \sum\nolimits_{\boldsymbol{m}'} \sum\nolimits_{i} u_{\boldsymbol{m}'}^{(a')}\left(\hat{\vec{r}}_i\right) \left[\boldsymbol{S}^{-1}\right]_{\boldsymbol{m}',\boldsymbol{m}}^{(a,a')}$$

which is a one-electron operator localized around nucleus a. The overlap integrals $\boldsymbol{S}$ appear in the Franck–Condon treatment of molecular transitions and have been thoroughly analyzed using harmonic oscillator operators, to derive recurrence relations in the m indices [35]. Interaction integrals I and

J within A can be obtained in compact form introducing the inverse function representation

$$r^{-1} = \frac{2}{\pi^{\frac{1}{2}}} \int_0^\infty d\kappa \exp(-\kappa^2 r^2),$$

using the property that a product of Gaussian functions gives a new Gaussian at a shifted position, and using translation properties of the harmonic oscillator functions within A. The resulting integrals are similar to the ones extensively employed in calculations of electronic structure with Gaussian basis sets [36]. Expansions for B are done similarly with a basis of functions $u_{\boldsymbol{n}}^{(b)} = u_{n_x}^{(b)}(x) u_{n_y}^{(b)}(y) u_{n_z}^{(b)}(z)$ localized at $\vec{r}_b$.

Interaction energies between A and B involve oscillator functions of $\vec{r}$ and $\vec{s}$ localized around $\vec{r}_a$ and $\vec{r}_b$ in each structure. Their products overlap in space at electron positions $\vec{\rho}$ and the products of Gaussian functions for A and B contain the relation

$$\exp\left(-\alpha\left|\vec{\rho}-\vec{r}_a\right|^2\right)\exp\left(-\beta\left|\vec{\rho}-\vec{r}_b\right|^2\right)$$
$$= \exp\left(-\frac{\alpha\beta}{\alpha+\beta}\left|\vec{r}_a-\vec{r}_b\right|^2\right)\exp\left[-(\alpha+\beta)\left(\left|\vec{\rho}-\vec{r}_I\right|^2\right)\right]$$

where $\vec{r}_I = (\alpha\vec{r}_a + \beta\vec{r}_b)/(\alpha+\beta)$ is an intermediate location. The first exponential to the right is very small when fragments of A and B are well separated, indicating that their interaction energies can be discarded. This can greatly decrease the number of integrals required to achieve accuracy.

An alternative is to use distributed multipoles with the density operator for A given as sums of terms localized around positions $\vec{r}_a$ [15],

$$\hat{c}_A^{(el)}(\vec{r}) = \sum\nolimits_a \sum\nolimits_{n,l,m} \hat{c}_{a,nlm}^{(el)} L_{a,nlm}(\vec{r})$$

$$L_{a,nlm}(\vec{r}) = A_{nlm} r_a^l \left[\frac{4\pi}{(2l+1)}\right]^{1/2} Y_{lm}(\vartheta_a, \varphi_a) \exp\left(-\alpha_n\left|\vec{r}-\vec{r}_a\right|^2\right)$$

where A_{nlm} is a normalization constant, and (ϑ_a, φ_a) are angular variables for $\vec{r}-\vec{r}_a$. The $L_{a,nlm}(\vec{r})$ functions contain solid spherical functions with selected exponential parameters α_n. They have known translational and rotational properties, and can be located at nuclear positions and along bonds. This expansion allows decomposition of large molecules into molecular fragments selected to contain certain nuclei.

Returning to expansions with the oscillator functions $u_{\boldsymbol{m}}^{(a)}(\vec{r})$, they introduce localized electronic density operators $\hat{c}_a^{(el)}(\boldsymbol{m})$ and $\hat{c}_b^{(el)}(\boldsymbol{n})$ into the Hamiltonian operator $\hat{H}_{AB}^{(int)} = \hat{H}_{AB}^{(nn)} + \hat{H}_{AB}^{(ne)} + \hat{H}_{AB}^{(ee)}$. This leads to expressions for electrostatic, induction, and dispersion energies in terms of ground state electronic charge densities $\langle 0_A | \hat{c}_{\boldsymbol{a}}^{(el)}(\boldsymbol{m}) | 0_A \rangle = c_0^{(a)}(\boldsymbol{m}) = C_e \rho_{0,\boldsymbol{m}}^{(a)}$ for A and $C_e \rho_{0,\boldsymbol{n}}^{(b)}$ for B. The dynamical susceptibility for A is a double sum over locations a and a',

$$\chi_0^{(A)}(\vec{r}, \vec{r}'; i\omega) = \sum_{a \in A} \sum_{\boldsymbol{m}=1}^{N_a} \sum_{a' \in A} \sum_{\boldsymbol{m}'=1}^{N_{a'}} u_{\boldsymbol{m}}^{(a)}(\vec{r}) \chi_0^{(aa')}(\boldsymbol{m}, \boldsymbol{m}'; i\omega) u_{\boldsymbol{m}'}^{(a')}(\vec{r}')$$

with small terms when the distance $|\vec{r}_a - \vec{r}\,'_a|$ is large for molecular fragments far away, so that many of them can be discarded in calculations. A similar expression applies to the susceptibility of B using $u_{\boldsymbol{n}}^{(b)}(\vec{s})$.

The two expansions replaced in induction and dispersion energy integrals over space variables contain products of summations over $(a, \boldsymbol{m})$ and $(b, \boldsymbol{n})$, and here again many terms can be discarded when $|\vec{r}_a - \vec{r}_b|$ is large.

7.3.4 Expansion in a Basis Set of Plane Waves

An expansion in a basis set of plane waves is done enclosing the whole system in a volume $\Omega = L_x L_y L_z$, and introducing periodic boundary conditions, which restrict the wavevectors in expansion functions $\langle \vec{r} | \vec{q}_{\boldsymbol{n}} \rangle = \Omega^{-1/2} \exp(-i\vec{q}_{\boldsymbol{n}}.\vec{r}) = \langle \vec{r} + \vec{L} | \vec{q}_{\boldsymbol{n}} \rangle$ to a denumerable set $\{\vec{q}_{\boldsymbol{n}}\}$ with $\boldsymbol{n} = (n_x, n_y, n_z)$ a triplet of integers, and $q_{n_\xi} = n_\xi 2\pi / L_\xi, \xi = x, y, z$. The choice $L_\xi = N_\xi l_\xi$, with l_ξ a distance over which densities change, and $|n_\xi| \leq N_\xi$ generates a grid with a desired density of points in the reciprocal space of $\vec{q}_{\boldsymbol{n}}$. The charge density operator for A or B is Fourier transformed from space to reciprocal space using that $\langle \vec{q}_{\boldsymbol{n}} | \vec{q}_{\boldsymbol{n}'} \rangle = \Omega^{-1} \int_\Omega d^3 r \exp(-i\vec{q}_{\boldsymbol{n}}.\vec{r}) \exp(i\vec{q}_{\boldsymbol{n}'}.\vec{r}) = \delta_{\boldsymbol{n}\boldsymbol{n}'}$, and that within the volume Ω, this basis set satisfies $\sum_{\boldsymbol{n}} |\vec{q}_{\boldsymbol{n}}\rangle\langle \vec{q}_{\boldsymbol{n}}| = \hat{I}$ so that

$$\hat{c}(\vec{r}) = \sum_{\boldsymbol{n}} \exp(i\vec{q}_{\boldsymbol{n}}.\vec{r}) \hat{c}_{\boldsymbol{n}},$$

$$\hat{c}_{\boldsymbol{n}} = \int_\Omega \frac{d^3 r}{\Omega} \exp(-i\vec{q}_{\boldsymbol{n}}.\vec{r}) \hat{c}(\vec{r}) = \frac{1}{\Omega} \sum_I C_I \exp\left(-i\vec{q}_{\boldsymbol{n}}.\hat{\vec{r}}_I\right)$$

and the coefficients $\hat{c}_{\boldsymbol{n}}$ are used in interactions energy operators. This expansion is common in treatments of solid state properties, and is particularly useful for polymers, surfaces, and solids with periodic atomic structure [37]. Products of operators like $\hat{c}(\vec{r})\hat{c}(\vec{r}')$ are likely to contribute little to interactions when the phase difference $\vec{q}_{\boldsymbol{n}}.\vec{r} - \vec{q}_{\boldsymbol{n}'}.\vec{r}'$ is large, in which case the exponential functions in the products show fast oscillations that add up to nearly zero in integrals.

The expansion in the plane waves $\Omega^{-1/2} \exp(-i\vec{q}_{\boldsymbol{m}}.\vec{r})$ gives for the dynamical susceptibility

$$\chi_0^{(A)}(\vec{r}, \vec{r}'; i\omega) = \sum_{\boldsymbol{m}} \sum_{\boldsymbol{m}'} \exp(i\vec{q}_{\boldsymbol{m}}.\vec{r}) \chi_0^{(A)}(\vec{q}_{\boldsymbol{m}}, \vec{q}_{\boldsymbol{m}'}; i\omega) \exp(i\vec{q}_{\boldsymbol{m}'}.\vec{r}')$$

These summations can be replaced in the space integrals for induction and dispersion energies of interaction between A and B, containing also $\chi_0^{(B)}(\vec{q}_n,\vec{q}_{n'};i\omega)$, and can be transformed as done in Section 3.2.5.

The charge density operator of all the charges I in species A distributed over space can be re-expressed in terms of its Fourier components $Q_A(\vec{k})$ as in

$$\hat{c}_A(\vec{r}) = \sum\nolimits_I C_I \delta\left(\vec{r}-\hat{\vec{r}}_{IA}\right) = \sum\nolimits_m \exp(i\vec{q}_m.\vec{r})\hat{Q}_A(\vec{q}_m)$$

with the inverse $\hat{Q}_A(\vec{q}_m) = \int_\Omega d^3 r \exp(-i\vec{q}_m.\vec{r})\hat{c}_A(\vec{r})/\Omega = \sum\nolimits_I C_I \exp\left(-i\vec{q}_m.\hat{\vec{r}}_I\right)/\Omega$ and similarly for species B.

The interaction energy operator $\hat{H}_{AB}^{(Coul)}$ can be re-expressed in terms of Fourier components using here that $|\vec{r}-\vec{s}|^{-1} = \int \left[d^3k/(2\pi)^3\right]\exp\left[i\vec{k}.(\vec{r}-\vec{s})\right]k^{-2}$ to obtain, using that $\delta(\vec{q}_m+\vec{q}_n) = \Omega\delta_{m,-n}$ and $\vec{q}_{-m} = -\vec{q}_m$, the result

$$\hat{H}_{AB}^{(Coul)} = \frac{\Omega}{4\pi\varepsilon_0}\sum\nolimits_m \hat{Q}_A(\vec{q}_m)\hat{Q}_B(-\vec{q}_m)|q_m|^{-2}$$

to be incorporated into the first- and second-order perturbation expansions.

The electrostatic interaction energy is, for the ground electronic states of A and B,

$$E_{els}^{(AB)} = \frac{\Omega}{4\pi\varepsilon_0}\sum\nolimits_m Q_0^{(A)}(\vec{q}_m)Q_0^{(B)}(-\vec{q}_m)|q_m|^{-2}$$

where $Q_0^{(A)}(\vec{q}_m) = \langle 0_A|\hat{Q}_A(\vec{q}_m)|0_A\rangle$ is the total charge density component of A.

The induction energy is $E_{ind}^{(AB^*)} + E_{ind}^{(A^*B)}$, containing the charge density $c_0^{(A)}(\vec{r})$ and static susceptibility $\chi_0^{(B)}(\vec{s},\vec{s}')$ for the first term, and similarly for B in the second term. They can be obtained from the charge density components $\hat{Q}_A(\vec{q}_m)$ and from the transform $\tilde{\chi}_0^{(B)}(\vec{k},\vec{k}')$ of the susceptibility into reciprocal space variables. Doing the integrals over space variables first, followed by integrals over d^3k and d^3k' appearing in the Fourier transforms of $|\vec{r}-\vec{s}|^{-1}$, we find

$$E_{ind}^{(AB^*)} = -\frac{1}{2(4\pi\varepsilon_0)^2}\sum\nolimits_m\sum\nolimits_{m'} Q_0^{(A)}(\vec{q}_m)|q_m|^{-2}\tilde{\chi}_0^{(B)}(-\vec{q}_m,-\vec{q}_{m'})Q_0^{(A)}(\vec{q}_{m'})|q_{m'}|^{-2}$$

$$\tilde{\chi}_0^{(B)}(\vec{k},\vec{k}') = \frac{1}{\Omega^2}\int d^3s\int d^3s'\exp\left(-i\vec{k}.\vec{s}\right)\exp\left(-i\vec{k}'.\vec{s}'\right)\chi_0^{(B)}(\vec{s},\vec{s}') =$$
$$= \frac{2}{\hbar}\sum\nolimits_{K\neq 0}\omega_K^{-1}\left\langle 0|\hat{Q}_B(\vec{k})|K\right\rangle\left\langle K|\hat{Q}_B(\vec{k}')|0\right\rangle$$

This is a double sum over grid points involving the transformed static susceptibility in reciprocal space, for excited states K and excitation frequencies ω_K. A similar expression applies for $E_{ind}^{(A^*B)}$ reversing the A and B labels.

Using the Casimir–Polder procedure to write the dispersion energy in terms of dynamical susceptibilities at imaginary-valued frequencies, and with $\hat{Q}_A(\vec{q}_m)^\dagger = \hat{Q}_A(-\vec{q}_m)$, we find that

$$E_{dsp}^{(AB)} = -\left(\frac{\Omega}{4\pi\varepsilon_0}\right)^2 \sum\nolimits_m \sum\nolimits_n \frac{1}{|q_m|^2}\frac{1}{|q_n|^2}\frac{\hbar}{2\pi}\int_0^\infty d\omega \tilde{\chi}_0^{(A)}(\vec{q}_m,\vec{q}_n; i\omega)\tilde{\chi}_0^{(B)}(-\vec{q}_m,-\vec{q}_n; i\omega)$$

$$\tilde{\chi}_0^{(A)}(\vec{q}_m,\vec{q}_n; i\omega) = \frac{2}{\hbar}\sum\nolimits_{J\neq 0} \frac{\omega_J}{\omega_J{}^2+\omega^2}\langle 0 \,|\, \hat{Q}_A(\vec{q}_m) \,|\, J\rangle\langle J \,|\, \hat{Q}_A(\vec{q}_n) \,|\, 0\rangle$$

with $\hbar\omega_J = E_J^{(A)} - E_0^{(A)}$, and similarly for B. This gives the dispersion energy as a double sum over grid points in reciprocal space, and avoids an expansion in multipolar components of the density.

7.4 Models of Hydrocarbon Chains and of Excited Dielectrics

7.4.1 Two Interacting Saturated Hydrocarbon Compounds: Chains and Cyclic Structures

As an example we consider the long-range interaction of two saturated hydrocarbon chains $CH_3(CH_2)_{N-2}CH_3$, each containing a large number N_C of carbon atoms in a linear conformation. The two chain axes are taken to be parallel and at a distance D between them. The dispersion forces can be approximately calculated assuming that electronic polarizations occur at the C—C bonds, and that the resulting bond polarizabilities are additive. This can be physically justified and tested numerically by application of a partitioning of the chain charge density distribution into components labeled by the bond positions along the chain. Local components of dynamical polarizabilities can be extracted from electronic structure treatments of the whole chain. These local polarizabilities are not the same as the polarizabilities of a single bond, because the exchange of bond electrons along the chain leads to charge density interactions among bonds, so that the local polarizabilities must be parametrized from calculations or measurements. In more detail, the chain is made up of fragments or units $HC(sp^3)$—$C(sp^3)H$, with each shared C atom counting as a half atom, and with a C—C bond length $l_{bnd} = 0.127\ nm$ [38, 39].

For each bond $A(a)$ at location $\vec{R}_a$ along chain A we can use an expansion in localized functions $u_m^{(a)}(\vec{r})$ centered midway between the two C atoms in the bond, to introduce a localized bond charge density operator

$$\hat{c}_a^{(el)}(\vec{r}) = \sum\nolimits_{m=1}^{N_a} \hat{c}_{a,m}^{(el)} u_m^{(a)}(\vec{r})$$

with the total density given by $\hat{c}_A^{(el)}(\vec{r}) = \sum_{a \in A} \hat{c}_a^{(el)}(\vec{r})$. Neglecting bond–bond couplings insofar the localized functions have little overlap, we can introduce related bond susceptibilities $\chi_a(\omega)$, which (for now) can be taken as isotropic averages. Similarly chain B is broken into components $B(b)$ at $\vec{R}_b$ with their susceptibilities $\chi_b(\omega)$. The bond–bond induction energies are then

$$E_{ind}^{(AB^*)} = \sum_{a \neq b} -\frac{1}{2} \int d^3r \int d^3r' \int d^3s \int d^3s' \frac{c_0^{(a)}(\vec{r})\, \chi_0^{(b)}(\vec{s}, \vec{s}')\, c_0^{(a)}(\vec{r}')}{(4\pi\varepsilon_0)^2 |\vec{r} - \vec{s}||\vec{r}' - \vec{s}'|}$$

$$= \frac{1}{2} \sum_{a \neq b} E_{0,0}^{(ind)}(ab^*)$$

where $c_0^{(a)}(\vec{r})$ is the total (nuclear plus ground state electronic) charge density at bond a, and similarly for $E_{0,0}^{(ind)}(A^*B)$. The integrals can be simplified considering that functions in the numerator are localized at $\vec{R}_a$ and $\vec{R}_b$. The integrand can be expanded in powers of the displacements $\vec{r} - \vec{R}_a$ and $\vec{s} - \vec{R}_b$, keeping only terms of zero and first order corresponding to total charge and dipole. To lowest order, induction energies result from the charge $C^{(a)}$ at a multiplying the static dipolar polarizability $\alpha_1^{(b)}(0)$ at b, with $E_{0,0}^{(ind)}(ab^*) = -(4\pi\varepsilon_0)^{-2}\left(C^{(a)}\right)^2 \alpha_1^{(b)}(0)/\left(2R_{ab}^4\right)$, as shown in Chapter 3. This is zero in the present example insofar the net charge of bond a is zero here. The same applies to $E_{0,0}^{(ind)}(a^*b)$. Bond–bond induction energies appear, however, when there are permanent bond dipoles and quadrupoles of a, and bond quadrupole polarizability at b.

The dispersion energy follows from a similar analysis leading to

$$E_{dsp}^{(A^*B^*)} = \frac{1}{2} \sum_{a \neq b} E_{0,0}^{(dsp)}(a^*b^*)$$

$$E_{0,0}^{(dsp)}(a^*b^*) = -\frac{\hbar}{2\pi} \int d^3r \int d^3r' \int d^3s \int d^3s' \int_0^\infty d\omega \frac{\chi_0^{(a)}(\vec{r}, \vec{r}'; i\omega)\, \chi_0^{(b)}(\vec{s}, \vec{s}'; i\omega)}{(4\pi\varepsilon_0)^2 |\vec{r} - \vec{s}||\vec{r}' - \vec{s}'|}$$

Here again it is possible to expand the integrand around $\vec{R}_a$ and $\vec{R}_b$ and to keep to lowest order the bond dynamical dipole polarizabilities, which give $E_{0,0}^{(dsp)}(a^*b^*) = -C_6^{(dsp)}(a^*b^*)/\left[(4\pi\varepsilon_0)^2 R_{ab}^6\right]$ with the $C_6^{(dsp)}$ coefficient expressed in terms of dipolar bond polarizabilities $\alpha_1^{(a)}(\omega)$ and $\alpha_1^{(b)}(\omega)$ as shown in Chapter 3. These polarizabilities can be chosen to be isotropic averages in the simplest case.

In the present example, the dispersion coefficients for bonds in the two chains are equal to a value $C_6^{(AB)}$ and one finds for chains with $N_{chn} = N_C - 1$ bonds,

$$E_{dsp}^{(A^*B^*)} = -\frac{C_6^{(AB)}}{(4\pi\varepsilon_0)^2}\sum_{a=1}^{N_{bnd}}\sum_{b=1}^{a} R_{ab}^{-6}$$

with $R_{ab}^2 = D^2 + l_{bnd}^2(a-b)^2$. The two summations can be rewritten as a sum over $a = b$ and a sum over $k = |a-b|$, of which there are $2(N_{chn} - k)$ at distances $\left[D^2 + l_{bnd}^2 k^2\right]^{1/2}$. This gives

$$E_{dsp}^{(A^*B^*)} = -\frac{C_6^{(AB)}}{(4\pi\varepsilon_0)^2}\left[\frac{N_{chn}}{D^6} + \sum_{k=1}^{N_{chn}-2}\frac{2(N_{chn}-k)}{\left[D^2 + l_{bnd}^2 k^2\right]^3}\right]$$

This allows calculation of limits when the chain length $L_{chn} = N_{chn}l_{bnd}$ is much larger or much smaller than the chain-to-chain distance D. When $N_{chn} \geq 10$, the quotient $y = k/N_{chn}$ varies almost continuously and the summation can be replaced by the integral $N_{chn}\int dy\ldots$. Doing the integration with $\rho = L_{chn}/D$ as a parameter, one finds [38]

$$E_{dsp,chn}^{(A^*B^*)} = -\frac{C_6^{(AB)}}{(4\pi\varepsilon_0)^2}\left\{\frac{N_{chn}}{D^6} + \frac{\rho}{4\,l_{bnd}{}^2 N_{chn}{}^4}\left[3\,\mathrm{atan}(\rho) + \frac{\rho}{1+\rho^2}\right]\right\}$$

When $D \gg L_{chn}$ (or $\rho \ll 1$) and the chains are distant, one simply finds $E_{0,0}^{(dsp)} = -C_6^{(AB)}N_{chn}^2/\left[(4\pi\varepsilon_0)^2 D^6\right]$, for pairs of bonds with energy varying as the inverse-six distance. When instead the chains are close and $D \ll L_{chn}$ (or $\rho \ll 1$), however, the interaction of pairs does not reach far, and one finds $E_{dsp,chn}^{(A^*B^*)} = -C_6^{(AB)}3\pi N_{chn}/\left[(4\pi\varepsilon_0)^2 8 l_{bnd}D^5\right]$, linear in the number of bonds and varying as the inverse-five distance. The value of the coefficient $C_6^{(AB)}$ can be fit to theory or experiment, and has a magnitude around 0.561 kJmol^{-1} nm^6 [38].

A similar procedure has been followed to describe the interaction of two identical cyclic compounds on parallel planes. They are taken to be circular with aligned centers and a circumference of length $L_{cycl} = N_C l_{bnd}$, at a distance D between planes. The dispersion energy is found to be [38]

$$E_{dsp,cycl}^{(A^*B^*)} = -\frac{C_6^{(AB)}}{(4\pi\varepsilon_0)^2}\frac{3\pi^2\rho'^2}{8\,l_{bnd}{}^2 D^4\left(\rho'^2+1\right)^{\frac{1}{2}}}\left[1 + \frac{2}{3}\frac{1}{\left(\rho'^2+1\right)} + \frac{1}{\left(\rho'^2+1\right)^2}\right]$$

where now $\rho' = L/(\pi D)$. The linear chain result is recovered at distances D small compared with L.

More accurately, the bond can be described by anisotropic parallel and perpendicular bond polarizabilities $\alpha_{\parallel}^{(a)}(\omega)$ and $\alpha_{\perp}^{(a)}(\omega)$ obtained in a reference frame attached to the bond, such that for a bond making an angle θ with a reference z-axis in a space fixed frame, the oriented bond polarizability is

$$\alpha_{\theta}^{(a)}(\omega) = \cos^2(\theta)\alpha_{\parallel}^{(a)}(\omega) + \sin^2(\theta)\alpha_{\perp}^{(a)}(\omega)$$

with an average over orientations $\alpha_{\mathrm{Av}}^{(a)}(\omega) = \left[\alpha_{\parallel}^{(a)}(\omega) + 2\alpha_{\perp}^{(a)}(\omega)\right]/3$. Also, a molecular fragment can be chosen to include the C—H bonds so that the polarizability can be constructed along a given axis as $\alpha\left(\mathrm{H\ddot{C}-\ddot{C}H}\right) = \alpha(\mathrm{C-C}) + 2\alpha(\mathrm{C-H})$ introducing expansions in basis sets $u_{\boldsymbol{m}}^{(a)}(\vec{r})$ located at the midpoints of all three atom–atom bonds, and using oriented bond polarizabilities. Values of static polarizabilities for a large number of molecular fragments have been collected [40, 41] and can be used to construct dynamical polarizabilities with interpolation formulae as shown in Chapter 2.

7.4.2 Two Interacting Conjugated Hydrocarbon Chains

Intermolecular forces between two extended systems containing delocalized electrons involve not just charges and dipoles, but also higher multipoles whose interactions are not simple functions of the distance between the systems. A general treatment can be based on charge density operators and their matrix elements between many-electron states, which are functions of density locations. The interaction energies require integrations over two or four space variables and these can best be done expanding the density operators of A and B with a common set of Fourier basis functions $u(\vec{r};\vec{q}_{\boldsymbol{n}}) = \Omega^{-1/2}\exp(-i\vec{q}_{\boldsymbol{n}}.\vec{r})$, to transform space integrals into sums over Fourier components. These density components are one-electron operators, and their matrix elements between many-electron states, as they appear in interaction energies, can be given in basis sets of delocalized electronic orbitals $\varphi_j^{(A)}(\vec{r})$ and $\varphi_k^{(B)}(\vec{s})$. The Fourier expansion is obtained putting the interacting chains A and B within a periodic supercell of sides L_x, L_y and L_z large enough to include the electronic densities of both species, with wavenumbers $q_{n\xi} = n_\xi \pi / L_\xi$, $n_\xi = 1, \ldots N_{\xi,\, max}$, and with the total number $N_G = N_{x,\, max} N_{y,\, max} N_{z,\, max}$ of grid points as needed for accuracy.

We consider the case of electrons in two identical linear chains of conjugated polyenes such as $CH_2 = [CH - CH=]_k CH_2$ with k units of alternating single and double C—C bonds of lengths l_1 and l_2, respectively, each with a frame in the (xz)-plane and the chain axis parallel to the z-axis, with chain A at the origin of coordinates and a parallel chain B displaced by a distance D between them along the x-axis. A simple model separates sigma- and pi-electron orbitals on account of their different symmetry on reflection of orbitals on the plane of the chain frame. The electronic charge density operator is then $\hat{c}^{(el)}(\vec{r}) = \hat{c}^{(\sigma)}(\vec{r}) + \hat{c}^{(\pi)}(\vec{r})$. The sigma-density is localized at bonds and can be expanded as done before. The total ion and electronic charges are equal and cancel in a neutral chain. The pi-electron density in a chain with N_C carbon atoms and $N_\pi = N_C$ pi electrons is instead delocalized and requires expansions in a set of delocalized MOs for a chain of length $L_{chn} = (N_C + 1)(l_1 + l_2)/2$.

Expansion of the two electronic densities $\hat{c}_A^{(\pi)}(\vec{r})$ and $\hat{c}_B^{(\pi)}(\vec{r})$ in the same $u(\vec{r};\vec{q}_n)$ basis set provides density components $\hat{Q}_S^{(\pi)}(\vec{q}_n), S = A, B$, and its matrix elements between pi-many-electron states $|J_S\rangle$ give static densities and susceptibility components $\chi_\pi^{(S)}(\vec{q}_n, \vec{q}_{n'}; i\omega)$. To calculate the susceptibilities, it is necessary to introduce electronic states and energies of the structures. The model here is at the simplest level just a collection of independent pi-electrons described by a one-electron Hamiltonian $\hat{H}_\pi$ and by many-electron determinantal functions containing pi-electron molecular orbitals $\varphi_j^{(\pi)}(\vec{r})$ perpendicular to the chain frame, from

$$\hat{H}_\pi \varphi_j^{(\pi)}(\vec{r}) = \varepsilon_j^{(\pi)} \varphi_j^{(\pi)}(\vec{r})$$

The matrix elements of one-electron operators between many-electron states are given by sums of one-electron integrals containing spin-orbitals $\varphi_j^{(\pi)}(\vec{r})\vartheta(\zeta)$ [42]. Susceptibilities involve the pi-electron transition densities $\rho_{jj'}^{(A)}(\vec{r}, \vec{r}\,') = \varphi_j^{(\pi)}(\vec{r})\varphi_{j'}^{(\pi)}(\vec{r}\,')^*$ of A, interacting with $\rho_{kk'}^{(B)}(\vec{s}, s')$ of B, and again include all electric multipole components of the chain. The space integrals appearing in the interaction energies can be expressed in terms of density Fourier components as described in Section 7.3, for more efficient computational work.

The orbitals can easily be constructed in two approximations:

a) From particle-in-a-box states for a box with the dimensions of the chain, and positive charges of the atomic ions spread and assumed to be homogeneous inside the box on the average. The electron densities are obtained from orbitals $\varphi^{(\pi)}(\vec{r};\vec{k}_j)$ of electrons in the box. The total ion and electronic charges are equal and cancel in a neutral chain.
b) Alternatively and more accurately, the delocalized orbitals can be constructed as LCAOs from 2pπ atomic orbitals $\xi_\mu(\vec{r})$ orthogonal to the plane of the chain frame, with $\varphi_j^{(\pi)} = \sum_\mu c_{\mu j}\xi_\mu$ as pi-MO-LCAOs.

Matrix elements $\langle J_A | \hat{Q}_A^{(\pi)}(\vec{q}_n) | J_A' \rangle$ of the one-electron density operator components between determinantal states are different from zero only if the states are identical or different by only one spin-orbital (the Condon-Slater rules) [42]. They are given by one-electron integrals $\langle \boldsymbol{j} | C_e \exp(-i\vec{q}_n \cdot \vec{r})/\Omega | \boldsymbol{j}' \rangle$ between orbitals $\langle \vec{r} | \boldsymbol{j} \rangle = \varphi_j^{(\pi)}(\vec{r})$. Ground state expectation values of electronic density components follow from the two locations of the chains separated by a distance D, with $\langle 0_A | \hat{Q}_A^{(\pi)}(\vec{q}_n; 0) | 0_A \rangle$ and $\langle 0_B | \hat{Q}_B^{(\pi)}(\vec{q}_n; D) | 0_B \rangle$ giving the electrostatic interaction energy. The dynamical susceptibility $\chi_{\pi}^{(A)}\left(\vec{q}_n,\vec{q}_{n'};i\omega\right)$ of A can be calculated from integrals $\langle 0_A \mid \hat{Q}_A^{(\pi)}\left(\vec{q}_n;0\right) \mid J_A\rangle$ with orbitals $\boldsymbol{j}$ initially occupied in the ground state and excited to unoccupied $\boldsymbol{j}'$, and similarly for B. A change of integration variables from x to $x-D$ gives the relation

$$\left\langle \boldsymbol{j} \mid c_e \exp\left(-i\vec{q}_n.\hat{\vec{r}}\right)/\Omega \mid \boldsymbol{j}'\right\rangle_B = \exp\left(\frac{-in_x\pi D}{L_x}\right)\left\langle \boldsymbol{j}\middle| c_e \exp\left(\frac{-i\vec{q}_n.\hat{\vec{r}}}{\Omega}\right)\middle|\boldsymbol{j}'\right\rangle_A$$

which displays the dependence of matrix elements on the distance D.

They provide electrostatic, induction, and dispersion interaction energies. The electronic term in the electrostatic interaction energy is

$$E_{els,\pi}^{(AB)}(D) = \left(\frac{\Omega}{4\pi\varepsilon_0}\right)\sum\nolimits_{n}\exp\left(\frac{in_x\pi D}{L_x}\right)Q_{0,\pi}^{(A)}\left(\vec{q}_n;0\right)Q_{0,\pi}^{(A)}\left(-\vec{q}_n;0\right)|q_n|^{-2}$$

with wavenumbers $q_\xi = n_\xi\pi/L_\xi$, $n_\xi = 1, \dots N_{\xi,\,max}$. The total electrostatic interaction energy must also include interactions with nuclear charges, and the net result is zero if the chains are neutral and far apart. Induction and dispersion energies are obtained from susceptibilities. The dynamical susceptibility $\chi_{\pi}^{(A)}\left(\vec{q}_n,\vec{q}_{n'};i\omega\right)$ of A can be calculated from integrals $\langle 0_A \mid \hat{Q}_A^{(\pi)}\left(\vec{q}_n;0\right) \mid J_A\rangle$ and excitation energies $\Delta\varepsilon_{j'j}^{(\pi)} = \varepsilon^{(\pi)}(\boldsymbol{j}') - \varepsilon^{(\pi)}(\boldsymbol{j})$ appearing in denominators, which are smallest for the transition between highest occupied (HO) and lowest unoccupied (LU) energy levels. Similar quantities are needed for B displaced by D along x with the dependence on the distance D appearing in the integrals for the Fourier components of the density.

In the box model, two chains can be located one inside a box A with boundaries $0 \le x \le L_\perp$, $0 \le y \le L_\perp$,and $0 \le z \le L_\| = L_{chn}$ and the other inside a similar box B parallel to the first with its origin displaced along the x-axis by a distance D so that a new boundary is $D \le x \le D + L_\perp$. In a particle-in-a-box model, the electronic states are null outside the boxes and at their edges. For A, they are

$$\varphi_{\pi}^{(A)}\left(\vec{r};\vec{k}_j\right) = \left(\frac{2}{L_\perp}\right)^{\frac{1}{2}}\sin\left(k_x x\right)\left(\frac{2}{L_\perp}\right)^{\frac{1}{2}}\sin\left(k_y y\right)\left(\frac{2}{L_\|}\right)^{\frac{1}{2}}\sin\left(k_z z\right)$$

where $k_\xi = j_\xi\pi/L_\perp$, $\xi = x, y$ and $k_z = j_z\pi/L_\|$ for integers j_ξ. This assumes that the spacing of grid points is dense enough, with the total system in a box with $L_\xi \gg L_\perp, L_\|$, allowing choices like $L_{x,\,y} = 100L_\perp$ and $L_z = 100L_\|$ to properly describe the interaction of the chains for varying distance D between them. The box energies are $\varepsilon^{(\pi)}(\vec{k}_j) = \left(\hbar^2/2m_e\right)\left|\vec{k}_j\right|^2, 1 \le j_\xi < \infty$, and the ground state energy has quantum numbers $\boldsymbol{j} = (111)$. The orbitals $\varphi_{\pi}^{(B)}(\vec{r};\vec{k}_l)$ for B are similarly obtained for the identical displaced chain along x, with boundaries $D \le x \le D + L_\perp$, $0 \le y \le L_\perp$, and $0 \le z \le L_\|$ and the factor $(2/L_\perp)^{\frac{1}{2}}\sin\left[k_x(x-D)\right]$ giving the x-dependence.

The present model assumes that the pi electrons occupy orbitals and are excited only along the z-axis, but remain in the lowest orbital along x- and y-axes, so that here $j_x = j_y = 1$.

The one-electron integrals $\langle \boldsymbol{j} | c_e \exp\left(-i\vec{q}_{\boldsymbol{n}}.\hat{\vec{r}}\right)/\Omega | \boldsymbol{j}' \rangle$ between box orbitals $\langle \vec{r} | \boldsymbol{j} \rangle = \varphi_{\pi}^{(A)}(\vec{r};\vec{k}_j)$ are products of integrals over the x, y, and z variables. Each factor contains products of *sin* and *cos* functions, readily integrated and giving negligible values for very large j_z. For example, the z-integrals for the density component $\hat{Q}_A^{(el)}(q_z)$ are

$$\left\langle j_z \middle| \exp\left(-\frac{in_z\pi\hat{z}}{L_z}\right) \middle| j_z' \right\rangle = \frac{2}{L_\parallel}\int_0^{L_\parallel} dz \sin\left(j_z\frac{\pi}{L_\parallel}z\right)\exp\left(-\frac{in_z\pi z}{L_z}\right)\sin\left(j_z'\frac{\pi}{L_\parallel}z\right)$$

$$= -\frac{iL_\parallel}{\pi L_z}n_z\left[\frac{1+\exp\left(in_z\pi L_\parallel/L_z\right)(-1)^{j_z-j_z'}}{\left(j_z-j_z'\right)^2-n_z{}^2\left(L_\parallel/L_z\right)^2}+\frac{1-\exp\left(in_z\pi L_\parallel/L_z\right)(-1)^{j_z+j_z'}}{\left(j_z+j_z'\right)^2-n_z{}^2\left(L_\parallel/L_z\right)^2}\right]$$

which can be further simplified for $L_z \gg L_\parallel$, showing then that the matrix elements change linearly with $L_\parallel/L_z$. Integrals over $0 \le x \le L_\perp$ and $0 \le y \le L_\perp$ have corresponding forms, also changing linearly with $L_\perp/L_\xi$. All integrals decrease as j_ξ^{-2} for large orbital quantum numbers, and as n_ξ^{-1} for the density operator Fourier components. Matrix elements $\langle \boldsymbol{k} | C_e \exp\left(-i\vec{q}_{\boldsymbol{n}}.\hat{\vec{r}}\right)/\Omega | \boldsymbol{k}' \rangle$ for B involve the same integrals over y and z as for A, while the integral over x has new limits $\le x \le D + L_\perp$. Integrals have the same form as given, but with the additional factor $\exp(-in_x\pi D/L_x)$ from the integration over $x - D$.

Ground state expectation values of electronic density components follow from $\langle 0_A | \hat{Q}_A^{(\pi)}(\vec{q}_{\boldsymbol{n}};0) | 0_A \rangle = Q_{0,x}^{(A)}(\vec{q}_{\boldsymbol{n}};0)Q_{0,y}^{(A)}(\vec{q}_{\boldsymbol{n}};0)Q_{0,z}^{(A)}(\vec{q}_{\boldsymbol{n}};0)$, where the two first factors are given by the above integral with $j_x = j_x' = 1$, $j_y = j_y' = 1$ and

$$Q_{0,z}^{(A)}(\vec{q}_{\boldsymbol{n}};0) = \left(\frac{C_e}{\Omega}\right)\sum_{j_z=1}^{j_z=N_\pi/2} 2\left\langle j_z \middle| \exp\left(-\frac{in_z\pi\hat{z}}{L_z}\right) \middle| j_z \right\rangle$$

for an even number of carbon atoms in the chain, and orbitals with double spin occupancy. The corresponding expectation value for B is $\langle 0_B | \hat{Q}_B^{(\pi)}(\vec{q}_{\boldsymbol{n}};D) | 0_B \rangle = Q_{0,\pi}^{(B)}(\vec{q}_{\boldsymbol{n}};D) = \exp(-in_x\pi D/L_x)Q_{0,\pi}^{(A)}(\vec{q}_{\boldsymbol{n}};0)$.

The dynamical susceptibility $\chi_\pi^{(A)}(\vec{q}_{\boldsymbol{n}},\vec{q}_{\boldsymbol{n}'};i\omega)$ of A can be calculated from integrals $\langle 0_A | \hat{Q}_A^{(\pi)}(\vec{q}_{\boldsymbol{n}};0) | J_A \rangle$ and excitation energies $\Delta\varepsilon_{j'j}^{(\pi)} = \varepsilon^{(\pi)}\left(\vec{k}_{j'}\right) - \varepsilon^{(\pi)}\left(\vec{k}_j\right) = \left(\hbar^2/2m_e\right)\left(\pi/L_\parallel\right)^2\left(j_z'^2 - j_z^2\right)$ and similar quantities for box B displaced by D along x, all of which are known for the particle-in-a-box model. These susceptibilities give the right qualitative dependences with the length of chains, but their numerical values must be corrected introducing more accurate excitation energies $\Delta\varepsilon_{j'j}^{(\pi)} + \Delta\varepsilon_G^{(A)}$ with a HO–LU energy gap parameter $\Delta\varepsilon_G^{(A)}$. Related induction and dispersion energies are $E_{ind}^{(\pi\pi*)}(AB^*)$, $E_{ind}^{(\pi*\pi)}(A^*B)$ and $E_{dsp}^{(\pi*\pi*)}(A^*B^*)$, which depend on D and the box dimensions of the chain.

The interaction energies contain not just the dipole–dipole interaction but all the multipolar contributions in the charge density and in the susceptibilities, and their interactions in the electrostatic, induction, and dispersion energies of the chains. When the distance D between A and B is much larger than the chain dimensions, it is possible to expand in the density components as $\exp(-in_\xi\pi\xi/L_\xi) \approx 1 - in_\xi\pi\xi/L_\xi$ within the one-electron transition integrals to obtain transition dipoles and the dipolar polarizabilities of A and B. Matrix elements of $\hat{H}_{AB}^{(Coul)}$ vary as n_ξ^{-2} and $E_{dsp}^{(\pi*\pi*)}$ varies as n_ξ^{-6} for large n_ξ. Adding over the n_ξ index for density components, the dispersion energy recovers in this limit the van der Waals form corresponding to interacting dipole fluctuations, and a D^{-6} dependence on the distance between the chains.

In the model using pi-MO-LCAOs $\varphi_j^{(\pi)}(\vec{r}) = \sum_\mu c_{\mu j}\,\xi_\mu(\vec{r})$, coefficients and orbital energies can be obtained from a one-electron pi-Hamiltonian treatment where $\mu = 1\ to\ N_C$ gives the location of a pi-AO, and orbital energies are given in terms of Coulomb (or atom) and resonance (or bond) energy parameters α and β, as [42]

$$c_{\mu j} = \left(\frac{2}{N_C + 1}\right)^{\frac{1}{2}} \sin\left(\frac{j.\mu.\pi}{N_C + 1}\right)$$

$$\varepsilon_j = \alpha + 2\beta\cos\left(\frac{j.\pi}{N_C + 1}\right)$$

Matrix elements $\langle J_A \mid \hat{Q}_A^{(\pi)}(\vec{q}_n) \mid J'_A \rangle$ are given by sums over one-electron atom–atom integrals $\langle \mu \mid C_e \exp(-i\vec{q}_n.\hat{\vec{r}})/\Omega \mid \mu' \rangle$ with $\langle \vec{r} \mid \mu \rangle = \xi_\mu(\vec{r})$. These integrals are large only for pi-AOs at the same or near-neighbor locations. This means that we need the Fourier transform of only two densities, $|\xi_1(\vec{r})|^2$ and $\xi_1(\vec{r})\,\xi_2(\vec{r})^*$, for a generic pair of pi AOs. Matrix elements can be obtained from Fourier integrals $\langle \mu \mid c_e \exp(-i\vec{q}_n.\hat{\vec{r}})/\Omega \mid \mu \rangle = \langle 1 \mid C_e \exp(-i\vec{q}_n.\hat{\vec{r}})/\Omega \mid 1 \rangle = (C_e/\Omega) I_{atm}(\vec{q}_n)$ for an atom in chain A and $\langle \mu \mid C_e \exp(-i\vec{q}_n.\hat{\vec{r}})/\Omega \mid \mu + 1 \rangle = \langle 1 \mid C_e \exp(-i\vec{q}_n.\hat{\vec{r}})/\Omega \mid 2 \rangle = (C_e/\Omega) I_{bnd}(\vec{q}_n)$ for a bond in A, and corresponding integrals $I_{atm,bnd}$ for B which follow by displacing integrands by D along the x-axis. These integrals have explicit forms in the Fourier variables $\vec{q}_n$ for given pi-AOs functions and their Fourier transforms.

This model can be made more accurate allowing for changes of parameters with bond lengths. In particular, excitation energies are more accurate when allowance is made for changes of the β parameter between short and long bond lengths, allowing calculation of excitation energies to within an error of 10% [42]. Electrostatic, induction, and dispersion energies can be given in terms of the parameters α, β, I_{atm}, and I_{bnd}, or α, β_1, β_2, I_{atm} and $I_{bnd,\,1}$, $I_{bnd,\,2}$ when allowance is made for two bond lengths.

To complete the treatment of conjugated hydrocarbons, one must also calculate the energies $E^{(ind)}_{\sigma\pi}(AB^*)$, $E^{(ind)}_{\sigma\pi}(A^*B)$, and $E^{(dsp)}_{\sigma\pi}(A^*B^*)$ from expansions of one-electron density operators $\hat{c}^{(\pi)}(\vec{r})$ and $\hat{c}^{(\sigma)}(\vec{r})$, respectively in the $u(\vec{r};\vec{q}_n)$ and $u^{(a)}_m(\vec{r})$ basis sets. Furthermore, structures of the two chains other than parallel conformations can be considered along similar lines, for example, with the two chains in a "T" conformation, as done already in early work [40].

7.4.3 Electronic Excitations in Condensed Matter

Light absorbed by a molecular liquid or a molecular crystal creates electronically excited molecules in strong interactions with their neighbors. Examples are crystalline benzene and the simpler Ar crystal.We consider a system containing only one species M, as an aggregate, which can be generalized to include liquid and solid solutions of two components A and B. In a first approximation, the state of the many-molecule system can be expressed in terms of the ground and excited states of individual molecules M(m), m = 1 to N_{mol}, assuming that electron exchange among them is not important. This allows decomposition of the Hamiltonian $\hat{H}$ of the system into sums of individual molecular Hamiltonians $\hat{H}_m$ plus their interactions $\hat{H}^{(int)}_{mn}$ for each molecular pair (mn). This is a sum

$$\hat{H}^{(int)}_{mn} = \hat{H}^{(Coul)}_{mn} + \hat{H}^{(rep)}_{mn} + \hat{H}^{(vibr)}_{mn}$$

containing the Coulomb energy, repulsion energy at short distances, and vibronic interaction energy $\hat{H}^{(vibr)}_{mn}$ resulting from couplings of electrons and nuclear vibrations.

The Coulomb molecule–molecule interaction term is

$$\hat{H}^{(Coul)}_{mn} = (4\pi\varepsilon_0)^{-1}\int d^3r \int d^3s\, \hat{c}(\vec{r};\mathbf{Q}_m)\left|\vec{r}-\vec{s}\right|^{-1}\hat{c}(\vec{s};\mathbf{Q}_n)$$

in terms of molecular charge densities $\hat{c}(\vec{r};\mathbf{Q}_m)$, and can be further decomposed into contributions from electrons and nuclei (or ion cores) for each molecule. Electronic states $|\Phi_J\rangle, \mathbf{J} = (J_1,\ldots,J_{N_{mol}})$ for the many-molecule system can be constructed from electronic states $|J_m\rangle$ of molecule m, obtained from $\hat{H}_m|J_m\rangle = E^{(m)}_J|J_m\rangle$. The ground state is $|\Phi_0\rangle, \mathbf{0} = (0_1,\ldots,0_{N_{mol}})$, while singly-excited total states corresponding to all molecules except m being in a ground state $|0_m\rangle$ and m being in excited state $|J_m\rangle \neq |0_m\rangle$ are

$$|\Phi_{mJ}\rangle = |J_m\rangle \prod\nolimits_{n(\neq m)} |0_n\rangle$$

The interaction among molecules leads to electronic energy transfer so that states with any one molecule in the excited state $|J\rangle$ is a superposition $|\Phi(J)\rangle = \sum_m c_m(J)|\Phi_{mJ}\rangle$, with coefficients and corresponding aggregate energy $E_{aggr}(J)$ obtained from the eigenstates of the total Hamiltonian of the many-molecule system,

$$\hat{H}_{aggr} = \sum_m \hat{H}_m + \frac{1}{2}\sum_{m \neq n} \hat{H}_{mn}^{(int)}$$

for $J \neq 0$ as

$$\hat{H}_{aggr}|\Phi(J)\rangle = E_{aggr}(J)|\Phi(J)\rangle$$

These states and energies characterize the properties of excitons, which are defined as electronic excitations delocalized over all the molecules. They have been extensively studied for molecular solids and for aggregates including organic macromolecules [35, 43–46].

The total Hamiltonian contains one- and two-molecule terms and its matrix elements between many-molecule states have similarities with matrix elements between many-electron states [9]. Here we provide a simplified version assuming no electron exchange between molecules and using a basis set of molecular states satisfying the approximate orthonormalization $\langle J_m | J_n' \rangle = \delta_{mn}\delta_{JJ'}$ which is accurate when electronic overlaps between molecules are small. Matrix elements of $\hat{H}_m$ and $\hat{H}_{mn}$ can be obtained in the basis set $\{|\Phi_0\rangle, |\Phi_{mJ}\rangle\}$ and to simplify, we consider here only one excited molecular state $J = 1$.

Excitonic energies can be obtained from first-order perturbation theory, as

$$\Delta E_1 = E_{aggr}(1) - E_{aggr}(0) \cong \langle \Phi_{m1} | \hat{H}_{aggr} | \Phi_{m1} \rangle - \langle \Phi_0 | \hat{H}_{aggr} | \Phi_0 \rangle$$

Matrix elements $E_{JJ'}^{(m)} = \langle J_m | \hat{H}_m | J_m' \rangle$ of $\hat{H}_m$ with $J = 0, 1$ are nonzero only for $E_{JJ}^{(m)} = E_J^{(m)}$, $E_{0,1}^{(m)}$, or $E_{1,0}^{(m)}$. Matrix elements of $\hat{H}_{mn}$ are $V_{JJ',KK'}^{(mn)} = \langle J_m K_n | \hat{H}_{mn} | J_m' K_n' \rangle$ and nonzero only for $J' = J = 1$, $K = K' = 0$, and $K' = K = 1$, $J = J' = 0$, corresponding to molecular excitation transfers $1 \rightarrow 0$ and $0 \rightarrow 1$.

Here the Coulomb interaction involves matrix elements $\langle J_m | \hat{c}(\vec{r}; \mathbf{Q}_m) | J_m' \rangle$ of the density operators, and the density–density interaction contains not only the dipole–dipole interaction but also all higher multipoles. A formalism for interacting excitons beyond the first-order perturbation treatment can be based on a model Hamiltonian with linear, bilinear, and quartic terms in the excitation (and de-excitation) bosonic operators $\hat{B}_m^{\dagger} = |1_m\rangle\langle 0_m|$ (and $\hat{B}_m = |0_m\rangle\langle 1_m|$) [47].

More accurately, one must solve for the eigenenergies $E_{aggr}(J)$, $J = 0, 1$. This can be conveniently done when the aggregate is a crystal structure or periodic chain, taking advantage of the periodicity of the molecular lattice. For a cubic lattice with molecules at locations $\vec{R}_{\mathbf{m}} = \sum_\xi m_\xi \vec{a}_\xi, \xi = x, y, z$, with $-M_\xi \leq m_\xi \leq M_\xi$ an integer, and unit cells of sides $\vec{a}_\xi$, a choice of periodic boundary conditions require invariance under the change $\vec{R}_m \rightarrow \vec{R}_{m\pm M}$. The expansion coefficients in the eigenstates $|\Phi(J)\rangle$ can then be written as $c_m(J) = \exp(i\vec{k}_j.\vec{R}_m)b_j(J)$ with wave-vectors $k_\xi = j_\xi 2\pi/(M_\xi a_\xi)$, and integers j_ξ giving lattice symmetry labels, which can be used to identify energy bands of the aggregate. Periodicity allows a simpler calculation of the coefficients $b_j(0)$ and $b_j(1)$ and energies $E_{aggr}(0)$ and $E_{aggr}(1)$ from the eigenvalue equation for $|\Phi(J)\rangle$ [43].

7.5 Density Functional Treatments for All Ranges

7.5.1 Dispersion-Corrected Density Functional Treatments

Dispersion energy functions of intermolecular and intramolecular position variables can be combined with near-range functionals obtained from DFT and containing all energies (electrostatic, exchange, semilocal correlation, long-range induction) except for vdW energies to be added. This is a very active area of research where several treatments have been proposed and tested by comparison with accurate results from CCSD(T) or SAPT calculations. Here the focus is on methods applicable to large classes of chemical systems, described as many-atom structures A and B containing sets of atoms $\{a\}$ and $\{b\}$ with their centers of mass at a distance R in a body-fixed reference frame, and with internal variables $\boldsymbol{Q} = \{\vec{R}_a, \vec{R}_b\}$ referred to their C-of-M.

Although DFT is a theory that should in principle provide accurate results for all intermolecular distances, in practice it has been implemented with semilocal functional of the electron density and its space derivatives. As such, energy results contain electrostatic and charge induction terms but do not include dispersion energies, which physically originate in the quantal coupling of charge density fluctuations in A and in B. We have seen that the dispersion energy can be separately treated and calculated from the dynamical susceptibilities of the two species, or from their multipolar polarizabilities. The aim here is to add them to the semilocal energies while avoiding double counting of long-range interactions, and to ascertain how important these are for development of potential energy surfaces valid at all distances.

This has led recently to treatments where the semilocal DFT energies $E_{sl}(R, \boldsymbol{Q})$ at near-range distances, and dispersion (van der Waals) energies $E_{dsp}(R, \boldsymbol{Q})$ at large distances are combined to generate results for all intermolecular distances, while restricting the dispersion contribution to the long range by multiplying it times a suitable damping function $f_d(R)$, where as before $f_d(R) \approx 1$, for $R \rightarrow \infty$ and $f_d(R) \approx 0$, for $R \rightarrow 0$, so that

$$E_{tot}(R, \boldsymbol{Q}) = E_{sl}(R, \boldsymbol{Q}) + E_{dsp}(R, \boldsymbol{Q}) f_d(R)$$

This expression must be generalized when one of the species, such as B, surrounds A and requires damping not only as a function of the distance R between centers of mass, but also damping when atoms of A and B come too close, with a new $f_d(R, \mathbf{Q})$ as a function also of interatomic distances.

The interaction of two many-atom molecules has been similarly treated adding atom-pair dispersion energies with pair damping functions. This assumes that atom-pair dispersion energies are approximately additive, which can be shown to be correct for many-atom interactions to second order of perturbation theory, as previously discussed. The schemes developed so far have made different choices for the construction of E_{dsp} from atomic structures, as a sum over (ab) atom-pair interactions $E_{dsp}^{(ab)}(R_{ab})$ with $R_{ab} = \left|\vec{R} + \vec{R}_a - \vec{R}_b\right|$ multiplying instead a pair damping function $f_d^{(ab)}(R_{ab})$, so that

$$E_{tot}(R, \mathbf{Q}) = E_{sl}(R, \mathbf{Q}) + \sum_{a,b} E_{dsp}^{(ab)}(R_{ab}) f_d^{(ab)}(R_{ab})$$

where $\vec{R}$ is the relative position vector of the molecular centers of mass.

Much effort has been directed to the identification of the correct methods for calculating the atom-pair terms. They must be constructed from atomic properties of the atoms in the medium provided by neighboring atoms, which differ from free atomic properties. They must account for changes in size, shape, and electrical charge of each bonded atom, for its bonding coordination, and for changes in its dynamical response to fluctuating electric fields in the same molecule and from the other interacting molecule. Several treatments have been recently reviewed and provide many details of alternatives being explored [33, 48]. The above damping function is somewhat arbitrary, and can be chosen to make the energy a smooth function of position coordinates, or instead to make forces smoothly varying.

The DFT-D approach constructs $E_{dsp}^{(ab)}$ from atomic components with polarizabilities obtained from empirical data or more fundamentally, in the so-called D3 version [48], from time-dependent DFT calculations of dynamical polarization of the hydrides of each atom a by an oscillating electric field of frequency ω. This is separately done for the atom a in a medium where N_a bonds to nearest neighbors have been modeled by a corresponding number of single bondings to N_a hydrogen atoms. It provides dynamical polarizabilities for distorted atoms with coordination numbers, called CN^A for an atom a [48], given by functions $\alpha_a^{(N_a)}(\omega)$ in our notation, to be used in the Casimir–Polder expression for the dispersion coefficient obtained from $\alpha_a^{(N_a)}(\omega)$ and $\alpha_b^{(N_b)}(\omega)$. The preferred damping function for the long-range dispersion energy going as R^{-n} for large R is $f_d^{(n)}(R) = R^n/(R^n + R_d{}^n)$, containing a parameter R_d to be chosen [49]. It follows that the dispersion energy, containing $R^{-n} f_d^{(n)}(R)$, goes to a constant at $R = 0$, which agrees with theoretical results [50] and provides a smoothly varying force along R. Results of DFT-D3 calculations of potential energy surfaces done for several functional, including PBE-D3, B3LYP-D3, and TPSS-D3, have been compared with accurate values of equilibrium energies obtained from the CCSD(T) treatment. The comparisons for two interacting Ar atoms, $\pi - \pi$ stacking energy of two coronenes in a dimer, and the lattice energy of the benzene crystal versus unit cell volume, all shown marked improvements when the D3 procedure is applied [48].

An alternative treatment of the atoms a in a medium introduces distributed densities $\rho_a(\vec{r})$ and polarizabilities that are coupled to account for medium effects and many-body shifts of frequencies, in a new so-called many-body dispersion (MBD) treatment also recently applied and tested [33]. An averaged (or scaled) static atomic polarizability $\alpha_a^{(0)}(0) = \nu_a \alpha_a^{(free)}(0)$ of an atom in a medium is taken to be proportional to its value $\alpha_a^{(free)}(0)$ when free, with a scaling factor $\nu_a = V_a^{(0)}/V_a^{(free)}$ obtained from the volumes quotient $V_a^{(0)}/V_a^{(free)} = \int d^3r\,\rho_a^{(0)}(\vec{r})\left|\vec{r}-\vec{R}_a\right|^3 / \int d^3r\,\rho_a^{(free)}(\vec{r})\left|\vec{r}-\vec{R}_a\right|^3$ of bonded and free atoms with atomic densities $\rho_a^{(0)}$ and $\rho_a^{(free)}$, respectively. A starting partition of the total density $\rho(\vec{r})$ is obtained by means of approximate weights $w_a(\vec{r}) = \rho_a^{(free)}(\vec{r}) / \left[\sum_a \rho_a^{(free)}(\vec{r})\right]$ [32] calculated from free atomic densities to begin with, and then improved to allow for changes in atomic size and charge, reflecting electronic rearrangements in molecule A as mentioned in the section on partitioned densities.

The static polarizabilities are used to construct dynamical polarizabilities $\alpha_a^{(0)}(\omega)$ containing an effective excitation energy $\hbar\bar{\omega}_a$ as described in Chapter 2 on molecular properties and in Chapter 3. These are then entered in an equation for interacting induced dipoles [15] that couples the atomic polarizability of a to nearest neighbors c through a short-range form $t_{\xi\eta}^{(sr)}(R_{ac}) = [1 - f_d(R_{ac})]\, t_{\xi\eta}(R_{ac})$ of the cartesian tensor dipole–dipole coupling factor $t_{\xi\eta}(R_{ac})$ introduced in Chapter 3. Effective atomic polarizabilities $\alpha_a^{(eff)}(\omega)$, or $\alpha_a^{(N_a)}(\omega)$ in our notation, shaped by their environment, are obtained in tensor detail from

$$\alpha_{a,\xi\eta}^{(N_a)}(\omega) = \alpha_{a,\xi\eta}^{(0)}(\omega) + \sum_{c=1}^{N_c} \sum_{\zeta,\lambda} \alpha_{a,\xi\zeta}^{(0)}(\omega)\, t_{\zeta\lambda}^{(sr)}(R_{ac})\, \alpha_{c,\lambda\eta}^{(N_c)}(\omega)$$

and used in a treatment of coupled charge density fluctuations.

The energy of a collection of fluctuating charges in A and B is obtained from a Hamiltonian operator for the oscillating effective polarizabilities coupled through a long-range dipole–dipole factor $t_{\xi\eta}^{(lr)}(R_{ab}) = f_d(R_{ab})\, t_{\xi\eta}(R_{ab})$, of the form

$$\hat{H}_{MBD} = -\frac{1}{2}\sum\nolimits_{\mu=a,b}\nabla_\mu{}^2 - \frac{1}{2}\sum\nolimits_{\mu=a,b}\bar{\omega}_\mu{}^2\vec{u}_\mu{}^2 + \frac{1}{2}\sum\nolimits_{a\neq b}\bar{\omega}_a\,\bar{\omega}_b\left[\alpha_a^{(N_a)}(0)\alpha_b^{(N_b)}(0)\right]^{1/2}\sum\nolimits_{\xi,\eta}u_{a,\xi}\,t_{\xi\eta}^{(lr)}\,u_{b,\eta}$$

with atomic mass weighted displacements such as $u_{a,\xi} = M_a{}^{\frac{1}{2}}(\xi_a - R_{a,\xi})$ oscillating with frequencies $\bar{\omega}_{a,\xi}$. Its eigenvalue equation $\hat{H}_{MBD}F_p(\{u\}) = \hbar\Omega_p F_p(\{u\})$ is solved to obtain many-body dispersion (MBD) energies $E_{dsp}^{(MBD)}(R,\boldsymbol{Q})$ from a sums over $N_{AB}^{(df)} = 3(N_A + N_B) - 3$ eigenfrequency shifts $\Omega_p(R, \boldsymbol{Q}) - \Omega_p(\infty, \boldsymbol{Q})$, given that there are N_A atoms in A and N_B in B. This provides

$$E_{tot}(R,\boldsymbol{Q}) = E_{sl}(R,\boldsymbol{Q}) + E_{dsp}^{(MBD)}(R,\boldsymbol{Q})$$

in a DFT-MBD treatment and has been applied usually combined with the PBE density functional, to generate results for several benchmark sets of small and intermediate size compounds (S66 and S22), crystals (X23 set), and large molecular complexes (S12L). They have been compared with accurate results from the CCSD(T) treatment with generally very good agreement [33].

7.5.2 Long-range Interactions from Nonlocal Functionals

The adiabatic connection fluctuation-dissipation (ACFD) relation gives a formally exact expression for the electron–electron correlation energy in terms of the dynamical susceptibility, which can be decomposed into semilocal (*sl*) and nonlocal (*nl*) terms as $E_c(R,\boldsymbol{Q}) = E_c^{(sl)}(R,\boldsymbol{Q}) + E_c^{(nl)}(R,\boldsymbol{Q})$ with the first term coming from a semilocal DFT functional and the second term given by an ACFD nonlocal functional. This second term can be constructed for large distances between A and B from their susceptibilities, as known functionals of the electronic density so that no parametrization from molecular properties is needed.

The vdW interaction energy $E_{Ae,Be}^{(2)}$, in a notation indicating that only electrons are being described, can be written in terms of the electronic dynamical susceptibility tensors introduced in Chapter 2, with $\tilde{\chi}_{Se}^{(+)}(\vec{r},\vec{r}\,';\omega)_{\xi\eta} = \tilde{\chi}_S^{(el)}(\vec{r},\vec{r}\,';\omega)_{\xi\eta}, S = A,B; \xi,\eta = x,y,z,$ in a new notation here for two separate charge distributions. It takes the form

$$E_{Ae,Be}^{(2)} = -\frac{\hbar}{2\pi}\int_A d(1)\int_A d(1')\int_B d(2)\int_B d(2')\frac{c_e^2}{r_{12}}\frac{c_e^2}{r_{1'2'}}\int_0^\infty d\omega\tilde{\chi}_A^{(el)}(1,1';i\omega)\tilde{\chi}_B^{(el)}(2,2';i\omega)$$

where integrals extend to nonoverlapping regions around A and B as shown. Here (1) stands for $\vec{r}_1$ and the integral over its element of volume extends only over a nonoverlapping region around A and contains the density of species A as shown, $i\omega$ is the imaginary-valued frequency that appears in the Casimir–Polder transformation as described in Chapter 3, and an assumed isotropy of the electronic interaction has been used to replace susceptibility tensors by their isotropic components,

$$\tilde{\chi}_S^{(el)}\left(\vec{r},\vec{r}\,';\omega\right)=\frac{1}{3}\sum_{\xi}\tilde{\chi}_S^{(el)}\left(\vec{r},\vec{r}\,';\omega\right)_{\xi\xi}$$

The susceptibility function is given by the time correlation of density fluctuations around equilibrium values at each position $\vec{r}$, and the above $E^{(2)}_{Ae,Be}$ therefore describes all multipolar interactions. The dipole–dipole vdW interaction energy proportional to R^{-6} appears when C_e^2/r_{12} is expanded in powers of r_{A1}/R and r_{B2}/R to first order in each variable. It can be expressed in terms of the tensor $t_{\xi\eta}$ in Chapter 3 and the dipolar polarizabilities as given in Chapter 2,

$$\tilde{\alpha}_S^{(el)}\left(\vec{r},\vec{r}\,';\omega\right)_{\xi\xi'}=C_e^2\xi_S\xi'_S\tilde{\chi}_S^{(el)}\left(\vec{r},\vec{r}\,';\omega\right)_{\xi\xi'}$$

after using that, insofar density fluctuations integrate to zero, one finds

$$\int_S d(1)\tilde{\chi}_S^{(el)}(1,1';i\omega)=\int_S d(1')\tilde{\chi}_S^{(el)}(1,1';i\omega)=0$$

The relation with DFT is done assuming that the polarizabilities are local. The original literature by several authors has been updated in recent reviews that also include more general formalisms [51–54].

The present equations are simplified first choosing the local form $\tilde{\alpha}_S^{(el)}\left(\vec{r},\vec{r}\,';i\omega\right)=\delta(\vec{r}-\vec{r}\,')\gamma_S^{(el)}\left(\vec{r};i\omega\right)$, and using that for electron distributions (1) and (2), respectively, at molecules A and B far from each other, their leading dipole–dipole interaction energy goes as $\left|\vec{r}_1-\vec{r}_2\right|^{-3}$, which gives

$$E^{(dsp)}_{Ae,Be}=-\frac{3\hbar}{\pi}\int_0^\infty d\omega\int_A d(1)\int_B d(2)\gamma_A^{(el)}\left(\vec{r}_1;i\omega\right)\frac{1}{\left|\vec{r}_1-\vec{r}_2\right|^6}\gamma_B^{(el)}\left(\vec{r}_2;i\omega\right)$$

and applies to either A and B at a relative distance R, or to A surrounded by B. In the first case, the dispersion energy at large R takes the form

$$E^{(dsp)}_{Ae,Be}(R)=-\frac{C_{AB^{(6)}}}{R^6},C^{(6)}_{AB}=\frac{3\hbar}{\pi}\int_0^\infty d\omega\Gamma_A^{(el)}(i\omega)\Gamma_B^{(el)}(i\omega)$$

$$\Gamma_S^{(el)}(i\omega)=\int_S d(1)\tilde{\gamma}_S^{(el)}(1;i\omega)$$

When in addition, the local polarizability function is given a form suggested by plasmon theory [55] as

$$\gamma_S^{(el)}\left(\vec{r}; i\omega\right) = \frac{c_e^2}{m_e}\frac{\rho_e\left(\vec{r}\right)}{\omega_0\left(\vec{r}\right)^2 + \omega^2}$$

dependent on the local electronic density $\rho_e\left(\vec{r}\right)$ and a local excitation frequency $\omega_0\left(\vec{r}\right)$, the dispersion energy is found to be

$$E_{Ae,Be}^{(dsp)} = -\frac{3\hbar}{2}\left(\frac{c_e^2}{m_e}\right)^2 \int_A d(1) \int_B d(2) \frac{1}{\left|\vec{r}_1 - \vec{r}_2\right|^6} \frac{\rho_e(\vec{r}_1)}{\omega_0\left(\vec{r}_1\right)^2} \frac{1}{\left[\omega_0\left(\vec{r}_1\right) + \omega_0\left(\vec{r}_2\right)\right]} \frac{\rho_e(\vec{r}_2)}{\omega_0\left(\vec{r}_2\right)^2}$$

and can be obtained from local properties of an inhomogeneous many-electron system in a variety of conformations for interacting A and B.

Local polarizabilities can be obtained from models of the species electronic structure, such as from the inhomogeneous electron gas, or from a spherical electronic distribution with an electronic density of states displaying an energy gap $\hbar\omega_0$. One of the proposed forms in the literature is a polarizability $\gamma_{DF}^{(el)}$ obtained from the inhomogeneous electron gas, with $\omega_{0,DF} = (3m_e)^{-1}\hbar k_F^2 (1 + 0.22\,s^2)^2$, where $k_F\left(\vec{r}\right) = \left[3\pi^2\rho_e\left(\vec{r}\right)\right]^{1/3}$ is a local Fermi wavenumber, and $s\left(\vec{r}\right) = |\nabla\rho_e| / \left[2k_F\rho_e\left(\vec{r}\right)\right]$ is a dimensionless function of the electronic density gradient [56]. Using this for species A and B, the coefficient $C_{AB}^{(6)}$ was calculated to reproduce a set of 34 $C_{AA}^{(6)}$ vdW values to an accuracy of about 10%, for pairs containing identical closed-shell atoms, diatomics, and polyatomic molecules.

This long-range dispersion energy can be combined with results from semi-local DFT treatments to obtain an exchange-correlation functional for all distances, using

$$E_{xc}(R) = E_{xc}^{(sl)}\left[\rho_e\left(\vec{r}\right), \ldots; R\right]\{1 - f_d(R)\} + E_{Ae,Be}^{(dsp)}\left[\rho_e\left(\vec{r}\right); R\right] f_d(R)$$

where $f_d(R)$ is a damping function going to zero at small distances. The damping factors can be dropped if the functionals have been chosen to avoid double counting of vdW forces, but must otherwise be included. Recent calculations [57, 58] have used a hybrid version of DFT for $E_{xc}^{(sl)}$ and has included higher order multipolar dispersion interactions to compare results with accurate CCSD(T) values for a set of compounds, showing excellent agreement.

Alternatively, a long-range correlation (*lrc*) energy, derived from the ACFD formula given in the previous chapter, can be constructed so that it is valid for all distances and does not lead to double counting of dispersion energies. It must also be constructed so that for large interatomic distances, it reproduces the form of $E_{Ae,Be}^{(dsp)}$ given above within the plasmon model. A form that satisfies these conditions is the nonlocal functional [56]

$$E_{lrc}^{(nl)}\left[\rho_e(\vec{r}),\rho_e(\vec{r}')\right] = \frac{\hbar}{2}\int d^3r \int d^3r' \rho_e(\vec{r}) \frac{3\,c_e^4}{2\,m_e^2 g(\vec{r}) g(\vec{r}')\left[g(\vec{r}) + g(\vec{r}')\right]}\rho_e(\vec{r}')$$

$$g\left(\vec{r};\, |\vec{r}-\vec{r}'|\right) = \omega_{0,VV}(\vec{r})\,|\vec{r}-\vec{r}'|^2 + \kappa(\vec{r})$$

obtained from a model of a spherical inhomogeneous electron distribution with an energy band gap given by

$$\omega_{0,VV}(\vec{r}) = \left(\frac{\omega_p^2}{3} + c\frac{\hbar^2\, k_F^4}{m_e^2}s^4\right)^{\frac{1}{2}}$$

where $\omega_p^2 = 4\pi\rho_e c_e^2/m_e = \left(4c_e^2/3\pi m_e\right)k_F^3$ is a local plasmon frequency squared, and where $\kappa = b(\hbar k_F)^2/\left(m_e^2\omega_p\right)$ with an adjustable parameter b, which controls the damping of g at short distances $|\vec{r}-\vec{r}'|$. The term containing s prevents a divergence in $\gamma_S^{(el)}$ when both $\omega \to 0$ and $\rho_e \to 0$.This choice, named VV09/10, has given values for 58 $C_{AA}^{(6)}$ vdW coefficients to an accuracy within 15%, obtained for a test set of pairs including identical closed- and open-shell atoms, diatomics, and polyatomic molecules, with the value of the numerical parameter $c = 0.149$ multiplying the s^4 term adjusted to minimize standard deviations from accurate values of $C_{AA}^{(6)}$ [56]. The above volume integrals over d^3r and d^3r' extend over the whole space containing both A and B species and have been constructed so that the dispersion energy vanishes at short distances for a homogeneous electron gas. An example of results obtained this way is shown for the methane dimer in Figure 7.2 from [56].

The good accuracy of the described nonlocal density functional $E_{lrc}^{(nl)}\left[\rho_e(\vec{r}),\rho_e(\vec{r}')\right]$ for $C_{AB}^{(6)}$ coefficients, and its physical behavior for all distances, allows a treatment combining the more standard semilocal exchange-correlation functionals $E_{xc}^{(sl)}\left[\rho_e(\vec{r}),\ldots\right]$, already tested for interaction energies at short and intermediate distances, with the nonlocal form $E_{lrc}^{(nl)}$ to construct intermolecular energies valid over all distances R as

$$E_{xc}(R) = E_{xc}^{(sl)}\left[\rho_e(\vec{r}),\ldots;R\right] + E_{lrc}^{(nl)}\left[\rho_e(\vec{r}),\rho_e(\vec{r}');R\right]$$

now without a damping factor.

7.5.3 Embedding of Atomic Groups with DFT

Large systems can be treated isolating groups of atoms to be described as clusters with suitably saturated peripheral bonds, or can be more generally treated with embedding methods where a cluster is self-consistently coupled to its environment. This is also related to the *QM/MM* approaches as described in

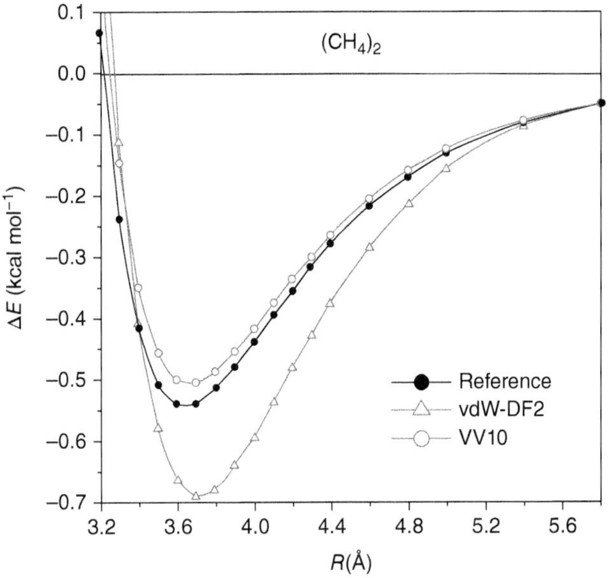

Figure 7.2 Methane dimer interaction energy versus distance R. *Source:* from Ref. [56]. Reproduced with permission of American Institute of Physics. Here R is the distance between the centers of mass of the monomers. VV10 and vdW-DF2 results were obtained self-consistently with the aug-cc-pVTZ basis set.

Section 7.2, but can be more generally done using efficient methods from DFT for both cluster and environment [59–61]. Embedding DFT methods start with a partition of the physical system into molecular fragments K containing selected atomic cores $\kappa = 1$ to $N_K^{(at)}$ and assigned N_K electrons, with this number allowed to change as interfragment electronic rearrangement occurs. This number of electrons can be fractional and interpreted as resulting from a statistical average over an ensemble of fragments with varying integer number of electrons in each one. In a simple version, the total system is partitioned into two: a cluster of selected atoms with $K = A$ and an environment with $K = B$.

The total energy E_{tot} can be obtained to begin with by means of a density functional embedding theory (or *DFET*), which describes the whole system, to be corrected by accurate calculation of the embedded cluster (or *emb cls*) fragment, for example, with a correlated many-electron wavefunction (or *CW*), by adding a correction term $\Delta E_{emb\,cls} = E_{emb\,cls}^{(CW)} - E_{emb\,cls}^{(DFT)}$ to $E_{tot}^{(DFT)}$. This is similar to the mentioned subtraction scheme in the *QM/MM* approach, and assumes that the environment is not changed by atomic rearrangements in the cluster.

A better description follows from the alternative decomposition of the total electron density and total (ionic plus electronic) energy into cluster, environment, and interaction terms, as [61]

$$E_{tot} = E_{cls}[\rho_{cls}] + E_{env}[\rho_{env}] + E_{int}[\rho_{cls}, \rho_{env}]$$

with the densities ρ_{cls} and ρ_{env} constructed so that the total electronic density is $\rho_{tot} = \rho_{cls} + \rho_{env}$. The functional for the interaction energy changes with densities of cluster and environment, providing embedding potential energy functions through the functional derivatives in

$$\frac{\delta E_{int}}{\delta \rho_{cls}(\vec{r})} = V_{emb}^{(cls)}(\vec{r}), \frac{\delta E_{int}}{\delta \rho_{env}(\vec{r})} = V_{emb}^{(env)}(\vec{r})$$

Equilibration between the two fragments requires that the two embedding potentials should be equal and gives the constraint $V_{emb}^{(cls)}(\vec{r}) = V_{emb}^{(env)}(\vec{r}) = V_{emb}(\vec{r})$, which is a unique embedding potential for the total system. This is also the potential that has been introduced in a general density functional formalism for chemical reactivity [62].

Given a choice of fragment atomic cores and of number of electrons for A and B, and a constructed density functional at any of the many available levels of accuracy, it is possible to variationally optimize the energy as a function of atomic positions, to obtain the potential energy function $E_{tot}(\mathbf{Q}_A, \mathbf{Q}_B)$. This can be done writing the total energy as a functional of the densities $\rho_A(\vec{r})$ and $\rho_B(\vec{r})$, and of the embedding potential $V_{emb}(\vec{r})$ as

$$E_{tot} = E_A[\rho_A, V_{emb}] + E_B[\rho_B, V_{emb}] + E_{int}[\rho_A, \rho_B, V_{emb}]$$

with a straightforward generalization to several fragments K. A potential functional embedding theory (or PFET) has been developed and applied to molecular and materials systems [61]. The computational procedure starts with a guess for $V_{emb}(\vec{r})$, N_A, and N_B. The fragment functionals $E_K[\rho_K(\vec{r}), V_{emb}(\vec{r}); \{N_K\}, \{\mathbf{Q}_K\}], K = A, B$, are constructed and the $\rho_K[V_{emb}; N_A, N_B]$ are found for fixed atom positions. The energy is minimized to find the embedding potential, calculating first $\delta E_{tot}/\delta V_{emb}(\vec{r}) = 0$ to search for a minimum for fixed $\{N_K\}$, followed by the calculation of the fragment number of electrons from $\partial E_K/\partial N_K = \mu$, a chemical potential common to A and B. An iteration of this sequence leads to the final potential energy surface $E_{tot}(\mathbf{Q}_A, \mathbf{Q}_B)$ for the total system. Some applications are covered in the following chapter on interactions at surfaces.

An alternative that avoids the issue of possibly finding unphysical fractional electronic charges in the fragments, as they separate into $A + B$ at large distances in a dissociation, can be based on the introduction of an energy E_{tot} written as a functional of electron orbitals as explained in Section 6.4, for the total system, to be obtained now within an optimized effective potential procedure generalized for a collection of interacting fragments. The functional $E_{tot}\left[\left\{\psi_j^{(K)}\left(\vec{r},\zeta\right)\right\}\right]$ of a set of $j = 1$ *to* N_K spin-orbitals $\psi_j^{(K)}$ for each fragment K can be constructed to describe electrons moving in an effective potential $V_{emb}\left(\vec{r}\right)$, with the orbitals given as functionals of $V_{emb}\left(\vec{r}\right)$. This can be derived from the relation $\delta E/\delta V_{emb}\left(\vec{r}\right) = 0$, and optimized by energy minimization, to obtain an *optimized embedding effective potential* (OEEP) $V_{emb}^{(OEEP)}\left(\vec{r}\right)$ and the corresponding PES satisfying $E_{tot}(\mathbf{Q}_A, \mathbf{Q}_B) \approx E_A(\mathbf{Q}_A) + E_B(\mathbf{Q}_A)$ with integer number of electrons in each dissociation species [60].

7.6 Artificial Intelligence Learning Methods for Many-Atom Interaction Energies

The energy of two interacting structures containing very many atoms, such as those involving quantum dots, solids, proteins, or biological membranes, presents special challenges for modeling with computational methods. Potential energy functions depend on very many atomic position variables and even when decomposed into fragments, these are too numerous for standard computational methods. This can be a barrier to discovery of new compounds with desirable properties. However, as large numbers of compounds are studied, providing their conformations and energies, the accumulated data can be used to identify new compounds with data mining and machine learning (ML) procedures applied to molecular structures and their interactions.

New methods are being developed making use of artificial intelligence to train neural networks in AI-NN algorithms to provide desired results. This has been done as briefly explained in Chapter 4 for a many-atom system in its ground electronic state, to generate potential energy surfaces for a given structure, as $V(\mathbf{Q}) = \sum_a E_a^{(emb)}[\mathbf{G}(\mathbf{Q})]$, a sum over atomic components with each one dependent on a set of global variables $\mathbf{G}(\mathbf{Q})$ containing neighbor atomic locations. The decomposition is applicable to both bonding and nonbonding interactions and can be validated comparing interpolated PES values with results from selected previous sets of calculations. It requires specifying for each atom what neighboring atoms are present within a certain range, which defines its embedding environment. The embedded atom energies contain parameters found by fitting procedures and can be obtained training an AI neural network with an AI-NN procedure as described in Chapter 4.

That treatment, designed to calculate details of a potential energy surface for a given compound, can be extended to allow for calculation of properties of new compounds and materials from known properties of their fragments [63, 64]. The extended treatments can be applied to calculation of the bonding energy of new structures from a large supply of data on related structures already known, by decomposing the energies of the many compounds and materials in the data sets into molecular fragments, each with embedded-atom components. The data sets contain not only the atomic environment in a single compound but also its variations for the same molecular fragment as present in many different compounds.

The applicability of the AI-NN method for calculations of the energetics of new components depends on the extent of the data sets available for the NN training. A large set of compounds including first and second row elements of the periodic table is sufficient to generate the energetics of many organic compounds, utilizing, for example, the ANI-1 potential data base [64]. As the available data sets of compounds are expanded to include atoms of the third and higher rows, new organic and inorganic compounds can be investigated.

The general procedure is illustrated in Figure 7.3 in the very simple test case where the potential energy surface of the water molecule for varying bond variables is obtained from the known energetics of other compounds containing hydrogen and oxygen atoms bonding to each other. Here the three atoms making the water molecule have a conformation given by internal coordinates $\boldsymbol{Q} = (q_1, q_2, q_3)$. A set of values of global symmetry variables $\boldsymbol{G}^{\mathrm{A}}(\boldsymbol{Q})$for each atom A are inputs to atomic NNs, which are trained using data for many molecular species containing H and O atoms, with its output giving the embedded atom energy. The atomic energies are added to obtain the total energy, in a procedure detailed in reference [64].

The available AI methods for the energetics of new compounds and for molecular interactions are being extended to make them both more efficient and more encompassing. The structure of neural networks can be optimized with respect to the number of layers and nodes, exploring their best combination for accuracy and efficiency, insofar the NNs do not necessarily give better results as those numbers increase. Also the NN training procedures can be made faster utilizing alternative forms for the symmetry variables $\boldsymbol{G}^{\mathrm{A}}(\boldsymbol{Q})$ describing radial and angular atomic distributions, with fewer numerical operations required for training. A by-product of the AI-NN procedure for energetics is the availability of calculated interaction forces needed in classical molecular dynamics treatments, so that an AI-NN procedure can be combined with time integration of equations of motion for atomic positions, where forces are needed at each of many time steps along trajectories and for many initial conditions. This has been applied to the unimolecular decomposition of vinyl bromide, and later on to the reaction dynamics of H + HBr [65].

The accuracy of derived energetics for new compounds depends on the accuracy of the values in the data sets. Starting with a DFT-trained network, using lots of DFT data, and then correcting the network training with a much smaller data set from accurate CCSD(T)/CBS calculations of input compounds, has been found to give more accurate energetics than the original results using only the less accurate DFT calculations [66].

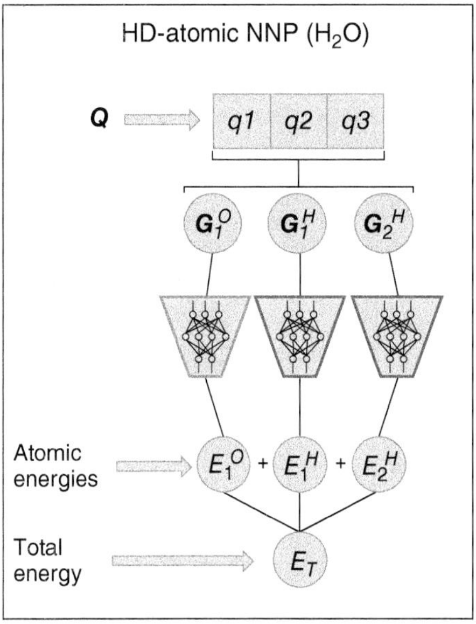

Figure 7.3 Combined atomic neural network (NN) architectures used to obtain high-definition NN potentials in a HD-Atomic NNP procedure, here for the three atoms making the water molecule with its conformation given by internal coordinates $\mathbf{Q} = (q_1, q_2, q_3)$. A set of values of global symmetry variables $\mathbf{G}^A(\mathbf{Q})$ for each atom A are inputs to atomic NNs, which are trained using data for many molecular species containing H and O atoms, with its output giving the embedded atom energy. The atomic energies are added to obtain the total energy. *Source:* adapted from Ref. [64]. Reproduced with permission of Royal Society of Chemistry.

Another methodological improvement consists of finding an alternative way to select fewer components of a data set with active learning (or AL). It goes beyond ML to incorporate knowledge gained in the NN training about which subsets of data are more reliable and are sufficient for sampling the space of new compounds [67]. This is done by discarding redundant data points with small standard deviation of their calculated atomic energies relative to reference energy values for those data points, while keeping data points that contain relevant new information.

Extensions of the AI-NN procedure to include many-atom systems such as solid surfaces, complex materials, and biomolecules are being done, with the applicability to new structures dependent on how many data points are available in related data sets for NN training. Some of the applications have already been mentioned in Chapter 4. In more detail and as an example, intermolecular interactions involved in bond breaking at a solid surface can be evaluated in test cases where an analytical PES, constructed from hundreds of energy points from DFT calculations, is known to begin with and is reproduced by NN training to verify the learning procedure.

This has been done for H_2 adsorbed on the (2 × 2)S/Pd(100) adsorbate surface and with the H—H bond stretched and breaking in dissociation. The original location of the diatomic was chosen to be at the fourfold hollow, bridge, and on-top locations of the diatomic center of mass. The AI-NN procedure involved nine global variables accounting for the periodicity of the surface, with their values as input for a three-layer NN, and training done with hundreds of epochs (each a cycle of data inputs). The results have shown accuracy errors smaller than 100 meVs in comparison with the original PES, and have shown smooth behavior, so that molecular dynamics simulations of dissociation could be accurately done [68]. This can also be done for many other PESs of interactions at solid surfaces, relevant to a wide range of applications to the discovery of new materials.

The majority of applications of AI-NN learning have been done to generate the energies of a many-atom system in its ground electronic state. However, the ML procedure can also be used for nonadiabatic processes to generate energies, PES forces and momentum couplings for a set of coupled ground and excited states, as needed for nonadiabatic dynamics calculations, provided enough information is available to construct data sets to be used for training and verification. This has recently been shown to be doable for conical intersections appearing in the photoisomerization of CH_3NH [69] and for coupled ground and excited states of the reactive triatomic LiFH [70].

Another related application of AI-NN learning of relevance to the development of new medical drugs involves the calculation of receptor–ligand scoring indices for small molecular ligands interacting with large biomolecular receptors. Neural network scoring functions have been found to reproduce results from state-of-the-art docking programs based on atomistic models, with NN computational speeds up to two orders of magnitude larger than speeds for the available scoring procedures [71].

Development of new methods and applications for AI-NN machine learning is very active at present. These AI methods are complementary to the ones based on atomistic models and detailed calculations of electronic structure, because AI methods can be used only when large numbers of data points can be collected and organized in data sets to be used for NN training and verification. Organization of the data sets require the introduction of molecular structure descriptors based on physical and chemical considerations, a subject that needs much attention. A good choice of descriptors can lead to the desired results in a short time, while bad descriptor choices may yet lead to needed values but after a much longer computational time. When the data sets are available, the AI-NN procedures are computationally very efficient. Many improvements in accuracy and efficiency of AI-NN machine learning methods can be expected in the near future.

References

1 Longuet-Higgins, H.C. (1956). The electronic states of composite systems. *Proc. R. Soc. A* 235: 537.

2 Dzyaloshinskii, I.E., Lifshitz, E.M., and Pitaevskii, L.P. (1961). The general theory of van der Waals forces. *Adv. Phys.* 10: 165.

3 Longuet-Higgins, H.C. and Salem, L. (1961). The forces between polyatomic molecules. I. Long-range forces. *Proc. R. Soc. Lond.* A259: 433.

4 McLahlan, A.D., Gregory, R.D., and Ball, M.A. (1964). Molecular interactions by the time-dependent Hartree method. *Mol. Phys.* 7: 119.

5 Linderberg, J. (1964). Dispersion energy and electronic correlation in molecular crystals. *Arkiv f. Fysik* 26: 323.

6 Linder, B. (1967). Reaction-field techniques and their applications to intermolecular forces. *Adv. Chem. Phys.* 12: 225.

7 Linder, B. and Rabenold, D.A. (1972). Unified treatment of van der Waals forces between two molecules of arbitrary sizes and electron delocalization. *Adv. Quantum Chem.* 6: 203.

8 Lowdin, P.O. (1955). Quantum theory of many-particle systems. II. Study of the ordinary Hartree-Fock approximation. *Phys. Rev.* 97: 1490.

9 McWeeney, R. (1989). *Methods of Molecular Quantum Mechanics*, 2e. London, England: Academic Press.

10 Hirschfelder, J.O., Brown, W.B., and Epstein, S.T. (1964). Recent developments in perturbation theory. *Adv. Quantum Chem.* 1: 255.

11 Goscinski, O. (1968). Upper and lower bounds to polarizabilities and van der Waals forces. I General theory. *Int. J. Quantum Chem.* 2: 761.

12 Langhoff, P., Gordon, R.G., and Karplus, M. (1971). Comparison of dispersion force bounding methods with applications to anisotropic interactions. *J. Chem. Phys.* 55: 2126.

13 Mavroyannis, C. and Stephen, M.J. (1962). Dispersion forces. *Mol. Phys.* 5: 629.

14 Gordon, M.S., Freitag, M.A., Bandyopadhyay, P. et al. (2001). The effective fragment potential method: a QM-based MM approach to modeling environmental effects in Chemistry. *J. Phys. Chem. A* 105: 293.

15 Stone, A.J. (2013). *The Theory of Intermolecular Forces*, 2e. Oxford, England: Oxford University Press.

16 Jeziorski, B., Moszynski, R., and Szalewicz, K. (1994). Perturbation theory approach to intermolecular potential energy surfaces of van der Waals complexes. *Chem. Rev.* 94: 1887.

17 Szalewicz, K. (2012). Symmetry-adapted perturbation theoryof intermolecular forces. *WIREs Comput. Mol. Sci.* 2: 254.

18 Gordon, M.S., Fedorov, D.G., Pruitt, S.R., and Slipchenko, L.V. (2011). Fragmentation methods: A route to accurate calculations on large systems. *Chem. Rev.* 112: 632.

19 Gordon, M.S., Smith, Q.A., Xu, P., and Slipchenko, L.V. (2013). Accurate first principles model potentials for intermolecular interactions. *Annu. Rev. Phys. Chem.* 64, Palo Alot, CA, USA, Annual Reviews Inc.: 553.

20 Gao, J., Truhlar, D.G., Wang, Y. et al. (2014). Explicit polarization: a quantum mechanical framework for developing next generation force fields. *Acc. Chem. Res.* 47: 2837.

21 Xie, W., Song, L., Truhlar, D.G., and Gao, J. (2008). Variational explicit polarization potential and analytical first derivative of energy: towards a new generation force field. *J. Chem. Phys.* 128: 234108.
22 Jensen, F. (1999). *Introduction to Computational Chemistry*. New York: Wiley.
23 Ufimtsev, I.S. and Martinez, T.J. (2008). Quantum chemistry on graphical processing units. 1. Strategies for two-electron integral evaluation. *J. Chem. Theory Comp.* 4: 222.
24 Warshel, A. and Levitt, M. (1976). QM/MM. *J. Mol. Biol.* 103: 227.
25 Bash, P.A., Field, M.J., and Karplus, M. (1987). Free energy perturbation method for chemical reactions in the condensed phase: a dynamic approach based on a combined quantum and molecular mechanics potential. *J. Am. Chem. Soc.* 109: 8092.
26 Field, M.J., Bash, P.A., and Karplus, M. (1990). A combined quantum mechanical and molecular mechanical potential for molecular dynamics simulations. *J. Comput. Chem.* 11: 700.
27 Maseras, F. and Morokuma, K.I. (1995). A new integrated Ab Initio+ molecular mechanics geometry optimization scheme of equilibrium structures and transition states. *J. Comput. Chem.* 16: 1170–1179.
28 Friesner, R.A. and Guallar, V. (2005). Ab initio quantum chemical and mixed quantum mechanics/molecular mechanics (QM/MM) methods for enzymatic catalysis. *Annu. Rev. Phys. Chem.* 56: 389.
29 Lin, H. and Truhlar, D.G. (2007). QM/MM: what have we learned, where are we, and where do we go from here? *Theor. Chem. Accounts* 117: 185.
30 Senn, H.M. and Thiel, W. (2009). QM/MM methods for biomolecular systems. *Angew. Chem. Int. Ed.* 48: 1198.
31 Brunk, E. and Rothlisberger, U. (2015). Mixed quantum mechanical/molecular mechanical molecular dynamics simulations of biological systems in ground and electronically excited states. *Chem. Rev.* 115: 6217.
32 Hirschfeld, F.L. (1977). Bonded-atom fragments for describing molecular charge densities. *Theor. Chim. Acta* 44: 129.
33 Hermann, J., DiStasio, R.A.J., and Tkatchenko, A. (2017). First-principles models for van der Waals interactions in molecules and materials: concepts, theory, and applications. *Chem. Rev.* 117: 4714.
34 Cohen-Tanoudji, C., Diu, B., and Laloe, F. (1977). *Quantum Mechanics*, vol. 1, Chap. IV. New York: Wiley.
35 May, V. and Kuhn, O. (2000). *Charge and Energy Transfer Dynamics in Molecular Systems*. Berlin: Wiley-VCH.
36 Shavitt, I. (1963). The Gaussian function in calculations of statistical mechanics and quantum mechanics. In: *Methods of Computational Physics*, vol. 2 (eds. B. Alder, S. Fernbach and M. Rotenberg), 1. New York: Academic Press.
37 Martin, R.M. (2004). *Electronic Structure: Basic Theory and Practical Methods*. Cambridge, England: Cambridge University Press.

38 Salem, L. (1962). Attractive forces between long saturated chains at short distances. *J. Chem. Phys.* 37: 2100.
39 Zwanzig, R. (1963). Two assumptions in the theory of atttractive forces between long saturated chains. *J. Chem. Phys.* 39: 2251.
40 Hirschfelder, J.O., Curtis, C.F., and Bird, R.B. (1954). *Molecular Theory of Gases and Liquids*. New York: Wiley.
41 Israelachvili, J. (1992). *Intermolecular and Surface Forces*. San Diego, CA: Academic Press.
42 Levine, I.N. (2000). *Quantum Chemistry*, 5e. Upper Saddle River: Prentice-Hall.
43 Davydov, A.S. (1971). *Theory of Molecular Excitons*. New York: Plenum.
44 Nitzan, A. (2006). *Chemical Dynamics in Condensed Phases*. Oxford, England: Oxford University Press.
45 Kittel, C. (2005). *Introduction to Solid State Physics*, 8e. Hoboken, NJ: Wiley.
46 Mukamel, S. (1995). *Principles of Nonlinear Optical Spectroscopy*. Oxford, England: Oxford University Press.
47 Mukamel, S. and Abramovicius, D. (2004). Many-body approaches for simulating coherent non-linear spectroscopies of electronic and vibronic excitons. *Chem. Rev.* 104: 2073.
48 Grimme, S., Hansen, A., and Brandenburg, J.G.C. (2016). Dispersion corrected mean-field electronic structure methods. *Chem. Rev.* 116: 5105.
49 Johnson, E.R. and Becke, A.D. (2005). A post-Hartree-Fock model of intermolecular interactions. *J. Chem. Phys.* 123: 024101.
50 Koide, A. (1976). A new expansion for dispersion forces and its applications. *J. Phys. B Atomic Mol. Phys.* 9: 3173.
51 Langreth, D.C., Lundqvist, B.I., Chakarova-Kaeck, S.D. et al. (2009). A density functional for sparse matter. *J. Phys. Condens. Matter* 21: 084203.
52 Dobson, J.F. and Gould, T. (2012). Calculation of dispersion energies. *J. Phys. Condens. Matter* 24: 073201.
53 Dobson, J.F. (2012). Dispersion (van der Waals) forces and TDDFT. In: *Fundamentals of Time-Dependent Density Functional Theory* (eds. M.A.L. Marques, N.T. Maitra, F.M.S. Nogueira, et al.), 417. Berlin: Springer-Verlag.
54 Berland, K., Cooper, V.R., Lee, K. et al. (2015). van der Waals forces in density functional theory: A review of the vdW-DF method. *Rep. Prog. Phys.* 78: 066501.
55 Ashcroft, N.W. and Mermin, N.D. (1976). *Solid State Physics*. London, England: Thomson.
56 Vydrov, O.A. and Van Voohis, T. (2010). Nonlocal van der Waals density functional: the simpler the better. *J. Chem. Phys.* 133: 244103–244101.
57 Sato, T. and Nakai, H. (2009). Density functional method including weak interactions: Dispersion coefficients based on the local response approximation. *J. Chem. Phys.* 131: 224104.
58 Sato, T. and Nakai, H. (2010). Local response dispersion method. II. Generalized multicenter interactions. *J. Chem. Phys.* 133: 194101.

59 Huang, P. and Carter, E.A. (2008). Advances in correlated electronic structure methods for solids, surfaces, and nanostructures. *Annu. Rev. Phys. Chem.* 59: 261.
60 Huang, C. and Carter, E.A. (2011). Potential-functional embedding theory for molecules and materials. *J. Chem. Phys.* 135: 194104–194101.
61 Libisch, F., Huang, C., and Carter, E.A. (2014). Embedded correlated wavefunction schemes: theory and applications. *Acc. Chem. Res.* 47: 2768.
62 Cohen, M.H. and Wassermann, A. (2007). On the foundations of chemical reactivity theory. *J. Phys. Chem. A* 111: 2229.
63 Ramakrishnan, R., Dral, P.O., Rupp, M., and von Lilienfeld, O.A. (2015). Big data meets quantum chemistry approximations: the delta-machine learning approach. *J. Chem. Theory Comput.* 11: 2087.
64 Smith, J.S., Isayev, O., and Roitberg, A.E. (2017). ANI-1: an extensible neural network potential with DFT accuracy at force-field computational cost. *Chem. Sci.* 8: 3192.
65 Raff, L.M., Malshe, M., Hagan, M. et al. (2005). Ab initio potential energy surfaces for complex, multichannel systems using modified novelty sampling and feedforward neural networks. *J. Chem. Phys.* 122: 084104–084101.
66 Smith, J.S., Nebgen, B.T., Zubatyuk, R. et al. (2018). Outsmarting quantum chemistry through transfer learning. *ChemRxiv* https://doi.org/10.26434/chemrxiv.6744440.v1.
67 Smith, J.S., Nebgen, B., Lubbers, N. et al. (2018). Less is more: sampling chemical space with active learning. *J. Chem. Phys.* 148: 241733.
68 Lorenz, S., Scheffler, M., and Gross, A. (2006). Description of surface chemical reactions using a neural network representation of the potential-energy surface. *Phys. Rev. B* 73: 115431.
69 Chen, W.-K., Liu, X.-Y., Fang, W.-H. et al. (2018). Deep learning for nonadiabatic excited-state dynamics. *J. Phys. Chem. Lett.* 9: 6702.
70 Guan, Y., Zhang, D.H., and Yarkony, D.R. (2018). representation of coupled adiabatic potential energy surfaces using neural network based quasi-adiabatic Hamiltonians: 1,2 2A' states of LiFH. *Phys. Chem. Chem. Phys.* https://doi.org/10.1039/c8cp06598e.
71 Durrant, J.D. and McCammon, J.A. (2011). NNScore 2.0: a neural network receptor-ligand scoring function. *J. Chem. Inf. Model.* 51: 2897.

8

Interaction of Molecules with Surfaces

CONTENTS

8.1 Interaction of a Molecule with a Solid Surface

8.1.1 Interaction Potential Energies at Surfaces

The interaction of a molecule with a solid surface depends on the atomic structure of the solid, which is usually classified as: (i) a covalent structure, (ii) an ionic structure, (iii) a weak-bonded (van der Waals) structure, (iv) a hydrogen-bonded structure or, (v) a metal. The first four involve localized electrons

Molecular Interactions: Concepts and Methods, First Edition. David A. Micha.

and can be treated as dielectrics, while metals (and the related semiconductors) contain delocalized electrons and require a different treatment. Surfaces of these solids have corresponding electronic properties, which affect the way the molecule interacts with the surface. Furthermore, the solid can be regular with a periodic lattice structure or amorphous. Rearrangement of atomic positions and of electronic distributions occur at a surface when this is formed in a solid, which further affects its interaction energy with a molecule.

The potential energy of interaction of a molecule near a solid surface can be constructed as a sum of terms for the molecule interacting with unit cells for a periodic lattice or with atom groups for amorphous surfaces. This provides a starting description of more complicated situations, where many molecules may be present as the solid surface is in contact with a gas, liquid, or another solid. At short distances between an adsorbed molecule and a surface, the interaction leads to *physisorption* if it is weak and nonbonding, while a strong interaction involving electronic rearrangement and bond breaking or formation leads to *chemisorption*. The interaction between two adsorbed molecules is both direct and indirect through surface effects, and their interaction strength depends on whether they are physisorbed or chemisorbed.

A clean and regular surface of a crystal can be defined by a cut through its three-dimensional lattice, giving a plane with atomic or molecular units in a periodic structure. An introduction to the nomenclature and definitions for lattices can be found in book chapters on the physics and chemistry of solids [1, 2] and in more detailed treatments of clean surfaces [3] and surfaces with adsorbates [4]. In brief, a plane through a crystal lattice formed by periodic translation of a unit cell with side vectors (primitive translations) $\left(\vec{a}, \vec{b}, \vec{c}\right)$ is defined by *Miller indices* (h, k, l), which are the inverse of intersection distances along the three axes, in units of primitive side lengths. Examples for a cubic lattice are the (001) plane, cutting the $\vec{c}$ axis and parallel to the $\left(\vec{a}, \vec{b}\right)$ plane, (011) cutting $\vec{b}$ and $\vec{c}$ axes, and (111) cutting all three.

For a given planar lattice, the primitive translations $\left(\vec{a}, \vec{b}\right)$ define five Bravais lattice types (oblique, square, hexagonal, rectangular, and centered rectangular), which combine with 10 crystalline point group symmetries to give 17 two-dimensional space groups. Translation vectors between cell units on a plane are of the form $\vec{T}_s = m\vec{a} + n\vec{b}$ with m and n integers. Lines on the plane are defined by two Miller indices (h, k) and the distance between two lines, when $\vec{a}.\vec{b} = 0$, is $d(h, k) = [(h/a)^2 + (l/b)^2]^{-1/2}$.

Electronic densities in the planar lattice must satisfy periodicity conditions, and they determine the periodic potential energies of interaction with a nearby molecule. Periodicity can be imposed introducing the primitive reciprocal lattice vectors $\left(\vec{A}, \vec{B}\right)$, which define a Brillouin lattice in reciprocal space, and are given by $\vec{A}.\vec{a} = 2\pi$, $\vec{A}.\vec{b} = 0$, and $\vec{B}.\vec{a} = 0$, $\vec{B}.\vec{b} = 2\pi$. Translations in the reciprocal lattice are given by the reciprocal lattice vectors $\vec{G} = h\vec{A} + k\vec{B}$.

Considering structures where the $\vec{c} = c\vec{n}_z$ primitive vector is perpendicular to the surface plane and along a z-axis with a unit vector $\vec{n}_z$ pointing into the vacuum, an origin of coordinates can be chosen at one of the lattice units. The interaction potential energy $V(\vec{R})$ between an atom or molecule with its center of mass at location $\vec{R} = \vec{R}_S + Z\vec{n}_z$, where $\vec{R}_S$ has components (X, Y) on the surface, with $Z \geq 0$, and a planar regular lattice must satisfy the periodicity condition $V(\vec{R} + \vec{T}_s) = V(\vec{R})$. This can be imposed with the choice

$$V(\vec{R}) = \sum_{\vec{G}} V_{\vec{G}}(Z)\exp(i\vec{G}.\vec{R}_s)$$

$$V_{\vec{G}}(Z) = \frac{1}{A_S}\int_{A_s} d^2R_s\, V(\vec{R})\exp(-i\vec{G}.\vec{R}_s)$$

where A_s is the area of a unit surface cell and the $V_{\vec{G}}$ are Fourier components of the potential energy at each distance Z. Periodicity follows from the fact that $\vec{G}.\vec{T}_s = 2\pi(hm + kn)$ with the factor in the parenthesis being an integer.

The potential energy $V(\vec{R})$ is composed of long-range, intermediate-, and short-range interactions of the molecule with the surface and depends also on their internal atomic positions ($\mathbf{Q}_A$, $\mathbf{Q}_S$) for the molecule and surface. Long-range interactions can be obtained by analogy with the procedure for molecule–molecule systems, with the difference that molecule–surface interactions may extend over a large region at the solid surface and also inside it, so that the overall interaction results from sums of local interactions over the surface and volume of the solid.

This strongly modifies the dependence of the potential energy on the distance Z between a molecular center-of-mass and a surface plane. The van der Waals interaction energy, which goes as R^{-6} for two molecules, becomes a function going as Z^{-3} after addition over the solid's atomic components, with a coefficient expressed in terms of the molecular and solid dynamical susceptibilities. The induction energy between a charged molecule and induced anisotropic dipoles extending over the solid surface similarly changes from R^{-4} to Z^{-1}, with a coefficient given in terms of the solid polarizability per unit volume. And the interaction energy between a surface dipole layer and a charge at the molecule becomes a constant independent of Z. This will be shown in more detail in what follows.

Intermediate- and short-range interaction energies can be obtained from functionals of the electronic densities of molecule and solid surface, similarly to what is done for molecule–molecule interactions. The treatment, however, must be adapted to the degree of delocalization of electrons in the solid surface. When the electronic charge density is well represented by a sum over atomic components, as is the case for dielectrics, intermediate- and short-range interactions follow from the treatments in previous chapters for interactions in many-atom systems, applied to the interaction of the molecule with atomic groups at the surface. In the case of metals, however, electrons are delocalized and energies depend on the electronic charge distribution over the whole solid surface, which must be calculated in the presence of the molecule. These two situations are separately described in what follows.

The combination of long-, intermediate- and short-range forces lead to the formation of attraction wells in potential energy functions and to adsorption of the molecule to the surface. Depending on the strength of the binding energy at the adsorbate site and on the extent of electronic rearrangement, the attraction leads to physisorption or chemisorption. Physisorption involves small electronic rearrangement and weak binding of the species, with van der Waals binding energies of the order of 20 kJ mol^{-1} (about 200 meV), typical distances between adsorbate and substrate around 350 pm, where the adsorbed species maintain their properties so that the total system electronic density can be constructed from the densities of separate molecule and solid surface, possibly weakly perturbed by their interaction. Chemisorption involves larger binding energies, of the order 200 kJ mol^{-1} (about 2 eV) corresponding to formation or breaking of chemical bonds with extensive electronic charge redistribution, or to electron transfer, with adsorbate–surface bond distances around 150 pm, and with rearrangement of atomic components of the molecule and surface. This can be the result of electronic charge transfer at the surface, molecular dissociation, or bonding between molecular and surface atoms. Physisorption and chemisorption can both be present at a surface as the distance between A and S changes, with a transition (or activation) energy linking them as shown in Figure 8.1a. Physisorption and chemisorption will be further considered, with their treatment emphasizing general concepts and computational methods. Figure 8.1 shows typical potential energy surfaces for a species A physisorbed on a surface S, for chemisorption of A on S due to electronic charge transfer, and for a species AB undergoing dissociation into $A + B$ due to its interaction with the surface S.

Electron transfer involves two or more potential energy surfaces and their crossings, as treated in the previous chapter on intermolecular interactions. The same arguments apply here about adiabatic or nonadiabatic potential functions and about potential couplings due to atomic displacements leading to electronic rearrangement. Dissociation at a surface is a simple case of a chemical reaction and can be treated introducing potential energy functions of several variable distances to describe dissociation energy barriers, similarly to what is done for atom–diatom interactions leading to bond breaking and formation.

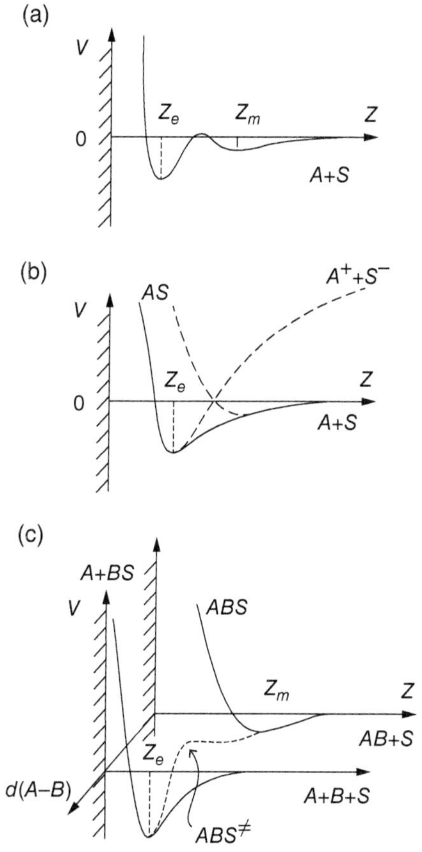

Figure 8.1 Potential energy $V(Z)$ functions of the distance Z between a molecular center-of-mass and a surface plane of a solid S for fixed internal atomic positions in the molecule and solid. (a) Physisorption of A on S at large distances around Z_m and chemisorption at short distances around Z_e, linked by an activation barrier; (b) electron-transfer chemisorption of A; (c) dissociative chemisorption of AB on S leading to $A + BS$, for variables Z and distance $d(A-B)$ between A and B, with a potential energy surface saddle point and reaction energy barrier shown as $ABS^{\neq}$.

The treatment in this chapter is focused on a single molecule interacting with the surface, but the same concepts and methods can be used to describe the interaction of two molecules at the surface, or the interaction of the surface with islands or layers of adsorbed molecules [4, 5].

In some detail, the calculation of the interaction potential energy between molecule A and solid surface S can be done by analogy with the developments in Chapter 7 on the interaction of molecules A and B, with B now replaced by the solid surface S. The interaction energy can be obtained from energy functionals of the densities of AS, A, and S, within a subtraction scheme of the form

$$V_{A,S}\left(\vec{R},\mathbf{Q}_A,\mathbf{Q}_S\right) = E^{(AS)}\left[\vec{R},\mathbf{Q}_A,\mathbf{Q}_S;\rho_{AS}\left(\vec{r}\right)\right] - E^{(A)}\left[\mathbf{Q}_A;\rho_A\left(\vec{r}\right)\right] - E^{(S)}\left[\mathbf{Q}_S;\rho_S\left(\vec{r}\right)\right]$$

This expression can be used in calculations to describe a situation where the molecule A and surface S undergo electronic rearrangement as the relative position $\vec{R}$ changes while the internal positions ($\boldsymbol{Q}_A$, $\boldsymbol{Q}_S$) remain close to their values at large relative distances.

However, a more general approach is needed when there is extensive atomic rearrangement at shorter distances, for example due to molecular dissociation of A, which affect not only A but also the substrate S. This situation can be described introducing new fragments A' and S' and their self-consistent coupling, with new densities $\rho_{A'}$ and $\rho_{S'}$, and are functions of new atomic internal positions $(\boldsymbol{Q}_{A'}, \boldsymbol{Q}_{S'})$, and the relative coordinates. The fragment A' includes A and neighboring surface atoms, while S' is the remaining indented surface structure. Their densities can be constructed so that the total density is $\rho_{AS} = \rho_{A'} + \rho_{S'}$, and the total energy of the AS system can be obtained from an additive scheme (or electronic embedding) writing instead that the total energy for A at relative location $\vec{R}$ over the surface is

$$E^{(AS)} = E^{(A')}\left[\vec{R}; \rho_{A'}(\vec{r})\right] + E^{(S')}\left[\vec{R}; \rho_{S'}(\vec{r})\right] + E_{int}^{(A'S')}\left[\vec{R}; \rho_{A'}(\vec{r}) + \rho_{S'}(\vec{r})\right]$$

which contains the interaction energy $E_{int}^{(AS)}$ between the two coupled subsystems. This is given by the quantal expectation value of the energy coupling operator between charges of A and S obtained from a many-electron state $\Phi^{(AS)}$, which allows for atomic rearrangement, calculated from the full Hamiltonian $\hat{H}_{AS} = \hat{H}_A + \hat{H}_S + \hat{H}_{AS}^{(Coul)} = \hat{H}_{A'} + \hat{H}_{S'} + \hat{H}_{A'S'}^{(Coul)}$. Alternatively $E^{(AS)}$ can be constructed as an approximate functional within DFT. The interaction energy can be given as

$$V_{A,S}\left(\vec{R}\right) = \Delta E^{(A')}\left[\vec{R}; \rho_{A'}(\vec{r})\right] + \Delta E^{(S')}\left[\vec{R}; \rho_{S'}(\vec{r})\right] + E_{int}^{(A'S')}\left[\vec{R}; \rho_{A'}(\vec{r}) + \rho_{S'}(\vec{r})\right]$$

where $\Delta E^{(A')} = E^{(A')} - E^{(A)}$ and $\Delta E^{(S')} = E^{(S')} - E^{(S)}$ are fragment adsorption and indentation energies changing with $\vec{R}$, in an embedding treatment.

The following sections provide details on electronic states, densities, and susceptibilities for a solid surface, as needed to consider its interaction with molecules.

8.1.2 Electronic States at Surfaces

The electronic structure of a solid with a surface separating it from a vacuum can be obtained from a model of atomic ion cores, containing localized electrons, interacting with delocalized electrons. The number N_e of delocalized electrons moves in the field of positive ion charges with the same but opposite total charge $|c_e| N_e$. They can be treated as independent electrons in spin orbitals $\psi_{\nu\sigma}(\boldsymbol{x})$, where ν and σ are orbital and spin quantum numbers and $\boldsymbol{x} = (\vec{r}, \zeta)$

includes position and spin variables, obtained from a one-electron effective Hamiltonian. Total energies follow from a sum of one-electron energies $\varepsilon_{\nu\sigma}$ of occupied orbitals, corrected by electron–electron correlation energies.

More accurately, energies follow from a many-electron treatment where N_e electrons are described by a many-electron Hamiltonian and wavefunctions $\Psi_J(\boldsymbol{x}_1,\ldots,\boldsymbol{x}_{N_e})$ accounting for electrostatic, electronic exchange, and correlation effects. In this case, a practical approach involves the introduction of a set of N_e Kohn–Sham spin orbitals, the solid electronic density $\rho(\boldsymbol{x})$, and an energy density functional.

For an electron moving in a lattice potential with a translational symmetry so that $v(\vec{r}+\vec{T}) = v(\vec{r})$, its orbitals $\varphi_{\vec{k},j}(\vec{r})$ for each band j and wavevector $\vec{k}$ must satisfy the Bloch theorem as $\varphi_{\vec{k},j}(\vec{r}+\vec{T}) = \exp(i\vec{k}.\vec{T})\varphi_{\vec{k},j}(\vec{r})$, with $\vec{k} = k_1\vec{A} + k_2\vec{B} + k_3\vec{C}$, the solutions of a one-electron Schrodinger equation with eigenenergies giving bands of energy $\epsilon_{\vec{k},j}$ [2]. Periodicity means that if $\epsilon_{\vec{k},j}$ is an eigenenergy, so is $\epsilon_{\vec{k}+\vec{G},j}$ with $\vec{G}$ a reciprocal lattice vector. The presence of a lattice potential leads to couplings of orbitals $\varphi_{\vec{k},j}$ and $\varphi_{\vec{k}+\vec{G},j}$ and to splitting of energy bands at edges in reciprocal space corresponding to the elementary reciprocal vector boundaries of a Brillouin zone [2].

The orbitals of dielectrics and semiconductors, including transition metal oxides, can be accurately treated within a tight-binding approach that combines atomic orbitals $\chi_\mu(\vec{r})$ to construct Bloch orbitals as

$$\varphi_{\vec{k},\mu}(\vec{r}) = \sum_{\vec{T}} \exp\left(i\vec{k}.\vec{T}\right)\chi_\mu\left(\vec{r}-\vec{T}\right)$$

Alternatively it is possible to first construct atomic orbitals as combinations of localized and plane wave functions and then form from them Bloch orbitals of the correct translational symmetry. For the example of solid Si, its atomic ground state configuration $3s^2 3p^2$ allows formation of four hybrids $|\chi_j\rangle = |sp_xp_yp_z\rangle_j$, $j = 1$ to 4, along tetragonal directions, and gives Bloch orbitals $\varphi_{\vec{k},j}(\vec{r})$ with the correct electronic distribution for the Si solid bulk and surface atoms.

More detailed treatments accounting for electrostatic interactions, exchange, and correlation can be done with the methods described in the chapter on many-electron approaches. Among these, a preferred one for solids and their surfaces is again DFT [6], which has been developed using a variety of functionals and basis sets. Introduction of atomic core pseudopotentials and a basis set of plane waves has allowed calculations of many solid structures with surface boundaries [7].

Many of the features of the surface electronic structure of metals and semiconductors, and related interaction with nearby atoms and molecules, can be qualitatively treated within a simple model where the lattice of positive ions is averaged into a homogeneous positive background of density $\rho_+(Z) = \overline{\rho_+}$, $Z \leq 0$, $\rho_+(Z) = 0$, $Z > 0$ per unit volume extending from a planar surface going through the uppermost layer of ions and perpendicular to the z-axis down into $-\infty \geq Z \geq 0$. A set of independent delocalized electrons of density $\rho(\vec{r})$ are bound to the substrate by its positive charge and form what is usually called the *jellium* model. They spill beyond the positive surface boundary to form an electronic surface dipole.

To allow for the electronic spread into the vacuum and to impose simple boundary conditions, the electrons are assumed to move in a well of potential energy $V(\vec{r})$ of relative value zero inside a semi-infinite region with $-\infty \leq z \leq D$, with D to be chosen so that the total charge of electrons plus lattice ions averages to zero, and with V infinitely repulsive for $z \geq D$ so that the delocalized electrons stay within $-\infty \leq z \leq D$. By symmetry, the electron density $\rho(\vec{r}) = \rho(z)$ and electron states $\varphi_{\vec{k}}(\vec{r})$ can be obtained from the solution for a particle with momentum $\hbar\vec{k}$ in a semi-infinite box as

$$\varphi_{\vec{k}}(\vec{r}) = A_{\vec{k}} \exp\left(i\vec{k}_{\|}.\vec{r}\right) \sin\left[k_{\perp}(z-D)\right],\ z \leq D,\ \varphi_{\vec{k}}(\vec{r}) = 0,\ z > D$$

where $\vec{k}_{\|}$ and $\vec{k}_{\perp}$ are parallel and perpendicular vector components relative to the surface plane, such that $k_{\|}^2 + k_{\perp}^2 = k^2$. At absolute-zero temperature, the electronic states satisfy the Fermi–Dirac distribution and are doubly occupied (with up and down spins) if their energies are below and up to the Fermi value $E_F = \hbar^2 k_F^2/(2m_e)$ [2].

The number of electronic states per unit volume of k-space in a homogeneous medium of large volume Ω is obtained from periodic boundary conditions as $\Omega/(2\pi)^3$, and the total number of occupied spin orbitals is $N_{st} = 2\left[\Omega/(2\pi)^3\right]$ $4\pi \int_0^{k_F} dk\, k^2 = k_F^3 \Omega/(3\pi^2)$. With one electron per occupied spin orbital, so that $N_{st} = N_{el}$, k_F is related to the electronic density $\rho_{el} = N_{el}/\Omega = k_F^3/(3\pi^2)$ or equivalently to the inverse of the electron *spread radius* r_s defined by the electronic volume $\Omega/N_{el} = 4\pi r_s^3/3 = 3\pi^2/k_F^3$. Values of the spread radius for metals are within $2.0 \leq r_s/a_0 \leq 6.0$, with r_s/a_0 equal to 2.07, 2.67, 3.02, and 3.93, respectively, for the common metals Al, Cu, Ag, and Na.

The total local density of electrons is $\rho(\vec{r}) = \int_{k \leq k_F} dk^3 \left|\varphi_{\vec{k}}(\vec{r})\right|^2 = \rho(z)$ and the orbital amplitudes $A_{\vec{k}}$ can be chosen all equal to A_0, to satisfy the neutrality condition $\rho \approx \rho_+$ for $z \to -\infty$ far inside the solid. Writing $\rho(z) = |A_0|^2 \int_{k \leq k_F} d^2\vec{k}_{\|}$ $dk_{\perp} \sin^2(k_{\perp} z')$, $z' = z - D$, and integrating gives

$$\rho(z) = |A_0|^2 \left[1 + \frac{3\cos(2k_F z')}{(2k_F z')^2} - \frac{3\sin(2k_F z')}{(2k_F z')^3}\right]$$

which has the limits $\rho \approx |A_0|^2 = \rho_+$ for $z \to -\infty$ and $\rho \approx 0$ for $z \to D_-$, and is zero for $z > D$. This shows that the density rises from $z = D$ inwards and oscillates along decreasing z away from the surface plane, with a length period $z_F = k_F/\pi$ which increases with the density of electrons, and with an amplitude, which decreases as the density increases. The parameter D can now be fixed requiring that deviations of the electronic density $\rho(z; D)$ from the value $-\overline{\rho_+}$, which neutralizes the positive background, should add up to zero, or $\int_{-\infty}^{D} dz[\rho(z;D) + \overline{\rho_+}] = 0$.

More details about the surface electronic structure can be obtained from a model of a solid with a surface perpendicular to the z-axis at location $z = 0$, with an electronic potential energy $V(z)$ of constant value $V_{in} < 0$ inside a semi-infinite region with $-\infty \le z \le 0$, and value $V_{out} = 0$ for $z \ge 0$ in a vacuum, so that now electrons stay mostly within $-\infty \le z \le 0$ but spill out into the vacuum. As before, the electron density $\rho(\vec{r}) = \rho(z)$ and electron states $\varphi_{\vec{k}}(\vec{r})$ can be obtained from the solution for a particle with momentum $\hbar\vec{k}$, but now they decay exponentially into the vacuum when their energy is $\varepsilon_{\vec{k}} = \hbar^2 k^2/2m_e + V_{in} < V_{out}$ so that

$$\varphi_{\vec{k}}(\vec{r}) = A_{\vec{k}}(z)\exp\left(i\vec{k}_{\parallel}.\vec{r}\right)\sin(k_{\perp}z + \gamma), z \le 0,$$

$$\varphi_{\vec{k}}(\vec{r}) = B_{\vec{k}}\exp\left(i\vec{k}_{\parallel}.\vec{r}\right)\exp(-\kappa z)\, z > 0$$

with $\hbar^2\kappa^2/2m_e = V_{out} - \varepsilon_{\vec{k}}$. Imposing continuity of $\varphi_{\vec{k}}$ and $\partial\varphi_{\vec{k}}/\partial z$ at $z = 0$ and solving the Schrodinger equation for $\varphi_{\vec{k}}(\vec{r})$ provide values for the constants γ and $B_{\vec{k}}$ and for the function $A_{\vec{k}}(z)$. This function can be obtained from boundary conditions, for a delocalized state extending all the way inside $-\infty \le z \le 0$, or as a bound state localized at the surface and normalizable. Surface states are found for $A_{\vec{k}}(z) = A_0\exp\left(|k'_{\perp,p}|z\right)$, and the $k'_{\perp,p}$, with p an integer, can be interpreted as the discretized imaginary part of a complex-valued $k_{\perp}$ wavevector [2, 8].

These features are maintained in more accurate treatments accounting for electronic exchange and correlation such as done within the density functional approach. Here we summarize results from the theory as given, for example, in Refs. [6, 9]. The functional form of $E[\rho]$ in Chapter 5 can be adapted to the present jellium model with the uniform positive background charge $\rho_+(z) = \overline{\rho_+}, z \le 0, \rho_+(z) = 0, z > 0$ adding also a related lattice potential energy per electron $v_+(\vec{r})$ due to the ion cores in the solid, giving the electron–nuclei interaction energy $\int d^3r\rho(\vec{r})v_+(\vec{r})$. Adding to this the electron–electron Coulomb energy and exchange and correlation energy functionals, one finds the effective potential energy functional per electron $v_{eff}[\vec{r};\rho(\vec{r})]$ appearing in the Kohn–Sham equations for the orbitals $\varphi_{\vec{k}}^{(KS)}(\vec{r})$. This effective potential energy $v_{eff} = v + v_{xc}$, containing an exchange–correlation term, is quite different from the electrostatic potential $v(\vec{r}) = v_+(\vec{r}) + v_{Coul}(\vec{r})$, and leads to larger attraction for electrons inside the metal.

Calculation of orbital energies, with the zero of energy at the bottom of the v_{eff} value as $z \to -\infty$, gives a corrected Fermi energy ε_F and a work function (the energy needed to remove an electron from the bulk inside the metal) $W_{blk} = v_{eff}(z \gg 0) - \varepsilon_F$, if one ignores the distortion of the ion lattice and electronic charges at the surface. However, the surface forces created by the surface dipole require an additional energy W_{srf} giving a corrected work function $W = W_{blk} + W_{srf}$, which can be obtained from density functional calculations.

To proceed further and obtain properties dependent on the lattice structure of the surface, such as work functions and surface densities as functions of the surface lattice Miller indices (h, k, l), it is necessary to introduce the ion lattice structure of the metal and its surfaces and to calculate electronic properties for them. A convenient model involves a periodic solid slab between two surface planes of Miller indices (h, k, l) perpendicular to the z-axis, with a group of ions in each unit cell at the positions $\vec{T}_{m,n,p}$, and electronic density $\rho(\vec{r})$ in each cell. The cell electrostatic potential at a location $\vec{r}$ is $\phi\left[\vec{r} - \left(\vec{d}_0 + \vec{T}_{m,n,p}\right); \rho(\vec{r})\right]$, a functional of the density, and the total electrostatic potential is the sum over cell units,

$$\phi_{h,k,l}(\vec{r}) = \sum\nolimits_{m,n,p}^{(h,k,l)} \phi\left[\vec{r} - \left(\vec{d}_0 + \vec{T}_{m,n,p}\right)\right]$$

This summation extends over all cell units contained within the two surface planes (h, k, l) defining the slabs and is different for each set of plane indices. Computational treatments and results of DFT for surfaces can be found, for example, in Ref. [7].

The potential energy of an electron in the solid can be calculated for this electrostatic potential and provides work functions $W(h, k, l)$, which depend on the surface indices. Taking the surface to be perpendicular to the z-axis, with an averaged (over x- and y-variables) electronic potential energy $\bar{v}(z)$, W can be obtained as the difference of electronic energies

$$W = [\bar{v}(\infty) + E_{N-1}] - E_N = \bar{v}(\infty) - \mu$$

where μ is the chemical potential of the electrons. For example, values for *fcc* Cu surfaces are 4.59, 4.48, and 4.98 eV (or 442.9, 432.3 and 480.5 kJ mol^{-1}) for surface planes (100), (110), and (111), respectively. These values are obtained from photoemission measurements, and DFT calculations show the same trends. Accurate calculations must account for distortion of the solid lattice near the surfaces. The work function for Si(111) is calculated to be about 4.15 eV, but that surface shows reconstruction into Si(111) (2 × 1) and (7 × 7) at varying temperatures, with a related small change in the work function for the (2 × 1) structure, but giving a work function lower by about 0.2 eV (19.3 kJ mol^{-1}) for (7 × 7).

8.1.3 Electronic Susceptibilities at Surfaces

A many-electron system bounded by a surface and driven by an external electric field, created by an interacting molecule or adsorbate, responds by showing electronic excitations, which differ for dielectrics or conductors, and contains effects of surface inhomogeneity. A general treatment of the response can be formally done for both dielectrics and conductors, starting from a model of the solid with its electrons moving in the field of its atomic ion cores. Ions vibrating around equilibrium positions are slow compared to electronic motions and can be assumed to stay fixed at given ionic conformations while calculating electronic response. The dynamics of electronic motions is described as done in Chapter 2 on properties of an extended molecule, in terms of dynamical susceptibilities.

A time-dependent driving electric field may vary over the lengths of a bounded solid, and it is necessary to consider a position- and time-dependent electric potential $\phi(\vec{r},t)$. For an extended system of charges I of electrons and nuclei (or ionic cores) at locations $\hat{\vec{r}}_I$, the total charge density operator per unit volume at space location $\vec{r}$ is $\hat{c}(\vec{r}) = \sum_I C_I \delta(\vec{r} - \hat{\vec{r}}_I)$ for charges C_I and operator-valued positions $\hat{\vec{r}}_I$. The treatment for an extended inhomogeneous system is conveniently done in terms of the expectation value of the total charge density, with the potential energy of coupling between the solid and field written as $\hat{V}_{SF} = \int dr^3 \hat{c}(\vec{r})\phi(\vec{r},t)$, and by working directly with the charge density operator.

For a time-dependent electric potential $\phi(\vec{r},t)$, the average charge density in a many-electron state a is a sum $c_a(\vec{r},t) = c_a^{(0)}(\vec{r}) + c_a'(\vec{r},t)$ of a permanent static value $c_a^{(0)}$ plus an induced charge density changing over time, with the latter obtained from the density operators $\hat{c}(\vec{r},t)$ with a many-electron state $\Phi_a(t)$. It is convenient to proceed within a formal operator approach to calculate susceptibilities without expanding in the usually unknown excited states for the extended system. The average induced charge density $c_a'(\vec{r},t)$ arises in response to the applied potential $\phi(\vec{r},t')$ present from time $t = 0$ and at all earlier times $0 \le t' \le t$, and can be expressed in terms of a delayed response function $\chi_a^{(1)}(\vec{r},t;\vec{r}_1,t_1)$ for phenomena linear in the applied field, written here using response functions of time and positions [10–12]. For a system initially in a time independent unperturbed state a, it is a function of the difference $t' = t - t_1$ and it is convenient to introduce the retarded susceptibility $\chi_a^{(+)}(\vec{r},\vec{r}_1;t') = \theta(t')$ $\chi_a^{(1)}(\vec{r},t';\vec{r}_1,0)$ where $\theta(t')$ is the step function null for negative arguments. This provides the useful expression

$$c_a'(\vec{r},t) = \int_{-\infty}^{\infty} dt' \int d^3 r_1 \chi_a^{(+)}(\vec{r},\vec{r}_1;t')\phi(\vec{r}_1,t-t')$$

which is a convolution form with a Fourier transform from time to frequency giving $\tilde{c}'_a(\vec{r},\omega) = \sum_\eta \int d^3r_1 \tilde{\chi}_a^{(+)}(\vec{r},\vec{r}_1;\omega)\tilde{\phi}(\vec{r}_1,\omega)$ in terms of the transforms of the three expressions. The dynamical susceptibility $\tilde{\chi}_a^{(+)}(\omega)$ can be obtained from a perturbation calculation of the average polarization. As shown in Chapter 2, the charge polarization is found to be

$$c'_a(\vec{r},t) = \int_{-\infty}^{\infty} d\tau \int d^3r' \chi_{cc,a}^{(+)}(\vec{r},\vec{r}',\tau)\phi(\vec{r}',t-\tau)$$

$$\chi_{cc,a}^{(+)}(\vec{r},\vec{r}',\tau) = -\left(\frac{i}{\hbar}\right)\left\langle \Psi_a^{(0)}(0) \mid \left[\hat{c}(\vec{r},\tau)\hat{c}(\vec{r}',0) - \hat{c}(\vec{r}',0)\hat{c}(\vec{r},\tau)\right] \mid \Psi_a^{(0)}(0)\right\rangle$$

for $\tau > 0$ and $\chi_{cc}^{(+)}(\tau) = 0$ for $\tau < 0$. Here the time-dependent density operator $\hat{c}(\vec{r},t) = \hat{U}^{(0)}(t)^\dagger \hat{c}(\vec{r})\hat{U}^{(0)}(t)$, with $\hat{U}^{(0)}(t)$ the time-evolution operator of the unperturbed system, appears at two different times in an expectation value with the unperturbed state, and the bracket is shown as a function of only the difference $\tau = t - t'$. The two sides can be Fourier transformed into $\tilde{c}'_a(\omega) = \tilde{\chi}_{cc,a}^{(+)}(\omega)\tilde{\phi}(\omega)$, which are complex-valued functions of frequency. The dynamical response can be alternatively treated in terms of the dielectric function $\varepsilon_a(\omega) = 1 + \tilde{\chi}_{cc,a}^{(+)}(\omega)$, which also depends on the electronic positions $\vec{r}, \vec{r}'$.

The response function of a bounded solid can be calculated in a variety of ways, using expansions of operators in a known basis set, or numerically generating a solution over time. It can also be constructed semiempirically to incorporate known features of the response. The response functions can be calculated introducing a convenient basis set $\{\langle \vec{r} \mid \mu \rangle\}$ of states to expand the charge operators, given a reference unperturbed state a, in which case responses can be expressed in terms of the amplitudes $\langle \mu \mid \hat{c}(t) \mid \mu' \rangle = c_{\mu\mu'}(t)$ and susceptibility matrices $\tilde{\chi}_{\mu\mu',a}^{(+)}(\omega)$.

Excitations in dielectrics can be described as transitions between many-electron states, similarly to what is done for molecular electronic excitations. They may be electron–hole pairs localized at ionic or molecular sites, or delocalized electron–hole pairs in periodic lattices, forming excitons [13, 14]. Total electrical susceptibilities arise from the distortion of valence electron shells and from the displacement of atomic ion cores [15]. For ionic dielectrics such as NaCl, or dielectrics composed of noble gas atoms or molecules like CO, a simple model of the susceptibility involves electronic polarizabilities as well as atomic displacement polarizabilities for each atomic ion core n, and sums over all the species in a solid bounded by a surface perpendicular to the z-axis at position $Z = 0$. This gives a polarization $P(\vec{r},t)$ in the solid, which can be averaged over variables in $\vec{r}$ to obtain $\bar{P}(t)$ inside the solid and $\bar{P}(t) = 0$ outside.

The electric field $\mathcal{E}_{int}(t)$ inside the solid is related to an external field $\mathcal{E}_{ext}(t)$ outside it by $\mathcal{E}_{int}(t) = \mathcal{E}_{ext}(t) + \epsilon_0{}^{-1}\bar{P}(t)$. Insofar the polarization is a delayed response to the applied external field, its value is $\bar{P}(t) = \int d\tau \chi^{(+)}(\tau)\mathcal{E}_{ext}(t-\tau)$, with $\chi^{(+)}(t)$ a susceptibility in the solid, null outside it. The dielectric displacement function $\mathcal{D}(t) = \int d\tau \varepsilon^{(+)}(\tau)\mathcal{E}_{ext}(t-\tau)$ defines the solid dielectric function $\varepsilon^{(+)}(t) = \epsilon_0 \varepsilon_r^{(+)}(t)$ and equals $\mathcal{D}(t) = \epsilon_0 \mathcal{E}_{int}(t)$ inside the solid. Given an external oscillating field of frequency ω, taking Fourier transforms over times leads to $\tilde{P}(\omega) = \chi(\omega)\tilde{\mathcal{E}}_{ext}(\omega)$, $\tilde{\mathcal{D}}(\omega) = \tilde{\varepsilon}^{(+)}(\omega)\tilde{\mathcal{E}}_{ext}(\omega) = \epsilon_0 \tilde{\mathcal{E}}_{int}(\omega)$, and to $\tilde{\varepsilon}_r^{(+)}(\omega) = \varepsilon_r(\omega) = 1 + \chi(\omega)$.

A local field $\mathcal{E}_{loc}(t)$ at a chosen location is obtained in more detail as a sum adding to the external field a depolarization field from physical surface charges, a field from the surface charges of a model cavity around the location, and the field of the all the species J inside the cavity, with average density $\bar{\rho}_J$. Letting $\alpha_J(\omega)$ be the total dynamical polarizability of each species J, the dielectric function in the solid is given for the many-electron state a by [2]

$$\frac{\varepsilon_r(\omega) - 1}{\varepsilon_r(\omega) + 2} = \frac{1}{3\epsilon_0} \sum\nolimits_J \bar{\rho}_J \alpha_J(\omega)$$

This expression can also be used to account for changes across a surface introducing a molecular density $\bar{\rho}_J(Z)$, which varies when averaging its value over a surface perpendicular to the Z-axis, and gives a varying dielectric function $\varepsilon_r(Z, \omega)$. It can further be simplified by introducing parametrized forms with frequencies and intensities obtained from independent calculations or measurements. Letting $\alpha_J(\omega) = \alpha_{J,el}(\omega) + \alpha_{J,ion}(\omega)$ to include electronic and displaced ion core terms, these polarizabilities can be written as [2]

$$\alpha_{J,el}(\omega) = \frac{c_e^2/m_e}{\omega_{J,0}{}^2 - \omega^2}, \quad \alpha_{J,ion}(\omega) = \frac{c_J^2/M_J}{\omega_{J,T}{}^2 - \omega^2}$$

where $\hbar\omega_{J,\,0}$ is the lowest electronic excitation energy of species J, while c_J, M_J, and $\hbar\omega_{J,\,T}$ are the charge, mass of ion J, and the transversal optical vibrational energy of the lattice of J ions. Typical energy values for dielectrics are $\hbar\omega_{J,\,0} \cong$ 10 eV (or $\cong 10^3$ kJ mol^{-1}) and $\hbar\omega_{J,\,T} \cong 10^{-1}$ to 10^{-2} eV (or $\cong$ 10 to 1.0 kJ mol^{-1}) so that for photons in the visible, near-IR and near-UV spectral regions, with $\omega^2 \ll \omega_{J,\,0}{}^2$, the electronic polarizability can frequently be taken to be a constant independent of frequencies.

8.1.4 Electronic Susceptibilities for Metals and Semiconductors

The dielectric function $\varepsilon(Z, \omega)$ of a metal or semiconductor driven by a homogeneous external oscillating field $\mathcal{E}_{ext}(t;\omega)$ must instead be derived keeping in mind that electrons are delocalized in these solids, so that it is convenient to work with its spatial Fourier transform from Z to the wavenumber q within a slab of thickness L,

$$\varepsilon_r(q,\omega) = \frac{1}{L}\int_L dZ \exp(-iqZ)\tilde{\varepsilon}_r(Z,\omega)$$

a function of the wavenumber q, which accounts for the components of the induced electric field with wavelengths $2\pi/q$. Two relevant limits are $\varepsilon_r(0, \omega)$, for phenomena over very long wavelengths such as occur for molecules interacting from large distances to the surface, and the static limit $\varepsilon_r(q, 0)$, which can be obtained from static external and induced charges [2].

A simple treatment of the electronic dielectric function $\epsilon_{r,el}(\omega) = \tilde{\varepsilon}_{el}(0,\omega)/\epsilon_0$ of metals and semiconductors driven by an external oscillating field $\mathcal{E}_{ext}(t;\omega)$ is based on a classical treatment of a homogeneous electronic density ρ_e displaced a distance $u(t;\omega)$ by the field with respect to a positive homogeneous charge background representing the averaged lattice ions. It is also assumed that the wavelength of the external field is large compared to the thickness of the solid, so that the field is homogeneous inside the solid. The displacement creates surface charges $\sigma = \pm u\rho_e c_e$ perpendicular to the field and a resulting induced field $\mathcal{E}_{ind}(t;\omega) = -|u(t;\omega)\rho_e c_e|/\epsilon_0$ in between them [2]. The equation of motion for the displacement of electrons is $m_e\rho_e d^2u/dt^2 = \rho_e c_e \mathcal{E}_{ext}$, which can readily be solved for a harmonic field $\mathcal{E}_{ext}(t) = \mathcal{E}_0\cos(\omega t)$ with the choice $u(t) = u_0\cos(\omega t)$ giving $u(t) = c_e\mathcal{E}_{ext}(t)/(m_e\omega^2)$. The induced field is then $\mathcal{E}_{ind} = -{\omega_p}^2\mathcal{E}_{ext}/\omega^2$, with ${\omega_p}^2 = \rho_e {c_e}^2/(\epsilon_0 m_e)$ the square of the frequency of the collective oscillation of the electrons, forming the plasmon. The electronic dielectric function follows from

$$\epsilon_{r,el}(\omega) = \frac{\mathcal{E}_{ext} + \mathcal{E}_{ind}}{\mathcal{E}_{ext}} = 1 - \frac{{\omega_p}^2}{\omega^2}$$

which gives propagating waves for positive values, as found from the Maxwell equations, when $\omega > \omega_p$.

In more detail, the susceptibilities of conductive materials, whether semiconductors like Si or TiO_2, or metals like Cu or Ag, can be obtained from their atomic structure expanding charge density operators in a basis set of stationary many-electron states $\Psi_K(\boldsymbol{x}_1,\ldots,\boldsymbol{x}_{N_e})$ and deriving the dynamical susceptibility $\tilde{\chi}^{(+)}_{cc,a}(\omega)$, with $K = a$, from its expression given above and involving the time-correlation of charge density operators, as described in Chapter 2 for dynamical polarizabilities. A useful but more approximate treatment derives the susceptibility from self-consistent field or DFT one-electron treatments with a basis set of Bloch orbitals $\varphi_{\vec{k},j}(\vec{r}) = \varphi_\lambda(\vec{r})$ giving matrices $\tilde{\chi}^{(+)}_{\lambda\lambda',a}(\omega)$, a susceptibility $\tilde{\chi}^{(+)}_a(\omega) = \chi_a(\omega)$ and $\tilde{\varepsilon}^{(+)}_{r,a}(\omega) = \varepsilon_{r,a}(\omega) = 1 + \chi_a(\omega)$.

As explained in the treatment of solids [2, 15], the electronic dielectric function $\tilde{\epsilon}_{el}(0,\omega) = \epsilon_0\epsilon_{r,el}(\omega)$ of an isotropic metal with electronic density ρ_e can be written as a sum over electronic state-to-state transitions $\Lambda = (\lambda\lambda')$, for fixed nuclei, so that

$$\epsilon_{r,el}(\omega) = 1 + \frac{\rho_e c_e^{\,2}}{\epsilon_0 m_e}\sum\nolimits_\Lambda \frac{f_\Lambda}{(\omega_\Lambda^{\,2} - \omega^2) - i\omega\Gamma_\Lambda}$$

where ω_Λ, f_Λ, and Γ_Λ are transition frequencies, strengths, and rates. In more detail, the summation must be written as a double integral over densities of band state $\lambda = \left(\vec{k},j\right)$ and $\lambda' = \left(\vec{k}',j'\right)$ per unit energy. The quotient $\rho_e c_e^{\,2}/(\epsilon_0 m_e) = \omega_p^{\,2}$ is the square of the plasmon frequency for the free electron gas of the given density, with typical plasmon energies $\hbar\omega_p$ of the order of 10 eV (about 10^3 kJ mol^{-1}) for metals. At zero frequency, the static dielectric constant is then $\epsilon_{r,el}(0) = 1 + \omega_p^{\,2}\sum_\Lambda f_\Lambda/\omega_\Lambda^{\,2} = 1 + \omega_p^{\,2}/\omega_0^{\,2}$ with ω_0 a frequency parameter. At large frequencies, provided transitions have small strengths f_Λ for $\omega^2 \gg \omega_\Lambda^{\,2}$, it is

$$\epsilon_{r,el}(\omega) = 1 - \frac{\omega_p^{\,2}}{\omega^2},\ \text{for}\ \omega > \omega_p = \left[\frac{\rho_e c_e^{\,2}}{(\epsilon_0 m_e)}\right]^{\frac{1}{2}}$$

as already found with the simple displacement model. The total dielectric function including ionic contributions can be obtained within a model of oscillating ion charges with a restoring (transverse optical) frequency ω_T and ionic plasmon frequency $\omega_{p,\,ion}$ from $\omega_{p,\,ion}^{\;\;2} = \rho_{ion}c_{ion}^{\;\;2}/(\epsilon_0 M_{ion})$ giving a dielectric function for each type of ion in the solid metal as

$$\epsilon_{ion}(\omega) = 1 + \omega^2_{\;p,ion}/\left(\omega_T^{\,2} - \omega^2\right)$$

at low frequencies $\omega < \omega_T$, and $\epsilon_{ion}(\omega) = \epsilon_{ion}(\infty)$ at the high frequencies when the ions do not respond to a rapidly oscillating field. The constant parameter $\omega_T^{\,2}$ can be re-expressed in terms of a measurable $\epsilon_{ion}(0)$. The total relative dielectric function can be written as

$$\epsilon_r(\omega) = \epsilon_{r,el}(\omega)\epsilon_{ion}(\omega) = \epsilon_{ion}(\omega) - \bar{\omega}_p(\omega)^2/\omega^2$$

with $\bar{\omega}_p^{\,2} = \epsilon_{ion}(\omega)\omega_p^{\,2}$ [15], at frequencies $\omega > \omega_p$. Typical values for the static dielectric function $\epsilon_r(0)$ are about 12 for Si, 11 for Al, and 6 for Cu.

The presence of a surface in a conductor alters its collective electronic oscillations, and leads to the appearance of slower collective oscillations tangential to the surface with surface plasmon frequencies $\omega_s = \omega_p/\sqrt{2}$. This follows from consideration of the total fields inside and outside the surface created by an oscillating charge localized at the surface, and by imposition of continuity of the perpendicular electric displacement fields inside and out, at the surface

[2]. These displacement fields are $\tilde{\mathcal{D}}_z(\omega) = \tilde{\varepsilon}(\omega)\tilde{\mathcal{E}}_z(\omega)$, with the electric field $\tilde{\mathcal{E}}_z(\omega)$ of surface charges equal but opposite inside and outside, so that $\tilde{\varepsilon}_{r,ins}(\omega)\tilde{\mathcal{E}}_z(\omega) = -\tilde{\varepsilon}_{r,out}(\omega)\tilde{\mathcal{E}}_z(\omega)$ gives $1 - \omega_p{}^2/\omega^2 = -1$ from which $\omega_p{}^2/\omega_s{}^2 = 2$. Some typical surface plasmon energies are 2.25 eV (217.1 kJ mol^{-1}) for Cs and 10.6 eV (1022.7 kJ mol^{-1}) for Al and Si.

In more detail, the dependence of the dielectric function on locations Z near the surface can be extracted from the Maxwell equations and external and induced electronic densities $\rho_{ext}(\vec{r},t)$ and $\rho_{ind}(\vec{r},t)$. To simplify, consider only functions of distances along the z-direction. The Maxwell divergence relations give $\partial\mathcal{E}/\partial Z = (\rho_{ext} + \rho_{ind})/\epsilon_0$ and $\partial\mathcal{D}/\partial Z = \rho_{ext}$. Taking Fourier transforms to q and ω from Z and t give $q\tilde{\mathcal{E}}(q,\omega) = [\tilde{\rho}_{ext}(\omega) + \tilde{\rho}_{ind}(q,\omega)]/\epsilon_0$ and $q\tilde{\mathcal{D}}(q,\omega) = \tilde{\rho}_{ext}(\omega)$ and from these $\tilde{\mathcal{D}}(q,\omega)/\left[\epsilon_0\tilde{\mathcal{E}}(q,\omega)\right] = \tilde{\varepsilon}_r(q,\omega) = 1 - \tilde{\rho}_{ind}(q,\omega)/[\tilde{\rho}_{ext}(\omega) + \tilde{\rho}_{ind}(q,\omega)]$, which can be expanded around $q = 0$. For an isotropic medium and absence of the term linear in q, this gives the dispersive dielectric function

$$\tilde{\varepsilon}_r(q,\omega) = 1 - \frac{\tilde{\rho}_{ind}(0,\omega)}{\tilde{\rho}_{ext}(\omega) + \tilde{\rho}_{ind}(0,\omega)} - A(\omega)\,q^2$$

where $A(\omega)$ depends on $\tilde{\rho}_{ext}(\omega)$ and the second term equals $\omega_p{}^2/\omega^2$ for $\omega > \omega_p$. An even more detailed treatment of surface effects on dielectric functions can be derived introducing an inhomogeneous perturbing potential energy $V_s(\vec{r},t)$ for the surface, and its Fourier transform $\tilde{V}_s(\vec{q},\omega)$, and calculating the linear response to this perturbation by the electrons in the solid. This provides a more general dielectric function $\varepsilon_r(\vec{q},\omega)$, which accounts for spatial as well as temporal effects [8].

8.2 Interactions with a Dielectric Surface

8.2.1 Long-range Interactions

Basic insight on the interaction energy of a molecule at or near the surface of a solid with a dielectric function of frequencies, $\varepsilon(\omega)$, can be obtained from a model with point charges outside a semi-infinite solid and with their image charges created inside the solid by electronic rearrangement. The dielectric function is related to the total dynamical susceptibility $\chi(\omega) = P(\omega)/[\varepsilon_0\mathcal{E}(\omega)] = \chi_{el}(\omega) + \chi_{ion}(\omega) = \varepsilon(\omega) - 1$ of the bounded solid, containing the polarizability $P(\omega)$ response of its electrons and atomic cores (ions) to an oscillating electric field $\mathcal{E}(\omega)$ as presented in the previous section. A simple treatment locates a charge C at a distance d above the surface of a semi-infinite solid with a static dielectric constant (or permittivity) $\varepsilon = \varepsilon_r\varepsilon_0$ as shown in

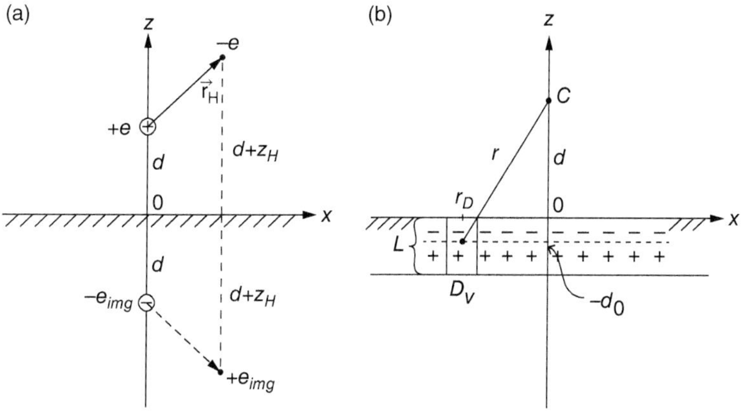

Figure 8.2 (a) Atomic nucleus and electron point charges $\pm e$ above a dielectric and their images $\pm e_{img}$ inside it; (b) A charge C above a surface dipole layer of a dielectric, with a dipole density D_V per unit volume.

Figure 8.2a, leading to rearrangement of the electronic density near the surface and to appearance of an image charge C_{img} and an electric potential $\phi\left(\vec{R}\right)$, which can be derived from C and its image charge C_{img} at distance $-d$ inside the solid. Its magnitude is determined by continuity of the gradient of the electric potential at the surface.

A single positive point charge C at a distance d from a slab surface perpendicular to the z-axis attracts electrons in the slab toward the surface, and this creates the same effect as a charge image C_{img} of opposite sign inside the slab at a distance d from its surface [16]. The electrical potential $\phi\left(\vec{R}\right)$ at a point $\vec{R} = \vec{R}_s + Z\vec{n}_z$ in the vacuum due to the charge and its image is

$$\phi^{(vac)}\left(\vec{R}\right) = \frac{1}{4\pi\epsilon_0}\left(\frac{C}{\left|\vec{R}+d\vec{n}_z\right|} + \frac{C_{img}}{\left|\vec{R}-d\vec{n}_z\right|}\right)$$

The force between the charge at distance Z from the surface and its image is $F_{img} = CC_{img}/(4\pi\varepsilon_0 2Z)^2$, which gives by integration over $d \le Z \le \infty$ a charge-image potential energy

$$E_{img}^{(vac)} = CC_{img}/(4\pi\epsilon_0 4d)$$

Inside the solid, the electric potential $\phi^{(sol)}$ has the same form as $\phi^{(vac)}$ but with ε_0 replaced by $\varepsilon_r\varepsilon_0$.

The distribution $\sigma\left(\vec{R}_s\right)$ of surface charges per unit area at $\vec{R}_s$ induced by the point charge C outside the solid follows from the potential gradient at the surface as $\sigma\left(\vec{R}_s\right) = -(1/4\pi)\left(\partial\varphi^{(vac)}/\partial Z\right)_{\vec{R}_s}$. A special case is that of a metal with perfect conductivity, in which case one finds that $C_{img} = -C$, $E_{img} = -C^2/(4\pi\varepsilon_0 4d)$, and

$$\sigma\left(\vec{R}_s\right) = -\frac{C}{8\pi^2\varepsilon_0}\frac{d}{\left|\vec{R}_s + d\vec{n}_z\right|^3}$$

which integrates over the surface to a charge $-C/(4\pi\varepsilon_0)$.

More generally, continuity of the electric field $\mathcal{E}\left(\vec{R}_s\right)$, given by the gradients of the potentials inside and outside the solid at the surface location, provides the relation between the charges as [17]

$$C_{img} = -C\,(\varepsilon_r - 1)/(\varepsilon_r + 1)$$

with the two charges equal and opposite in the special case of a perfectly conducting metal where $\varepsilon_r \rightarrow \infty$.

As a simple application, consider a H atom near the surface with its nucleus of charge $C_p = e$ at a distance d and its electron of charge $C_e = -e$ at a position $\vec{r}_H$ relative to the nucleus. Assuming that these two charges instantly interact with their opposite images $\pm e\,(\varepsilon_r - 1)/(\varepsilon_r + 1)$, the interaction potential energy operator is, for proton-(image proton) and electron-(image electron) attractions plus twice the electron-(other proton) repulsions, from Figure 8.2b,

$$\hat{H}_{H,img}\left(\vec{r}_H\right) = \frac{e^2}{4\pi\varepsilon_0}\frac{\varepsilon_r - 1}{\varepsilon_r + 1}\left(-\frac{1}{4d} - \frac{1}{4|d + z_H|} + 2\frac{1}{2\left|2\vec{d} + \vec{r}_H\right|}\right)$$

where z_H is the z-component of $\vec{r}_H$. This can be expanded in powers of the electron coordinates in $\vec{r}_H$. The zeroth-order term is zero for the neutral atom, the first order gives forces that add up to zero, and the second-order term provides the leading average electronic energy. In detail,

$$\hat{H}_{H,img}\left(\vec{r}_H\right) = -\frac{e^2}{4\pi\varepsilon_0 16d^3}\frac{\varepsilon_r - 1}{\varepsilon_r + 1}\left(\left|\vec{r}_H\right|^2 + {z_H}^2\right) + \cdots$$

and upon averaging over the H atom ground state $|g\rangle_H$, with $\langle g\,|\,{\xi_H}^2\,|\,g\rangle_H = \left\langle g\,|\,\left|\vec{r}_H\right|^2\,|\,g\right\rangle_H/3 = a_0$ for $\xi = x, y, z$, this gives a long-range electrostatic potential energy

$$V_{H,img}^{(els)}(d) = \left\langle\hat{H}_{H,img}\right\rangle_g = -\frac{e^2}{4\pi\varepsilon_0 12d^3}\frac{\varepsilon_r - 1}{\varepsilon_r + 1}\left\langle g\,|\,\left|\vec{r}_H\right|^2\,|\,g\right\rangle_H$$

that shows its distance dependence and trend with changing dielectric permittivity. The interaction is attractive insofar $\varepsilon_r > 1$, it decreases as $\varepsilon_r \to 1$ and reaches an attraction maximum when $\varepsilon_r \to \infty$, in pure metals.

Along the same lines, another application of the images model can be made for a polar molecule A with a permanent dipole $\vec{D}_A$ at an angle ϑ with the surface normal $\vec{n}_z$. From a model of two opposite charges making the dipole, and their images one finds

$$V_{dip,img}^{(els)}(d,\vartheta) = -\frac{1}{4\pi\varepsilon_0 12 d^3}\frac{\varepsilon_r - 1}{\varepsilon_r + 1}\left|\vec{D}_A\right|^2\left(1 + \cos^2\vartheta\right)$$

for the electrostatic interaction energy. Here again the energy varies as d^{-3} with distance, and reaches a maximum for a perfectly metallic solid.

A more general treatment for electrostatic, induction, and dispersion interactions of a molecule near a surface can be done constructing an interaction Hamiltonian from the collection of all charges and their images, and calculating first- and second-order perturbation energies of interaction. Results can be obtained from charge distributions and the dynamical polarizabilities of the molecule and solid. The dispersion interaction energy appears in a second-order perturbation treatment and involves, similarly to the expression for two interacting molecules, the polarizability $\alpha_A(\omega)$ of the species A, which is a 3×3 tensor, and a dielectric function of frequency, or relative permittivity $\varepsilon_r(\omega) = \varepsilon_S(\omega)$ of the solid (or a slab). Insofar a transient dipole $D_A(\omega)$ creates an image dipole $D_{A,img}(\omega) = D_A(\omega)(\varepsilon_S - 1)/(\varepsilon_S + 1)$, one expects (and finds) that such quotient appears in the molecule–surface dispersion energy. More accurately, the treatment also must account for retardation effects in the very long-range interaction of adsorbate charges with surface charges very far away. This can be done within a quantal dielectric response treatment to second order in the strength of the interaction Hamiltonian, alternatively using quantum field theory [18], the theory of fluctuating electromagnetic fields [16], response theory [19], and a quantal perturbation treatment [20], all of which lead to the same results, with the response treatment being simpler. The main relations are summarized as follows.

The atomic structure of the whole system can be described by the distance d between the center-of-mass of the molecule and the surface plane and by the internal atomic positions $\boldsymbol{Q}_A$ of the molecule and $\boldsymbol{Q}_S$ of the surface. The polarizability α_A and dielectric function ϵ_S depend on these internal positions. The result for the molecule–surface dispersion potential energy, omitting fixed internal positions and with α_A given by its the tensor zz-component, is [19]

$$V_{A,S}^{(dsp)}(d) = -\frac{1}{4\pi\varepsilon_0}\frac{\hbar}{2\pi}\int_0^\infty d\omega\, \alpha_A(i\omega)\frac{\epsilon_S(i\omega) - 1}{\epsilon_S(i\omega) + 1}$$
$$\times \left[\frac{4}{(2d)^3} + \frac{4\omega}{c(2d)^3} + \frac{2\omega^2}{c^2(2d)^3}\right]\exp\left(-2\frac{d.\omega}{c}\right)$$

where c is the speed of light. This expression contains retardation effects that arise for very long-range interactions. For relatively small distances $d \ll c/\omega$, compared to radiation wavelengths, the dispersion energy becomes

$$V_{A,S}^{(dsp)}(d) = -\frac{1}{4\pi\varepsilon_0}\frac{\hbar}{4\pi d^3}\int_0^{\infty} d\omega\, \alpha_A(i\omega)\frac{\epsilon_S(i\omega)-1}{\epsilon_S(i\omega)+1}$$

which is usually employed to calculate the energy from parametrized polarizabilities and dielectric functions. This shows a d^{-3} dependence on the distance to the surface, which we'll find below to also result from the van der Waals R^{-6} dependence on interatomic distances, after averaging over atomic positions in the solid lattice. The solid dielectric permittivity function $\epsilon_S(i\omega)$ can be obtained as shown in Section 8.1.3 in terms of the sum $F(\omega) = \sum_J \bar{\rho}_J \alpha_J(\omega)/3\epsilon_0$, which leads to

$$\frac{\epsilon_S(i\omega)-1}{\epsilon_S(i\omega)+1} = \frac{3F(i\omega)}{2+F(i\omega)}$$

to be obtained from the polarizabilities of species in the dielectric solid.

The quotient containing the solid dielectric function in the above integral over frequencies is the one that arises in the response of a dielectric to perturbation by an oscillating dipole far from the surface. It can be reinterpreted in terms of an effective susceptibility of the solid. Introducing the susceptibility $\chi_S(\omega) = \epsilon_S(\omega) - 1$, one finds that the quotient in the integral for $V_{A,S}^{(dsp)}$ is equal to $\bar{\chi}_S(\omega) = \chi_S(\omega)/[2+\chi_S(\omega)]$ a renormalized solid susceptibility, which accounts for medium effects and gives a $\bar{\chi}_S(i\omega)$ smaller than the original $\chi_S(i\omega)$. In this notation,

$$V_{A,S}^{(dsp)}(d) = -\frac{C_{A,S}}{d^3}, C_{A,S} = \frac{1}{4\pi\varepsilon_0}\frac{\hbar}{4\pi}\int_0^{\infty} d\omega\, \alpha_A(i\omega)\bar{\chi}_S(i\omega)$$

with the dispersion coefficient $C_{A,S}$ a function of internal coordinates $\boldsymbol{Q}_A$ and $\boldsymbol{Q}_S$. This coefficient can be evaluated from the molecular polarizability $\alpha_A(\omega)$ as described in Chapter 2 and from models of the solid dielectric susceptibility $\chi_S(\omega) = 3F(\omega)/[1 - F(\omega)]$ as given in the previous section. As done for molecules, we can introduce the function $\alpha_A(i\omega) = \beta^{(A)}(\omega)$. As before, we have for $\omega \to \infty$ the asymptotic value $\omega^2\beta_0^{(A)} \approx (c_e^2\hbar^2/m_e)\, N_{el}^{(A)*}$, with $N_{el}^{(A)*}$ the number of active electrons of A in its ground electronic state. This together with the static isotropic average value $\beta_0^{(A)}(0) = \bar{\alpha}_0^{(A)}(0)$ at zero frequency gives the previous interpolation function $\beta_0^{(A)}(\omega) = \bar{\alpha}_0^{(A)}(0)\left[1 + \left(\omega/\bar{\varpi}^{(A)}\right)^2\right]^{-1}$ with the frequency parameter $\bar{\varpi}^{(A)}$ satisfying $\bar{\alpha}_0^{(A)}(0)\left(\bar{\varpi}^{(A)}\right)^2 = (c_e^2\hbar^2/m_e)\, N_{el}^{(A)*}$ and of the order $0.02 \le \bar{\varpi}^{(A)} \le 2.00$ atomic units [21]. Similar expressions can be used for the species J in the solid dielectric, to parametrize $V_{A,S}^{(dsp)}$ using static polarizabilities $\bar{\alpha}_0^{(J)}(0)$ of atoms and molecules and their frequency constants $\bar{\omega}^{(J)}$.

A more detailed analysis of the coupling of electronic charge density fluctuations at the surface with fluctuations in a molecule at position $\vec{R}$, with coordinates (X, Y, Z) relative to the solid surface, leads to a correction needed at short distances. It replaces Z^{-3} with the form $(Z - Z_0)^{-3}$ where Z_0 is the position of a reference plane and depends on the surface electronic density [20]. Without going into details, the procedure is to express the second-order interaction energy $E^{(2)}$ as a sum over excited states, with state-to-state transition integrals for the Coulomb electronic interaction Hamiltonian containing $1/r$ functions written in cylindrical coordinates $(\vec{\rho}, z)$ there, doing a two-dimensional Fourier transform from $\vec{\rho}$ to a wavevector $\vec{q}_{\|}$, and re-expressing the sum over excited states as an integral over frequencies containing the dynamical susceptibilities of molecule and solid.

This leads to an expression $E^{(2)} = (-2/\pi)\int_0^\infty d\omega\, \alpha_A(i\omega)F_S\left(i\omega;\vec{q}_{\|},Z\right)$ where Z is the distance from A to the surface location defined by the first layer of atomic positions. Derivation of $F_S\left(\omega;\vec{q}_{\|},Z\right)$ and its expansion in powers of $\vec{q}_{\|}$ for large Z leads to the expansion of $E^{(2)}$ as $V_{A,S}^{(dsp)}(Z) = -(A/Z^3)(1 + 3Z_0/Z + \cdots) \approx -A/(Z - Z_0)^3$ with an explicit form for Z_0 in terms of susceptibilities [20]. This correction to the Z^{-3} dependence is relevant at short distances, but a more detailed treatment is needed to account for the transition from long distances d to short ones. For a molecule interacting with a surface lattice, the coefficient A depends also on the coordinates (X, Y) of molecule A above the surface, with the form

$$V_{A,S}^{(dsp)}\left(\vec{R}\right) = -A(X,Y)/(Z - Z_0)^3$$

and the coefficient A dependent also on internal coordinates $\boldsymbol{Q}_A$ of the molecule and $\boldsymbol{Q}_S$ of the solid, appearing in the molecular and solid susceptibility functions.

8.2.2 Short and Intermediate Ranges

The interaction energies between a dielectric surface and a molecule at intermediate and short distances can be treated with the same methods used for molecule–molecule interactions, insofar the electrons in the dielectric are localized. This means that many of the concepts and tools already introduced in previous chapters would also help here. They rely, in semiempirical treatments, on the parametrization of a near-range (short and intermediate range) potential energy function of atomic positions arising from the overlap of electronic charges of molecule and surface atomic groups and on smooth fitting to long-range energies. A more fundamental treatment involves calculations using density functional theory and also more accurate many-electron methods [22–24]. These treatments provide interaction potentials $V_{A,S}\left(\vec{R}\right)$ for a molecular species A containing a group of atoms $\{a\}$ interacting with a solid surface S containing lattice groups of atoms $\{b\}$. These atomic groups may be in a cell periodically repeated forming a crystal, or may form an amorphous lattice.

Dielectric surfaces resulting from cuts through dielectric lattices, such as in Ar or in LiF crystals, have electronic density distributions approximately given by sums of densities of their unit cell components at location $\vec{T}_{m,n,p} = m\vec{a} + n\vec{b} + p\vec{c}$ at or below the surface. We consider a molecular unit B in the cell made up of atoms at positions $\vec{d}_b$. Potentials of interaction at short and intermediate distances, which are functionals of electronic densities of the molecule and the surface, can be given to good accuracy as sums of potential energies $V^{(Ab)}_{m,n,p}\left(\vec{R}\right) = V^{(Ab)}\left(\vec{R}-\vec{R}_\beta\right)$ with $\vec{R}_\beta = \vec{d}_b + \vec{T}_{m,n,p}, \beta = (m,n,p,b)$, between the molecule A and each atom b in cell component (m,n,p). The total potential energy of A near the surface S, with both systems in their ground electronic state, and using a no-dispersion label (nd), is given by

$$V^{(nd)}_{A,S}\left(\vec{R}\right) = \sum\nolimits_\beta V^{(Ab)}\left(\vec{R}-\vec{R}_\beta\right)$$

Each term can be assumed to contain short- and intermediate-range terms accounting for electrostatic (including charge-induced polarization), exchange and correlation energies, as well as long-range electrostatic and induction terms, and can be constructed as described in the chapter on model potentials. However, parameters in these potential terms must account for medium effects on the properties (such as charge, size, polarizability) of the cell atoms b, with values that may be quite different from those of free atoms. The long-range interactions describing dispersion forces must be added with a proper damping function at short distances, $f_d\left(\vec{R}\right)$, going from 1.0 at large distances Z to zero at $Z = 0$, giving

$$V_{A,S}\left(\vec{R}\right) = V^{(nd)}_{A,S}\left(\vec{R}\right) + V^{(dsp)}_{A,S}\left(\vec{R}\right) f_d\left(\vec{R}\right)$$

The total potential must satisfy the condition $V_{A,S}\left(\vec{R}+\vec{T}_s\right) = V_{A,S}\left(\vec{R}\right)$, and its Fourier components $V_{\vec{G}}(Z)$ can be extracted from the sum over atomic positions β given above.

For example, the potential energy for He in empty space at location (X, Y, Z) interacting with the surface LiF(001) of the ionic solid, with the LiF molecular unit in a f.c.c. cell of side length $a = 0.402$ nm, can be constructed from interactions of He with ions Li^+ at location (0,0,0) and F^- at (1/2,1/2,0) in the primitive cell for the (001) surface. Its form after adding over all pair interactions can be given as

$$V_{He,S}(X,Y,Z) = V_0(Z) + V_1(Z)Q(X,Y)$$

with the function Q satisfying $Q(X+a, Y) = Q(X, Y+a) = Q(X, Y)$. It has been used in calculations of cross sections for low-energy elastic scattering of the atom by the ionic surface, which requires introduction of both attractive and repulsive potential energies along the z-direction. A useful parametrization is given by [25]

$$V_0(Z) = D\exp[\alpha.(Z_m - Z)]\{\exp[\alpha.(Z_m - Z)] - 2.0\}$$

$$V_1(Z) = -2\beta D\exp[2\alpha.(Z_m - Z)]$$

$$Q(X,Y) = \cos\left(\frac{2\pi X}{a}\right) + \cos\left(\frac{2\pi Y}{a}\right)$$

with parameters $a = 0.284$ nm, $\alpha = 11.0\ \text{nm}^{-1}$, $D = 7.63$ meV (0.7362 kJ/mol), $\beta = 0.04$ to 0.10, and $Z_m = 0.10$ nm, with the range of β values available to fit experimental results [26]. This gives a potential well along Z, and the same form can be used for other rare gas atoms with suitable parameters. Addition of the dispersion energy to this $V^{(nd)}_{He,S}\left(\vec{R}\right)$ would only give a small correction, compared with the attractive induction energy due to polarization of the atom by the ions. This potential energy function has been used in calculations of resonance energies and shapes, in low-energy collisions of noble gas atoms with the ionic dielectric surface.

Better agreement with scattering experimental results were obtained for He + LiF(001) with a semi-(ab initio) potential energy obtained from a surface-projectile sum $V^{(He,S)}\left(\vec{R}\right)$ of He-(ion pair p) potential energies $v^{(He,p)}\left(\vec{R} - \vec{R}_\beta\right)$ with each term including a long-range van der Waals attraction as well as a short-range repulsion, plus an additional term $V^{(ind)}\left(\vec{R}\right)$ describing the induction attraction of a polarized He atom to the lattice of ions in the solid [27]. The induction term is obtained as $V^{(ind)}\left(\vec{R}\right) = -(\alpha_{He}/2)\left|\vec{\mathcal{E}}^{(latt)}\left(\vec{R}\right)\right|^2$, where the electric field $\vec{\mathcal{E}}^{(latt)} = -\vec{\nabla}\,\phi^{(latt)}$ is the gradient of the electric potential $\phi^{(latt)}\left(\vec{R}\right)$ of the lattice ions interacting with He, with the potential itself a sum over all lattice charges C_b at $\vec{R}_\beta$. This lattice potential is calculated for LiF(001) as a sum over Yukawa-type potentials for each ion, decomposed as a Fourier series, and taken to the limit of the Coulomb potentials. The results is $V_{He,S}\left(\vec{R}\right) = V^{(He,S)}\left(\vec{R}\right) + V^{(ind)}\left(\vec{R}\right)$, or in detail

$$V_{He,S}\left(\vec{R}\right) = \sum\nolimits_{\vec{G}} V_{\vec{G}}^{(He,S)}(Z)\exp\left(i\vec{G}.\vec{R}_s\right) - \frac{\alpha_{He}}{2} f_d(Z)\left|\sum\nolimits_{\vec{G}} \vec{\mathcal{E}}_{\vec{G}}^{(latt)}(Z)\exp\left(i\vec{G}.\vec{R}_s\right)\right|^2$$

with $\vec{\mathcal{E}}_{\vec{G}}^{(latt)}$ as derived in [27] in terms of charges and their positions for the given lattice. The induction interaction is cut-off at short distances with a damping factor $f_d(Z)$ going from 1.0 at large distances to zero at $Z = 0$.

More accurately, the long-range induction and dispersion energies can be obtained from the charge distributions and dynamical polarizabilities of the atom and the solid surface, using methods of response theory [19].

8.3 Continuum Models

8.3.1 Summations Over Lattice Cell Units

When the molecule A near the surface interacts with many substrate lattice cell units, the total potential energy can be approximated in a continuum model where the positions $\vec{T}_{m,n,p}$ of cells are considered to be changing smoothly, and sums over cell indices (m, n, p) are approximately calculated as integrals. To simplify, consider a surface where there is one atom B per cubic unit cell, with its position in the primitive cell at its center $\vec{d}_0 = -d_0\vec{c}$, and its interaction potential with A given by $v^{(A)}$, and omit the dependence of parameters on internal coordinates of A and of the solid. The total interaction potential, given as a sum over all the cells, or

$$V_{AS}\left(\vec{R}\right) = \sum\nolimits_{m,n,p} v^{(A)}\left[\vec{R} - \left(\vec{d}_0 + \vec{T}_{m,n,p}\right)\right]$$

is approximated as an integral for A located on the Z-axis, and with atoms B forming a lattice with vector position variables $S_x = ma$, $S_y = nb$, and $S_z = Z - pc + d_0$ for fixed Z. The indices m, n, p can be assumed to be continuously changing as $-\infty < m, n < \infty$, $-\infty < p \leq 0$, with the summation changed into integrals, giving

$$V_{AS}\left(\vec{R}\right) \cong \frac{1}{a.b.c}\int_{-\infty}^{\infty} dS_x \int_{-\infty}^{\infty} dS_y \int_{Z+d_0}^{\infty} dS_z\, v^{(A)}\left(\vec{S}\right)$$

For a cell species B such that the interaction with A has axial symmetry around the z-axis, with $a = b$, integration variables (S_x, S_y) can be replaced by the length $S_s = (S_x{}^2 + S_y{}^2)^{1/2}$ and an axial angle. The interaction potential becomes

$$V_{AS}(Z) \cong \frac{2\pi}{a^2 c}\int_0^{\infty} dS_s S_s \int_{Z+d_0}^{\infty} dS_z\, v^{(A)}\left[\left(S_s{}^2 + S_z{}^2\right)^{1/2}\right]$$

which is now only a function of the distance Z from A to the surface, and can be calculated for a variety of A–B pair interactions where the distance of A to the species B is given by $(S_s{}^2 + S_z{}^2)^{1/2}$.

Three useful examples, with $a = b = c$, for isotropic potentials and with bars over their parameters signifying values adjusted for medium effects, are [26]:

a) An inverse power potential energy $v^{(A)} = -W(R_0/R)^m$, giving

$$V_{A,S}(Z) = -\frac{\pi}{a^3}\int_0^{\infty} d\left(S_s^2\right)\int_{Z+d_0}^{\infty} dS_z \frac{W R_0^m}{\left(S_s{}^2 + S_z{}^2\right)^{\frac{m}{2}}}$$

$$= -\frac{2\pi W}{(m-3)(m-2)}\left(\frac{R_0}{a}\right)^3\left(\frac{R_0}{Z+d_0}\right)^{m-3}$$

which in particular shows that the R^{-6} atom–atom dispersion interaction becomes a $(Z + d_0)^{-3}$ atom–surface interaction.

b) The Lennard-Jones (12,6) potential $v^{(A)} = \bar{\epsilon}\left[(\bar{R}_m/R)^{12} - (\bar{R}_m/R)^6\right]$, giving

$$V_{A,S}(Z) = W\left[\left(\frac{z_m}{Z+d_0}\right)^9 - 3\left(\frac{z_m}{Z+d_0}\right)^3\right]$$

with $W/\bar{\epsilon} = \left(5^{\frac{1}{2}}\pi/9\right)(\bar{R}_m/a)^3$ and $z_m/\bar{R}_m = 5^{-1/6}$, showing how the well depth and position relate to the original L–J (12,6) potential parameters.

c) An exponential repulsion $v^{(A)} = W exp[-\alpha.\,(R - R_0)]$ giving

$$V_{A,S}(Z) = \frac{4\pi W}{a^3\alpha^3}\left[1 + \frac{\alpha.(Z+d_0)}{2}\right]\exp[-\alpha.(Z + d_0 - R_0)]$$

which is a shifted repulsion in Z with the same relative slope.

8.3.2 Surface Electric Dipole Layers

A continuum model can also be used to describe interactions of a molecule with a surface electric dipole layer formed by the rearrangement of atomic positions and electronic charge at the surface. The surface dipole appears even in clean surfaces in a vacuum due to the spilling of electronic charge from the solid into the vacuum, which creates a layer of surface dipoles oriented along the

z-direction pointing into the solid, as shown in Figure 8.2b. A dipolar layer also appears when a layer of polar molecules is adsorbed at the solid surface. For a collection of charges C_I at distances d_I from the surface, contained in a polar molecule, the interaction energy can be obtained by first locating each single positive charge C in the vacuum at location (0, 0, d) above the surface plane, interacting with a layer of surface electric dipoles of thickness l divided into cells with dipole density D_V per unit volume. The dipoles are assumed to be perpendicular to the surface with the dipolar plane centered at $-d_0 = -l/2$, as shown in Figure 8.2b.

The interaction energy per unit volume for the charge at distance $R = [S_D{}^2 + d_D{}^2]^{1/2}$ from a cell dipole located on the surface at distance S_D from the z-axis, with $d_D = d + d_0$ the distance from C to the plane of the dipole layer is given, including the cosine d_D/R of the angle between dipole and charge position, by

$$u_{C,S}(R) = -C\left[\frac{D_V}{(4\pi\varepsilon_0 R^2)}\right]\frac{d_D}{R}$$

which can be integrated over a surface area of radius L_D with elements of volume $l\,2\pi S_D dS_D$. The integral over $0 \leq S_D \leq L_D$ gives a total electrostatic energy

$$V_{C,S}^{(els)}(d) = -\frac{CD_V l}{4\pi\varepsilon_0}2\pi\left[1 - \frac{d_D}{\left(L_D{}^2 + d_D{}^2\right)^{1/2}}\right] \approx -2\pi\frac{CD_V l}{4\pi\varepsilon_0}\left(1 - \frac{d_D}{L_D}\right)$$

for small d_D/L_D. It provides an attractive force for a positive charge and shows a small decrease of the interaction energy as the distance d of the charge C to the surface increases, for a fixed radius of the dipole layer. It also shows increasing attraction at a given distance as the radius of the dipolar layer increases. This result can be extended to a collection of charges C_I at distances d_I contained in a polar molecule, adding over charges in the presence of a surface with a dipolar layer. More generally, the energy can be obtained for any charge density distribution at the solid surface, solving the Poisson equation for the resulting electrostatic potential [17].

The previous result can be applied as an example to the calculation of the electrostatic interaction of a hydrogen atom with the planar surface dipole distribution. Locating the hydrogen proton with charge $C_p = |c_e| = +e$ at a distance $Z_D = Z + d_0$ from the dipole layer, and the electron at position $\vec{r}_e$ relative to the proton, the interaction Hamiltonian of the two atomic charges and the surface dipoles is

$$\hat{H}_{H,S}(Z, z_e) = -2\pi\frac{C_p D_V l}{4\pi\varepsilon_0}[\gamma(Z_D) - \gamma(Z_D + z_e)]$$

with $\gamma(Z) = 1 - Z/(L_D{}^2 + Z^2)^{1/2}$. This can be expanded up to second order in powers of z_e insofar $z_e/Z_D \ll 1$ away from the surface, and the electrostatic potential energy can be calculated from $V_{H,S}^{(els)}(Z_D) = \langle g\,|\,\hat{H}_{H,S}\,|\,g\rangle_H$ to first order in this coupling Hamiltonian for the H ground state $|g\rangle_H$ using that $\langle g\,|\,z_e\,|\,g\rangle_H = 0$, with the result

$$V_{H,S}^{(els)}(Z) = 2\pi \frac{C_P D_V l}{4\pi\varepsilon_0} \langle g\,|\,r_e^{\,2}\,|\,g\rangle_H \frac{Z_D}{\left(L_D^{\,2} + Z_D^{\,2}\right)^{3/2}},\; Z_D = Z + d_0$$

where furthermore $\langle g\,|\,z_e^{\,2}\,|\,g\rangle_H = \langle g\,|\,r_e^{\,2}\,|\,g\rangle_H/3 = a_0$, which gives a repulsive force for the chosen dipole orientation, and is seen to go as a Z^{-2} at large distances $Z_D \gg L_D$ between the H atom and a dipolar surface layer with a finite radius.

8.3.3 Adsorbate Monolayers

Two molecules adsorbed on a solid surface interact directly between them and also indirectly through the surface. When the surface is that of a dielectric or a semiconductor solid, and the molecular adsorption is weak and does not strongly change the surface electronic density, it is possible to describe the interaction energy of the pair of adsorbed molecules in terms of molecule–molecule and molecule–surface interactions using the electronic densities and properties of the isolated molecules and of the clean solid. This is the case for adsorbed closed-shell molecules and atoms, such as Ar atoms adsorbed on graphite or N_2 adsorbed on NaF(s). Adsorption of many molecules or atoms leads to formation of a surface adsorbate lattice or of an adsorbate island. Depending on whether the adsorbed species are bound to locations or moving around, the adsorbate system can be treated as a lattice or as a fluid, by analogy with the treatments presented in the chapter on model potentials, but done here for planar structures.

A simple treatment describes an adsorbed monolayer as a continuum with a given density. A finite adsorbate lattice or an adsorbate island of surface area A, with its electronic density concentrated between distances d and $d + l$ along a z-axis perpendicular to an infinitely large substrate surface, is attracted to the substrate with an energy derived from the long-range molecule–surface interaction energy. The molecule–surface dispersion energy changes as Z^{-3} as shown above. Taking the limit of a continuous distribution of adsorbate molecules with a number density $\rho_{ads} = N_{ads}/(Al)$, the overall adsorbate interaction energy per unit of surface area A is found integrating over Z the inverse power Z^{-m+3} in the model potential energy given above in 8.3.1, between limits d and $d + l$ for an adsorbate layer of thickness l, so that

$$V^{(ads)}(d)/A = -\frac{2\pi W}{(m-3)(m-2)} \left(\frac{R_0}{a}\right)^3 \rho^{(ads)} \int_d^{d+l} dZ \left(\frac{R_0}{Z}\right)^{m-3}$$

$$= -\frac{2\pi W}{(m-4)(m-3)(m-2)} \left(\frac{R_0}{a}\right)^3 \rho^{(ads)} R_0 \left(\frac{R_0}{d}\right)^{m-4} \left[1 - \left(\frac{d}{d+l}\right)^{m-4}\right]$$

using the results in Section 8.3.1.Therefore the dispersion energy, where $m = 6$ for the atom–atom interaction, goes here as d^{-2} for large l, for a layer of A species interacting with an infinite substrate surface made up of B atoms. To this one must add a repulsive interaction energy function of d at short distances, including a damping factor for the long-range term.

Returning to the description of an adsorbate layer as a collection of many interacting adsorbed molecules, a lattice model of an adsorbate layer undergoing phase transformations at a temperature T (for instance, in a bath under an external inert gas) can be described with a Hamiltonian containing a sum of (single molecule)–surface interactions, plus molecule–molecule interactions, and also adding a three-molecule term if the density of adsorbates is large. The substrate surface can be divided into M two-dimensional cells labeled by two indices in $\boldsymbol{j} = (j_1, j_2)$, with cell dimensions large enough to accommodate one molecule. For N_{ads} adsorbed molecules A at an average distance Z_{ads} from the substrate and with small bonding energy $\varepsilon_{ads} = v^{(A)}(Z_{ads})$ to its lattice sites $\boldsymbol{j}$, among which the molecules can jump, the Hamiltonian can be written in terms of lattice site occupation operators $\hat{c}_j = 1,0$ respectively for occupied or empty lattice site $\boldsymbol{j}$, as

$$\hat{H}_{ads} = -\varepsilon_{ads}\sum\nolimits_{1\le j\le M}\hat{c}_j - \sum\nolimits_{1\le j<k\le M}\varphi(\boldsymbol{j},\boldsymbol{k})\hat{c}_j\hat{c}_k$$

where $\varphi(\boldsymbol{j}, \boldsymbol{k}) = v^{(\boldsymbol{jk})}(Z_{ads})$ is the $(\boldsymbol{j},\boldsymbol{k})$ molecule–molecule interaction energy.

Thermal states of the adsorbate can be described in terms of the temperature and its density ρ_{ads}, to obtain the statistical coverage density $\theta_{ads} = \sum_{j=1}^{M}\langle\hat{c}_j\rangle_T/M$ with $\langle\hat{c}_j\rangle_T$ an average over a thermal distribution [28]. This treatment is mathematically analogous to the Ising model of up- and down-spins in two dimensions, as can be displayed with the introduction of spin variables $\hat{s}_j = \pm 1$ and the correspondence $\hat{c}_j = (1-\hat{s}_j)/2$. Conclusions from the Ising model about phase transitions, for example, for the heat capacity of He on graphite and for X-ray intensity versus light wavelength profiles of inert gases on graphite, have been confirmed by experiments [28].

When the adsorbate molecules A are mobile, it is convenient to instead describe the properties of a monolayer or of an adsorbate island as a liquid of molecules bound by the attraction of the B species in the substrate. The molecules A move at an average distance Z_{ads} from the substrate, and are found at adsorbate locations $\vec{S} = (S_x, S_y)$. A treatment by analogy to what was done in the previous chapter on interaction energies in liquids constructs the thermodynamical internal energy $U_{ads}(\rho_{ads}, \beta)$ with $\beta = 1/(k_B T)$ from A–A pair interaction energies $v(S; Z_{ads})$, where $S = (S_x^2 + S_y^2)^{1/2}$ in terms of the two-dimensional pair-distribution function $g_{ads}(S; \rho_{ads}, \beta)$, as

$$U_{ads}(\rho_{ads},\beta) = \frac{N_{ads}}{\beta} + \frac{N_{ads}}{2}\rho_{ads}\int_0^{\infty} dS\,2\pi S\,v(S;Z_{ads})g_{ads}(S;\rho_{ads},\beta) - \varepsilon_{ads}M\theta_{ads}(\rho_{ads},\beta)$$

Here the first term contains the thermal kinetic energy in two dimensions, and the last term accounts for the bonding energy of the adsorbate to the substrate, with coverage density $\theta_{ads}(\rho_{ads}, \beta)$.

8.4 Nonbonding Interactions at a Metal Surface

8.4.1 Electronic Energies for Varying Molecule–Surface Distances

Metals such as Na, Al, Cu, or Ni contain localized electrons bound in cores of lattice ions, and electrons delocalized over the lattice framework. At a metal surface, such as at Na(011), Al(111), Cu(001), or Ni(001), delocalized electrons spread beyond the surface ions further into the vacuum as compared with their distribution inside the solid. This creates local net charge densities and formation of a surface dipole. Metal atoms at the surface also have unsaturated bonds available for adsorption of other atoms or of molecules. A full treatment of interactions of a molecule A near or at a metal surface M requires modeling of the interaction energy dependence on the position $\vec{R}$ of the center of mass of A relative to the surface and on the location of atoms in the molecule and at the metal surface, $V_{A,M}\left(\vec{R}, \mathbf{Q}_A, \mathbf{Q}_M\right)$. This can be done starting from a collection of all atomic ion cores and all delocalized electrons for the metal and also for the atoms in the molecule.

Physisorption or chemisorption are special cases where the molecule is adsorbed at the surface at a location $\vec{R}_{eq}$ around which it vibrates. The potential energy surface (PES) at equilibrium is then $V_{A,M}\left(\vec{R}_{eq}, \mathbf{Q}_A, \mathbf{Q}_M\right) = V_{AM}^{(ads)}(\mathbf{Q}_A, \mathbf{Q}_M)$ a function of internal atomic positions, showing small binding energies of the order of 20 kJ mol^{-1} (about 200 meV) for physisorption of A and large binding energies, of the order 200 kJ mol^{-1} (about 2 eV), for chemisorption of A or its fragments. Section 8.4 deals with nonbonding interactions and physisorption due to weak attraction forces. As examples, the physisorption energies at equilibrium (not including vibrational zero-point energy) for He on Cu(001) is about 5.6 meV (0.5403 kJ mol^{-1}), and for H_2 on Cu(001), it is in the range of 40–80 kJ mol^{-1} depending on the location of the adsorbate at the surface. The physisorption energies depend also on the surface Miller indices and for H_2, they are smaller at Cu(111) and larger at Cu(110) compared to Cu(001).

A self-consistent field model of $N_e^{(AM)} = N_e^{(A)} + N_e^{(M)}$ delocalized electrons in $A + M$ introduces an effective potential energy $v_{AM}^{(eff)}(\boldsymbol{x})$ derived by a mean-field treatment of the electron motions, such as the Hartree–Fock or Kohn–Sham treatments with an effective Hamiltonian $\hat{F} = \sum_m \hat{f}_m$ for independent electrons $m = 1$ *to* N_e, with $\hat{f}_m = \hat{h}_m + \hat{u}_m$ a one-electron effective Hamiltonian containing an averaged electron–electron interaction $\hat{u}_m$. The delocalized electrons move in the field of pseudopotentials for positive ions in a structure with a total charge $|c_e|N_e^{(AM)}$, that add to the electron kinetic energy in $\hat{h}_m$. To this one must also add a residual mean-field correlation energy term $\hat{U}^{(res)} = \sum_{m<n} v_{mn} - \sum_m \hat{u}_m$ to obtain the full Hamiltonian $\hat{H}_{AM} = \hat{F}_{AM} + \hat{U}_{AM}^{(res)}$ for $A + M$. In some detail, electron spin orbitals are solutions of

$$\left\{ -\left[\hbar^2/(2m_e)\right]\nabla^2 + v_{AM}^{(eff)}(\boldsymbol{x}) \right\}\psi_{\nu\sigma}(\boldsymbol{x}) = \varepsilon_{\nu\sigma}\psi_{\nu\sigma}(\boldsymbol{x})$$

and total energies are obtained adding all one-electron energies for spin orbitals with occupation number $n_{\nu\sigma}$ and subtracting an electron–electron residual correlation energy $E_{res}^{(AM)}$ after accounting for electronic correlation, to avoid double counting of correlation energy, so that the total energy is [29, 30]

$$E_{tot}^{(AM)} = \sum_{\nu\sigma} n_{\nu\sigma}\varepsilon_{\nu\sigma} - E_{res}^{(AM)}$$

from which one can obtain $V_{AM}(\vec{R}) = E_{tot}^{(AM)}(\vec{R}) - E_{tot}^{(A)}(\vec{R}) - E_{tot}^{(M)}$ with subtraction of similar expressions for $E_{tot}^{(A)}$ and $E_{tot}^{(M)}$. This is also a function of all the internal coordinates of atoms in A and M.

Simple models for molecules near metal surfaces help to extract physical insight into the many aspects of surface interactions. A one-electron treatment relies on a piecewise potential energy along the z-direction and expands the expressions introduced in Section 8.1 for a pure metal M to include the potential energy at the location of an adsorbed or nearby molecule A, and the interaction of each electron with the atomic ions in A and M in the field of all the other $N_e - 1$ electrons screening the ion charges. The electronic structure of $A + M$ can be obtained from a one-electron *jellium* model of the metal with a surface perpendicular to the z-axis at location $Z = 0$, with an electronic potential energy $v_M(z)$ of constant value $v_{in} < 0$ inside a semi-infinite region with $-\infty \le z \le 0$, and value $v_{out} = 0$ for $z \ge 0$ in a vacuum, plus an attractive molecular potential $v_A(\vec{r})$ for an isolated A at location Z so that now orbitals of electrons staying mostly within $-\infty \le z \le 0$ spill out to overlap the orbitals of A. One must add an interaction potential energy $v_{AM}^{(int)}(\vec{r}, Z)$ to account for electron delocalization between A and M.

As A approaches M to a distance Z, each screened core ionic charge C_a in A creates an image charge $C_{a,\,img} = -C_a(\varepsilon_r - 1)/(\varepsilon_r + 1)$ in M, and each electron in A interacts with all the ion charges and their images, and also with its own image charge. This creates a long-range electronic potential energy operator $v_{AM}^{(a,img)}(\vec{r}, Z, \mathbf{Q}_A)$ for each charge a, which add up to give an electrostatic potential energy $v_{AM}^{(img)}(\vec{r}, Z, \mathbf{Q}_A)$ between the electron and all other charges. By analogy with the treatment of a hydrogen atom near a surface in Section 8.2.1, it is found that the leading interaction term at large distances between a neutral molecule A and the surface comes from the electron at location $\vec{r}_A = \vec{r} - Z\vec{u}_z$ interacting with all charges, so that

$$v_{AM}^{(img)}\left(\vec{r}, Z\right) = -\frac{C_{el}^2}{4\pi\epsilon_0 16Z^3}\frac{\varepsilon_r - 1}{\varepsilon_r + 1}\left(\left|\vec{r}_A\right|^2 + {z_A}^2\right)$$

plus higher order terms in $1/Z$, with $C_{el} = eN_e^{(AM)} = \sum_a C_a$. Further addition of a near-range repulsive electronic potential energy $v_{AM}^{(nr)}$ gives an one-electron interaction potential energy $v_{AM}^{(int)}\left(\vec{r}, Z\right) = v_{AM}^{(img)}\left(\vec{r}, Z\right) + v_{AM}^{(nr)}\left(\vec{r}, Z\right)$, and a total one-electron potential energy operator

$$v_{AM}^{(eff)}\left(\vec{r}, Z\right) = v_M(z) + v_A\left(\vec{r}, Z\right) + v_{AM}^{(int)}\left(\vec{r}, Z\right)$$

which also depends on internal positions $\boldsymbol{Q}_A$.

A one-electron description of the interaction valid for all distances can be based on orbitals for electrons moving in the $v_{AM}^{(eff)}\left(\vec{r}, Z\right)$ potential energy, constructed from linear combinations of molecular and metal orbitals, $\varphi_\mu\left(\vec{r}\right)$ and $\varphi_{\vec{k}}\left(\vec{r}\right)$ with energies ε_μ and $\varepsilon_{\vec{k}}$, respectively, and normalizations $\left\langle \varphi_\mu \mid \varphi_{\mu'} \right\rangle = \delta_{\mu\mu'}$ and $\left\langle \varphi_{\vec{k}} \mid \varphi_{\vec{k}'} \right\rangle = \delta\left(\vec{k} - \vec{k}'\right)$, as

$$\varphi_\lambda\left(\vec{r}; Z\right) = \sum_\mu c_{\mu\lambda}(Z)\varphi_\mu\left(\vec{r}\right) + \int d^3k\, c_{\vec{k}\lambda}(Z)\varphi_{\vec{k}}\left(\vec{r}\right)$$

where λ is a continuous orbital label, and the dependence on $\boldsymbol{Q}_A$ has been omitted. These orbitals are obtained from solutions of the one-electron Schrodinger equation for fixed atomic positions, with a normalization $\langle \varphi_\lambda \mid \varphi_{\lambda'} \rangle = \delta(\lambda - \lambda')$ suitable for delocalized orbitals. The one-electron energies go asymptotically as Z^{-3}, as follows from the state averages $\left\langle \varphi_\lambda \mid \hat{v}_{AM}^{(int)}(Z) \mid \varphi_\lambda \right\rangle$ of the interaction potential. Total energies are given by the sum of all one-electron energies of occupied orbitals corrected by subtracting the residual electronic correlation energy $E_{res}^{(AM)}(Z)$ so that

$$E^{(AM)}(Z) = \int d\varepsilon\, \varepsilon\, n_{AM}(\varepsilon, Z) g_{AM}(\varepsilon, Z) - E_{res}^{(AM)}(Z)$$

with $n_{AM}(\varepsilon, Z)$ and $g_{AM}(\varepsilon, Z)$ the population and density of levels.

The coupling of localized molecular orbitals with delocalized metal orbitals leads to broadening and shifting of the molecular energy levels with values dependent on the distance Z, and requires special treatment. The density of electronic states per unit energy $g_{AM}(\varepsilon; Z) = \int d\lambda\, \delta[\varepsilon - \varepsilon_\lambda(Z)]$ changes with the distance Z. To simplify, we consider only the highest occupied and lowest unoccupied molecular orbitals and metal orbitals filled up to the Fermi level. At large distances with no interaction between A and M, it is given for each electron spin state by

$$g_{AM}(\varepsilon) = \int_{|\vec{k}| \le k_F} d^3k\,\delta\left(\varepsilon - \varepsilon_{\vec{k}}\right) + \delta(\varepsilon - \varepsilon_{HO}) + \delta(\varepsilon - \varepsilon_{LU})$$

where the first term is $g_M(\varepsilon)$ with k_F the Fermi wavenumber, and the last two terms give $g_A(\varepsilon)$. At a shorter distance Z, the interaction shifts and broadens ε_{HO} by amounts $\Delta\varepsilon_{HO}(Z)$ and $\gamma_{HO}(Z)$, and ε_{LU} by amounts $\Delta\varepsilon_{LU}(Z)$ and $\gamma_{HO}(Z)$. The density of states is then of the form

$$g_{AM}(\varepsilon; Z) = \int_{|\vec{k}| \le k_F'} d^3k\,\delta\left(\varepsilon - \varepsilon_{\vec{k}}\right) + \frac{\gamma_{HO}}{\pi\left[\left(\varepsilon - \varepsilon'_{HO}\right)^2 + \gamma_{HO}^2\right]} + \frac{\gamma_{LU}}{\pi\left[\left(\varepsilon - \varepsilon'_{LU}\right)^2 + \gamma_{LU}^2\right]}$$

with a new Fermi wavenumber $k_F'(Z) = k_F + \Delta k_F$, $\varepsilon'_{HO} = \varepsilon_{HO} + \Delta\varepsilon_{HO}$, and $\varepsilon'_{LU} = \varepsilon_{LU} + \Delta\varepsilon_{LU}$ and level widths γ_{HO} and γ_{LU}. In cases of physisorption, where the interaction of molecule and surface is weak, the level shifts and widths can be obtained from perturbation theory and are found to be of second order in the coupling $v_{AM}^{(int)}$.

The perturbed density of states can be used to account for the population $N_{AM}(T; Z)$ of states at a given temperature T introducing the Fermi–Dirac energy distribution $n_{FD}(\varepsilon, T)$, with $N_{AM}(T; Z) = \int d\varepsilon g_{AM}(\varepsilon; Z) n_{FD}(\varepsilon, T)$ [15]. Corresponding combinations of orbitals can be bonding between A and M, and delocalized, or antibonding and localized near the surface. Energies of interaction $V_{AM}(Z, \mathbf{Q}_A)$ follow from independent electron models for $A + M$, A, and M under the assumption that $E_{res}^{(AM)} \cong E_{res}^{(A)} + E_{res}^{(M)}$ for the weak interactions typical of physisorption, as

$$V_{AM}(Z, \mathbf{Q}_A) = \int d\varepsilon\varepsilon.\left[n_{AM}(\varepsilon; Z, \mathbf{Q}_A) g_{AM}(\varepsilon; Z, \mathbf{Q}_A) - n_A(\varepsilon) g_A(\varepsilon) - n_M(\varepsilon) g_M(\varepsilon)\right]$$

where $n_{AM}(\varepsilon; Z, \mathbf{Q}_A)$ is the electronic energy level population in AM. It shows that the interaction energy dependence on $(Z, \mathbf{Q}_A)$ comes from the density of one-electron energies of $A + M$ and also from the broadening and shift of molecular levels.

This simple treatment provides some insight on the dependence of interaction energies as the distance Z between A and M is varied. In particular, insofar $v_{AM}^{(img)}(\vec{r}, Z) \approx -Q(\vec{r}) Z^{-3}$, the sum of one-electron energies gives an attractive potential energy changing as $1/Z^3$, and inasmuch level shifts and broadenings involve the square of the coupling matrix elements $\left\langle \varphi_\mu \,|\, \hat{v}_{AM}^{(int)}(Z) \,|\, \varphi_{\vec{k}} \right\rangle$, it is found that at large distances, $\Delta\varepsilon_\lambda$ and γ_λ contribute a repulsive energy changing as $1/Z^6$.

A more detailed treatment can be done introducing a Hartree–Fock or Kohn–Sham effective Hamiltonian operator $\hat{F}$ and its orbitals in an approximation adding electron correlation effects, to obtain energy shifts and widths from energy and overlap matrix elements [31]. Such treatment provides insight on trends of adsorption energies for several adsorbate atoms on a given metal surface, and has been extended to account for the atomic structure of the substrate and adsorbate site [8], in which case V_{AM} is obtained as a function of all atomic positions in $\left(\vec{R}, \boldsymbol{Q}_A, \boldsymbol{Q}_M\right)$. It gives density of states and energies that can be analyzed in weak A–M coupling and also in strong coupling cases, and provides variations of physisorption energies for different transition metal surfaces. Results show the right trends but are usually inaccurate for values of adsorption energies.

Details of physisorption energies and of interaction potential energies versus atomic structure, and accurate parametrizations, require many-electron treatments such as those provided by many-atom cluster models or by generalized density functional treatments. A variety of theoretical treatments of energies of physisorption have been extensively compared to experimental results [32]. We show in what follows some of the related concepts and methods.

8.4.2 Potential Energy Functions and Physisorption Energies

In a treatment of molecule–metal dispersion energies using polarizabilities, valid for light of wavelengths large compared to surface interatomic structures, we set the wavenumbers introduced in Section 8.2.1 to $q = 0$, so we can write for the metal permittivity $\varepsilon_r(0, \omega) = \varepsilon_M(\omega)$, and for the dispersion interaction energy between A and M at the relative distance Z between A and the surface plane, the potential energy

$$V_{A,M}^{(dsp)}(Z) = -\frac{1}{4\pi\varepsilon_0}\frac{\hbar}{4\pi Z^3}\int_0^\infty d\omega\, \alpha_A(i\omega)\frac{\epsilon_M(i\omega)-1}{\epsilon_M(i\omega)+1} = -\frac{C_{A,M}^{(dsp)}}{Z^3}$$

with $C_{A,M}^{(dsp)}$ a dispersion coefficient dependent on internal coordinates of molecule and solid metal. As done for the interaction of molecules with dielectric solids, we can introduce the function $\alpha_A(i\omega) = \beta^{(A)}(\omega)$, and also define a new function $\epsilon_M(i\omega) = \eta_M(\omega)$. Similarly for the solid metal, we have $\omega^2\eta_M(\omega) \approx \omega^2 + {\omega_p}^2$ for $\omega \to \infty$, $\eta_M(\omega) = 1 + {\omega_{p,\,ion}}^2/({\omega_T}^2 + \omega^2)$ for $\omega < \omega_T$, and the static value $\eta_M(0) = \epsilon_M(0)$ as a parameter. A suggested interpolation formula for a metal with a single-atom composition is

$$\eta_M(\omega) = \left[1 + \frac{\omega_{p,ion}^2}{{\omega_T}^2 + \omega^2}\right][1 - g_d(\omega)] + \left[1 - \frac{{\omega_p}^2}{\omega^2}\right] g_d(\omega)$$

with $g_d(\omega)$ a damping function equal to one at large frequencies and decaying to zero over a transition region $\omega_T < \omega < \omega_p$. This can be used in the approximate dispersion coefficient

$$C_{A,M}^{(dsp)} = \frac{1}{4\pi\varepsilon_0}\bar{\alpha}_0^{(A)}(0)\frac{\hbar}{4\pi}\int_0^{\infty} d\omega \frac{1}{1+\left(\omega/\bar{\omega}^{(A)}\right)^2}\frac{\eta_M(\omega)-1}{\eta_M(\omega)+1}$$

which can be calculated from the static polarizability and the known number of valence electrons of A contained in $\bar{\omega}^{(A)}$, and from the static dielectric constant and plasmon frequency of the solid. This shows that the solid dielectric function contributes to the integral mostly when $\omega < 2\bar{\omega}^{(A)}$, and should give reasonable results for noble gases on metal surfaces provided transitions $\Lambda = (\lambda' \lambda)$ there, as introduced in Section 8.1.3, have small oscillator strengths f_Λ where $\omega^2 \gg \omega_\Lambda{}^2$ [18].

The effect of variations of the surface electronic charge distribution $\rho_e\left(\vec{r};\vec{R},\mathbf{Q}_A,\mathbf{Q}_M\right)$, for fixed atomic positions, on the dispersion energy can be approximately calculated introducing a plasmon frequency dependent on the location $\vec{r}$ across the surface by means of $\omega_p\left(\vec{r}\right)^2 = \rho_e\left(\vec{r}\right)c_e{}^2/(\epsilon_0 m_e)$ and integrating over $\vec{r}$ the functional $V_{A,S}^{(dsp)}\left[\vec{R};\rho_e\left(\vec{r}\right)\right]$ to average it across the metal surface. This can provide trends for the interaction energy dependence on the distances over which the metal electron density ρ_e decays outside the metal surface. An alternative is to introduce the dispersion dependence of ϵ_r on the wavevector $\vec{q}$ of light, derived from a Boltzmann-equation treatment of the electronic fluid and parametrized as $\epsilon_{r,el}\left(\vec{q},\omega\right) = 1-\omega_p{}^2/\left[\omega(\omega-i\Gamma)-(\alpha q)^2\right]$ [33], to be combined with a parametrized form for the polarizability of the molecule to obtain an energy density $v_{A,S}^{(dsp)}\left(\vec{R},q\right)$, and to integrate this over a distribution of q values.

More generally, the energy of physisorption of a molecule A at all distances Z from the surface shows a distance dependence with a minimum at a well position Z_m, which results from combined long-range electrostatic, induction, and dispersion attraction and near-range electrostatic, exchange, and correlation forces giving repulsion. Long-range dispersion energies $V_{A,M}^{(dsp)}\left(\vec{R},\mathbf{Q}_A,\mathbf{Q}_M\right)$ are obtained from polarizabilities of A and M and go as Z^{-3}. Near-range potential energies $V_{A,M}^{(nr)}\left(\vec{R},\mathbf{Q}_A,\mathbf{Q}_M\right)$, containing also electrostatic and induction long-range forces for ground electronic states, must be obtained from the combined molecule–surface electronic charge distributions where they overlap. They are repulsive at short range due to the Pauli exclusion effect between electrons at the molecule and surface, imposed by the antisymmetry of joint electronic wavefunctions, and typically decay exponentially at short distances, while electrostatic and induction components go as powers of Z^{-1} away from the surface.

Near-range (nr) potential energies can be obtained in a variety of approximations, many of them based on DFT. Constructing a model of the adsorbate A and substrate M and choosing a density functional $E\left[\rho\left(\vec{r}\right)\right]$, the interaction energy at position $\vec{R}$ of the molecule relative to the surface is obtained from the difference between energies of the complex and its components as

$$V_{A,M}^{(nr)}\left(\vec{R}\right) = E_{A,M}\left[\vec{R};\rho_{A,M}\left(\vec{r}\right)\right] - E_A\left[\vec{R};\rho_A\left(\vec{r}\right)\right] - E_M\left[\rho_M\left(\vec{r}\right)\right]$$

where care must be taken in calculations to subtract large numbers in an accurate way, by using the same computational treatment for all terms, and by carefully subtracting energies calculated at the same interatomic distances. The full interaction potential energy is

$$V_{A,M}\left(\vec{R}\right) = V_{A,M}^{(nr)}\left(\vec{R}\right) + f_d\left(\vec{R}\right)V_{A,M}^{(dsp)}\left(\vec{R}\right)$$

with f_d a damping function in the Z variable and showing surface lattice periodicity in the (X, Y) variables. The two terms also depend on the internal atomic coordinates $(\boldsymbol{Q}_A, \boldsymbol{Q}_M)$.

Results of near-range DFT calculations containing electrostatic, exchange, correlation, and polarization terms can be fitted to parametrized forms, and in the simplest case have been fitted in the literature to functions decreasing exponentially with Z. However, DFT can provide some of the attraction forces through polarization and a more general potential function fit should include a potential well. Combined near-range and long-range potential energies can be parametrized for A at position $\vec{R}$ outside the solid S, with functions such as a near-range Morse potential with a minimum coming from the induction attraction, as

$$V_{A,M}\left(\vec{R}\right) = W(X,Y)\exp[-\alpha(X,Y)(Z-Z_m)].\{\exp[-\alpha(X,Y)(Z-Z_m)]-2.0\}$$
$$-f_d(Z)C_{A,M}^{(dsp)}(X,Y)(Z-Z_0)^{-3}$$

where the parameters W, α, and C_{AM} are shown to depend on the location of species A along the surface, to account for changes of its interaction energies along atoms in the surface lattice, and $f_d(Z)$ is a damping function at short distances. The parameters are periodic functions of lattice cell distances a and b for crystalline surfaces. This dependence can be averaged integrating over the surface positions when the distance Z is much larger than interatomic distances at the surface.

Insofar physisorption does not involve electronic rearrangement at the surface, insight can be extracted about $V_{A,M}^{(nr)}\left(\vec{R}\right)$ even when using in functionals a simple additive form of total electron density as $\rho_{A,M}\left(\vec{r};\vec{R}\right)=\rho_A\left(\vec{r};\vec{R}\right)+\rho_M\left(\vec{r}\right)$, with mutually consistent functionals for components and total system involving kinetic energy, exchange, and correlation, and with an energy subtraction procedure similar to what has been used for nonbonded molecule–molecule interactions [34].

An alternative based on physical considerations corrects for density overlaps and introduces instead an effective medium treatment where the physisorption is modeled calculating first the energy of the adsorbate immersed in a medium of electrons with a density $\bar{\rho}_M$ of the substrate metal, obtained averaging over angles around the adsorbate. This is followed by a perturbation treatment to account for the energy resulting from the distortion charge density $\Delta\rho_{AM}\left(\vec{r}\right)=\rho_{AM}\left(\vec{r}\right)-\bar{\rho}_M$, due to the addition of the adsorbate, in the presence of the electric potential change $\Delta\phi\left(\vec{r}\right)$ due to the substrate density change $\rho_M\left(\vec{r}\right)-\bar{\rho}_M$ [22, 35]. Features of physisorption binding and of interaction potential energy of a closed-shell species can be described in many cases introducing such a model, which has also provided insight on chemisorption of atomic species such as H and O on metals.

Early results for noble gas atoms on metals were obtained from DFT variational calculations of $V_{A,M}^{(nr)}\left(\vec{R}\right)$ with the *jellium* model of the metals [9]. These treatments appear to provide correct trends as structures change with different adsorbates at a given metal surface, or with different substrate surfaces for a given adsorbate.

Calculations of the electronic density for physisorption systems, for example, Ar adsorbed on *jellium* with $r_s = 3\ au$ (near the value for the density of Ag), show that some electronic charge moves from the surface toward the atom at a distance d from the surface, to provide an attraction force and an interaction energy minimum for varying distance d [36]. Results are shown in Figure 8.3a for the potential energy curve versus distance, and in Figure 8.3b as isocontours of electronic density changes on a plane perpendicular to the surface containing the atom, with solid lines for positive changes and dashed lines for negative changes. Changes are more pronounced at shorter distances, and the increase of electronic density is noticeable at equilibrium and in repulsive regions. This leads to long-range attraction due to induced polarization, and to repulsion at short distances due to the Pauli electronic exclusion.

Understanding of more subtle aspects, such as the change of the interaction energy of the adsorbate with the lattice Miller indices (h, k, l) of a given metal surface, or of the adsorbate location of lowest energy (on top of a surface atom or at a bridge or hollow), must be done introducing the atomic structure of the whole system, with more accurate treatments of the electronic distribution and long-range interaction energies. One such accurate treatment is based on symmetry-adapted perturbation theory (SAPT) as described in Chapter 6. This is

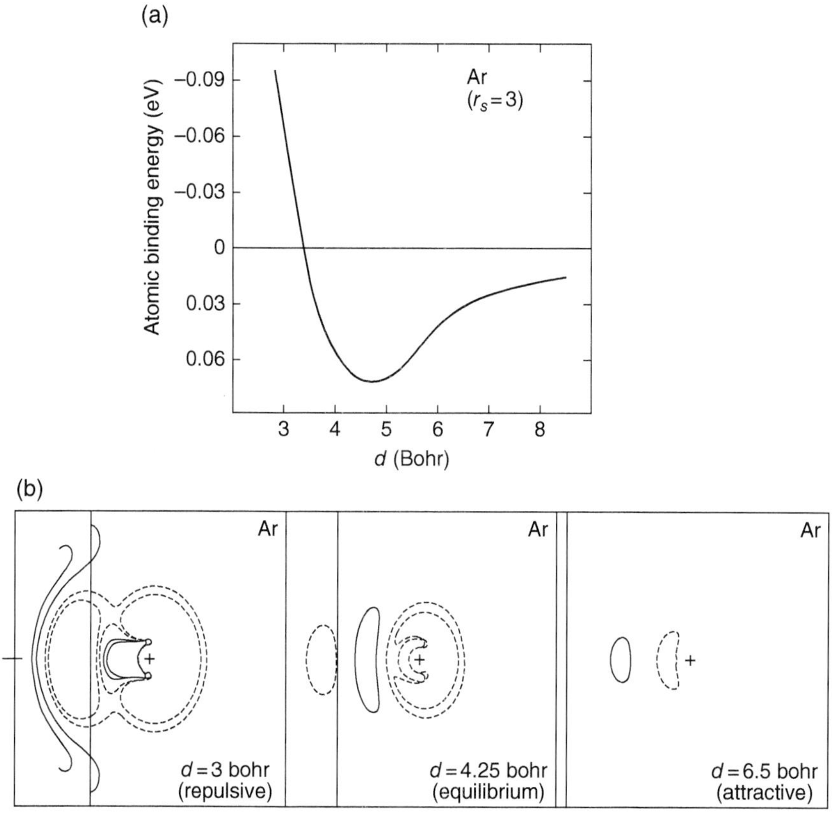

Figure 8.3 (a) Potential energy versus distance for Ar interacting with the surface of a metal with electronic density corresponding to a spread radius $r_s = 3$ *au* (near the value for the density of Ag). (b) Electronic density changes for three distances, with full and dashed lines for positive and negative changes, respectively. *Source:* from Ref. [36]. Reproduced with permission of American Physical Society.

very demanding of computational times, but has nevertheless been applied to several physisorption cases such as a noble gas atom interacting with a surface of $TiO_2(110)$ [37] and to several noble gas atoms interacting with a coronene, graphene, or graphite surface [38] modeled both by a cluster embedded in a medium and by an adsorbate on a surface with periodic boundary conditions. These calculations provide accurate standards to compare with a variety of approximations based on density functional methods and including dispersion energies in different models.

Early treatments using the *jellium* model and the effective medium approximation have been superseded by detailed DFT treatments with a variety of semilocal and hybrid density functionals. Work was done, for example, using semilocal density functionals for systems such as Na adsorbed on the Al(001) and Al(111) surfaces, and Xe adsorbed on Cu(111) and other metal surfaces, to obtain binding energies and bond distances at equilibrium conformations, generally showing correct trends as compared with experimental results [24, 39, 40]. Results using a generalized gradient density functional show that noble gas atoms have their largest physisorption binding energy at locations on top of metal atoms instead of locating at bridges or hollows, because the repulsion energies at the top positions are relatively smaller while the long-range attraction energies are very similar over the surface [40]. This somewhat surprising conclusion has been found to be in agreement with reinterpreted experimental results, and has also been reached in more recent theoretical work with density functionals including van der Waals forces [41, 42]. The introduction of dispersion energies allows more accurate calculation of binding energies, which vary between different site locations, and give more accurate bond distances between adsorbed atoms and surface atoms.

Atomistic models combined with DFT treatments are useful for calculations of structures and binding energies at surfaces, provided the long-range van der Waals interactions are added. Addition of the dispersion energy changes binding energies and bond distances and can be done with recent treatments of the type vdW-DFT [41, 43, 44] and DFT-D [45, 46]. The DFT-D treatments start with a calculation of energies using a selected semilocal functional, and these are then corrected by addition of dispersion energies that multiply damping functions as described in the Chapter 7. This appears to improve agreement with experimental values of properties at equilibrium, and gives potential energy surfaces, which can be used to treat the molecular dynamics of atomic scattering at the surfaces. Earlier treatments have been replaced in recent years by more accurate treatments with semilocal and nonlocal density functionals, and hybrid functionals of Kohn–Sham electronic orbitals, which provide short- and intermediate-range interaction energies [7, 24]. They can be combined with long-range van der Waals functions to cover all distances, such as done with the DFT-D3 treatment [45] and the DFT-MBD treatment [46].

The alternative vdW-DFT approach starts from the formally exact expression of the exchange-correlation energy in terms of the two-electron dynamical susceptibility, obtained from DFT and implemented so that it correctly reproduces the long-range dispersion energy [44]. As explained in Chapters 6 and 7, the adiabatic connection fluctuation–dissipation (ACFD) relation gives a formally exact expression for the electron–electron correlation energy in terms of the dynamical susceptibility, which can be decomposed into semilocal (*sl*) and nonlocal (*nl*) terms. For species A at a relative position $\vec{R}$ from a surface plane, the correlation energy is $E_{A,M}^{(c)}\left(\vec{R},\mathbf{Q}\right)=E_{A,M}^{(slc)}\left(\vec{R},\mathbf{Q}\right)+E_{A,M}^{(nlc)}\left(\vec{R},\mathbf{Q}\right)$ with the first term coming from a semilocal DFT correlation functional and the second term given by an ACFD nonlocal correlation functional. This second term can be constructed for large distances between A and M from their susceptibilities given as known functionals of the electronic density, so that no parametrization from molecular properties is needed [47–49]. This approach has been extensively applied to the interaction of the noble gases with metal surfaces, such as Xe on Cu(111) and Cu(110), and to organic molecules with metal surfaces, such as butane (C_4H_{10}) adsorbed on Cu(111) and benzene on graphite [43, 44].

Accurate values for short-range and intermediate-range energies can be obtained from the coupled-cluster method or from symmetry-adapted perturbation theory as described in the chapter on many-electron treatments, with models using clusters or periodic boundary conditions for the substrate. Insight can also be extracted from extended atomic cluster models, which include the adjacent substrate atoms near the adsorbate location, with atoms added to saturate the peripheral (dangling) cluster bonds. These models have been recently extended in embedding treatments as briefly presented in the next subsection.

8.4.3 Embedding Models for Physisorption

The embedding treatment described in Chapter 7 and Section 8.1.1 is suitable for studies of physisorption, choosing the atomic group A' to be a cluster (or *cls*) including the adsorbate species A and neighboring atoms in the substrate, and choosing $B = M'$ to be the indented substrate or environment (*env*), usually modeled as a slab sufficiently thick to avoid confinement effects on the physisorption properties. The total electron density is constructed so that $\rho_{tot} = \rho_{cls} + \rho_{env}$ and the total (ionic plus electronic) energy decomposes into cluster, environment, and interaction terms, as [50] $E_{tot} = E_{cls}[\rho_{cls}] + E_{env}[\rho_{env}] + E_{int}[\rho_{cls}, \rho_{env}]$. The functionals and densities depend on the atomic positions in $\left(\vec{R}, \mathbf{Q}_A, \mathbf{Q}_M\right)$, and in the case of physisorption, the relative position is chosen to be around an equilibrium value $\vec{R}_{eq}$ while the remaining positions are varied, so that $E_{tot}^{(ads)}(\mathbf{Q}_A, \mathbf{Q}_M) = E_{cls}\left(\vec{R}_{eq}, \mathbf{Q}_A, \mathbf{Q}_M\right)$ gives the adsorption energy as the molecular structure is varied. The embedding procedure can be implemented for physisorption so that the number of electrons N_{cls} and N_{env} are fixed and do not change if the adsorbate conformations are varied, while the indented substrate can be constructed so that it has the same atomic positions as before adsorption.

Insofar the environment does not change its atomic structure or electronic charge distribution, it is possible to obtain the total energy using the subtraction procedure where E_{tot} is described within a density functional treatment for the whole system, giving $E_{tot}^{(DFT)}$. Some electronic rearrangement, however, occurs near the adsorption location, and it can be accounted for by accurate calculation of the energy of the embedded cluster (or *emb cls*) fragment, for example with a correlated many-electron wavefunction (or *CW*) Φ_{cls}, by adding a term $\Delta E_{emb\,cls} = E_{emb\,cls}^{(CW)} - E_{emb\,cls}^{(DFT)}$ to $E_{tot}^{(DFT)}$. This is similar to the previously mentioned subtraction scheme in the *QM/MM* approach. The assumption to begin with, that the environment is not changed by atomic displacements in the cluster, can be tested by enlarging the cluster size and verifying that the environment is unchanged.

The functional for the interaction energy varies with the densities of cluster and environment, providing embedding potential energy functions $V_{emb}^{(cls)}(\vec{r})$ and $V_{emb}^{(env)}(\vec{r})$ through the functional derivatives in

$$\frac{\delta E_{int}}{\delta \rho_{cls}(\vec{r})} = V_{emb}^{(cls)}(\vec{r}), \frac{\delta E_{int}}{\delta \rho_{env}(\vec{r})} = V_{emb}^{(env)}(\vec{r})$$

Equilibration between the cluster and substrate requires that the two embedding potentials should be equal and gives the constraint $V_{emb}^{(cls)}(\vec{r}) = V_{emb}^{(env)}(\vec{r}) = V_{emb}(\vec{r})$, which is a unique embedding potential for the total system, and is illustrated in Figure 8.4, adapted from Ref. [51].

Choosing the cluster to contain all the atoms likely to move due to interactions between adsorbate and surface, the internal atomic coordinates of the environment can be fixed at their equilibrium values, as $\mathbf{Q}_{env} = \mathbf{Q}_{env}^{(eq)}$, while the atomic positions of the cluster change with the position $\vec{R}$ of the center of mass of the adsorbate with respect to the surface, with variables $\left(\vec{R}, \mathbf{Q}_{cls}\right)$. The potential energy function of the variable positions is $E_{tot}^{(DFT)}\left(\vec{R}, \mathbf{Q}_{cls}\right)$ and can be written as a functional of the densities $\rho_{cls}(\vec{r})$ and $\rho_{env}(\vec{r})$, and also of the embedding potential $V_{emb}(\vec{r})$ in

$$E_{tot}^{(DFT)} = E_{cls}^{(DFT)}\left[\rho_{cls}, V_{emb}\right] + E_{env}^{(DFT)}\left[\rho_{env}, V_{emb}\right] + E_{int}^{(DFT)}\left[\rho_{cls}, \rho_{env}, V_{emb}\right]$$

The computational procedure starts with a guess for ρ_{cls}, ρ_{env}, and $V_{emb}(\vec{r})$. The fragment functionals $E_K\left[\rho_K(\vec{r}), V_{emb}(\vec{r})\right], K = cls, env$, are constructed for fixed atom positions. The energy is minimized to find the densities $\rho_K(\vec{r})$ for

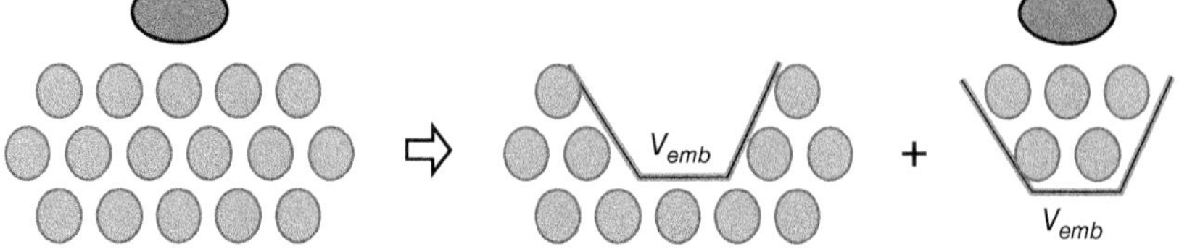

Figure 8.4 Embedding components for physisorption of an adsorbate *A* on a substrate (left), decomposed into its environment (center) and a cluster (right). The embedding potential V_{emb} constrains surface atoms to their positions, near equilibrium values for physisorption. *Source*: adapted from Ref. [51]. Reproduced with permission of American Physical Society.

the given embedding potential, and this is updated by calculating $\delta E_{tot}/\delta V_{emb}(\vec{r})$ to search for a minimum for fixed $\{\rho_K(\vec{r})\}$ [52]. An iteration of this sequence leads to the final potential energy surface $E_{tot}^{(DFT)}$ for the total system. This must be corrected with the addition of $\Delta E_{emb\,cls} = E_{emb\,cls}^{(CW)} - E_{emb\,cls}^{(DFT)}$, where $E_{emb\,cls}^{(DFT)}$ differs from $E_{cls}^{(DFT)}$ insofar the first term accounts for the boundary effect of $V_{emb}(\vec{r})$ present in the Hamiltonian for the cluster. The equations for the combined DFT and CW treatment including the embedding potential are

$$\frac{\delta E_{cls}^{(DFT)}}{\delta \rho_{cls}(\vec{r})} + V_{emb}(\vec{r}) = \mu_{emb\,cls}$$

$$\left[\hat{H}_{cls} + \sum_{j=1}^{N_{cls}} V_{emb}(\vec{r}_j)\right]\Phi_{cls} = E_{emb\,cls}^{(CW)}\Phi_{cls}$$

with $\hat{H}_{cls}$ the many-electron Hamiltonian (excluding the embedding potential) of the cluster for a chosen structure. Similarly, $E_{cls}^{(DFT)}$ is the energy functional for the cluster excluding the embedding potential, and solving the first equation provides the optimized density $\rho_{emb\,cls}(\vec{r})$ to be used in

$$E_{emb\,cls}^{(DFT)} = E_{cls}^{(DFT)}[\rho_{emb\,cls}] + \int d^3r\, V_{emb}^{(cls)}(\vec{r})\rho_{emb\,cls}(\vec{r})$$

The embedding treatment with the subtraction scheme has been applied to physisorption of atoms and molecules.

For example, embedding has been used to obtain the conformation and vibrational spectra of CO adsorbed on a Cu surface. Binding sites and energies and vibrational spectra of this adsorbate have been obtained for the Cu(001) [53] and Cu(111) surfaces. In the case of CO adsorbed on Cu(111), cluster and slab models left doubts as to whether the CO would have larger binding energy on top of a Cu atom or in a hollow of the surface, with different conclusions depending on the quality of the DFT functional. It was reconsidered with the embedded cluster treatment [54], using DFT for the environment and a multi-configuration wavefunction with single and double excitations for the cluster, which lead to the conclusion that CO is perpendicular to the surface and sits on top of a Cu(111) atom, as known from experiments.

8.5 Chemisorption

8.5.1 Models of Chemisorption

Chemisorption is the result of extensive electronic rearrangement when a molecule or atom interacts with a surface, leading to electron transfer or to bond breaking and formation [4, 8, 22, 28, 55]. Treatments of chemisorption energies depend on whether the solid surface refers to a dielectric surface S with localized electrons, to a semiconductor surface, or to a metal (using then $S = M$) with delocalized electrons.

Chemisorption at a dielectric or semiconductor surface can be accurately described with a cluster model, which includes the adsorbate and nearby atoms in the substrate, with boundary bonds atomically saturated. The whole many-atom system can be treated as a supermolecule with many-electron wavefunctions including electron correlation as described in Chapter 6, to allow for both short-range and long-range interaction energies. A cluster model can also be treated within DFT-D formulations [45, 46], allowing for dispersion energies as done for two interacting many-atom systems in Chapter 7. This is less demanding of computing times and can be applied to quite large clusters. The DFT-D treatments can also be applied to models of crystalline solid surfaces such as slabs with extended periodic lattices. This avoids problems with structural and bonding properties dependent on the size of the cluster models that may converge only for very large cluster sizes.

Chemisorption on metal surfaces brings in new aspects due to the delocalization and high susceptibility of electrons in the metal, and must be treated differently. This requires models with extended lattice structures for the metal surfaces, with their electronic properties usually described within DFT. It can be done including dispersion energies with a version of DFT-D that, however, must go beyond sums over atom-pair interactions to account for many-atom effects. Some of these extensions have been covered in Chapter 7 and in reviews [45, 46].

An alternative to the DFT-D approaches has been provided by the vdW-DF treatment [43, 44, 49] derived from the adiabatic connection theorem described in Chapters 6 and 7, which also includes dispersion energies in a seamless way between short and long ranges. Its introduction of a nonlocal expression for the electronic correlation energy has been implemented for both localized and delocalized electron distributions and therefore is suitable for calculations of chemisorption energies for atoms and molecules on dielectric, semiconductor, and metal surfaces.

Reviews of published calculations have compared results for chemisorption with a variety of DFT functional also adding dispersion interaction energies, and the general conclusion is that dispersion contributions are important and needed to obtain correct chemisorptions binding energies and structural bond distances [44–46]. Some examples are mentioned in what follows.

Physical insight can be obtained from one-electron treatments that account for strong coupling of orbitals in the molecule and metal and that incorporate electronic correlation through density functionals [8, 31]. The coupling leads to shifts of molecular levels and to their broadening, depending on the relations between the ionization and affinity energies of the molecule, and the work function and energy band shapes of the metal. This section considers features of chemisorption common to all solid surfaces, whether dielectric, semiconducting, or metallic, which can be described in terms of energy functionals. The following sections deal with chemisorption on metal surfaces, due to electronic charge transfer or to bond breaking and formation.

Electronic charge transfer can happen when an open-shell atomic or molecular adsorbate A interacts with surface atoms, whether the surface is a dielectric or metal, as electronic charge of magnitude δc_e is transferred between the adsorbate and the atoms of the surface lattice S, going from $A + S$ to $A^{\pm} + S^{\mp}$, with the direction of charge transfer determined by relative values of the ionization and affinity energies of A compared to the electronic work function of S. The charge transfer can be described with different models depending on whether the surface contains localized electrons or delocalized ones. Using energy functionals, the interaction potential energy is

$$V_{A,S}^{(el-tr)}\left(\vec{R},\mathbf{Q}_A,\mathbf{Q}_S\right) = E^{(A^{\pm}S^{\mp})}\left[\vec{R},\mathbf{Q}_A,\mathbf{Q}_S;\rho_{A^{\pm}S^{\mp}}\left(\vec{r}\right)\right] - E^{(A)}\left[\mathbf{Q}_A;\rho_A\left(\vec{r}\right)\right]$$
$$- E^{(S)}\left[\mathbf{Q}_S;\rho_S\left(\vec{r},\mathbf{Q}_S\right)\right]$$

where $\left(\vec{R},\mathbf{Q}_A,\mathbf{Q}_S\right)$ are the relative position of A with respect to S, the internal atomic positions of A, and those of S. Electronic charge transfer is likely to affect the atomic structure of the solid surface near the adsorbate.

As described in Section 8.1, these adsorbates can be treated calculating the functional $E^{(A^{\pm}S^{\mp})}$ as a sum over energies of the charged fragments with internal atomic positions $\mathbf{Q}_{A^{\pm}}$ and $\mathbf{Q}_{S^{\mp}}$ and with densities $\rho_{A^{\pm}}$ and $\rho_{S^{\mp}}$, plus their interaction energy in an embedding treatment, letting

$$E^{(A^{\pm}S^{\mp})}\left[\vec{R},\mathbf{Q}_A,\mathbf{Q}_S;\rho_{A^{\pm}S^{\mp}}\left(\vec{r}\right)\right] = E^{(A^{\pm})}\left[\vec{R},\mathbf{Q}_{A^{\pm}};\rho_{A^{\pm}}\left(\vec{r}\right)\right] + E^{(S^{\mp})}\left[\vec{R},\mathbf{Q}_{S^{\mp}};\rho_{S^{\mp}}\left(\vec{r}\right)\right]$$
$$+ E_{int}^{(A^{\pm}S^{\mp})}\left[\vec{R},\mathbf{Q}_{A^{\pm}},\mathbf{Q}_{S^{\mp}};\rho_{A^{\pm}S^{\mp}}\left(\vec{r}\right)\right]$$

with $\rho_{A^{\pm}S^{\mp}} = \rho_{A^{\pm}} + \rho_{S^{\mp}}$ by construction. Calculations can proceed as previously described for physisorption, with a chosen indented surface and cluster, and their atoms held in place by an embedding potential, but allowing here for new locations of atomic cores in the ionic fragments resulting from displacements in the original A and S structures. Unlike the case of physisorption, the indented surface atomic structure must be allowed to change from the original structure, and the total potential energy function $V_{A,S}^{(el-tr)}\left(\vec{R},\mathbf{Q}_A,\mathbf{Q}_S\right)$ can be obtained from the charge transfer energies $\Delta E^{(A^{\pm})} = E^{(A^{\pm})} - E^{(A)}$ and $\Delta E^{(S^{\mp})} = E^{(S^{\mp})} - E^{(S)}$ plus the interaction energy of the fragments.

An example is provided by the chemisorption of Ag_2 on the $TiO_2(011)$ surface. The diatomic can be located on top of a Ti atom, or at a bridge oxygen, or in a hollow of the surface. Calculations of interaction potential energies have been

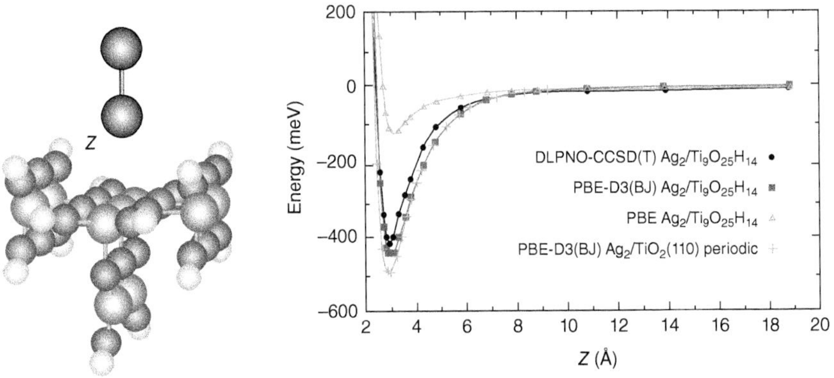

Figure 8.5 (a) To the left, cluster model $Ag_2/Ti_9O_{25}H_{14}$ of the perpendicular adsorbate and nearby surface atoms, with boundary bonds saturated with hydrogen atoms. (b) To the right, potential energy versus distance *Z* from a Ag atom to a nearby Ti surface atom, for the cluster model calculated from CCSD, DFT and DFT-D, and also for a periodic slab with four layers, from DFT-D. *Source:* from Ref. [56]. Reproduced with permission of Royal Society of Chemistry.

done for a cluster model $Ag_2/Ti_9O_{25}H_{14}$ [56] as shown in Figure 8.5a for the adsorbate in a perpendicular orientation including nearby surface atoms (with boundary bonds saturated with hydrogen atoms) with accurate many-electron CCSD(T) wavefunctions, and also with a DFT-D treatment including long-range dispersion energies for the same cluster. The addition of the dispersion energy has been done as explained in Chapter 7, with a damping function. The DFT-D treatment has also been implemented within a model of Ag_2 adsorbed on a slab of the solid, with a periodic lattice. Results for potential energies are shown in Figure 8.5b for the on-top Ti location. They indicate excellent agreement for the binding energies between CCSD(T) and PBE-D3 results, but large errors if the dispersion energy correction is ignored.

Chemisorption also happens for closed-shell molecules such as H_2, CO, N_2, H_2O, CO_2, and C_6H_6 on solid surfaces, when electronic rearrangement near the surfaces weakens the molecular bonds of a species *AB* and results in dissociation into fragments *A* and *B*. These can both bond to surface atoms to form adsorbates *AS* and *BS*, or one or both fragments can go free away from the surface, depending on the bond strength or repulsion force between each fragment and the surface atoms.

The dissociation of a molecule *AB* as it approaches a surface *S* and breaks into $AS + BS$ occurs on a potential energy function $E^{(ABS)}\left[\vec{R}, \mathbf{Q}_A, \mathbf{Q}_B, \mathbf{Q}_S; \rho_{ABS}(\vec{r})\right]$, with the relative position of the center of mass of *AB* near the surface reaching a value $\vec{R}_{ads}$ where the dissociation of *AB* occurs. The fragments then move on a

PES $V_{AB,S}\left(\vec{R}_{ads}, \mathbf{Q}_A, \mathbf{Q}_B, \mathbf{Q}_S\right) = V_{AB,S}^{(diss)}(\mathbf{Q}_A, \mathbf{Q}_B, \mathbf{Q}_S)$, a dissociation potential energy function obtained from

$$V_{AB,S}^{(diss)}(\mathbf{Q}) = E^{(ABS)}\left[\vec{R}_{ads}, \mathbf{Q}; \rho_{ABS}(\vec{r})\right] - E^{(AB)}\left[\mathbf{Q}_A, \mathbf{Q}_B; \rho_{AB}(\vec{r})\right] - E^{(S)}\left[\mathbf{Q}_S; \rho_S(\vec{r})\right]$$

with $\mathbf{Q} = (\mathbf{Q}_A, \mathbf{Q}_B, \mathbf{Q}_S)$. The functional $E^{(ABS)}$ can be obtained for different atomic arrangements along the dissociation path from embedding treatments, with clusters A', B', and AB' including, respectively, A, B, and AB and also neighbor atoms at the surface, and for a surface S' with an indentation large enough so that all three clusters can be accommodated in the surface hole. At the short distance $\vec{R}_{ads}$ between the adsorbate AB and the surface, the energy isocontours of $E^{(ABS)}$ show a minimum for the equilibrium structure of AB', and a dissociation path leading to $A' + B'$.

The energy function $E^{(ABS)}\left(\vec{R}, \mathbf{Q}_A, \mathbf{Q}_B, \mathbf{Q}_S\right)$ describes more generally dissociation for varying approach distances and orientation of AB with respect to the surface, with energy isocontours showing a valley as AB approaches the surface along a variable distance Z_{AB-S} and as AB' breaks into $A' + B'$, which separate along R_{AS-BS} with an activation energy along a path into an energy valley for $AS + BS$. The potential energy function $V_{AB,S}^{(ads)}(Z_{AB-S}, R_{AS-BS})$ can be fit to a London–Eyring–Polanyi–Sato (or LEPS) function of the two position variables for a surface rearrangement, as derived from a valence-bond description of bond breaking and forming and described in Chapter 4. However, one must keep in mind that here pair bonding and repulsion-potential functions are affected by the presence of the remaining surface atoms.

An example of molecular dissociative adsorption on a dielectric surface is found in the calculation of the adsorption fragmentation of a benzene C_6H_6 molecule on a Si(001)-(2 × 1) surface obtained within the vdW-DF approach, which shows two possible fragmentations. The dispersion energy contributions appear to be relevant to clarification of which dissociation structures are the most stable [43].

The rearrangement reaction $A + BC \rightarrow AB + C$ on the surface S, for all species already adsorbed on the surface, involves the reaction potential energy function

$$V_{ABC,S}^{(rct)}(\mathbf{Q}) = E^{(ABC,S)}\left[\mathbf{Q}; \rho_{ABC,S}(\vec{r})\right] - E^{(A+BC,S)}\left[\mathbf{Q}; \rho_{A+BC,S}(\vec{r})\right]$$

with $\mathbf{Q} = (\mathbf{Q}_A, \mathbf{Q}_B, \mathbf{Q}_C, \mathbf{Q}_S)$ the collection of all atomic positions in a reference frame attached to the surface. The energies to the right must be obtained as functions of the positions of A, B, and C on the surface S. In a treatment using embedding in an indented surface S', the energies of fragments A', B', and C' containing the species A, B, and C and neighbor surface atoms can be obtained along reaction paths leading from $A' + BC'$ to $AB' + C'$. Surface atoms in the fragments must be allowed to rearrange too, but the atomic structure of the indented surface S' can be kept fixed if the fragments and indentation are large enough.

8.5.2 Charge Transfer at a Metal Surface

Electron transfer between a molecule and the surface of an insulator or semiconductor involves localized surface spin orbitals ψ_s and spin orbitals ψ_μ of the molecule to form localized ionic bonds at the surface, similarly to what is found in intermolecular bonding. The interaction with delocalized electrons at a metal surface is fundamentally different in that the metal provides a large number of orbitals to be combined with adsorbate orbitals. This involves the coupling of molecular spin orbitals ψ_μ of well-defined atomic energies ϵ_μ labeled by space and spin quantum numbers μ, and metal energy band spin orbitals ψ_j with a continuum of energy levels ϵ_j, where $j = \left(b, \vec{k}, \sigma\right)$ stands for the indices of a band b, electronic wavevector $\vec{k}$, and electron spin quantum number σ. For solid surfaces of transition metals such as Ni or W, an intermediate situation arises involving narrow energy bands from states of both localized d- or f-orbitals and broad energy bands of delocalized s-orbitals, competing for bonding with the adsorbate atom orbitals. Bonding here shows some of the features (bond distances and binding energies) of ligand-field theory [1, 29] for transition element compounds.

For an atom adsorbed on a metal surface, the interaction of an atomic orbital with metal orbitals leads to shifts and broadening of the atomic energy levels and to charge rearrangement. A one-electron picture provides insight on the charge transfer process. At large distances, the electron transfer can be understood comparing the ionization $I = -E_{ion}$ and affinity $A = -E_{afn}$ energies of an atomic electron with the solid's work function W, leading to an electron moving from the atom to the solid if $W > I$ or to electron attraction from the solid to the atom if $A > W$. At shorter distances, as described in Section 8.2, the electron in the atom interacts with its opposite-sign image in the solid and this creates an attractive potential energy $C_e C_{e,img}/(4\pi\varepsilon_0 4Z)$, with $C_e C_{e,img} < 0$, on the electron toward the solid, which decreases as the electron is removed away from the surface. Therefore, the effective ionization energy for the atom at distance Z from the surface is given by $I_{eff}(Z) = I + C_e C_{e,img}/(4\pi\varepsilon_0 4Z)$, a smaller value than I. Bringing instead an electron to the molecule from far away and considering its positive image charge gives an effective affinity $A_{eff}(Z) = A - C_e C_{e,img}/(4\pi\varepsilon_0 4Z)$, a larger value. Comparing these to the work function W of the solid, it follows that when $W > I_{eff}(Z)$, it is energetically favorable for an electron to move from the adsorbate to the solid to gain a binding energy $W - I_{eff}(Z)$, while when $A_{eff}(Z) > W$, the electron moves from the solid to the adsorbate to form a negative ion. These arguments also apply to a molecule interacting with the metal surface, with its center of mass at a distance Z from the surface.

Within a many-electron description, the ground potential energy as a function of the distance Z between A and M shows a rapid change at a crossing point Z_c between a rapidly changing Coulomb attraction potential energy of the two charged species and a slowly changing potential function for the neutral species, shaped by electronic exchange and correlation, as shown in Figure 8.1b. The first case, of an electron moving into the solid to form $A^+ + M^-$ at intermediate distances where $W > I_{eff}(Z)$, is usual for the interaction of an alkali atom with the surface. As the distance Z increases, the Coulomb potential energy between the two ionic species goes as $V^{(A^+M^-)}(Z) = -C_e^2/4\pi\varepsilon_0 Z + W - I_{eff}(Z)$, which increases towards $W - I$ at large distances. However, it is likely to intersect a potential energy $V^{(AM)}(Z)$ for the two neutral species at a crossing distance Z_c where $V^{(A^+M^-)}(Z_c) = V^{(AM)}(Z_c)$ and the electron moves back to the molecule. Taking $V^{(AM)}(Z_c) \approx 0$, this gives

$$Z_c = \frac{C_e^2 - (C_e C_{e,img}/4)}{4\pi\varepsilon_0 (W - I)} = \frac{C_e^2}{4\pi\varepsilon_0} \frac{3(\varepsilon_r + 2)}{4(\varepsilon_r + 1)} \frac{1}{W - I}$$

where as before, ε_r is the static dielectric constant of the solid. At short distances, repulsion forces take over, while at distances larger than Z_c, the potential energy is made up of induction and dispersion terms. A potential can be constructed over all distances using a crossing (or switching) function around Z_c, as done in the chapter on model potentials.

The reverse charge transfer situation involves $A^- + M^+$ present at intermediate distances, going into $A + M$ at large distances, for example for halide atoms interacting with the surface. Transfer happens for increasing distances at the crossing value determined instead by the difference $A - W$ in $V^{(A^-M^+)}(Z_c) = -C_e^2/4\pi\varepsilon_0 Z_c + A_{eff}(Z_c) - W \approx 0$, past which the electron moves back to the metal.

A one-electron model extends the treatment described in Section 8.4 for physisorption, to allow here for strong coupling of molecular and metal orbitals. The strong coupling leads to large changes in molecular orbital energy shift and broadening. These can be obtained using a partitioning treatment, which leads to energy-dependent level shifts and widths, and several different cases for chemisorption due to electron transfer [8, 31]. The molecular and metal orbitals φ_μ and φ_j, $j = \left(b, \vec{k}, \sigma\right)$, for given electron spin quantum number, can be taken to be orthonormalized to begin with, so that $\langle \varphi_\mu | \varphi_{\mu'} \rangle = \delta_{\mu\mu'}$ and $\left\langle \varphi_j | \varphi_{j'} \right\rangle = \delta_{jj'}$. Molecular and metal orbitals are nonorthogonal with overlap $\langle \varphi_\mu | \varphi_j \rangle = S_{\mu j}$. A transformation of molecular orbitals from the set $\{\varphi_\mu\}$ to a new basis set $\{\varphi_\nu\}$ can, however, be done by symmetrical orthogonalization [57], to obtain new molecular and metal orbitals so that $\langle \varphi_\nu | \varphi_j \rangle = \delta_{\nu j}$, and to simplify a partitioning treatment. In practice, the original overlap integrals are small and the new functions φ_ν can be physically related to the old φ_μ and assigned to atomic positions. Partitioning relies on the introduction of projection operators

$$\hat{\mathcal{P}}_A = \sum\nolimits_\nu |\varphi_\nu\rangle\langle\varphi_\nu|,\ \hat{\mathcal{P}}_M = \sum\nolimits_j |\varphi_j\rangle\langle\varphi_j|$$

satisfying $\hat{\mathcal{P}}_A + \hat{\mathcal{P}}_M = \hat{I}$, $\hat{\mathcal{P}}_A^2 = \hat{\mathcal{P}}_A$, $\hat{\mathcal{P}}_M^2 = \hat{\mathcal{P}}_M$, and $\hat{\mathcal{P}}_A.\hat{\mathcal{P}}_M = 0$. The density of electronic states per unit energy $g_{AM}(\varepsilon)$ follows from the resolvent operator $\hat{G}(\varepsilon)$ in the equation for electron m,

$$\left(\varepsilon + i\eta - \hat{f}_m\right)\hat{G}_m(\varepsilon) = \hat{I}$$

with $\hat{f}_m = \hat{h}_m + \hat{u}_m$ the effective Hamiltonian containing a mean-field averaged electron–electron interaction $\hat{u}_m$, $\hat{G}_m$ is the resolvent operator for the equation, and the limiting parameter $\eta \to 0+$. For a given electron, projecting the above equation with $\hat{\mathcal{P}}_A$ to the right, then with $\hat{\mathcal{P}}_A$ to the left and also with $\hat{\mathcal{P}}_M$ to the left,

$$\left(\varepsilon + i\eta - \hat{f}_{AA}\right)\hat{G}_{AA}(\varepsilon) - \hat{f}_{AM}\hat{G}_{MA}(\varepsilon) = \hat{\mathcal{P}}_A$$

$$-\hat{f}_{MA}\hat{G}_{AA}(\varepsilon) + \left(\varepsilon + i\eta - \hat{f}_{MM}\right)\hat{G}_{MA}(\varepsilon) = 0$$

where, for example, $\hat{f}_{AM} = \hat{\mathcal{P}}_A\hat{f}\hat{\mathcal{P}}_M$. Solving formally for $\hat{G}_{MA}$ from the second line and substituting in the first line, one obtains

$$\hat{G}_{AA}(\varepsilon) = \left[\varepsilon + i\eta - \hat{f}_{AA} - \hat{f}_{AM}\frac{1}{\varepsilon + i\eta - \hat{f}_{MM}}\hat{f}_{MA}\right]^{-1}$$

Then in the limit $\eta \to 0+$, we can use the relation

$$\frac{1}{E + i\eta} = \mathcal{P}\frac{1}{E} - i\pi\delta(E)$$

with $\mathcal{P}$ indicating the principal value of an integral, to obtain a reduced resolvent displaying energy shift $\hat{\Delta}_A$ and width $\hat{\Gamma}_A$ operators in

$$\hat{G}_{AA}(\varepsilon) = \left[\varepsilon + i\eta - \hat{f}_{AA} - \hat{\Delta}_A(\varepsilon) + i\hat{\Gamma}_A(\varepsilon)\right]^{-1}$$

$$\hat{\Delta}_A(\varepsilon) = \left(\hat{f}_{AM}\frac{\mathcal{P}}{\varepsilon - \hat{f}_{MM}}\hat{f}_{MA}\right)$$

$$\hat{\Gamma}_A(\varepsilon) = \pi\hat{f}_{AM}\delta\left(E - \hat{f}_{MM}\right)\hat{f}_{MA}$$

The density of states at A follows from

$$g_{A,M}(\varepsilon) = \sum\nolimits_{\nu} \left\langle \varphi_{\nu} \mid \delta\left(\varepsilon - \hat{f}\right) \mid \varphi_{\nu} \right\rangle = -\frac{1}{\pi} \sum\nolimits_{\nu} Im\langle \varphi_{\nu} \mid \hat{G}_{AA}(\varepsilon) \mid \varphi_{\nu} \rangle$$

$$\left\langle \varphi_{\nu} \mid \hat{G}_{AA}(\varepsilon) \mid \varphi_{\nu} \right\rangle = \left[\varepsilon + i\eta - f_{\nu\nu} - \Delta_{\nu}(\varepsilon) + i\Gamma_{\nu}(\varepsilon)\right]^{-1}$$

and its imaginary part as

$$g_{A,M}(\varepsilon) = \frac{1}{\pi} \sum\nolimits_{\nu} \frac{\Gamma_{\nu}(\varepsilon)}{\left[\varepsilon - f_{\nu\nu} - \Delta_{\nu}(\varepsilon)\right]^2 + \Gamma_{\nu}(\varepsilon)^2}$$

$$\Gamma_{\nu}(\varepsilon) = \pi \sum\nolimits_{j} \left|f_{\nu j}\right|^2 \delta\left(\varepsilon - f_{jj}\right),$$

$$\Delta_{\nu}(\varepsilon) = \mathcal{P} \sum\nolimits_{j} \frac{\left|f_{\nu j}\right|^2}{\varepsilon - f_{jj}} = \mathcal{P} \int_{-\infty}^{\infty} d\varepsilon' \frac{\Gamma_{\nu}(\varepsilon)}{\varepsilon - \varepsilon'}$$

where the summation over $j = \left(b, \vec{k}\right)$ involves also integration over values of the wavevector $\vec{k}$. This density of states can be considered for each separate state φ_{ν} of energy $f_{\nu\nu}$ to find how the molecular energy levels have changed due to interactions with the metal. The density of states function $g_{A,M}(\varepsilon)$ can be used to calculate a potential energy function $V_{AM}\left(\vec{R}, \mathbf{Q}_A, \mathbf{Q}_M\right)$ similar to the one given in Section 8.4.1 within a one-electron treatment of interaction energies at surfaces, possibly including now the change in residual correlation energies, by means of

$$V_{AM} = \int d\varepsilon\varepsilon.\left[n_{AM}(\varepsilon) g_{AM}(\varepsilon) - n_A(\varepsilon) g_A(\varepsilon) - n_M(\varepsilon) g_M(\varepsilon)\right] + E_{res}^{(AM)} - E_{res}^{(A)} - E_{res}^{(M)}$$

calculated for varying atomic positions. The partitioning treatment has also been developed for time-dependent equations that arise in the description of electron transfer dynamics and has been applied to calculations of their probabilities in collisions of ions with metal surfaces [58, 59].

Given the forms of chemisorption shifts and widths $\Delta_{\nu}(\varepsilon)$ and $\Gamma_{\nu}(\varepsilon)$ as they follow from the location and widths of energy bands in the metal and from the strength of the molecule–metal coupling energies $f_{\nu j}$, it is possible to classify the changes of A energy levels due to chemisorption as giving either sharp isolated levels or shifted and broadened ones that are overlapping [8, 31]. For isolated levels, it is convenient to find the roots ε_{ν} of the equation $\varepsilon - f_{\nu\nu} - \Delta_{\nu}(\varepsilon) = 0$, and to expand the density of states around each root value. Letting ε_1 and ε_2 be the lower and higher limits of the range of $\Gamma_{\nu}(\varepsilon)$, and considering first weak coupling and only one root ε_{ν} near $f_{\nu\nu}$, the new level $\varepsilon'_{\nu} = f_{\nu\nu} + \Delta_{\nu}(\varepsilon_{\nu})$ has no width if $\varepsilon'_{\nu} < \varepsilon_1$ or $\varepsilon'_{\nu} > \varepsilon_2$, and has a small width $\Gamma_{\nu}(\varepsilon_{\nu})$ if $\varepsilon_1 \leq \varepsilon'_{\nu} \leq \varepsilon_2$.

When the coupling is strong and level widths overlap, an alternative treatment is to first combine an adsorbate orbital with one or more substrate orbitals, such as a $3s$- or $3p$- orbital of Si and a d-orbital of Ni(001) for Si adsorbed on Ni, to form surface (bonding and anti-bonding) orbitals φ_s, $s = a, b$ and to calculate their energies ε_s. Allowing for their interaction with the remaining energy band orbitals of the metal substrate, the partitioning procedure provides their shifts Δ_s and widths Γ_s. This is a generalization of tight-binding models, valid for adsorbates and their energies near metal surfaces [7, 8].

In applications, the matrix elements of $\hat{f}_m$ can be parametrized at the Hartree–Fock level using ionization and affinity energies for the molecule and energy bandwidths for the metal, to reach qualitative conclusions about chemisorption energies. Parametrizations of chemisorption energy expressions containing the coupling energy between adsorbate and substrate orbitals have also been developed in semiempirical models and have been used in treatments of catalysis, at surfaces of simple and transition metals [23].

Some of the considered systems have been hydrogen, oxygen, and alkali atoms adsorbed on light metals and transition metals and CO adsorbed on transition metals [8, 60]. An example is provided by the chemisorption of O atoms on the Ni(111) and Ni(001) surfaces. The coordination number of O with a Ni surface atom changes from three to four between the surfaces, respectively, with a change of the amount of electronic charge transfer between O and Ni, in a one-electron description. This can qualitatively explain the corresponding changes in binding energy and bond distance for the two surfaces [8, 55]. Correct qualitative conclusions on chemisorption energy trends can be obtained with one-electron treatments, but accurate values of chemisorption energies and potential energy changes with the relative position of the molecule near the surface require a many-electron treatment, or a DFT treatment including dispersion energies.

At intermediate and shorter distances, the potential energy is shaped by electron exchange and correlation, while large distances involve electrostatic, induction, and dispersion energies. Many-electron treatments can be done within DFT extended to include long-range dispersion energies, or more accurately with atomic lattice models for the metal slab and adsorbate. The roles of ionization and affinity electron energies were found in early DFT calculations, with electronic density distributions for Li, Si, and Cl adsorbed on a metal with the density of Al corresponding to a spread radius $r_s = 2.0$ au, and the *jellium* model for the metal, and are shown in Figure 8.6 [61]. Electronic rearrangement increases that electronic density for Li and Si in the region between atom and surface, and decreases it between Cl and the surface. Potential energies of interaction follow from the subtraction of energy density functionals, as previously given, and display bonding minima and equilibrium chemisorption distances due to the attraction effect of long-range induction forces.

The main features of the DFT results for chemisorption can be reproduced by an effective-medium treatment of chemical binding of an atom A to a surface M with a realistic inhomogeneous electronic distribution $\rho_{AM}(\vec{r})$ [35], embedding

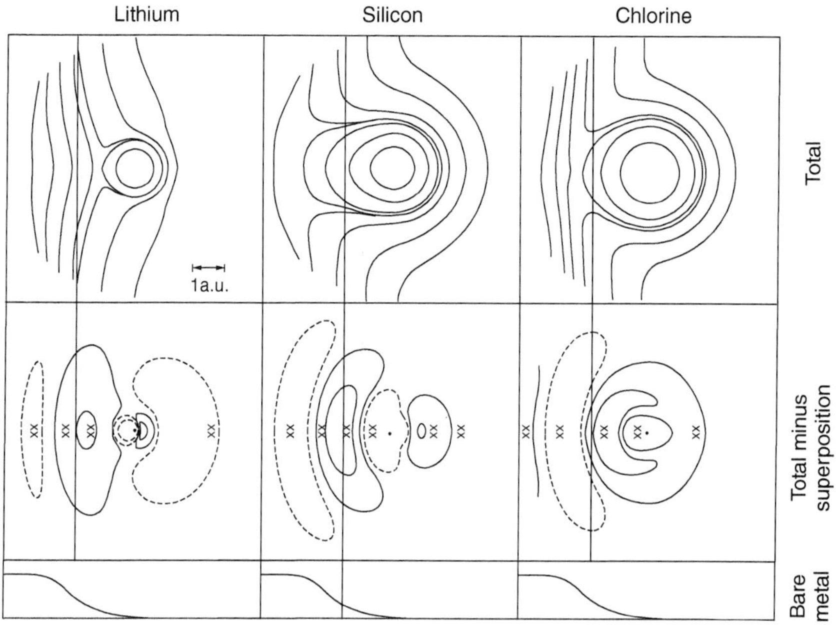

Figure 8.6 Electron density contours for chemisorption of Li, Si, and Cl on a substrate with the electron density of Al, corresponding to a spread radius $r_s = 2.0$ au (or 0.03 electrons/bohr^3). The middle panels (the second row) show the total densities minus the superposition of the atom and metal densities and the extent of electronic charge transfer, with full lines indicating a net increase in density and dashed lines a decrease. The bottom panels show the bare metal electron density near its surface. *Source:* from Ref. [61]. Reproduced with permission of Elsevier.

the atom in a homogeneous electron gas of a density $\bar{\rho}_S$ equal to the average density of the host surface, to calculate its energy. Simple first-order and second-order corrections to the energy account for binding changes due to the host charge inhomogeneity and related distortion charge density $\Delta\rho_{AM}(\vec{r}) = \rho_{AM}(\vec{r}) - \bar{\rho}_M$. More recently, density functionals including gradient corrections, hybrid forms with partial exact exchange energy, and with added dispersion energies, have been used to generate accurate PESs not only for electron transfer, but also for adsorbate dissociation and reactions.

8.5.3 Dissociation and Reactions at a Metal Surface from Density Functionals

The PES for a diatomic molecule AB brought to the metal surface at a relative position $\vec{R}_{ads}$ from which it dissociates into $A + B$ is given by

$$V_{AB,M}^{(diss)}(\mathbf{Q}) = E^{(ABM)}\left[\mathbf{Q};\rho_{ABM}(\vec{r})\right] - E^{(AB,M)}\left[\mathbf{Q};\rho_{AB,M}(\vec{r})\right] - E^{(M)}\left[\mathbf{Q}_M;\rho_M(\vec{r})\right]$$

with $\boldsymbol{Q} = (\boldsymbol{Q}_A, \boldsymbol{Q}_B, \boldsymbol{Q}_M)$ containing atomic positions in a reference frame attached to the surface. The first term to the right contains all active atoms of *AB* and of the substrate metal, the second term is the energy of the *AB* molecule initially adsorbed at the metal surface, and the third term is the energy of the pure metal. The set of all metal and adsorbate atomic positions can be separated into a small subset for the active atoms that undergo large displacements, and a larger subset for the remaining atoms, which stay close to their original equilibrium positions.

Chemisorption of a closed-shell adsorbate on a metal surface usually involves bond breaking in the adsorbate, with interaction energies, which depend on its orientation and the surface structure as defined by its Miller indices [39]. The location and orientation of a diatomic molecule near a surface are fully given by six position variables, which specify the surface location (X, Y) under the diatomic center of mass (or CM), the distance Z between a surface plane and the CM, the distance R between the diatomic atoms, and two angles (ϑ, φ) for the orientation of the diatomic axis. A detailed treatment of interaction energies involves consideration of the six variables and also their changes with the Miller indices of the surface, a very extensive effort that has been undertaken in only a few cases.

Some insight can be extracted from a simple example, for H_2 on the Ni(001) surface leading to its dissociation and to adsorption of the two hydrogen atoms [55, 62]. In a simple description, isocontours of the potential energy $V(Z, R)$ versus the distance Z of the center of mass of H_2 to the surface, oriented parallel to it and to a row of surface atoms, and versus the distance R between the two hydrogen atoms, shows an energy valley as Z decreases and R increases from the equilibrium R_{HH} distance, with an activation energy along a path going into a products valley where R is large and Z goes to a constant at the equilibrium Z_{HNi} distance. This is also found for N_2 dissociating on *W*. Adsorption of fragments following dissociation frequently occurs at transition metal substrate solids such as Ni or W with their localized *d*- and *f*-electron surface orbitals in open shells available for bonding with the fragments. In contrast, H_2 oriented parallel to a Cu surface does not show dissociation upon adsorption because the Cu atom *d*-shell is closed and the H + Cu interaction is weaker, but H_2 is physisorbed there.

However, chemisorption is found in more detailed treatments involving calculations of interaction energies as functions of all the six coordinates of the adsorbed H_2 and as the Cu surfaces change with their Miller indices. They have been done for H_2 on Cu(001) with a cluster model and *ab initio* configuration interaction wavefunctions [63], and also with a slab model and a GGA DFT functional [64], and have been done for H_2 on Cu(111) with DFT functionals [65]. These calculations show reaction paths with activation barriers toward dissociation of H_2 into two adsorbed H atoms, with barrier energies that can be overcome in collisional events by the molecular translational energy or by its vibrational energy. The addition of dispersion energies to these results, not included in the cited work, may, however, be needed to obtain accurate values for activation barriers.

Another example is given by the diatomic dissociation and atomic adsorption of O_2 on an aluminum surface, for which one wants to establish whether there is an activation energy and if so what is its origin, insofar it involves in principle electronic charge and spin rearrangements. The orientation of O_2 is again fully specified by six atomic coordinates, and the aluminum surface has different shapes for different Miller indices. A full treatment of the energetics of the dissociation would be very laborious, but insight can be extracted considering only approaches of O_2 with its axis parallel or perpendicular to the surface of Al(111) and at locations with chosen surface symmetries such as on-top, fcc, hcp, or bridge of nearby surface Al atoms.

Figure 8.7 shows isocontours of energy versus O – O and O – Al distances obtained by an embedding treatment [51, 52], with a cluster including O_2 and several Al atoms embedded in an environment provided by an indented Al surface. Results are presented in Figure 8.7a for an O_2 perpendicular to the surface and for the fcc symmetry, described with the PBE density functional, while Figure 8.7b shows the results for the cluster and a CW treatment implemented with CASSCF + PT2 calculations [50], which include long-range interactions. The isocontours differ, with the CW results showing an activation

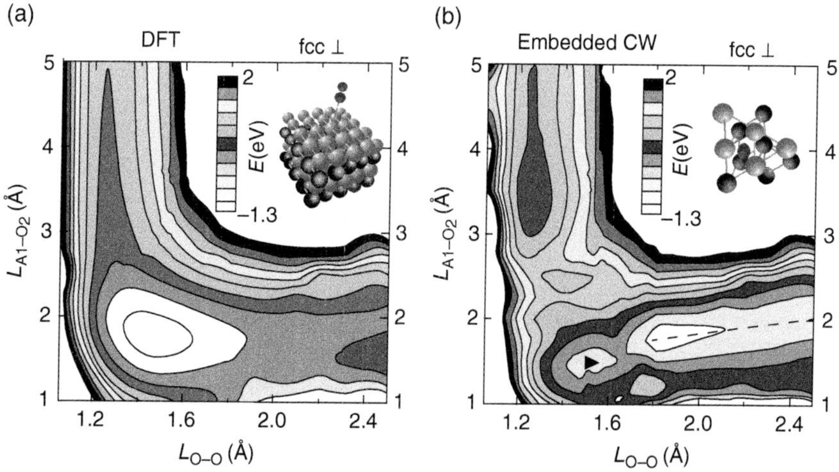

Figure 8.7 Isocontours of energy versus O – O and O – Al distances obtained by an embedding treatment with a cluster including O_2 and several Al atoms embedded in an environment consisting of an indented Al surface. Results in Figure 8.7a are for an O_2 perpendicular to the surface and for the fcc symmetry, described with the PBE density functional, while Figure 8.7b shows the results for a correlated wavefunction treatment implemented with CASSCF + PT2 calculations. *Source*: From Ref. [50].

energy due to electronic charge rearrangement, which varies with the location on the surface and is around 600 meV (57.891 kJ mol^{-1}) for the on-top position with $R(\mathrm{O}-\mathrm{Al})\cong 280$ pm, in accordance with experimental results. Additional calculations of the PES for O_2 on Al(111) have been used to parametrize a flexible periodic LEPS potential energy function, suitable for reaction dynamics on a periodic surface structure. It is constructed as in the LEPS formulation from valence-bond Coulomb and exchange energies using Morse functions and parameters, which are, however, expanded in Fourier series with the periodicity of two-dimensional lattice distances corresponding to the surfaces' Miller indices [66].

A surface chemical reaction such as $A + BC \rightarrow AB + C$, with A, B, and C here taken to be atoms to simplify, starts with the two reactant species adsorbed at positions $\vec{R}_{A,ads}$ and $\vec{R}_{BC,ads}$ relative to the surface, and proceeds on a PES

$$V_{ABC,M}^{(rct)}(\mathbf{Q}) = E^{(ABCM)}\left[\mathbf{Q};\rho_{ABCM}(\vec{r})\right] - E^{(A,M)}\left[\mathbf{Q}_A,\mathbf{Q}_M;\rho_{A,M}(\vec{r})\right] - E^{(BC,M)}\left[\mathbf{Q}_B,\mathbf{Q}_C,\mathbf{Q}_M;\rho_{BC,M}(\vec{r})\right]$$

with $\mathbf{Q} = (\mathbf{Q}_A, \mathbf{Q}_B, \mathbf{Q}_C, \mathbf{Q}_M)$. The first term to the right depends on positions of all participating atoms including metal atoms, the second term is the energy of an adsorbed atom A, and the last term is the energy of an adsorbed BC diatomic. As in the case of dissociation, the set of all metal atom positions can be separated into a small subset for active atoms that undergo large displacements during reactions, and a larger subset for the remaining atoms that stay close to their original equilibrium positions. These two subsets define a cluster and its environment, respectively, and can be treated within embedding methods as described.

One of the concerns about using DFT for chemisorption energies is that the treatment can lead to fractional electronic charge transfer for each reaction fragment, with a related unphysical energy dependence on a continuously varying number of electrons, instead of physically meaningful fragments with integer numbers of electrons. As described in Chapters 6 and 7, an alternative treatment can be based on the introduction of an energy E_{tot} written as a functional of electron orbitals as explained in Section 6.4, for the total system, to be obtained within an optimized effective potential procedure for interacting adsorbate and substrate fragments K. The functional $E_{tot}\left[\left\{\psi_j^{(K)}(\vec{r},\zeta)\right\}\right]$ of a set of $j = 1$ *to* N_K spin orbitals $\psi_j^{(K)}$ for each fragment K can be constructed to describe electrons moving in an effective potential $V_{emb}(\vec{r})$, optimized by energy minimization, to derive an *optimized embedding effective potential* (OEEP) $V_{emb}^{(OEEP)}(\vec{r})$ suitable for chemisorption phenomena, with potential energy functions $V_{AB,M}^{(diss)}(\mathbf{Q})$ and $V_{ABC,M}^{(rct)}(\mathbf{Q})$ obtained as sums of energies for fragments with integer numbers of electrons in each one [52, 67].

8.6 Interactions with Biomolecular Surfaces

The interaction of molecules with biomolecular surfaces is a very extensive and very active area of research, where the same physical and chemical concepts can be used as done so far for interactions at inorganic surfaces. But the complex atomic composition of biological materials, plus the fact that they are soft matter presenting multiple structural shapes, pushes the available methodologies to their limits and requires novel approaches for theory and computational work. Progress has nevertheless been made using force fields for biosurfaces interacting with both small molecules and large molecules such as proteins, and using structural multiscale methods. Ongoing research involves development of new polarized force fields, and many-atom modeling of cohesive energies and of interaction energies between molecules and biosurfaces.

Biological surfaces can take spherical shape as in micelles, or can display extended surfaces as in membranes of biocells, in the presence of a water solvent. They are usually formed by assemblies of amphiphilic organic molecules, with hydrophilic electrically charged heads assembled to form the boundaries of the micelles or membranes in contact with water molecules, and long hydrophobic tails pointing away from the aqueous region. Molecules with short tails can form micelles while molecules with long tails form layers or bilayers [68].

For example, phospholipid molecules with a phosphoric acid group as a head and two hydrocarbon tails assemble to form a bilayer membrane, of thickness between 5 and 8 nm, with the heads containing phosphate groups arrayed as a surface, shown in Figure 8.8.

Molecules can interact with micelles and membranes to form biological complexes. Biological membranes are found to contain extrinsic proteins at their surfaces, or intrinsic proteins inside the membranes. Small molecules composed of a few atoms can cross biomembranes through pores moving along channels created by tubular proteins [5].

Because biomolecular surfaces display large displacements of their atoms, their fluid nature is more accurately described by treatments similar to those for liquids, in terms of atom pair distribution functions and force fields, as described in Sections 4.5 and 4.6.

The interaction energies between a molecule and a biosurface, in the presence of a solvent such as water, are similar to the ones found between a molecule and an extended many-atom solids, and they can be described in terms of force fields provided these are modified to account for extended polarization effects, which introduce dielectric constants in force fields, and must also incorporate solvent effects as treated in Chapter 7. Polarization effects can be included by parametrizing potential energies between atomic groups containing embedded charges and localized polarizabilities, as described in Section 4.6.

Force-field developments for modeling of lipid membranes have been recently reviewed [69–71]. They are based on parametrized potential energy functions

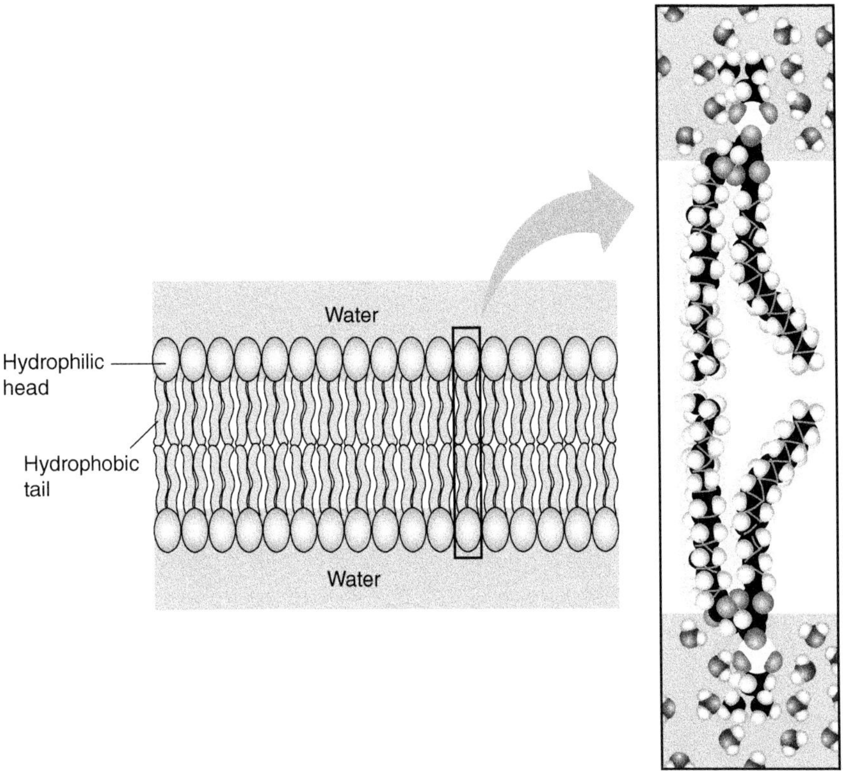

Figure 8.8 Biomembrane bilayer, to the left, formed by self-assembly of phospholipid molecules as seen to the right, with hydrophilic heads surrounded by water molecules. *Source:* Adapted from public domain figures.

containing bonding $V^{(B)}$ and nonbonding $V^{(NB)}$ terms as described in Section 4.6.2, with

$$V^{(B)} = V_{str} + V_{bnd} + V_{tor} + V_{bck} + V_{H-bnd}$$

where V_{str} is the bond stretch component involving sums over two atoms, V_{bnd} is the angle bending components for sets of three atoms, V_{tor} the torsion angle component with sums over four atoms, V_{bck} is the buckling component also for four atoms, and the last term is a hydrogen-bonding energy dependent on the positions of a hydrogen and two other atoms. The nonbonding potential energy is

$$V^{(NB)} = \sum\nolimits_{I,J} \epsilon_{IJ} \left[\left(\frac{\sigma_{IJ}}{R_{IJ}} \right)^{12} - 2 \left(\frac{\sigma_{IJ}}{R_{IJ}} \right)^{6} \right] + v_{IJ}^{(Coul)} \left(\{ R_{ij} \} \right) + v_{IJ}^{(ind)} \left(\{ R_{ij} \} \right)$$

$$v_{IJ}^{(Coul)}\left(\{R_{ij}\}\right) = \sum\nolimits_{i \in I, j \in J} \frac{C_i C_j}{4\pi\varepsilon_0\varepsilon_r R_{ij}} f_d\left(R_{ij}\right)$$

$$v_{IJ}^{(ind)}\left(\{R_{ij}\}\right) = -\sum\nolimits_{i \in I, j \in J} \frac{C_i\alpha_j + \alpha_i C_j}{\left(4\pi\varepsilon_0\varepsilon_r\right)^2 R_{ij}^{\,4}} f_d\left(R_{ij}\right)$$

where ε_r is the dielectric constant of the medium. The first term in the summation describes dispersion energies, and the Coulomb and induction terms have been given as sums of pair interactions between point charges C_k and polarizabilities α_k in fragments $\mathcal{K} = \mathcal{I}, \mathcal{J}$ located at the relative distance R_{ij}. They contain a damping function $f_d(R_{ij}) \approx 0$ for $R_{ij} \rightarrow 0$ at short distances, as described in Chapters 4 and 7. These point charges and polarizabilities are approximate values for the actual spatial distributions of charges and polarizabilities, which furthermore change as conformations vary. With reference to citations in [69], well-tested force fields (FFs) are available from GROMOS, where CH_n, $n = 1$–3, groups of atoms are treated as a single particle, as recently improved for lipids [72], and from the all-atoms CHARMM FF as recently improved for lipids [73], AMBER with improvements incorporated for lipids, MARTINI with a coarse-grained model [74], and other ongoing developments for polarizable FFs suitable for lipids.

The very large number of atoms to be considered in the calculation of structural and dynamical properties of a biomembrane interacting with a large molecule presents a challenge in the computational modeling of the complex. This can be dealt with grouping several atoms into chemically meaningful fragments, which become basic constituents, reducing the total number of degrees of freedom in the whole system. It requires parametrization of fragment properties independently calculated with many-electron treatments. Consistent ways to do this employ multiscale techniques based on a full atomic description and proceed from the ground-up by coarse-graining many-atom systems into fragments (or new particles), to describe condensed matter and biomolecular systems with fewer interacting particles [75–79]. Optimal ways to extract parameters for the fragment structures and their interaction, from extensive atomistic calculations, can be found using machine learning of artificial intelligence rules from big data collected for biomolecules [80–82].

A quantitative coarse-graining (CG) treatment of the modeling of a large many-atom system by a smaller collection of particles can be done starting from a mapping of the set $\mathbf{R}_A = \{r^{3N_A}\}$ of all the $3N_A$ atomic position coordinates into a set of $3N_P$ model particle position coordinates $\mathbf{R}_P = \{r^{3N_P}\}$ plus N_L model parameters $\boldsymbol{\lambda}$, such as point charges and polarizabilities, collected in the set of particle variables ($\Lambda = \mathbf{R}_P, \boldsymbol{\lambda}$). The mapping is of form

$$\Lambda = \mathcal{M}\left(\mathbf{R}_A\right)$$

where Λ and $\mathbf{R}_A$ are column matrices and $\mathcal{M}$ is a matrix of order $3N_A \times (3N_P + N_L)$. This describes a general case where the parameters $\boldsymbol{\lambda}$ are functions of the original atomic positions and change with the conformation of molecules. A simpler treatment is usually developed where the particle parameters do not change, in which case a simpler mapping is done between atomic positions and particle positions of the form $\mathbf{R}_P = \mathcal{M}_{PA}(\mathbf{R}_A; \boldsymbol{\lambda})$.

The construction of particles as groups of atoms must be done so that their interactions reproduce statistical properties of the original many-atom system. This can be enforced using the probability distribution $p_A(\mathbf{R}_A)$ of the atomic ensemble, to obtain the probability distribution of particles by means of

$$p_P(\Lambda) = \int d\mathbf{R}_A \, p_A(\mathbf{R}_A)\delta[\mathcal{M}(\mathbf{R}_A) - \Lambda]$$

in a compact notation signifying an integral over $3N_A$ variables. Alternatively, the model system can be described by a potential of mean force $U_P(\Lambda, \beta)$ when the original atomic system has been kept at a temperature $T = 1/(k_B\beta)$, defined by a Boltzmann-type distribution $p_P(\Lambda,\beta) = \exp[-\beta U_P(\Lambda)]/Z_P(\beta)$, with Z_P a partition function for the system of particles. This gives for the potential of mean force

$$U_P(\Lambda,\beta) = -\beta^{-1}\ln\left\{ Z_P(\beta) \int d\mathbf{R}_A \, p_A(\mathbf{R}_A,\beta)\delta[\mathcal{M}(\mathbf{R}_A) - \Lambda]\right\}$$

where $p_A(\mathbf{R}_A,\beta)$ is the original Boltzmann distribution for the atomic system. It simplifies to give a function $U_P(\mathbf{R}_P, \beta)$ when the parameters can be chosen to be independent of conformations. Construction of this potential energy starts with a calculation of $p_A(\mathbf{R}_A,\beta)$ possibly for a small atomic subsystem, to identify particles' structure and properties to be incorporated in the full system. A variety of ways to do this by iteration has been described in the reviews already cited, and also in a proposed treatment based on the minimization of a relative entropy function $S_{rel} = \sum_I p_P(I)\ln[p_P(I)/p_A(I)]$ where the summation is over configurations in the CG procedure [83].

An example of application of the CG procedure is found in a calculation where the potential of mean force is decomposed into sums of pairs of particle interactions, within a multiscale-CG (or MS-CG) procedure, which does the force-matching by linearization of equations with the original pair forces [84]. It has been applied to a model lipid bilayer composed of dimyristoilphosphatidilcholine (or DMPC) molecules, containing two alkane chains. This molecule, with nearly 50 atoms, was modeled by 13 particles associated with the choline (CH), phosphate (PH), glycerol (GL), ester groups (E1 and E2), six alkane particles (SM), and two tail ends (ST). Results from the MS-CG simulation reproduced the structural properties of the lipid bilayer as obtained from the atomistic simulation, quite accurately [84].

As more applications of coarse-graining treatments are done, and provide big data collections, it becomes possible to use machine learning of rules for choosing optimal mappings of the original many-atom system into a smaller system of particles, while reproducing relevant properties, such as total energies of changing conformations. Artificial intelligence tools are being developed and applied for that purpose [80–82], and are available for treatments of interactions at complex biomolecular surfaces, including the related important determination of the optimal location and binding of a small bioactive molecule on a membrane from among a very large number of possible sites.

References

1 Atkins, P. and De Paula, J. (2010). *Physical Chemistry*, 9e. New York: W. H. Freeman and Co.
2 Kittel, C. (2005). *Introduction to Solid State Physics*, 8e. Hoboken, NJ: Wiley.
3 Somorjai, G.A. (1972). *Principles of Surface Chemistry*. Englewood Cliffs, NJ: Prentice-Hall.
4 Somorjai, G.A. (1994). *Introduction to Surface Chemistry and Catalysis*. New York: Wiley-Interscience Publication.
5 Israelachvili, J. (2011). *Intermolecular and Surface Forces*, 3e. New York: Elsevier.
6 Lang, N.D. (1973). The density-functional formalism and the electronic structure of metal surfaces. In: *Solid State Physics*, vol. 28 (eds. H. Ehrenreich, F. Seitz and D. Turnbull), 225. New York: Academic Press.
7 Martin, R.M. (2004). *Electronic Structure: Basic Theory and Practical Methods*. Cambridge, England: Cambridge University Press.
8 Desjonqueres, M.C. and Spanjaard, D. (1995). *Concepts in Surface Physics*, 2e. Berlin, Germany: Springer-verlag.
9 Lang, N.D. (1983). Density functional approach to the electronic structure of metal surfaces and metal-adsorbate systems. In: *Theory of the Inhomogeneous Electron Gas* (eds. S. Lundqvist and N.H. March), 309. New York: Plenum Press.
10 May, V. and Kuhn, O. (2000). *Charge and Energy Transfer Dynamics in Molecular Systems*. Berlin: Wiley-VCH.
11 Mukamel, S. (1995). *Principles of Nonlinear Optical Spectroscopy*. Oxford, England: Oxford University Press.
12 Micha, D.A. (2015). Generalized response theory for a photoexcited many-atom system. In: *Adv. Quantum Chemistry*, vol. 71, *Chapter 8* (eds. J.R. Sabin and R. Cabrera-Trujillo), 195. New York: Elsevier.
13 Yu, P. and Cardona, M. (2005). *Fundamentals of Semiconductors: Physics and Materials Properties*, 3e. Berlin, Germany: Springer-Verlag.
14 Haug, H. and Koch, S.W. (2004). *Quantum Theory of the Optical and Electronic Properties of Semiconductors*. World Scientific.

15 Ashcroft, N.W. and Mermin, N.D. (1976). *Solid State Physics*. London, England: Thomson.
16 Landau, L.D. and Lifshitz, E.M. (1960). *Electrodynamics of Continuous Media*. London: Pergamon Press.
17 Jackson, J.D. (1975). *Classical Electrodynamics*. New York: Wiley.
18 Mavroyanis, C. (1963). The interaction of neutral molecules with dielectric surfaces. *Mol. Phys.* 6: 593.
19 McLachlan, A.D. (1964). Van der Waals forces between an atom and a surface. *Mol. Phys.* 7: 381.
20 Zaremba, E. and Kohn, W. (1976). van der Waals interaction between an atom and a solid surface. *Phys. Rev. B* 13: 2270.
21 Kramer, H.L. and Herschbach, D.R. (1970). Combination rules for van der Waals force constants. *J. Chem. Phys.* 53: 2792.
22 Bortolani, V., March, N.H., and Tosi, M.P. (eds.) (1990). *Interaction of Atoms and Molecules with Solid Surfaces*. New York: Plenum Press.
23 Greeley, J., Noerskov, J.K., and Mavrikakis, M. (2002). Electronic structure and catalysis on metal surfaces. *Annu. Rev. Phys. Chem.* 53: 319.
24 Huang, P. and Carter, E.A. (2008). Advances in correlated electronic structure methods for solids, surfaces, and nanostructures. *Annu. Rev. Phys. Chem.* 59: 261.
25 Wolken, G.J. (1976). Scattering of atoms and molecules from solid surfaces. In: *Dynamics of Molecular Collisions Part A* (ed. W.H. Miller), 211. New York: Plenum Press.
26 Goodman, F.O. and Wachman, H.Y. (1976). *Dynamics of Gas-Surface Scattering*. New York: Academic Press.
27 Celli, V., Eichenauer, E., Kaufhold, A., and Toennies, J.P. (1985). Pairwise additive semi ab initio potential for the elastic scattering of He atoms from the LiF(001) crystal surface. *J. Chem. Phys.* 83: 2504.
28 Zangwill, A. (1988). *Physics at Surfaces*. Cambridge, England: Cambridge University Press.
29 Levine, I.N. (2000). *Quantum Chemistry*, 5e. Upper Saddle River: Prentice-Hall.
30 Parr, R.G. and Yang, W. (1989). *Density Functional Theory of Atoms and Molecules*. Oxford, England: Oxford University Press.
31 Newns, D.M. (1969). Self-consistent model of hydrogen chemisorption. *Phys. Rev.* 178: 1123.
32 Maurer, R.J., Ruiz, V.G., Camarillo-Cisneros, J. et al. (2016). Adsorption structures and energetics of molecules on metal surfaces: bridging experiment and theory. *Prog. Surf. Sci.* 91: 72.
33 Landman, U. and Kleiman, G.G. (1977). Microscopic approaches to physisorption. In: *Surface and Defect Properties of Solids*, vol. 6 (eds. M.W. Roberts and J.M. Thomas), 1. London: Chemical Society.
34 Gordon, R.G. and Kim, Y.S. (1972). Theory for the forces between closed shell atoms and molecules. *J. Chem. Phys.* 56: 3122.

35 Norskov, J.K. and Lang, N.D. (1980). Effective-medium theory of chemical binding: application to chemisorption. *Phys. Rev. B* 21: 2131.
36 Lang, N.D. (1981). Interaction between closed-shell systems and metal surfaces. *Phys. Rev. Lett.* 46: 842.
37 deLara-Castells, M.P., Stoll, H., and Mitrushchenkov, A.O. (2014). Assessing the performance of dispersionless and dispersion-accounting methods: Helium interaction with cluster models of the TiO2 Surface. *J. Phys. Chem. A* 118: 6367.
38 deLara-Castells, M.P., Bartolomei, M., and Mitrushchenkov, A.O. (2015). Transferability and accuracy by combining dispersionless density functional and incremental post-Hartree-Fock theories: Noble gases adsorption on coronene/graphene/graphite surfaces. *J. Chem. Phys.* 143: 194701–194701.
39 Scheffler, M. and Stampfl, C. (2000). Theory of adsorption on metal surfaces. In: *Handbook of Surface Science: Vol. 2 Electronic Structure* (eds. K. Horn and M. Scheffler), 286. Amsterdam: Elsevier.
40 DaSilva, J.L.F., Stampfl, C., and Scheffler, M. (2005). Xe adsorption on metal surfaces: first principles investigations. *Phys. Rev. B* 72: 075424–075421.
41 Chen, D.-L., AlSaidi, W.A., and Johnson, J.K. (2012). The role of van der Waals interactions in the adsorption of noble gases on metal surfaces. *J. Phys. Condens. Matter* 24: 424211.
42 Maurer, R.J., Ruiz, V.G., and Tkatchenko, A. (2015). Many-body dispersion effects in the binding of adsorbates on metal surfaces. *J. Chem. Phys.* 143: 102808–102801.
43 Langreth, D.C., Lundqvist, B.I., Chakarova-Kaeck, S.D. et al. (2009). A density functional for sparse matter. *J. Phys. Condens. Matter* 21: 084203.
44 Berland, K., Cooper, V.R., Lee, K. et al. (2015). van der Waals forces in density functional theory: a review of the vdW-DF method. *Rep. Prog. Phys.* 78: 066501.
45 Grimme, S., Hansen, A., and Brandenburg, J.G.C. (2016). Dispersion corrected mean-field electronic structure methods. *Chem. Rev.* 116: 5105.
46 Hermann, J., DiStasio, R.A.J., and Tkatchenko, A. (2017). First-principles models for van der Waals interactionsin molecules and materials: concepts, theory, and applications. *Chem. Rev.* 117: 4714.
47 Dobson, J.F. (2012). Dispersion (van der Waals) forces and TDDFT. In: *Fundamentals of Time-Dependent Density Functional Theory* (eds. M.A.L. Marques, N.T. Maitra, F.M.S. Nogueira, et al.), 417. Berlin: Springer-Verlag.
48 Dobson, J.F. and Gould, T. (2012). Calculation of dispersion energies. *J. Phys. Condens. Matter* 24: 073201.
49 Vydrov, O.A. and Van Voohis, T. (2010). Nonlocal van der Waals density functional: the simpler the better. *J. Chem. Phys.* 133: 244103–244101.
50 Libisch, F., Huang, C., Liao, P. et al. (2012). Origin of the energy barrier to chemical reactions of O2 on Al(111). *Phys. Rev. Lett.* 109: 198303–198301.
51 Libisch, F., Huang, C., and Carter, E.A. (2014). Embedded correlation function schemes: theory and applications. *Acc. Chem. Res.* 47: 2768.

52 Huang, C. and Carter, E.A. (2011). Potential-functional embedding theory for molecules and materials. *J. Chem. Phys.* 135: 194104–194101.
53 Sudhyadhom, A. and Micha, D.A. (2006). Bonding and excitation in CO/Cu(001) from a cluster model and density functional treatments. *J. Chem. Phys.* 124: 101102.
54 Sharifzadeh, S., Huang, P., and Carter, E. (2008). Embedded configuration interaction description of CO on Cu(111): resolution of the site preference conundrum. *J. Phys. Chem. C* 112: 4649.
55 Norskov, J.K. (1990). Chemisorption at metal surfaces. *Rep. Prog. Phys.* 53: 1253.
56 de Lara Castells, M.P., Cabrillo, C., Micha, D.A. et al. (2018). Ab initio design of light absorption through silver atomic cluster decoration of TiO2. *Phys. Chem. Chem. Phys.* 20: 19110.
57 Jensen, F. (2001). *Introduction to Computational Chemistry*. New York: Wiley.
58 Feng, E.Q., Micha, D.A., and Runge, K. (1991). A time-dependent molecular orbital approach to electron transfer in ion-metal surface collisions. *Int. J. Quantum Chem.* 40: 545.
59 Micha, D.A. and Feng, E.Q. (1994). The calculation of electron transfer probabilities in slow ion-metal surface collisions. *Comput. Phys. Commun.* 90: 242.
60 Muscat, J.P. and Newns, D.M. (1978). Chemisorption on metals. *Prog. Surf. Sci.* 9: 1.
61 Lang, N.D. and Williams, A.R. (1978). Theory of atomic chemisorption on simple metals. *Phys. Rev. B* 18: 616.
62 Kresse, G. (2000). Dissociation and sticking of H2 on the Ni(111),(100), and (110) substrate. *Phys. Rev. B* 62: 8295.
63 Madhavan, P. and Whitten, J.L. (1982). Theoretical studies of chemisorption of hydrogen on copper. *J. Chem. Phys.* 77: 2673.
64 Wiesenekker, G., Kroes, G.J., and Baerends, E.J. (1996). An analytical six-dimensional potential energy surface for dissociation of molecular hydrogen on Cu(001). *J. Chem. Phys.* 18: 7344.
65 Hammer, B., Scheffler, M., Jacobsen, K.W., and Norskov, J.K. (1994). Multidimensional potential energy surface for H2 dissociation over Cu(111). *Phys. Rev. Lett.* 73: 1400.
66 Yin, R., Zhang, Y., Libisch, F. et al. (2018). Dissociative chemisorption of O2 on Al(111): Dynamics on a correlated wavefunction-based potential energy surface. *J. Phys. Chem. Lett.* 9: 3271.
67 Kummel, S. and Kronik, L. (2008). Orbital dependent density functionals: theory and applications. *Rev. Mod. Phys.* 80: 3.
68 Lehninger, A.L., Nelson, D.L., and Cox, M.M. (1992). *Principles of Biochemistry*, 2e (Chapter 10). New York: Worth.
69 Lyubartsev, A.P. and Rabinovich, A.L. (2016). Force field developments for lipid membrane simulations. *Biochim. Biophys. Acta* 1858: 2483.
70 Pluhackova, K., Kirsch, S.A., Han, J. et al. (2016). A critical comparison of biomembrane force fields: structure and dynamics of model DMPC, POPC, and POPE bilayers. *J. Phys. Chem. B* 120: 3888.

71 Sandoval-Perez, A., Pluhackova, K., and Boeckmann, R.A. (2017). Critical comparison of biomembrane force fields: Protein-Lipid interactions at the membrane interface. *J. Chem. Theory Comput.* 13: 2310.
72 Poger, D., VanGunsteren, W., and Mark, A.E. (2010). A new force field for simulating phosphatidylcholine bilayers. *J. Comput. Chem.* 31: 1117.
73 Klauda, J.B., O'Connor, J.W., Venable, R.M. et al. (*2010*). Update of the CHARMM all-atom additive force field for lipids: validation on six lipid types. *J. Phys. Chem. B* 114 (23): 7830–7843.
74 Marrink, S.J., Risselada, H.J., Yefimov, S. et al. (2007). The MARTINI force field: cgrained model for biomolecular simulations. *J. Phys. Chem. B* 111: 7812.
75 Voth, G.A. (2009). *Coarse-Graining of Condensed Phase and Biomolecular Systems.* Boca Raton, FL, USA: CRC Press.
76 Nielsen, S.O., Lopez, C.F., Srinivas, G., and Klein, M.L. (2004). Coarse grain models and the computer simulation of soft materials. *J. Phys. Condens. Matter* 15: R481.
77 Shih, A.Y., Arkhipov, A., Freddolino, P.L. et al. (2007). Assembly of lipids and proteins into lipoprotein particles. *J. Phys. Chem. B* 111: 11095.
78 Kamerlin, S.C.L., Vicatos, S., Dryga, A., and Warshel, A. (2011). Coarse-grained (multiscale) simulations in studies of biophysical and chemical systems. *Annu. Rev. Phys. Chem.* 62: 41.
79 Noid, W.G. (2013). Perspective: coarse-grained models for biomolecular systems. *J. Chem. Phys.* 139: 090901–090901.
80 Rupp, M. (2015). Machine learning for quantum mechanics in n Nutshell. *Int. J. Quantum Chem.* 115: 1058.
81 von Lilienfeld, O.A., Lins, R.D., and Rothlisberger, U. (2005). Variational particle number approach for rational compound design. *Phys. Rev. Lett.* 95: 153002.
82 Brandt, S., Sittel, F., Ernst, M., and Stock, G. (2018). Machine learning of biomolecular reaction coordinates. *J. Phys. Chem. Lett.* 9: 2144.
83 Shell, M.S. (2008). The relative entropy is fundamental to multiscale and inverse thermodynamical problems. *J. Chem. Phys.* 129: 144108.
84 Izvekov, S. and Voth, G.A. (2005). A multiscale coarse-graining method for biomolecular systems. *J. Phys. Chem. B* 109: 2469.

Index

Molecular Interactions: Concepts and Methods, First Edition. David A. Micha.

d

e

i